ENCYCLOPEDIA OF PLASTICS, POLYMERS, AND RESINS

VOLUME I

Compiled by

Michael and Irene Ash

Chemical Publishing Co.
New York, N.Y.

Encyclopedia of Plastics, Polymers, and Resins Volume 1

ISBN: 978-0-8206-0046-8

Chemical Publishing Company:
www.chemical-publishing.com
www.chemicalpublishing.net

First Edition:

© **Chemical Publishing Company, Inc.** - New York 1982

Second Impression:

Chemical Publishing Company, Inc. - 2011

Printed in the United States of America

PREFACE

This encyclopedia is an attempt to coordinate and unify practical information on plastic, polymer, and resin trademark products. These categories overlap, i.e., plastics are a subset of polymers, and resins are often, but not always, polymers. However, it has been our aim to give the user of this compilation fingertip availability to a large quantity of essential information about these products. This three-volume compendium has been made possible through the cooperation of the major national and international plastic, polymer, and resin manufacturers.

The information provided here has been gleaned from thousands of brochures, technical bulletins, and data sheets, but the extent of the information provided for each product has been limited by the amount of data given to us by the manufacturers. Whenever possible and/or appropriate, we have given the chemical description, applications, form and color, general, mechanical, thermal, and electrical properties of each product.

With the state of technological growth in these industries constantly increasing, this encyclopedia should serve as an important tool for chemists, engineers, and salespeople here and abroad.

We want to extend our thanks to Roberta Dakan for her tireless efforts in helping make this encyclopedia as accurate and consistent as possible.

It should be noted that when the temperatures of properties such as viscosity, density, solubility, etc. are not included, a standard temperature of 25 C is to be assumed. The information in this publication is reliable to the best of our knowledge. We would appreciate being informed of any errors or omissions so that these can be integrated into subsequent editions of this encyclopedia.

CONTRIBUTORS

Abbey Plastics Corp.
11 Brent Drive
Hudson, Mass. 01749

Acme Chem. & Insulation Co./
Div. of Allied Prod. Corp.
P.O. Box 1404
New Haven, Conn. 06505

AECI Ltd./Plastics Div.
P.O. Box 1122
Johannesburg, 2000, S.A.

Allied Chem., Fibers & Plastics Co.
1411 Broadway
New York, N.Y. 10018

American Cyanamid Co.
Stamford Research Laboratories
1937 W. Main St.
Stamford, Conn. 06904

American Hoechst Corp./
Plastics Div.
Leominster, Mass.

Ashland Chem. Co.
Box 2219
Columbus, Ohio 43216

Astro Chem. Co., Inc.
1205 Godfrey Lane
Schenectady, N.Y. 12309

Bacon Industries Inc.
192 Pleasant St.
Watertown, Mass. 02172

Bayer AG
Sitz der Gesellschaft
Leverkusen; Eintragung;
Amtsgericht Leverkusen
HRB 1122, Germany

Belding Chem. Industries
1430 Broadway
New York, N.Y. 10018

Blackhawk Resin & Chem. Co.
P.O. Box 838
Tooele, Utah 84074

Borg-Warner Chem. Corp.
International Center
Parkersburg, West Va. 26101

Cal Polymers, Inc.
2115 Gaylord St.
Long Beach, Calif. 90813

Celanese Corp.
Plastics and Specialties Co.
26 Main St.
Chatham, N.J. 07928

Chemapol
Ing. Scharff
Kodanska 40
Prague 10, Czechoslovakia

Chemfax Inc.
3 Rivers Road
Gulfport, Miss.

Chemplast, Inc.
150 Dey Road
Wayne, N.J. 07470

Chemplex Co.
Rolling Meadows, Ill. 60008

Ciech Plastofarb Div.
ul. Jasna 12
00950 Warsaw, Poland

Cincinnati Milacron Chem.
West St.
Reading, Ohio 45215

Compounding Technology, Inc.
3140 Pullman St.
Costa Mesa, Calif. 92626

Conoco Chem. Div.
Continental Oil Co.
5 Greenway Plaza East
P.O. Box 2197
Houston, Tex. 77001

Continental Polymers Inc.
2225 East Del Amo Boulevard
Compton, Calif. 90220

Cosden Oil & Chem. Co.
P.O. Box 2159
3350 N. Central Expressway
Dallas, Tex. 75221

Crosby Chem., Inc.
P.O. Box 460
Picayune, Miss. 39466

CY/RO Industries
697 Route 46
Clifton, N.J. 07015

Delta Plastics Co.
Sante Fe Srpings, Calif.

Diamond Shamrock/
Functional Polymers Div.
1100 Superior Ave.
Cleveland, Ohio 44114

Dow Chem. Co.
1703 S. Saginaw Road
Midland, Mich. 48640

Du Pont de Nemours, E.I. & Co.
Wilmington, Del. 19898

Durez Div./
Hooker Chem. & Plastics Corp.
North Tonawanda, N.Y. 14120

Eastman Chem. Products, Inc.
Coatings Chemicals Div.
Kingsport, Tenn. 37662

Emerson & Cuming/W.R. Grace & Co.
869 Washington St.
Canton, Mass. 02021

Emery Industries, Inc.
1300 Carew Tower
Cincinnati, Ohio 45202

Fabri-Coate Co.
P.O. Box 8835
4500 Roger B. Chaffee Dr.
Grand Rapids, Mich. 49508

Fiberite Corp.
501-559 W. Third St.
Winona, Minn. 55987

Firestone Plastics Co.
Chem. Div.
Pottstown, Pa. 19464

Firestone Synthetic Fibers Co.
Box 450 Main St.
Hopewell, Va. 23860

Freeman Chem. Corp.
Subsid. of H.H. Robertson Co.
P.O. Box 247
Port Washington, Wisc. 53074

GAF Corp.
140 W. 51st St.
New York, N.Y. 10020

General Electric Co.
Plastics Div.
One Plastics Ave.
Pittsfield, Mass. 01201

Georgia-Pacific Corp.
290 Ferry St.
Newark, N.J. 07105

M.C. Gill Corp.
4056 Easy St.
El Monte, Calif. 91731

B.F. Goodrich Co.
6100 Oak Tree Boulevard
Cleveland, Ohio 44131

Goodyear Tire & Rubber Co.
Chem. Div.
Akron, Ohio 44316

Hammond Plastics Div./
Carl Gordon Industries, Inc.
1001 Southbridge St.
Worcester, Mass. 01610

Hardman Inc.
600 Cortlandt St.
Belleville, N.J. 07109

Hercules Inc.
910 Market St.
Wilmington, Del. 19899

Hexcel Corp./Rezolin Div.
20701 Nordhoff St.
Chatsworth, Calif. 91311

Hooker Chem. & Plastics Corp./
Industrial Chem. Group
345 Third St.
Niagara Falls, N.Y. 14302

ICI United States Inc./
Specialty Chem. Div.
Wilmington, Del. 19897

Isochem Resins Co.
Cook St.
Lincoln, R.I. 02865

Koppers Co., Inc.
1350 Koppers Bldg.
Pittsburgh, Pa. 15219

Liquid Nitrogen Processing Corp.
412 King St.
Malvern, Pa. 19355

3M/Commercial Chem. Div.
3M Center
St. Paul, Minn. 55101

Mobay Chem. Corp.
Plastics and Coatings Div.
Penn Lincoln Pkwy. West
Pittsburgh, Pa. 15205

Mobil Chem. Co.
Polymers Group
Petrochemicals Div.
P.O. Box 726
Paramus, N.J. 07652

Monsanto Plastics & Resins Co.
800 N. Lindbergh Blvd.
St. Louis, Mo. 63166

Neville Chem. Co.
Neville Island
Pittsburgh, Pa. 15225

Polymer Research Corp. of America
2186 Mill Ave.
Brooklyn, N.Y. 11234

Reichhold Chem., Inc.
RCI Building
White Plains, N.Y. 10602

Rogers Corp.
Engineered Prod. Group
Atlanta Div.
5259 Minola Dr.
Lithonia, Ga. 30058

Rohm & Haas Co.
Independence Mall West
Philadelphia, Pa. 19105

Scott Bader Co. Ltd.
Wollaston, Wellingborough
Northamptonshire, NN9 7RL England

Solvay & Cie
Rue du Prince Albert 44-
1050 Brussels, Belgium

W.R. Grace & Co./Organic Chem. Div.
55 Hayden Ave.
Lexington, Mass. 02173

Witco Chem. Corp./
Argus Chem. Div.
633 Court St.
Brooklyn, N.Y. 11231

XCEL Corp.
290 Ferry St.
Newark, N.J. 07105

ABBREVIATIONS

@	at
ABS	acrylonitrile-butadiene-styrene
aq.	aqueous
ASTM	American Society for Testing Materials
avail	available
B.P.	boiling pt.
BS	British Standards
BS&D	backwashed, settled, and drained
Btu	British thermal unit
C	Centigrade
Cal	calories
cc	cubic centimeter
CC	closed cup
CD	cross direction
charact.	characteristics
chem.	chemical
cm	centimeter
COC	Cleveland Open Cup
coeff.	coefficient
comp.	compound
compr.	compressive
conc..	concentrated
conduct.	conductivity
const.	constant
cps	centipoise or cycles per second
defl. or deflect'n	deflection
dg.	decigram
dielec.	dielectric
DIN	Deutsche Industrie Normen
dissip.	dissipation
distort.	distortion
DMF	dimethylformamide
DOT	Department of Transportation
DWV	drain, waste, and vent
elec.	electrical
elong.	elongation

EPDM . . . . .ethylene propylene-diene monomer
esp.. . . . . especially
EVA . . . . ethylene vinyl acetate
exp. . . . . .expansion
F . . . . Fahrenheit
FAA . . . . Federal Aviation Agency
FB . . . . .free base
FDA . . . . Food and Drug Administration
flamm. . . . . flammability
flex. . . . . flexural
F.P. . . . . freezing point
ft . . . . .foot
g . . . . gram
G . . . . giga
gal . . . . .gallon
gen'l.. . . . . general
gr. . . . . gravity
GRP . . . . .glass-reinforced polyester
h . . . . hour
HAF . . . . high abrasion furnace carbon black
HB . . . . horizontal burning
HC . . . . hydrocarbon
hyd. . . . . hydroxyl
hydrog. . . . . hydrogenated
Hz . . . . hertz
ICC. . . . . Interstate Commerce Commission
ignit.. . . . . ignition
in. . . . . .inch
incl. . . . . including
insol.. . . . . .insoluble
insul.. . . . . insulation
J . . . . joule
k . . . . kilo
kg. . . . . .kilogram
l. . . . . .liter
lb . . . . pound
LDPE . . . . .low density polyethylene
m . . . . milli or meter
M . . . . mega
max.. . . . . maximum
MC . . . . .megacycle
MD. . . . . mold direction or machine direction

MDA. . . . . methylene disalicylic acid
mech. . . . . mechanical
med.. . . . . medium
MEK. . . . . methyl ethyl ketone
mfg. . . . . manufacturing
mg . . . . milligram
MIBK . . . . methyl isobutyl ketone
MIL . . . . Military Specifications
min. . . . . minute
min. . . . . minimum or mineral
misc.. . . . . miscellaneous
mm. . . . . millimeter
mo . . . . months
MOCA. . . . . methylene-bis-orthochloroaniline
mod.. . . . . modulus
M.P. . . . . melting point
m.w.. . . . . molecular weight
N . . . . Newton
no. . . . . number
NR . . . . natural rubber
OC . . . . open crucible
oz. . . . . ounce
Pa. . . . . Pascal
pbw . . . . parts by weight
pcf . . . . pounds per cubic foot
PET . . . . polyethylene terephthalate
PMCC . . . . Pensky-Martens Closed Cup
pH . . . . hydrogen-ion concentration
phr . . . . parts per hundred parts of resin or rubber
pkgs.. . . . . packages
PPD . . . . piperidine pentamethylene dithiocarbamate
psi . . . . pounds per square inch
pt. . . . . point
PU . . . . polyurethane
PVAc . . . . polyvinyl acetate
PVC . . . . polyvinyl chloride
PW . . . . potable water
quat.. . . . . quaternary
qt. . . . . quart
R&B . . . . Ring & Ball
ref.. . . . . refractive or reference
resist. . . . . resistance or resistivity

r.h.	relative humidity
rpm	revolutions per minute
R.T.	room temperature
s.	second
SAN	styrene-acrylonitrile polymer
sapon.	saponification
SBR	styrene-butadiene elastomer
soften.	softening
sol.	solubility or soluble
sol'n.	solution
sp.	specific
spec.	specification
SS	stainless steel
S/S	step by step
std.	standard
S/T	short time
str.	strength
surf.	surface
TCC	Taggart Closed Cup
TD	transverse direction
temp.	temperature
tens.	tensile
TDI	toluene diisocyanate
therm.	thermal
THF	tetrahydrofuran
TOC	Taggart Open Cup
trans.	transitional
UL	Underwriter's Laboratory
USDA	Unites States Department of Agriculture
uv.	ultraviolet
V	volt
visc.	viscosity
vol.	volume
W.E. or W. Elec.	Western Electric
wk	week
wt.	weight
yrs	years
$<$	less than
$>$	greater than
$\leqslant$	less than or equal to
$\geqslant$	greater than or equal to
$\approx$	approximately equal to

The following products which are included in this volume, are registered with the U.S. Patent Office.

Abalyn®
Acrylite®
Acryloid®
Adiprene®
Adtac®
Advastab®
Advawax®
A-Fax®
Alathon®
Ameripol®
BCI®
Butacite®
CAPA®
Carbopol®
Carstab®
Celcon®
Cellolyn®
Cereclor®
Chemplex®
Clorafin®
Clysar®
Cyanacryl®
Cycolac®
Cycoloy®
Cytor®
Daran®
Daratak®
Darex®
Dion®
Drapex®
Duolite®
Durez®
Dymerex®
Eccosil®
Elvace®
Elvanol®
Elvax®
Emerez®
Envex®
Epolene®
Epolite®
Epotuf®
Ertalon®
Estane®
Fluorel®
Fluorocomp®
Forel®
Fosta®
Fostafoam®
Fostalite®
Fostarene®
Fosta-Tuf-Flex®
Gafite®
Gantrez®
Geon®
Good-rite®

A

Abalyn. Hercules Inc.
Chem. Descrip.: Methyl ester of rosin
Category/Applications: Resin with wide compatibility, excellent surface-wetting properties, viscosity, and tack for use in lacquers, inks, paper coatings, varnishes, adhesives, sealing compounds, plastics, wood preservatives, and perfumes
Form: Viscous liquid; Color: Amber
Gen'l. Prop.: Sol.: Sol. in esters, ketones, alcohols, ethers, coal tar, petroleum hydrocarbons, and veg. and min. oils; insol. in water; Sp. Gr.: 1.03; Density: 1.03 kg/l; Visc. (Gardner-Holdt): Z_1; B.P.: 352–356 C; Acid No.: 6; Sapon. No.: 160; Stability: Good thermal stability Ref. Index: 1.5300 (20 C); Flash Pt.: 180 C (COC)
Std. Pkgs.: 205-kg metal drums (net), and in tank cars, tank trucks
Ref.: Product data No. 742-9

Abbey #100. Abbey Plastics Corp.
Chem. Descrip.: Conductive vinyl compound
Category/Applications: Thermoplastic material with outstanding electrical conductive properties for use in microphone cable, safety equipment, hospital O.R. equipment, test equipment, gasketing and tubing, radio, stereo, and TV cable, aircraft static eliminators, conductive calendered sheeting and tapes, high voltage conductor shielding and corona barrier, antistatic devices, solvent solutions
Gen'l. Prop.: Sp. Gr.: 1.39
Mech. Prop.: Tens. Str.: 1500 psi; Tens. Elong.: 125%; Tens. Mod.: 1450 psi (@ 100% elong.); Hardness (Shore A): 88
Elect. Prop.: Vol. Resist.: 6 ohm/cm^3

Abitol. Hercules Inc.
Chem. Descrip.: Primary, monohydric alcohol derived from hydrog. rosin acids
Category/Applications: Resinous plasticizer and tackifier in plastics, lacquers, inks, and adhesives; chemical intermediate in the production of oil-modified and oil-free alkyd resins
Form: Viscous liquid; Color: Colorless; Odor: Low
Composition: Avail. as viscous liquid (100% resin), and 90% solids toluene or xylene solutions
Gen'l. Prop.: Sol.: Sol. in alcohols, esters, ketones, chlorinated solvents, and in aliphatic, aromatic, or terpene hydrocarbons; insol. in water; Sp. Gr.: 1.008; Density: 8.4 lb/gal; Visc.: 400 poises (40 C); Acid No.: 0.3; Sapon. No.: 15; Hydroxyl: 4.75%; Stability: Resistant to oxidation; Ref. Index: 1.5262 (20 C); Flash Pt.: 185 C (COC)

Std. Pkgs.: 208-ℓ (195-kg net) metal drums
Ref.: Product data No. 750-9

ABS 124ESG. Mobil Chem. Co.
Chem. Descrip.: Acrylonitrile/butadiene/styrene copolymer
Category/Applications: General purpose molding grade with high tensile and flexural properties, and good heat resistance; used for spools and molded parts requiring high tensile modulus
Form: Solid; Color: Natural
Gen'l. Prop.: Sp. Gr.: 1.05 (23/23 C, D-792); Soften. Pt. (Vicat): 225 F (D1525)
Mech. Prop.: Tens. Str.: 7700 psi (D-638); Tens. Elong.: 10% (D638); Tens. Mod.: 340,000 psi (D638); Flex. Str.: 12,000 psi (D790); Flex. Mod.: 330,000 psi (D790); Impact Str. (Izod): 2.5 ft lb/in. notched (23 C, D256); Hardness (Rockwell): 112 R (D785)
Therm. Prop.: Distort. Temp.: 215 F (@ 264 psi, D648)

ABS 124ESGE. Mobil Chem. Co.
Chem. Descrip.: Acrylonitrile/butadiene/styrene copolymer
Category/Application: General purpose, semigloss, profile and sheet extrusion grade, for use in refrigerator breaker strips and profiles
Form: Solid; Color: Natural
Gen'l. Prop.: Sp. Gr.: 1.05 (23/23 C, D-792), Soften. Pt. (Vicat): 220 F (D-1525)
Mech. Prop.: Tens. Str.: 6000 psi (D-638); Tens. Elong.: 25% (D638); Tens. Mod.: 310,000 psi (D638); Flex. Str.: 9000 psi (D790), Flex. Mod.: 340,000 psi (D790); Impact Str. (Izod): 2.0 ft lb/in. notched (23 C, D256): Hardness (Rockwell): 110 R (D785)
Therm. Prop.: Distort. Temp.: 205 F (@ 264 psi, D648)

ABS 125M. Mobil Chem. Co.
Chem. Descrip.: Acrylonitrile/butadiene/styrene copolymer
Category/Applications: General purpose molding grade; high flow for thin sections and intricate moldings; easily decorated, good chemical resistance; used for toys, closures, small knobs and parts
Form: Solid; Color: Natural
Gen'l. Prop.: Sp. Gr.: 1.05 (23/23 C, D-792); Soften. Pt. (Vicat): 218 F (D1525)
Mech. Prop.: Tens. Str.: 5400 psi (D-638); Tens. Elong.: 30% (D638); Tens. Mod.: 310,000 psi (D638); Flex. Str.: 9000 psi (D790); Flex. Mod.: 355,000 psi (D790); Impact Str. (Izod): 1.5 ft lb/in. notched (23 C, D256); Hardness (Rockwell): 109 R (D785)
Therm. Prop.: Distort. Temp.: 205 F (@ 264 psi, D648)

ABS 224M. Mobil Chem. Co.
Chem. Descrip.: Acrylonitrile/butadiene/styrene copolymer
Category/Applications: General purpose molding grade copolymer; low luster with good toughness; used for tackle boxes, carrying cases
Form: Solid; Color: Natural
Gen'l. Prop.: Sp. Gr.: 1.05 (23/23 C, D-792); Soften. Pt. (Vicat): 220 F (D1525)
Mech. Prop.: Tens. Str.: 6500 psi (D-638); Tens. Elong.: 30% (D638); Tens. Mod.: 305,000 psi (D638); Flex. Str.: 10,000 psi (D790); Flex. Mod.: 360,000 psi (D790); Impact Str. (Izod): 2.5 ft lb/in. notched (23 C, D256); Hardness (Rockwell): 110 R (D785)
Therm. Prop.: Distort. Temp.: 205 F (@ 264 psi, D648)

ABS 236F. Mobil Chem. Co.
Chem. Descrip.: Acrylonitrile/butadiene/styrene copolymer
Category/Applications: General purpose

automotive grade copolymer with high impact and good flow; injection molding; used for automotive interior parts
Form: Solid; Color: Natural and colors
Gen'l. Prop.: Sp. Gr.: 1.05 (23/23 C, D-792); Soften. Pt. (Vicat): 218 F (D1525)
Mech. Prop.: Tens. Str.: 5600 psi (D-638); Tens. Elong.: 50% (D638); Tens. Mod.: 270,000 psi (D638); Flex. Str.: 8200 psi (D790); Flex. Mod.: 320,000 psi (D790); Impact Str. (Izod): 5.9 ft lb/in. notched (23 C, D256); Hardness (Rockwell): 103 R (D785)
Therm. Prop.: Distort. Temp.: 203 F (@ 264 psi, D648)

ABS 236MA. Mobil Chem. Co.
Chem. Descrip.: Acrylonitrile/butadiene/styrene copolymer
Category/Applications: Medium impact grade copolymer for profile extrusions for automotive and appliance applications
Form: Solid; Color: Natural
Gen'l. Prop.: Sp. Gr.: 1.05 (23/23 C, D-792); Soften. Pt. (Vicat): 220 F (D-1525)
Mech. Prop.: Tens. Str.: 5700 psi (D-638); Tens. Elong.: 35% (D638); Tens. Mod.: 285,000 psi (D638); Flex. Str.: 9500 psi (D790); Flex. Mod.: 360,000 psi (D790); Impact Str. (Izod); 4.5 ft lb/in. notched (23 C, D256); Hardness (Rockwell): 106 R (D785)
Therm. Prop.: Distort. Temp.: 208 F (@ 264 psi, D648)

ABS 236R. Mobil Chem. Co.
Chem. Descrip.: Acrylonitrile/butadiene/styrene copolymer
Category/Applications: General purpose automotive grade copolymer, low impact, high stiffness; for injection molding for automotive parts, package trays, console components
Form: Solid; Color: Natural and colors
Gen'l. Prop.: Sp. Gr.: 1.05 (23/23 C, D-792); Soften. Pt. (Vicat): 220 F (D1525)
Mech. Prop.: Tens. Str.: 7100 psi (D-638); Tens. Elong.: 11% (D638); Tens. Mod.: 310,000 psi (D638); Flex. Str.: 11,100 psi (D790); Flex. Mod.: 370,000 psi (D790); Impact Str. (Izod): 2.5 ft lb/in. notched (23 C, D256); Hardness (Rockwell): 114 R (D785)
Therm. Prop.: Distort. Temp.: 209 F (@ 264 psi, D648)

ABS 301K. Mobil Chem. Co.
Chem. Descrip.: Acrylonitrile/butadiene/styrene copolymer
Category/Applications: Injection molding, pipe fitting grade; DWV-NSF approved; high ESCR properties; fittings for DWV, sewer pipe, and conduit
Form: Solid; Color: Black
Gen'l. Prop.: Sp. Gr.: 1.05 (23/23 C, D-792); Soften. Pt. (Vicat): 212 F (D1525)
Mech. Prop.: Tens. Str.: 5500 psi (D-638); Tens. Elong.: 30% (D638); Tens. Mod.: 280,000 psi (D638); Flex. Str.: 9000 psi (D790); Flex. Mod.: 360,000 psi (D790); Impact Str. (Izod): 6.0 ft lb/in. notched (23 C, D256); Hardness (Rockwell): 103 R (D785)
Therm. Prop.: Distort. Temp.: 205 F (@ 264 psi, D648)

ABS 323MA. Mobil Chem. Co.
Chem. Descrip.: Acrylonitrile/butadiene/styrene copolymer
Category/Applications: Tough, high impact, automotive grade copolymer with good flow properties for molding low stress parts; automotive ducts and bezels
Form: Solid; Color: Natural
Gen'l. Prop.: Sp. Gr.: 1.05 (23/23 C, D-792); Soften. Pt. (Vicat): 212 F (D1525)
Mech. Prop.: Tens. Str.: 5600 psi (D-638); Tens. Elong.: 40% (D638); Tens.

Mod.: 275,000 psi (D638); Flex. Str.: 8300 psi (D790); Flex. Mod.: 348,000 psi (D790); Impact Str. (Izod): 6.0 ft lb/in. notched (23 C, D256); Hardness (Rockwell): 103 R (D785)
Therm. Prop.: Distort. Temp.: 205 F (@ 264 psi, D648)

ABS 400P. Mobil Chem. Co.
Chem. Descrip.: Acrylonitrile/butadiene/styrene copolymer
Category/Applications: General purpose molding grade copolymer; medium impact, semigloss; used for various appliance housings; UL 94 HB rating
Form: Solid; Color: Natural and colors
Gen'l. Prop.: Sp. Gr.: 1.05 (23/23 C, D-792); Soften. Pt. (Vicat): 220 F (D1525)
Mech. Prop.: Tens. Str.: 5500 psi (D-638); Tens. Elong.: 30% (D638); Tens. Mod.: 290,000 psi (D638); Flex. Str.: 9500 psi (D790); Flex. Mod.: 355,000 psi (D790); Impact Str. (Izod): 4.0 ft lb/in. notched (23 C, D256); Hardness (Rockwell): 106 R (D785)
Therm. Prop.: Distort. Temp.: 204 F (@ 264 psi, D648)

ABS 410ESG. Mobil Chem. Co.
Chem. Descrip.: Acrylonitrile/butadiene/styrene copolymer
Category/Applications: Profile and sheet extrusion grade copolymer, medium impact, semigloss, for use in appliance, cooler, and similar applications
Form: Solid; Color: Natural
Gen'l. Prop.: Sp. Gr.: 1.05 (23/23 C, D-792); Soften. Pt. (Vicat): 215 F (D-1525)
Mech. Prop.: Tens. Str.: 5100 psi (D-638); Tens. Elong.: 35% (D638); Tens. Mod.: 265,000 psi (D638); Flex. Str.: 8300 psi (D790); Flex. Mod.: 330,000 psi (D790); Impact Str. (Izod): 4.5 ft lb/in. notched (23 C, D256); Hardness (Rockwell): 104 R (D785)
Therm. Prop.: Distort. Temp.: 203 F (@ 264 psi, D648)

ABS 414MA. Mobil Chem. Co.
Chem. Descrip.: Acrylonitrile/butadiene/styrene copolymer
Category/Applications: High gloss, stiffness, automotive grade copolymer with easy flow; instrument panels
Form: Solid; Color: Natural
Gen'l. Prop.: Sp. Gr.: 1.05 (23/23 C, D-792); Soften. Pt. (Vicat): 224 F (D1525)
Mech. Prop.: Tens. Str.: 7200 psi (D-638); Tens. Elong.: 15% (D638); Tens. Mod.: 330,000 psi (D638); Flex. Str.: 10,500 psi (D790); Flex. Mod.: 320,000 psi (D790); Impact Str. (Izod): 1.0 ft lb/in. notched (23 C, D256); Hardness (Rockwell): 112 R (D785)
Therm. Prop.: Distort. Temp.: 204 F (@ 264 psi, D648)

ABS 420A. Mobil Chem. Co.
Chem. Descrip.: Acrylonitrile/butadiene/styrene copolymer
Category/Applications: Profile and sheet extrusion grade copolymer, high impact, matte finish, with good toughness for use for recreational vehicle interior parts, etc.
Form: Solid; Color: Natural
Gen'l. Prop.: Sp. Gr.: 1.05 (23/23 C, D-792); Soften. Pt. (Vicat): 217 F (D-1525)
Mech. Prop.: Tens. Str.: 3500 psi (D-638); Tens. Elong.: 70% (D638); Tens. Mod.: 225,000 psi (D638); Flex. Str.: 5500 psi (D790); Flex. Mod.: 214,000 psi (D790); Impact Str. (Izod): 5.0 ft lb/in. notched (23 C, D256); Hardness (Rockwell): 90 R (D785)
Therm. Prop.: Distort. Temp.: 205 F (@ 264 psi, D648)

ABS 424ESM. Mobil Chem. Co.
Chem. Descrip.: Acrylonitrile/butadiene/styrene copolymer
Category/Applications: Profile and sheet extrusion grade copolymer, medium impact, matte finish, for use in automotive, recreational, appliance, luggage cases, etc. applications
Form: Solid; Color: Natural
Gen'l. Prop.: Sp. Gr.: 1.05 (23/23 C, D-792); Soften. Pt. (Vicat): 219 F (D-1525)
Mech. Prop.: Tens. Str.: 4000 psi (D-638); Tens. Elong.: 60% (D638); Tens. Mod.: 235,000 psi (D638); Flex. Str.: 6100 psi (D790); Flex. Mod.: 230,000 psi (D790); Impact Str. (Izod): 3.0 ft lb/in. notched (23 C, D256); Hardness (Rockwell): 92 R (D785)
Therm. Prop.: Distort. Temp.: 207 F (@ 264 psi, D648)

ABS 500FR-1. Mobil Chem. Co.
Chem. Descrip.: Acrylonitrile/butadiene/styrene copolymer
Category/Applications: Flame retardant grade; high impact with good stability; easily reprocessed scrap trim (non-PVC); good heat distortion; used for aircraft, automotive, appliance parts; furniture, smoke detector housings; UL 94 V-O rating
Form: Solid; Color: Natural
Gen'l. Prop.: Sp. Gr.: 1.33 (23/23 C, D-792); Soften. Pt. (Vicat): 198 F (D1525)
Mech. Prop.: Tens. Str.: 4600 psi (D-638); Tens. Elong.: 20% (D638); Tens. Mod.: 255,000 psi (D638); Flex. Str.: 6100 psi (D790); Flex. Mod.: 270,000 psi (D790); Impact Str. (Izod): 5.0 ft lb/in. notched (23 C, D256); Hardness (Rockwell): 70 R (D785)
Therm. Prop.: Distort. Temp.: 187 F (@ 264 psi, D648)

ABS 504ESG. Mobil Chem. Co.
Chem. Descrip.: Acrylonitrile/butadiene/styrene copolymer
Category/Applications: Profile and sheet extrusion grade copolymer, high impact, glossy, for use in refrigerator liners and recreational uses
Form: Solid; Color: Natural
Gen'l. Prop.: Sp. Gr.: 1.04 (23/23 C, D-792); Soften. Pt. (Vicat): 220 F (D-1525)
Mech. Prop.: Tens. Str.: 5500 psi (D-638); Tens. Elong.: 20% (D638); Tens. Mod.: 280,000 psi (D638); Flex. Str.: 8500 psi (D790); Flex. Mod.: 295,000 psi (D790); Impact Str. (Izod): 6.5 ft lb/in. notched (23 C, D256); Hardness (Rockwell): 100 R (D785)
Therm. Prop.: Distort. Temp.: 205 F (@ 264 psi, D648)

ABS 506ESG. Mobil Chem. Co.
Chem. Descrip.: Acrylonitrile/butadiene/styrene copolymer
Category/Applications: Profile and sheet extrusion grade copolymer, glossy, very high impact grade with good low temp. performance, for use in camper, appliance, and housings
Form: Solid; Color: Natural
Gen'l. Prop.: Sp. Gr.: 1.03 (23/23 C, D-792); Soften. Pt. (Vicat): 216 F (D-1525)
Mech. Prop.: Tens. Str.: 4400 psi (D-638); Tens. Elong.: 40% (D638); Tens. Mod.: 225,000 psi (D638); Flex. Str.: 6700 psi (D790); Flex. Mod.: 235,000 psi (D790); Impact Str. (Izod): 7.5 ft lb/in. notched (23 C, D256); Hardness (Rockwell): 85 R (D785)
Therm. Prop.: Distort. Temp.: 204 F (@ 264 psi, D648)

ABS 510EC. Mobil Chem. Co.
Chem. Descrip.: Acrylonitrile/butadiene/styrene copolymer

Category/Applications: Pipe extrusion grade; WE and NSF approved; high modulus; used for conduit and sewer pipe
Form: Solid; Color: Black
Gen'l. Prop.: Sp. Gr.: 1.05 (23/23 C, D-792); Soften. Pt. (Vicat): 218 F (D1525)
Mech. Prop.: Tens. Str.: 6500 psi (D-638); Tens. Elong.: 15% (D638); Tens. Mod.: 300,000 psi (D638); Flex. Str.: 10,000 psi (D790); Flex. Mod.: 360,000 psi (D790); Impact Str. (Izod): 2.5 ft lb/in. notched (23 C, D256); Hardness (Rockwell): 110 R (D785)
Therm. Prop.: Distort. Temp.: 203 F (@ 264 psi, D648)

ABS 522ESM. Mobil Chem. Co.
Chem. Descrip.: Acrylonitrile/butadiene/styrene copolymer
Category/Applications: Profile and sheet extrusion grade copolymer, very high impact, matte finish, excellent low temp. performance; for use in luggage, ATV, camper, and other applications
Form: Solid; Color: Natural
Gen'l. Prop.: Sp. Gr.: 1.04 (23/23 C, D-792); Soften. Pt. (Vicat): 215 F (D1525)
Mech. Prop.: Tens. Str.: 3400 psi (D-638); Tens. Elong.: 70% (D638); Tens. Mod.: 200,000 psi (D638); Flex. Str.: 5300 psi (D790); Flex. Mod.: 194,000 psi (D790); Impact Str. (Izod): 7.0 ft lb/in. notched (23 C, D256); Hardness (Rockwell): 85 R (D785)
Therm. Prop.: Distort. Temp.: 203 F (@ 264 psi, D648)

ABS 550ES. Mobil Chem. Co.
Chem. Descrip.: Acrylonitrile/butadiene/styrene copolymer
Category/Applications: Profile and sheet extrusion grade copolymer, very high impact grade with high tensile and flexural properties; used for core laminates
Form: Solid; Color: Natural
Gen'l. Prop.: Sp. Gr.: 1.05 (23/23 C, D-792); Soften. Pt. (Vicat): 217 F (D1625)
Mech. Prop.: Tens. Str.: 5600 psi (D-638); Tens. Elong.: 15% (D638); Tens. Mod.: 272,000 psi (D638); Flex. Str.: 9000 psi (D790); Flex. Mod.: 321,000 psi (D790); Impact Str. (Izod): 9.0 ft lb/in. notched (23 C, D256); Hardness (Rockwell): 100 R (D785)
Therm. Prop.: Distort. Temp.: 206 F (@ 264 psi, D648)

ABS 606ED. Mobil Chem. Co.
Chem. Descrip.: Acrylonitrile/butadiene/styrene copolymer
Category/Applications: Pipe extrusion grade; DWV-NSF approved; used for DWV pipe
Form: Solid; Color: Black
Gen'l. Prop.: Sp. Gr.: 1.04 (23/23 C, D-792); Soften. Pt. (Vicat): 220 F (D1525)
Mech. Prop.: Tens. Str.: 5200 psi (D-638); Tens. Elong.: 25% (D638); Tens. Mod.: 280,000 psi (D638); Flex. Str.: 7900 psi (D790); Flex. Mod.: 280,000 psi (D790); Impact Str. (Izod): 8.5 ft lb/in. notched (23 C, D256); Hardness (Rockwell): 92 R (D785)
Therm. Prop.: Distort. Temp.: 209 F (@ 264 psi, D648)

ABS 707H. Mobil Chem. Co.
Chem. Descrip.: Acrylonitrile/butadiene/styrene copolymer
Category/Applications: General purpose molding grade; high flow, medium impact with good stiffness and gloss; used for appliances, housings, garden equipment; NSF approved for swimming pool valves in potable water
Form: Solid; Color: Natural
Gen'l. Prop.: Sp. Gr.: 1.04 (23/23 C, D-792); Soften. Pt. (Vicat): 212 F (D1525)

Mech. Prop.: Tens. Str.: 5500 psi (D-638); Tens. Elong.: 15% (D638); Tens. Mod.: 300,000 psi (D638); Flex. Str.: 9000 psi (D790); Flex. Mod.: 350,000 psi (D790); Impact Str. (Izod): 4.0 ft lb/in. notched (23 C, D256); Hardness (Rockwell): 107 R (D785)
Therm. Prop.: Distort. Temp.: 202 F (@ 264 psi, D648)

ABS 707K. Mobil Chem. Co.
Chem. Descrip.: Acrylonitrile/butadiene/styrene copolymer
Category/Applications: General purpose molding grade; medium impact, semi-gloss with good stiffness; used for housewares and household accessories, recreational equipment, housings
Form: Solid; Color: Natural
Gen'l. Prop.: Sp. Gr.: 1.04 (23/23 C, D-792); Soften. Pt. (Vicat): 208 F (D1525)
Mech. Prop.: Tens. Str.: 5600 psi (D-638); Tens. Elong.: 10% (D638); Tens. Mod.: 300,000 psi (D638); Flex. Str.: 9300 psi (D790); Flex. Mod.: 350,000 psi (D790); Impact Str. (Izod): 4.0 ft lb/in. notched (23 C, D256); Hardness (Rockwell): 105 R (D785)
Therm. Prop.: Distort. Temp.: 192 F (@ 264 psi, D648)

ABS 808K. Mobil Chem. Co.
Chem. Descrip.: Acrylonitrile/butadiene/styrene copolymer
Category/Applications: General purpose molding grade; high impact, glossy; excellent moldability and good low temp. properties; used for office and business machines, communications equipment, appliances, sporting equipment
Form: Solid; Color: Natural
Gen'l. Prop.: Sp. Gr.: 1.03 (23/23 C, D-792); Soften. Pt. (Vicat): 218 F (D1525)
Mech. Prop.: Tens. Str.: 5000 psi (D-638); Tens. Elong.: 11% (D638); Tens. Mod.: 235,000 psi (D638); Flex. Str.: 7700 psi (D790); Flex. Mod.: 270,000 psi (D790); Impact Str. (Izod): 6.0 ft lb/in. notched (23 C, D256); Hardness (Rockwell): 93 R (D785)
Therm. Prop.: Distort. Temp.: 203 F (@ 264 psi, D648)

ABS 910K. Mobil Chem. Co.
Chem. Descrip.: Acrylonitrile/butadiene/styrene copolymer
Category/Applications: Electroplating grade copolymer with excellent adhesion in thermocycling for exterior applications; automotive grilles, trim, and bezels
Form: Solid; Color: Gray
Gen'l. Prop.: Sp. Gr.: 1.06 (23/23 C, D-792); Soften. Pt. (Vicat): 220 F (D1525)
Mech. Prop.: Tens. Str.: 6200 psi (D-638); Tens. Elong.: 15% (D638); Tens. Mod.: 320,000 psi (D638); Flex. Str.: 10,000 psi (D790); Flex. Mod.: 350,000 psi (D790); Impact Str. (Izod): 3.0 ft lb/in. notched (23 C, D256); Hardness (Rockwell): 109 R (D785)
Therm. Prop.: Distort. Temp.: 208 F (@ 264 psi, D648)

ABS 920, 940, 960. Hammond Plastics Div./Carl Gordon Industries, Inc.
Chem. Descrip.: ABS
Form: Color: Avail. in natural and a full range of opaque colors
Mech. Prop.: Tens. Str.: 8200, 7000, 6000 psi respectively; Flex. Mod.: 390,000 psi, 346,000 psi, 330,000 psi respectively; Impact Str. (Izod): 2.0, 3.5, 4.8 ft lb/in. notch respectively; Hardness (Rockwell): R 118, 112, 108 respectively
Therm. Prop.: Distort. Temp.: 182 F, 182 F, 180 F respectively
Ref.: Data sheet

Abson 010. Mobay Chem. Corp.
Chem. Descrip.: Flame-retardant ABS resin containing brominated additive
Category/Applications: Thermoplastic injection molding resin offering easy processing, good toughness, and satin gloss, suitable for TV cabinets and business machines; continuous use temp. to 60 C (without impact, UL746B)
Form: Opaque, colorable
Gen'l. Prop.: Density: 1.17 mg/m^3
Mech. Prop.: Tens. Str.: 28 MPa (break); Tens. Elong.: 60% (break); Flex. Str.: 51 MPa (break, 3.2 mm); Flex. Mod.: 2.07 GPa (3.2 mm); Impact Str. (Izod): 134 J/m notch (23 C, 3.2 mm); Hardness (Rockwell): R 90
Therm. Prop.: Distort. Temp.: 74 C (@ 1.82 MPa, unannealed); Coeff. of Linear Exp.: 8.57 $\times$ 10^{-5} mm/mm/C; Flamm.: V-O (@ 3.05 mm, UL94)
Elec. Prop.: Arc Resist.: 24 s
Ref.: Product information

Abson 025. Mobay Chem. Corp.
Chem. Descrip.: Flame-retardant ABS/PVC alloy
Category/Applications: Comp. with excellent extrusion characteristics making it ideal for sheet which can in turn be thermoformed into business machine housings; continuous use temps. to 50 C (without impact, UL746B)
Form: Opaque, colorable
Gen'l. Prop.: Density: 1.20 mg/m^3
Mech. Prop.: Tens. Str.: 30 MPa (break); Tens. Elong.: 125% (break); Flex. Str.: 66 MPa (break, 3.2 mm); Flex. Mod.: 2.34 GPa (3.2 mm); Impact Str. (Izod): 641 J/m notch (23 C, 3.2 mm); Hardness (Rockwell): R 105
Therm. Prop.: Distort. Temp.: 85 C (@ 1.82 MPa, unannealed); Coeff. of Linear Exp.: 8.10 $\times$ 10^{-5} mm/mm/C; Flamm.: V-O (@ 2.54 mm, UL94)
Elec. Prop.: Arc Resist.: 33 s
Ref.: Product information

Abson 042. Mobay Chem. Corp.
Chem. Descrip.: Flame-retardant ABS resin containing a brominated additive
Category/Applications: Injection molding resin offering toughness, rigidity, high heat distortion, high gloss, and good flowability; used for large parts such as business machine housings or TV cabinetry
Form: Opaque, colorable
Gen'l. Prop.: Density: 1.22 mg/m^3
Mech. Prop.: Tens. Str.: 41 MPa (break); Flex. Str.: 70 MPa (break, 3.2 mm); Flex. Mod.: 2.41 GPa (3.2 mm); Impact Str. (Izod): 240 J/m notch (23 C, 3.2 mm); Hardness (Rockwell): R 97
Therm. Prop.: Distort. Temp.: 82 C (@ 1.82 MPa, unannealed); Flamm.: V-O (@ 1.6 mm, UL94)
Ref.: Product information

Abson 110, 120, 130, 140. Mobay Chem. Corp.
Chem. Descrip.: ABS resins
Category/Applications: Series of general purpose, injection molding compounds exhibiting high gloss and characterized by increasing toughness and impact resistance, and decreasing tensile stress and stiffness with the upward numerical change in nomenclature; continuous use temps. to 60 C for Abson 110 (without impact, UL746B)
Form: Opaque, colorable
All properties—Abson 110:
Gen'l. Prop.: Density: 1.06 mg/m^3
Mech. Prop.: Tens. Str.: 51 MPa (break); Tens. Elong.: 15% (break); Flex. Str.: 100 MPa (break, 3.2 mm); Flex. Mod.: 3.10 GPa (3.2 mm); Impact Str. (Izod): 107 J/m notch (23 C, 3.2 mm); Hard-

ness (Rockwell): R 115
Therm. Prop.: Distort. Temp.: 91 C (@ 1.82 MPa, unannealed); Coeff. of Linear Exp.: 4.30 X 10^{-5} mm/mm/C; Flamm.: HB (@ 1.52 mm, UL94)
Elec. Prop.: Arc Resist.: 50 s
Ref.: Product information

Abson 129. Mobay Chem. Corp.
Chem. Descrip.: Flame-retardant ABS/PVC alloy
Category/Applications: Comp. featuring high impact resistance and heat resistance for injection molding and extrusion applications, incl. telecommunication and electrical/electronic components; continuous use temps. to 80 C (without impact, UL746B)
Form: Opaque, colorable
Gen'l. Prop.: Density: 1.16 mg/m^3
Mech. Prop.: Tens. Str.: 41 MPa (break); Tens. Elong.: 25% (break); Flex. Str.: 76 MPa (break, 3.2 mm); Flex. Mod.: 2.28 GPa (3.2 mm); Impact Str. (Izod): 267 J/m notch (23 C, 3.2 mm); Hardness (Rockwell): R 107
Therm. Prop.: Distort. Temp.: 85 C (@ 1.82 MPa, unannealed); Coeff. of Linear Exp.: 8.30 X 10^{-5} mm/mm/C; Flamm.: V-1 (@ 3.18 mm, UL94)
Elec. Prop.: Arc Resist.: 95 s
Ref.: Product information

Abson 135. Mobay Chem. Corp.
Chem. Descrip.: ABS resin
Category/Applications: General purpose injection molding comp. with high impact, high gloss, and high flow performance to yield an excellent cost/performance balance; used in small appliances, toys, telecommunications, etc.; continuous use temps. to 75 C (without impact, UL746B)
Form: Opaque, colorable
Gen'l. Prop.: Density: 1.04 mg/m^3
Mech. Prop.: Tens. Str.: 39 MPa (break); Tens. Elong.: 15% (break); Flex. Str.: 72 MPa (break, 3.2 mm); Flex. Mod.: 2.30 GPa (3.2 mm); Impact Str. (Izod): 321 J/m notch (23 C, 3.2 mm); Hardness (Rockwell): R 106
Therm. Prop.: Distort. Temp.: 87 C (@ 1.82 MPa, unannealed); Coeff. of Linear Exp.: 5.00 X 10^{-5} mm/mm/C; Flamm.: HB (@ 1.5 mm, UL94)
Elec. Prop.: Arc Resist.: 114 s
Ref.: Product information

Abson 161. Mobay Chem. Corp.
Chem. Descrip.: ABS resin
Category/Applications: General purpose injection molding comp. with high gloss, very high impact resistance, and good retention of toughness at low temps. required by such demanding applications as protective headgear; continous use temps. to 60 C (without impact, UL-746B)
Form: Opaque, colorable
Gen'l. Prop.: Density: 1.02 mg/m^3
Mech. Prop.: Tens. Str.: 30 MPa (break); Tens. Elong.: 70% (break); Flex. Str.: 51 MPa (break, 3.2 mm); Flex. Mod.: 1.55 GPa (3.2 mm); Impact Str. (Izod): 481 J/m notch (23 C, 3.2 mm); Hardness (Rockwell): R 88
Therm. Prop.: Distort. Temp.: 89 C (@ 1.82 MPa, unannealed); Coeff. of Linear Exp.: 5.80 X 10^{-5} mm/mm/C; Flamm.: HB (@ 1.6 mm, UL94)
Elec. Prop.: Arc Resist.: 74 s
Ref.: Product information

Abson 171. Mobay Chem. Corp.
Chem. Descrip.: ABS resin
Category/Applications: Comp. offering comparable property performance to Abson 161 combined with excellent thermoformability, to produce a sheet

extrusion comp. for demanding applications, e.g., snowmobile shrouds; continuous use temps. to 60 C (without impact, UL746B)
Form: Opaque, colorable
Gen'l. Prop.: Density: 1.02 mg/m^3
Mech. Prop.: Tens. Str.: 31 MPa (break); Tens. Elong.: 110% (break); Flex. Str.: 52 MPa (break, 3.2 mm); Flex. Mod.: 1.58 GPa (3.2 mm); Impact Str. (Izod): 480 J/m notch (23 C, 3.2 mm); Hardness (Rockwell): R 92
Therm. Prop.: Distort. Temp.: 91 C (@ 1.82 MPa, unannealed); Coeff. of Linear Exp.: 5.80×10^{-5} mm/mm/C; Flamm.: HB (@ 1.6 mm, UL94)
Elec. Prop.: Arc Resist.: 74 s
Ref.: Product information

Abson 200. Mobay Chem. Corp.
Category/Applications: Engineering thermoplastic, blending resin for general purpose applications, featuring med. impact, low gloss
Mech. Prop.: Tens. Str.: 33 MPa (break); Flex. Mod.: 2.27 GPa (3.2 mm); Impact Str. (Izod): 107 J/m notch (23 C, 3.2 mm)
Therm. Prop.: Distort. Temp.: 88 C (@ 1.82 MPa); Flamm.: HB (@ 1.6 mm, UL94)
Ref.: Product information

A-C Copolymer 400A. Allied Chem., Fibers & Plastics Co.
Chem. Descrip.: Low m.w. ethylene-vinyl acetate copolymer
Category/Applications: Pigment dispersant; polystyrene color concentrates
Form: Powder
Ref.: Tech. data bulletin

A-C Copolymer 540A. Allied Chem., Fibers & Plastics Co.
Chem. Descrip.: Low m.w. ethylene-acrylic acid copolymer
Category/Applications: Plastics lubricant and processing aid, pigment dispersant; for internal lubricant PVC, nylon 6 processing aid, nylon color concentrates
Form: Powder
Ref.: Tech. data bulletin

A-C Polyethylene 6, 6A, 8, 8A, 9, 9A, 617, 617A. Allied Chem., Fibers & Plastics Co.
Chem. Descrip.: Low m.w. polyethylene
Category/Applications: Processing lubricant, melt index modifier, pigment dispersant, mold release aid; for external lubricant PVC, color concentrates, polyolefin flow modifiers
Form: "A" designates powdered form; all others prilled or diced
Ref.: Tech. data bulletin

A-C Polyethylene 629A. Allied Chem., Fibers & Plastics Co.
Chem. Descrip.: Oxidized low m.w. polyethylene
Category/Applications: Processing lubricant, mold release aid; balanced (external/internal) lubricant PVC
Form: Powder
Ref.: Tech. data bulletin

Aclar Films. Allied Chem., Fibers & Plastics Co.
Chem. Descrip.: Fluorocarbon thermoplastic films
Category/Applications: High moisture barrier, chemically inert flexibile film for packaging and container liners; packaging requiring exceptional barrier properties, pharmaceutical and cosmetic packaging, clean room and DOD applications, electronic applications as capacitors, printed circuit substrate
Form: Extruded films in various widths

and thicknesses from 0.75 mil; Color: Optically clear
Gen'l. Prop.: Stability: Resistant to acids, bases, and solvents
Toxicity/Handling: Avoid incineration
Std. Pkgs.: Extruded film on 6-in. cores in various widths
Ref.: Tech. data bulletin

Acralen AFR. Bayer AG
Chem. Descrip.: Acrylic acid ester, acrylonitrile
Category/Applications: Acrylic latex self-crosslinking binder for nonwoven fabrics, textile finishing and lamination
Form: Film color: Yellow, transparent
Compositon: 49.5–50.5% solids
Gen'l. Prop.: Density: 1.03 g/cc; Stability: Good resistance to ageing; pH: 6.5–7.0
Mech. Prop.: Hardness (Shore A): ≈ 30 (film)
Std. Pkgs.: 120-kg plastic drums, 600-kg plastic containers, 16–23 ton tank wagons
Ref.: Pamphlet KA 32125e

Acralen ATR. Bayer AG
Chem. Descrip.: Acrylic acid ester, styrene latex
Category/Applications: Acrylic latex self-crosslinking binder for nonwoven fabrics, lamination
Form: Film color: Colorless, transparent
Composition: 47.5–48.5% solids
Gen'l. Prop.: Density: 1.03 g/cc; Stability: Good resistance to ageing; pH: 4.7–5.2
Mech. Prop.: Hardness (Shore A): ≈ 50 (film)
Std. Pkgs.: See Acralen AFR
Ref.: Pamphlet KA 32125e

Acralen BS. Bayer AG
Chem. Descrip.: Butadiene, styrene, acrylonitrile latex
Category/Applications: Acrylic latex self-crosslinking binder for nonwoven fabrics, technical fabrics
Form: Film color: Yellow, transparent
Composition: 39.5–40.5% solids
Gen'l. Prop.: Density: 1.02 g/cc; Stability: Good resistance to ageing; pH: 7.5–8.5
Mech. Prop.: Hardness (Shore A): ≈ 93 (film)
Std. Pkgs.: See Acralen AFR
Ref.: Pamphlet KA 32125e

Acrylite GP, MS-2 Sheet. CY/RO Industries
Chem. Descrip.: Cast acrylic sheet
Category/Applications: Cast acrylic sheet for use in industrial plants and building products, e.g., window glazing, partitions, door panels, lighting fixture diffusers, skylights, fascia panels, tub and shower enclosures
Form: Sheet available in thicknesses from 0.030–4.25 in.; Color: Clear, 6 densities of translucent white, and many colors; Odor and Taste: None
Gen'l. Prop.: Sp. Gr.: 1.19 (for .250 in., D792-66); Stability: Good resistance to many chemicals incl. solutions of inorganic alkalies, e.g., ammonia, dilute acids, e.g., sulfuric acid, aliphatic hydrocarbons, e.g., hexane and VM&P naphtha; good weather, breakage, and heat resistance; Ref. Index: 1.49 (for .250 in., D542-50)
Mech. Prop. (for .250 in. thickness): Tens. Str.: 10,000 psi (D638-72); Tens. Elong.: 4.2% (D638-72); Tens. Mod.: 400,000 psi (D638-72); Flex. Str.: 16,500 psi (D790-71-Method 1); Flex. Mod.: 475,000 psi (D790-71-Method 1); Compr. Str.: 18,000 psi (D695-69); Compr. Mod.: 430,000 psi (D695-69); Shear Str.: 9000 psi (D732-46); Impact Str. (Izod): 0.4 ft lb/in. milled, notched (D256-73); Hardness (Rockwell): M94

(D785-65)
Therm. Prop. (for .250 in. thickness): Conduct.: 1.3 (h)(ft²)(F/in.) (Cenco-Fitch); Distort. Temp.: 210 F (@ 264 psi, D648-56); Coeff. of Linear Exp.: 0.000034 in./in./F (D696-70); Sp. Heat: 0.35 Btu (DuPont 900 Therm. An. Cal.); Flamm.: 1.2 in./min (1/8 in. thickness, D635-74); Self-Ignit. Temp.: 860 F (D-1929-68)
Elec. Prop. (for 0.250 in. thickness): Power Factor: 0.06 (60 cycles, D150-65T); Dielec. Str.: 430 V/mil (1/8 in. thickness, D149-64); Dielec. Const.: 3.5 (60 cycles, D150-65T); Vol. Resist.: 1.6×10^{16} ohm/cm (D257-66); Surf. Resist.: 1.9×10^{15} ohm (D257-66)
Toxicity/Handling: Combustible thermoplastic
Std. Pkgs.: Available in more than 50 std. sizes from 36 X 48 in. to 120 X 144 in.
Ref.: Fabrication manual

Acryloid A-10. Rohm & Haas Co.
Chem. Descrip.: Methyl methacrylate polymer
Category/Applications: Thermoplastic resin used for heat-resistant white enamels with good color retention; also used for vinyl clear coatings and printing inks (gravure, overprints); hardest and most resistant to solvents of Acryloid thermoplastics
Form: Liquid
Composition: 30% in Cellosolve acetate
Gen'l. Prop.: Density: 8.6 lb/gal; Visc. (Brookfield): 800–1200 cps; Flash Pt.: 128 F (TOC)
Ref.: Product pamphlet C-335

Acryloid A-11. Rohm & Haas Co.
Chem. Descrip.: Methyl methacrylate polymer
Category/Applications: Solid grade of Acryloid A-10
Form: Solid
Composition: 100% solids
Gen'l. Prop.: Sol.: Sol. in diacetone alcohol, most chlorinated hydrocarbons, most esters, dioxane, toluene, acetone, MEK
Ref.: Product pamphlet C-335

Acryloid A-21. Rohm & Haas Co.
Chem. Descrip.: Methyl methacrylate polymer
Category/Applications: Thermoplastic resin for automotive finishing and refinishing with excellent adhesion, hardness, and exterior durability; also for vinyl clear coatings, plastisol primers, specification lacquers, rail car and aircraft finishes
Form: Liquid
Composition: 30% solids in toluene/butanol (90/10)
Gen'l. Prop.: Density: 7.8 lb/gal; Visc. (Brookfield): 235–365 cps; Flash Pt.: 50 F (TOC)
Ref.: Product pamphlet C-335

Acryloid A-21LV. Rohm & Haas Co.
Chem. Descrip.: Methyl methacrylate polymer
Category/Applications: Thermoplastic resin with lower visc. and better cellulosics compatibility than Acryloid A-21; used in bake-sand-bake finishes, auto refinishes, aircraft and rail car finishes, vinyl clear coatings, plastisol primers
Form: Liquid
Composition: 30% solids in toluene/MEK/butanol (50/40/10)
Gen'l. Prop.: Density: 7.8 lb/gal; Visc. (Brookfield): 210–280 cps; Flash Pt.: 32 F (TOC)
Ref.: Product pamphlet C-335

Acryloid A-30. Rohm & Haas Co.
Chem. Descrip.: Methyl methacrylate polymer
Category/Applications: Thermoplastic resin with good abrasion resistance, toughness, and flexibility; used in vinyl topcoating with vinyl copolymers and homopolymers
Form: Powder
Composition: 100% solids
Ref.: Product pamphlet C-335

Acryloid A-101. Rohm & Haas Co.
Chem. Descrip.: Methyl methacrylate polymer
Category/Applications: Thermoplastic resin used for vinyl topcoating and printing inks (gravure, overprints); compatible with most vinyl copolymers
Form: Liquid
Composition: 40% solids in MEK
Gen'l. Prop.: Density: 7.9 lb/gal; Visc. (Brookfield): 700–1400 cps (@ 35% solids); Flash Pt.: 28 F (TOC)
Ref.: Product pamphlet C-335

Acryloid AT-50. Rohm & Haas Co.
Chem. Descrip.: Hydroxyl-type acrylic polymer
Category/Applications: Thermosetting resin for applications requiring ultimate hardness, and chemical, stain, and solvent resistance; used in product finishing (internal and external general purpose); light appliance finishing
Form: Liquid
Composition: 50% solids in xylene/butanol/methyl Cellosolve (60/22/18)
Gen'l. Prop.: Visc. (Brookfield): 625–1755 cps; Flash Pt.: 78 F (PMCC)
Curing Charact.: Typical Cure: 30 min @ 300 F
Ref.: Leaflet RB5-10

Acryloid AT-51. Rohm & Haas Co.
Chem. Descrip.: Hydroxyl-type acrylic polymer
Category/Applications: Thermosetting resin similar to Acryloid AT-50
Form: Liquid
Composition: 50% solids in xylene/butanol (78/22)
Gen'l. Prop.: Visc. (Brookfield): 1070–2265 cps; Flash Pt.: 76 F (PMCC)
Curing Charact.: Typical Cure: 30 min @ 300 F
Ref.: Leaflet RB5-10

Acryloid AT-56. Rohm & Haas Co.
Chem. Descrip.: Hydroxyl-type acrylic polymer
Category/Applications: Thermosetting resin with med. hardness, high gloss, long-term durability; used for automotive topcoats, exterior coil coating, interior and exterior general purpose product finishing
Form: Liquid
Composition: 50% solids in xylene/butanol (90/10)
Gen'l. Prop.: Visc. (Brookfield): 450–700 cps; Flash Pt.: 78 F (PMCC)
Curing Charact.: Typical Cure: 30 min @ 300 F (crosslinker required)
Ref.: Leaflet RB5-10

Acryloid AT-63. Rohm & Haas Co.
Chem. Descrip.: Hydroxyl-type acrylic polymer
Category/Applications: Thermosetting resin with fastest cure of Acryloid thermosets, med. hardness; used for general product finishing (interior, exterior, low temp. conversion), exterior coil coating; can react with isocyanates to make acrylic-urethane coatings
Form: Liquid
Composition: 50% solids in xylene

Gen'l. Prop.: Visc. (Brookfield): 1000–3000 cps; Flash Pt.: 76 F (PMCC)
Curing Charact.: Typical Cure: 30 min @ 250 F (crosslinker required)
Ref.: Leaflet RB5-10

Acryloid AT-64. Rohm & Haas Co.
Chem. Descrip.: Hydroxyl-type acrylic polymer
Category/Applications: Thermosetting resin with med.-high hardness for general use; used for appliance and product finishes needing chemical resistance (light appliance, hospital and lab equipment, interior and exterior product finishes, etc.)
Form: Liquid
Composition: 50% solids in xylene/butanol (90/10)
Gen'l. Prop.: Visc. (Brookfield): 1500–3300 cps; Flash Pt.: 74 F (PMCC)
Curing Charact.: Typical Cure: 30 min @ 300 F (crosslinker required)
Ref.: Leaflet RB5-10

Acryloid AT-70. Rohm & Haas Co.
Chem. Descrip.: Carboxyl-type acrylic polymer
Category/Applications: Thermosetting resin with excellent adhesion, hardness, flexibility, and color retention on overbaking; used for appliance enamels, and interior general purpose product finishing
Form: Liquid
Composition: 50% solids in xylene/Cellosolve acetate (75/25)
Gen'l. Prop.: Visc. (Brookfield): 1200–2500 cps; Flash Pt.: 82 F (PMCC)
Curing Charact.: Typical Cure: 20 min @ 350 F (crosslinker required)
Ref.: Leaflet RB5-10

Acryloid AT-71. Rohm & Haas Co.
Chem. Descrip.: Carboxyl-type acrylic polymer
Category/Applications: Thermosetting resin similar to Acryloid AT-70 but in slow solvent; used with Acryloid AT-75 for metal decorating needing maximum resistance to severe fabrication and dry heat; also for interior coil coating
Form: Liquid
Composition: 50% solids in Solvesso 150/Cellosolve acetate (75/25)
Gen'l. Prop.: Visc. (Brookfield): 2700–3600 cps; Flash Pt.: 116 F (PMCC)
Curing Charact.: Typical Cure: 20 min @ 350 F (crosslinker required)
Ref.: Leaflet RB5-10

Acryloid AT-75. Rohm & Haas Co.
Chem. Descrip.: Carboxyl-type acrylic polymer
Category/Applications: Soft thermosetting resin supplied in slow solvent for metal decorating
Form: Liquid
Composition: 50% solids in Solvesso 150/Cellosolve acetate (75/25)
Gen'l. Prop.: Visc. (Brookfield): 2700–3600 cps; Flash Pt.: 118 F (PMCC)
Curing Charact.: Typical Cure: 20 min @ 350 F (crosslinker required)
Ref.: Leaflet RB5-10

Acryloid B-44. Rohm & Haas Co.
Chem. Descrip.: Methyl methacrylate copolymer
Category/Applications: Thermoplastic resin with good hardness, flexibility, and adhesion; used for finishes for vacuum-metallized plastics; aluminum sash and chrome plate; luminescent finishes; metal and masonry clear coatings; farm and factory machinery; product finishing for light appliances, metal furniture, metal cabinets; specification lacquers; auto and aircraft finishes
Form: Liquid
Composition: 40% solids in toluene/methyl Cellosolve (95/5); also avail.

in 100% solids grade
Gen'l. Prop.: Density: 8.1 lb/gal; Visc. (Brookfield): 855–1700 cps; Flash Pt.: 38 F (TOC)
Ref.: Product pamphlet C-335

Acryloid B-48N. Rohm & Haas Co.
Chem. Descrip.: Methyl methacrylate copolymer
Category/Applications: Thermoplastic resin with good toughness, exterior durability, adhesion to bare metal; used for clear coating over copper, brass, bronze, zinc; aerosol coatings; plastic coatings; farm and factory machinery; product finishing for light appliances, metal furniture and cabinets; auto, rail car, and marine finishes
Form: Liquid
Composition: 45% solids in toluene
Gen'l. Prop.: Density: 8.1 lb/gal; Visc. (Brookfield): 6000–10,000 cps; Flash Pt.: 66 F (TOC)
Ref.: Product pamphlet C-335

Acryloid B-50. Rohm & Haas Co.
Chem. Descrip.: Methyl methacrylate copolymer
Category/Applications: Thermoplastic resin used for coatings for plastics, seamless floors, treated metals; aerosol coatings; farm and factory machinery; product finishing for light appliances, metal furniture, metal cabinets; auto, rail car, and marine finishes
Form: Liquid
Composition: 45% solids in toluene
Gen'l. Prop.: Density: 8.1 lb/gal; Visc. (Brookfield): 5000–9000 cps; Flash Pt.: 72 F (TOC)
Ref.: Product pamphlet C-335

Acryloid B-66. Rohm & Haas Co.
Chem. Descrip.: Methyl/butyl methacrylate copolymer
Category/Applications: Thermoplastic resin with fast air-dry and excellent compatibility with other film formers; used in specialty clear and pigmented coatings on metal, wood, plastic, and fabric; aerosols (metallics, clears, solid colors); farm and factory machinery; fluorescent pigment dispersion; printing inks (gravure, overprints); plastic coatings (barrier, top coats); product finishing for light appliances, metal furniture, metal cabinets; auto, rail car, and marine finishes; specification lacquers
Form: Liquid
Composition: 50% solids in toluene; also avail. in 100% solids grade
Gen'l. Prop.: Density: 8.2 lb/gal; Visc. (Brookfield): 2000–5000 cps; Flash Pt.: 62 F (TOC)
Ref.: Product pamphlet C-335

Acryloid B-67. Rohm & Haas Co.
Chem. Descrip.: Isobutyl methacrylate polymer
Category/Applications: Thermoplastic resin which improves properties of med. and long-oil alkyd and oleo-resinous varnishes; used in enamels with and without alkyd modification; aerosols (metallics, clears, solid colors); clear coatings (wood, masonry); farm and factory machinery; fluorescent pigment dispersion; plastic top coats; marine finishes
Form: Liquid
Composition: 45% solids in VM&P naphtha; also avail. in 100% solids grade
Gen'l. Prop.: Density: 7.3 lb/gal; Visc. (Brookfield): 900–1600 cps; Flash Pt.: 56 F (TOC)
Ref.: Product pamphlet C-335

Acryloid B-67MT. Rohm & Haas Co.
Chem. Descrip.: Isobutyl methacrylate polymer
Category/Applications: Similar to Acryloid B-67 but with slightly slower drying

speed and a higher flash point; used in plastic coating, printing ink, overprint varnishes, and refinish lacquers
Form: Liquid
Composition: 45% solids in mineral thinner; also avail. in 100% solids grade
Gen'l. Prop.: Density: 7.4 lb/gal; Visc. (Brookfield): 800–1600 cps; Flash Pt.: 120 F (TOC)
Ref.: Product pamphlet C-335

Acryloid B-72. Rohm & Haas Co.
Chem. Descrip.: Ethyl methacrylate copolymer
Category/Applications: Thermoplastic resin compatible with vinyl and silicone resins, with good stability, durability, and transparency; used in general purpose clear and pigmented coatings; aerosols (metallics, clears, solid colors); printing inks (flexographic, overprints); plastic coatings (barrier, top coats); product finishing (light appliances, metal furniture, metal cabinets); specification lacquers
Form: Liquid
Composition: 50% solids in toluene; also avail. in 100% solids grade
Gen'l. Prop.: Density: 8.3 lb/gal; Visc. (Brookfield): 3700–6200 cps; Flash Pt.: 60 F (TOC)
Ref.: Product pamphlet C-335

Acryloid B-82. Rohm & Haas Co.
Chem. Descrip.: Methyl methacrylate copolymer
Category/Applications: Low cost thermoplastic resin with typical acrylic properties; used in general purpose coatings; aerosols (metallics, clears, solid colors); clear coatings (masonry, cloth); flexographic printing inks; plastic coatings (barrier, top coats); product finishing (light appliances, metal furniture, metal cabinets); specification lacquers; aircraft finishes
Form: Liquid
Composition: 50% solids in toluene; also avail. in 100% solids grade
Gen'l. Prop.: Density: 8.4 lb/gal; Visc. (Brookfield): 4000–7000 cps; Flash Pt.: 68 F (TOC)
Ref.: Product pamphlet C-335

Acryloid B-99. Rohm & Haas Co.
Chem. Descrip.: Methyl methacrylate copolymer
Category/Applications: Thermoplastic resin used in pigment grind to improve gloss; improves short and med.-oil alkyd properties; used for grinding medium; refinish lacquer maintenance parts; aerosols (clears, solid colors); farm and factory machinery; conventional pigment dispersion; gravure printing inks; product finishing (light appliances, metal furniture, metal cabinets); auto refinishes; rail car and marine finishes
Form: Liquid
Composition: 50% solids in toluene; also avail. in 100% solids grade
Gen'l. Prop.: Density: 8.4 lb/gal; Visc. (Brookfield): 4000–7000 cps; Flash Pt.: 68 F (TOC)
Ref.: Product pamphlet C-335

Acryloid C-10LV. Rohm & Haas Co.
Chem. Descrip.: Methyl acrylate polymer
Category/Applications: Thermoplastic resin for brushing and dipping coating, thermoplastic adhesive; softest and most elastic of Acryloid resins
Form: Liquid
Composition: 40% in toluene
Gen'l. Prop.: Density: 8.2 lb/gal; Visc. (Brookfield): 38–82 cps; Flash Pt.: 40 F (TOC)
Ref.: Product pamphlet C-335

Acryloid F-10. Rohm & Haas Co.
Chem. Descrip.: Butyl methacrylate polymer

Category/Applications: Thermoplastic resin with good adhesion and elongation; used in clear and pigmented coatings on metal; luminescent pigment vehicles; aerosols (clears, solid colors); gravure printing inks
Form: Liquid
Composition: 40% solids in mineral thinner/Amsco F (90/10)
Gen'l. Prop.: Density: 7.4 lb/gal; Visc. (Brookfield): 1400-2600 cps; Flash Pt.: 116 F (TOC)
Ref.: Product pamphlet C-335

Acryloid NAD-10. Rohm & Haas Co.
Chem. Descrip.: Stable dispersion of acrylic polymer particles in VM&P naphtha
Category/Applications: Nonaqueous acrylic dispersion with fast solvent release, glossy films, good adhesion; used in rotogravure inks, fluorescent inks, general coatings
Form: Liquid
Composition: 40% solids in VM&P naphtha
Gen'l. Prop.: Density: 7.1 lb/gal; Visc. (Brookfield): 290–600 cps; Flash Pt.: 64 F (TOC)
Ref.: Product pamphlet C-335

Acryloid WR-97. Rohm & Haas Co.
Chem. Descrip.: Hydroxyl-type acrylic polymer
Category/Applications: Thermosetting acrylic solution resin reducible with water or solvent; crosslinkable with urea and melamine resins; universal pigment disperser; outstanding compatibility; also used in exterior coil coating, product and appliance finishing
Form: Liquid
Composition: 70% solids in isopropanol/butyl Cellosolve (83/17)
Gen'l. Prop.: Visc. (Brookfield): 14,000-17,000 cps; Flash Pt.: 67 F (PMCC)
Curing Charact.: Typical Cure: 30 min @ 300 F (crosslinker required)
Ref.: Leaflet RB5-10

Acrysol ASE-60. Rohm & Haas Co.
Chem. Descrip.: Crosslinked acrylic emulsion copolymer
Category/Applications: Used to suspend pigments and fillers in water-based paints, inks, or other coatings, and the abrasive particles in waxes or polishes; viscosity modifier for emulsion and latex compounds; temporary binder; thickener of aq. dispersions, polar solvents, surfactant solutions; its sodium salt is used for emulsification and emulsion stabilization
Form: Milky liquid
Composition: 28.0 ± 0.5% solids
Gen'l. Prop.: Sp. Gr.: 1.054; Visc. (Brookfield): 10 cps (as supplied, 12 rpm); pH: ≈ 3.5
Ref.: Leaflet RC-9

Acrysol G-110. Rohm & Haas Co.
Chem. Descrip.: Ammonium polyacrylate solution
Category/Applications: Thickening and stabilizing agent for certain synthetic latices; used in coatings, adhesives, making dipped, cast, and molded goods, cements for combining rug backing and pile fabrics, and in spraying, spreading, brushing, and extruding compounds
Form: Slightly hazy liquid; Color: Colorless
Gen'l. Prop.: Sp. Gr.: 1.06; Visc.: 90–170 cps (5% solids in water); Stability: Good; pH: 8.5–9.5
Ref.: Product information C-1

Acrysol I-94. Rohm & Haas Co.
Chem. Descrip.: Carboxyl-type aqueous colloidal acrylic
Category/Applications: Resin which

forms tough, durable, glossy, water-resistant films combining best properties of acrylic solution resins (pigment wetting and dispersion, easy transfer from a roller coater) and of emulsion polymers (toughness, water resistance, chemical resistance); can be further solubilized by the addition of a water-miscible solvent or an alkali; can be crosslinked with other water-sol. resins, e.g., melamine, urea; Acrysol I-94 is an excellent vehicle for printing inks, pigment grinding medium for all aq. coatings
Form: Liquid
Mech. Prop.: Hardness (Tukon): 12 KHN
Ref.: Leaflet RB5-10

Acrysol WS-12. Rohm & Haas Co.
Chem. Descrip.: Carboxyl-type aqueous colloidal acrylic
Category/Applications: See Acrysol I-94; WS-12 is a universal vehicle for air-dry (as supplied) or baking finishes (crosslinked with melamine resin), with good adhesion to glass, unprimed metal, plastics; fast dry; tough, durable
Form: Liquid
Mech. Prop.: Hardness (Tukon): 8 KHN
Ref.: Leaflet RB5-10

Acrysol WS-24. Rohm & Haas Co.
Chem. Descrip.: Carboxyl-type aqueous colloidal acrylic
Category/Applications: See Acrysol I-94; WS-24 has intermediate hardness, forms tough, glossy, water-resistant films at R.T.; can be crosslinked with other water-sol. resins at temps. as low as 250 F
Form: Liquid
Mech. Prop.: Hardness (Tukon): 6 KHN
Ref.: Leaflet RB5-10

Acrysol WS-32. Rohm & Haas Co.
Chem. Descrip.: Carboxyl-type aqueous colloidal acrylic
Category/Applications: See Acrysol I-94; WS-32 is a softer form of Acrysol WS-24; forms tacky films; improves flexibility and adhesion when added to harder resins
Form: Liquid
Mech. Prop.: Hardness (Tukon): 1 KHN
Ref.: Leaflet RB5-10

Acrysol WS-50. Rohm & Haas Co.
Chem. Descrip.: Carboxyl-type aqueous colloidal acrylic
Category/Applications: See Acrysol I-94; WS-50 is a slightly harder form of Acrysol WS-24; improves water resistance of uncrosslinked films
Form: Liquid
Mech. Prop.: Hardness (Tukon): 8 KHN
Ref.: Leaflet RB5-10

Adiprene Series. Du Pont Co.
Chem. Descrip.: Urethane rubber
Category/Applications: Reaction product of diisocyanate and polyalkylene ether glycol avail. in liquid and one solid grade; offers a combination of the resilience of rubber and the hardness of structural plastics, outstanding impact strength, load bearing capacity, and abrasion resistance; used for solid industrial tires, rolls, bearings, wear pads, die pads and other metal forming tools, diaphragms, seals and other mechanical goods, equipment coatings, linings, mining industry wear parts, and compounds for potting and encapsulation of electronic devices
Gen'l. Prop.: Stability: Good abrasion resistance and low temp. properties; excellent resistance to ozone, oxidation, and oils
Toxicity/Handling: Adiprene L rubbers contain small amounts of free toluene diisocyanate, an air contaminant, and Adiprene LW rubbers contain methylene-bis(4-cyclohexyl isocyanate), a

strong skin sensitizer and severe irritant to skin and mucous membranes; MOCA urethane curative may cause cancer and may be harmful if inhaled or absorbed through skin; Caytur 21 and 22 curing agents may cause eye and skin irritation and may be harmful if inhaled at processing temps.; other compounding ingredients, incl. MDA, may present hazards during handling
Ref.: Data sheet

Adiprene BL-16. Du Pont Co.
Chem. Descrip.: Urethane elastomer
Category/Applications: Liquid elastomer with blocked isocyanate curing sites which can be activated by heating and which allow protracted storage in the presence of moisture and active vulcanizing agents; produces tough, high modulus, highly extensible cured products in thin sections, such as cast films and coatings; well suited as a vehicle for abrasion-resistant coatings for fabric and rubber, and for solvent-resistant adhesives
Form: Viscous liquid; Color: Amber
Composition: 5.3–5.8/ blocked isocyanate content
Gen'l. Prop.: Sol.: Sol. in most common solvents, e.g., toluene, ethyl acetate, MEK, and trichloroethylene; limited sol. in aliphatic solvents and alcohols; Sp. Gr.: 1.06 ± 0.02 (25/4 C); Visc. (Brookfield): 7500–15,000 cps (50 C); Stability: Excellent raw polymer storage stability @ R.T.
Vulcanizate Properties with 11.8 phr of methylene dianiline curing agent (LD-2729):
Mech. Prop.: Tens. Str.: 37.2 MPa; Tens. Elong.: 455% (@ break); Tens. Mod.: 11.4 MPa (@ 100% mod.); Hardness (Durometer A): 92
Curing Charact.: Cure 1 h @ 130 C
Toxicity/Handling: During heat cure, the blocking agent (MEK), a flammable liquid, is evolved; tolylene diisocyanate may also be released; avoid inhalation of these vapors
Std. Pkgs.: 450-lb drums
Ref.: Data sheet AP-260.BL16

Adiprene CM. Du Pont Co.
Chem. Descrip.: Polyether urethane rubber
Category/Applications: Sulfur-curable urethane rubber offering outstanding resistance to abrasion, good low temp. characteristics and resistance to heat deterioration, ozone cracking, weathering, and swelling in oils or solvents; processable in conventional equipment to produce products with outstanding resistance to abrasion and water, e.g., footwear (soles and heels), electrical (abrasion-resistant cable jackets), oil well supplies (pipe protectors, pistons, valve inserts), belts (conveyor and V-belt covers), hose (sandblast and suction hose tubes, airdrill and jack hammer covers), coated fabrics (diaphragms, tarpaulins), rolls (textile cots, high-pressure rolls), solid tires, and misc. molded goods (packing, gaskets)
Form: Solid slabs; Color: Orange; Odor: Faint, characteristic
Gen'l. Prop.: Sol.: Sol. in tetrahydrofuran, MEK, and dimethylformamide; swollen by chlorinated solvents and petroleum fractions; Sp. Gr.: 1.06; Visc. (Mooney): ≈ 60 (MS-10 @ 100 C); Stability: Excellent storage stability
Mech. Prop.: Tens. Str.: 281–352 kg/cm^2; Tens. Elong.: Good; Tear Str.: 71.2–89 kg/cm); Hardness (Durometer A): 50–90
Therm. Prop.: Flamm.: Burns readily unless specifically compounded for flame resistance
Elec. Prop. (of typical insulation-type comp.): Dissip. Factor: 0.08 (100 cps);

Std. Pkgs.: 4 slabs to a 50-lb box
Ref.: Booklet MLD A-87069

Adiprene L-42. Du Pont Co.
Chem. Descrip.: Urethane rubber
Category/Applications: Liquid urethane rubber for fabrication of products of lower hardness with high flexibility at low temps.
Form: Liquid; Color: Honey
Composition: 2.8% NCO content
Gen'l. Prop.: Sol.: Sol. in aromatic and chlorinated hydrocarbons, ketones, and esters; limited sol. in aliphatic solvents; Sp. Gr.: 1.03 (25/4 C); Visc.: 17,000 cps (30 C)
Vulcanizate Properties with 8.8 phr of MOCA urethane curative:
Mech. Prop.: Tens. Str.: 20.7 MPa; Tens. Elong.: 800%; Tens. Mod.: 2.8 MPa (100% mod.); Tear Str.: 12.2 kN/m (D-470); Hardness (Durometer A): 80
Curing Charact.: Working Life: 7 min @ 100 C; Cure Schedule: Cure 3 h @ 100 C
Std. Pkgs.: 40-lb cans and 450-lb drums
Ref.: Data sheet AP-210.1 (R1)

Adiprene L-83. Du Pont Co.
Chem. Descrip.: Urethane rubber
Category/Applications: Liquid urethane rubber for fabrication of products in the 80–85 durometer A hardness range; esp. suitable for wheels, rolls, and mechanical goods
Form: Liquid; Color: Honey
Composition: 3.25% NCO content
Gen'l. Prop.: Sol.: See Adiprene L-42; Sp. Gr.: 1.05 (25/4 C); Visc.: 28,000 cps (30 C)
Vulcanizate Properties with 10.3 phr of MOCA urethane curative:
Mech. Prop.: Tens. Str.: 30.3 MPa; Tens. Elong.: 580%; Tens. Mod.: 4.8 MPa (@ 100% mod.); Tear Str.: 14.9 kN/m (D-470); Hardness (Durometer A): 83
Curing Charact.: Working Life: 5 min @ 100 C; Cure Schedule: 3 h @ 100 C
Std. Pkgs.: 40-lb cans and 450-lb drums
Ref.: Data sheet AP-210.1(R1)

Adiprene L-100. Du Pont Co.
Chem. Descrip.: Urethane rubber
Category/Applications: Liquid urethane rubber for fabrication of products of high hardness (up to 90A durometer), load-bearing capacity, and excellent resistance to abrasion, oils, oxidation, and ozone
Form: Liquid; Color: Honey
Composition: 4.1% NCO content
Gen'l. Prop.: Sol.: See Adiprene L-42; Sp. Gr.: 1.06 (25/4 C); Visc.: 18,000 cps (30 C)
Vulcanizate Properties with 12.5 phr of MOCA urethane curative:
Mech. Prop.: Tens. Str.: 31.0 MPa; Tens. Elong.: 450%; Tens. Mod.: 7.6 MPa (@ 100% mod.); Tear Str.: 13.1 kN/m (D-470); Hardness (Durometer A): 90
Curing Charact.: Working Life: 10 min @ 100 C; Cure Schedule: 1–3 h @ 100 C
Std. Pkgs.: 40-lb cans and 450-lb drums
Ref.: Data sheet AP-210.1(R1)

Adiprene L-167. Du Pont Co.
Chem. Descrip.: Urethane rubber
Category/Applications: Liquid urethane rubber for production of harder (to 50D durometer), higher modulus products than Adiprene L-100
Form: Liquid; Color: Honey
Composition: 6.3% NCO content
Gen'l. Prop.: Sol.: See Adiprene L-42; Sp. Gr.: 1.07 (25/4 C); Visc.: 6000 cps (30 C)
Vulcanizate Properties with 19.5 phr of MOCA urethane curative:
Mech. Prop.: Tens. Str.: 34.5 MPa; Tens. Elong.: 400%; Tens. Mod.: 12.4 MPa (@ 100% mod.); Tear Str.: 26.2 kN/m (D-470); Hardness (Durometer A): 95

Curing Charact.: Working Life: 5 min @ 85 C; Cure Schedule: 1 h @ 100 C
Std. Pkgs.: 40-lb cans and 450-lb drums
Ref.: Data sheet AP-210.1(R1)

Adiprene L-200. Du Pont Co.
Chem. Descrip.: Urethane rubber
Category/Applications: Liquid urethane rubber for production of harder (60D durometer), higher modulus products than Adiprene L-167
Form: Liquid; Color: Honey
Composition: 7.5% NCO content
Gen'l. Prop.: Sol.: See Adiprene L-42; Sp. Gr.: 1.09 (25/4 C); Visc.: 8000 cps (30 C)
Vulcanizate Properties with 23.2 phr of MOCA urethane curative:
Mech. Prop.: Tens. Str.: 57.2 MPa; Tens. Elong.: 320%; Tens. Mod.: 20.7 MPa (@ 100% mod.); Tear Str.: 23.6 kN/m (D-470); Hardness (Durometer D): 58
Curing Charact.: Working Life: 3 min @ 85 C; Cure Schedule: 1 h @ 100 C
Std. Pkgs.: 40-lb cans and 450-lb drums
Ref.: Data sheet AP-210.1(R1)

Adiprene L-213. Du Pont Co.
Chem. Descrip.: Urethane rubber
Category/Applications: Liquid urethane rubber which will give products of extreme hardness (to 75D durometer) and high impact resistance as well as load bearing capacity and resistance to physical abuse; specially designed for plastic-gum processing
Form: Liquid; Color: Honey
Composition: 9.3% NCO content
Gen'l. Prop.: Sol.: See Adiprene L-42; Sp. Gr.: 1.11 (25/4 C); Visc.: 10,000 cps (30 C)
Vulcanizate Properties with 26.6 phr of MOCA urethane curative:
Mech. Prop.: Tens. Str.: 60.7 MPa; Tens. Elong.: 240%; Tens. Mod.: 26.9 MPa (@ 100% mod.); Flex. Mod.: 800 MPa (D-790); Impact Str. (Izod): 800 J/m notch; Tear Str.: 25.4 kN/m (D470); Hardness (Durometer D): 73
Therm. Prop.: Distort. Temp.: 85 C
Curing Charact.: Working Life: 1.5 min @ 85 C; Cure Schedule: 1 h @ 100 C
Std. Pkgs.: 40-lb cans and 450-lb drums
Ref.: Data sheet AP-210.1(R1)

Adiprene L-300. Du Pont Co.
Chem. Descrip.: Urethane rubber
Category/Applications: Liquid urethane rubber for fabrication of products in the 87A durometer range, when cured with Caytur® 21; also processable with MBCA where rapid mold cycles are needed
Form: Liquid; Color: Honey
Gen'l. Prop.: Sp. Gr.: 1.06; Visc.: 20,000 cps
Std. Pkgs.: 40-lb cans and 450-lb drums
Ref.: Data sheet

Adiprene L-315. Du Pont Co.
Chem. Descrip.: Urethane rubber
Category/Applications: Similar to Adiprene L-213 with somewhat greater impact resistance, load-bearing capacity, and resistance to physical abuse
Form: Liquid; Color: Honey
Composition: 9.45% NCO content
Gen'l. Prop.: Sol.: See Adiprene L-42; Sp. Gr.: 1.11 (25/4 C); Visc.: 15,000 cps (30 C)
Vulcanizate Properties with 26.0 phr of MOCA urethane curative:
Mech. Prop.: Tens. Str.: 62.0 MPa; Tens. Elong.: 210%; Tens. Mod.: 33.4 MPa (@ 100% mod.); Flex. Mod.: 662 MPa (D-790); Impact Str. (Izod): 640 J/m notch; Tear Str.: 19.2 kN/m (D470); Hardness (Durometer D): 73
Therm. Prop.: Distort. Temp.: 140 C
Curing Charact.: Working Life: 1 min @ 77 C; Cure Schedule: 1–2 h @ 100 C
Std. Pkgs.: 40-lb cans and 450-lb drums
Ref.: Data sheet AP-210.1(R1)

Adiprene L-325. Du Pont Co.
Chem. Descrip.: Urethane rubber
Category/Applications: Liquid urethane rubber which will give products of extreme hardness (to 75D durometer) and exceptional impact resistance as well as load bearing capacity and resistance to physical abuse; allows more processing time and products have better hydrolysis resistance than Adiprene L-213 and L-315
Form: Liquid; Color: Honey
Composition: 9.15% NCO content
Gen'l. Prop.: Sol.: See Adiprene L-42; Sp. Gr.: 1.11 (25/4 C); Visc.: 20,000 cps (30 C)
Vulcanizate Properties with 25.8 phr of MOCA urethane curative:
Mech. Prop.: Tens. Str.: 64.3 MPa; Tens. Elong.: 260%; Tens. Mod.: 25.8 MPa (@ 100% mod.); Flex. Mod.: 586 MPa (D-790); Impact Str. (Izod): No break; Hardness (Durometer D): 72
Therm. Prop.: Distort. Temp.: 145 C
Curing Charact.: Working Life: 2.5 min @ 70 C; Cure Schedule: 6 h @ 100 C
Std. Pkgs.: 40-lb cans and 450-lb drums
Ref.: Data sheet AP-210.1(R1)

Adiprene L-367. Du Pont Co.
Chem. Descrip.: Urethane rubber
Category/Applications: Liquid urethane rubber which will produce harder (95A durometer), higher modulus products than Adiprene L-300, when cured with Caytur® 21; also processable with MBCA where rapid mold cycles are needed
Form: Liquid; Color: Honey
Gen'l. Prop.: Sp. Gr.: 1.07; Visc.: 6000 cps (30 C)
Std. Pkgs.: 40-lb cans, 50-lb drums
Ref.: Data sheet

Adiprene L-700. Du Pont Co.
Chem. Descrip.: Urethane rubber
Category/Applications: Liquid urethane rubber which will produce lower hardness (to 90A durometer) products with approx. the same modulus as L-367 when cured with Caytur® 21; also processable with MBCA curing agent where rapid mold cycles are needed
Form: Liquid; Color: Honey
Composition: 5.5% NCO content
Gen'l. Prop.: Sol.: See Adiprene L-42; Sp. Gr.: 1.07 (26/4 C); Visc.: 8500 cps (30 C)
Vulcanized Properties with 27.2 phr of Caytur® 21 curing agent:
Mech. Prop.: Tens. Str.: 32.4 MPa; Tens. Elong.: 430%; Tens. Mod.: 9.0 MPa (@ 100% mod.); Tear Str.: 17.5 kN/m (D-470); Hardness (Durometer A): 92
Curing Charact.: Working Life: 20+ min @ 70 C; Cure Schedule: 2 h @ 121 C
Std. Pkgs.: 40-lb cans and 450-lb drums
Ref.: Data sheet AP-210.1(R1)

Adiprene L-767. Du Pont Co.
Chem. Descrip.: Urethane rubber
Category/Applications: Liquid urethane rubber which will produce harder (to 50D durometer), higher modulus products than Adiprene L-700 when cured with Caytur® 21; also processable with MBCA where rapid mold cycles are needed
Form: Liquid; Color: Honey
Composition: 7.8% NCO content
Gen'l. Prop.: Sol.: See Adiprene L-42; Sp. Gr.: 1.09 (25/4 C); Visc.: 7000 cps (30 C)
Vulcanizate Properties with 38.6 phr of Caytur® 21 curing agent:
Mech. Prop.: Tens. Str.: 27.6 MPa; Tens. Elong.: 320%; Tens. Mod.: 12.4 MPa (@ 100% mod.); Tear Str.: 22.7 kN/m (D-470); Hardness (Durometer D): 53
Curing Charact.: Working Life: 25+ min @ 70 C; Cure Schedule: 2 h @ 121 C
Std. Pkgs.: 40-lb cans and 450-lb drums
Ref.: Data sheet AP-210.1(R1)

Adiprene LW-500. Du Pont Co.
Chem. Descrip.: Urethane rubber
Form: Liquid
Composition: 3.15% NCO content
Gen'l. Prop.: Sol.: See Adiprene L-42; Sp. Gr.: 1.01 (25/4 C); Visc.: 3300 cps (100 C)
Vulcanizate Properties with 7.6 phr of MDA curing agent:
Mech. Prop.: Tens. Str.: 37.9 MPa; Tens. Elong.: 430%; Tens. Mod.: 4.8 MPa (@ 100% mod.); Tear Str.: 9.6 kN/m (D-470); Hardness (Durometer A): 83
Curing Charact.: Working Life: 4 min @ 100 C; Cure Schedule: 20 h @ 100 C
Ref.: Data sheet AP-210.1(R1)

Adiprene LW-510. Du Pont Co.
Chem. Descrip.: Urethane rubber
Form: Liquid
Composition: 4.25% NCO content
Gen'l. Prop.: Sol.: See Adiprene L-42; Sp. Gr.: 1.03 (25/4 C); Visc.: 1700 cps (100 C)
Vulcanizate Properties with 10.1 phr of MDA curing agent:
Mech. Prop.: Tens. Str.: 37.9 MPa; Tens. Elong.: 400%; Tens. Mod.: 8.1 MPa (@ 100% mod.); Tear Str.: 12.3 kN/m (D-470); Hardness (Durometer A): 90
Curing Charact.: Working Life: 3–4 min @ 100 C; Cure Schedule: 20 h @ 100 C
Ref.: Data sheet AP-210.1(R1)

Adiprene LW-520. Du Pont Co.
Chem. Descrip.: Urethane rubber
Category/Applications: Liquid polymer based on an aliphatic diisocyanate for fabrication of products of high hardness (to 90A durometer), excellent resistance to abrasion and hydrolysis; curable with methylene dianiline using the plastic-gum technique
Form: Liquid; Color: Honey
Composition: 4.75% NCO content
Gen'l. Prop.: Sol.: See Adiprene L-42; Sp. Gr.: 1.03 (25/4 C); Visc.: 900 cps (100 C)
Vulcanizate Properties with 10.5 phr of MDA curing agent:
Mech. Prop.: Tens. Str.: 32.4 MPa; Tens. Elong.: 350%; Tens. Mod.: 7.8 MPa (@ 100% mod.); Tear Str.: 9.6 kN/m (D-470); Hardness (Durometer A): 90
Curing Charact.: Working Life: 4–6 min @ 100 C; Cure Schedule: 16 h @ 100 C
Std. Pkgs.: 40-lb cans and 450-lb drums
Ref.: Data sheet AP-210.1(R1)

Adiprene LW-550. Du Pont Co.
Chem. Descrip.: Urethane rubber
Form: Liquid
Composition: 5.55% NCO content
Gen'l. Prop.: Sol.: See Adiprene L-42; Sp. Gr.: 1.03 (25/4 C); Visc.: 900 cps (100 C)
Vulcanizate Properties with 13.0 phr of MDA curing agent:
Mech. Prop.: Tens. Str.: 40.0 MPa; Tens. Elong.: 340%; Tens. Mod.: 12.8 MPa (@ 100% mod.); Tear Str.: 16.6 kN/m (D-470); Hardness (Durometer D): 48
Curing Charact.: Working Life: 2–3 min @ 100 C; Cure Schedule: 20 h @ 100 C
Ref.: Data sheet

Adiprene LW-570. Du Pont Co.
Chem. Descrip.: Urethane rubber
Category/Applications: Liquid polymer based on an aliphatic diisocyanate and which will give products of extreme hardness (to 75D durometer) and high impact resistance as well as load bearing capacity requiring excellent hydrolytic stability; curable with methylene dianiline
Form: Liquid; Color: Honey
Composition: 7.5% NCO content
Gen'l. Prop.: Sol.: See Adiprene L-42; Sp. Gr.: 1.05 (25/4 C); Visc.: 900 cps (100 C)
Vulcanizate Properties with 16.8 phr of MDA curing agent:
Mech. Prop.: Tens. Str.: 43.3 MPa; Tens.

Elong.: 180%; Tens. Mod.: 29.5 MPa (@ 100% mod.); Flex. Mod.: 689 MPa (D-790); Impact Str. (Izod): 750 J/m notch; Tear Str.: 18.4 kN/m (D470); Hardness (Durometer D): 75
Curing Charact.: Working Life: 1.5 min @ 100 C; Cure Schedule: 20 h @ 100 C
Std. Pkgs.: 40-lb cans and 450-lb drums
Ref.: Data sheet AP-210.1(R1)

Adtac B10BHT. Hercules Inc.
Chem. Descrip.: Aliphatic hydrocarbon resin derived from monomers of petroleum origin
Category/Applications: Tackifier for natural rubber, most synthetic rubbers, and thermoplastic block copolymer elastomers used in adhesive compositions, esp. pressure sensitive systems; improves low-temp. flex properties of hot-melt adhesives
Form: Liquid; Color (Gardner): 5
Gen'l. Prop.: Sol.: Sol. in aliphatic and aromatic hydrocarbons, long-chain alcohols (ethylhexanol), MIBK, trichloroethylene; insol. in ketones, glycols, lower alcohols, and water; Density: 0.916 kg/l; Visc.: 125 cps (99 C); Soften. Pt. (R&B): 10 C; Acid No.: < 1; Stability: Good heat- and light-stability
Std. Pkgs.: 182-kg net heavy-gauge metal drums
Ref.: Data sheet 7233-3

Adtac B25BHT. Hercules Inc.
Chem. Descrip.: Aliphatic hydrocarbon resin derived from monomers of petroleum origin
Category/Applications: See Adtac B10BHT
Form: Semisolid; Color (Gardner): 5.5
Gen'l. Prop.: Sol.: See Adtac B10BHT; Density: 0.923 kg/l; Visc.: 310 cps (99 C); Soften. Pt. (R&B): 25 C; Acid No.: < 1; Stability: Good heat- and light-stability
Std. Pkgs.: 182-kg net heavy-gauge metal drums
Ref.: Data sheet 7233-3

Adtac L100 BHT. Hercules Inc.
Chem. Descrip.: Aliphatic hydrocarbon resin derived mainly from reactive olefin monomers
Category/Applications: Resin for use in adhesives, esp. rubber-based pressure sensitive and hot melt types; waterproofing agent for paper, paperboard, and textiles; paints, varnishes, and other coatings
Form: Solid; Color: Pale
Gen'l. Prop.: Stability: Resistance to moisture; good UV stability
Ref.: Products catalogue

Advastab LS-202. Cincinnati Milacron Chem.
Chem. Descrip.: Organotin comp.
Category/Applications: Lubricating stabilizer for high output PVC pipe production on multiple screw extruders
Form: Granulated waxy solid; Color: Off white to light yellow
Composition: 100% solid system
Gen'l. Prop.: Density: 485 g/l (bulk); Soften. Pt.: (65–119 C)
Toxicity/Handling: Avoid contact with skin and eyes, and prolonged breathing of vapors
Std. Pkgs.: 750 lb Gaylord boxes
Ref.: Tech. bulletin

Advastab T-52N Concentrate. Cincinnati Milacron Chem.
Chem. Descrip.: Butyltin carboxylate
Category/Applications: PVC and PVC-PVAc heat and light stabilizer, suggested for sol'n. resin systems employing conventional PVC and PVC-PVAc resins and for PVC-PVAc copolymer calendering and molding (rigid, semirigid, and flexible); gives good air release, optical clarity and brilliance in plastisols; ure-

thane catalyst for rigid foams and molded flexible systems made with a premix
Form: Liquid; Color (Gardner): 1
Composition: 20% tin content
Gen'l. Prop.: Sp. Gr.: 1.21 (75 F); Density: 10.0 lb/gal; Visc.: 150 cs (75 F)
Std. Pkgs.: 55 gal (450 lb net) drums
Ref.: Tech. bulletin

Advastab T-290. Cincinnati Milacron Chem.
Chem. Descrip.: Butyltin carboxylate
Category/Applications: Nonlubricating, nonplasticizing heat and light stabilizer for rigid polyvinyl chloride, esp. applications requiring high heat distort. temps.; provides excellent initial color; contains no sulfur; low volatility
Form: Fine powder; Color: White
Gen'l. Prop.: Sp. Gr.: 1.40 (75 F); Density: 6.95 lb/gal)
Std. Pkgs.: 300 lb net fiber drums
Ref.: Tech. bulletin

Advastab T-340. Cincinnati Milacron Chem.
Chem. Descrip.: Dibutyltin maleate
Category/Applications: Nonlubricating, nonplasticizing heat stabilizer for PVC compounds; imparts absolute clarity in rigid sheeting, moldings, and extrusions; contains no sulfur
Form: Crystalline solid; Color: White
Composition: 34% tin
Gen'l. Prop.: Sol.: Sol. in ethyl alcohol benzene, acetone; insol. in water and most common plasticizers, Sp. Gr.: 1.36–1.42 (75 F); Density: 11.50–11.82 lb/gal
Std. Pkgs.: 300 lb net fiber drums
Ref.: Tech. bulletin

Advastab TM-180. Cincinnati Milacron Chem.
Chem. Descrip.: Butyltin mercaptide
Category/Applications: Nonlubricating heat stabilizer, esp. suited for the prolonged and severe processing temps. of rigid PVC and PVC-PVA formualtions; recommended for solution vinyls containing reactive "ter-polymers," applications where NSF approvals are required, fluidized-bed powder coating applications, and where acrylic or chlorinated olefin modifiers are employed; contains sulfur
Form: Clear liquid; Color (Gardner): 4 max.
Gen'l. Prop.: Sp. Gr.: 1.110–1.130 (75 F); Density: 9.24–9.48 lb/gal; Visc.: 50 cps max. (75 F)
Std. Pkgs.: 55 gal (450 lb net) drums
Ref.: Tech. bulletin

Advastab TM-181. Cincinnati Milacron Chem.
Chem. Descrip.: Methyltin mercaptide
Category/Applications: Nonlubricating stabilizer for all rigid PVC processes: extrusion, calendering, injection and blow molding; provides good early color stabilization; suited for rigid PVC and PVC-PVA formulations having prolonged and severe processing temps.; compatible with most impact modifiers and processing aids; stabilizer for chlorinated PVC compounds; tin catalyst in polyurethane polymerization reactions
Form: Clear liquid; Color (APHA): 70
Gen'l. Prop.: Sp. Gr.: 1.175; Density: 9.8 lb/gal; Visc.: 50 cs
Toxicity/Handling: Toxic but not highly toxic by ingestion; avoid prolonged contact with skin; avoid contact with pump valves, tanks and lines of iron or steel during storage
Std. Pkgs.: 55 gal (500 lb net) drums
Ref.: Tech. bulletin

Advastab TM-185. Cincinnati Milacron Chem.
Chem. Descrip.: Methyltin mercaptide
Category/Applications: Sulfur-containing PVC stabilizer with outstanding ini-

tial color, color retention, and processing stability; recommended for critical clear rigid PVC bottle and sheet processing, injection molding applications, esp. high flow molding compounds, and for extruded profile, expanded rigid foams, siding, fluid-bed powder coating, and solution vinyl coil coating applications
Form: Clear liquid; Color (Gardner): 2–3
Gen'l. Prop.: Sp. Gr.: 1.193 (75 F); Density: 9.93 lb/gal; Visc.: 101.8 cp (75 F)
Std. Pkgs.: 55 gal (500 lb net) drums
Ref.: Tech. bulletin

Advastab TM-692. Cincinnati Milacron Chem.
Chem. Descrip.: Methyltin mercaptide
Category/Applications: Stabilizer for most rigid PVC applications with good initial color stability and during processing; recommended in single and twin screw extrusion of potable water, sewer, and irrigation pipe, conduit, and duct, and in most rigid PVC extrusion applications, e.g., siding, profile, foamed profile
Form: Liquid
Gen'l. Prop.: Sp. Gr.: 0.99; Density: 8.3 lb/gal; Visc.: 60 cs; Cloud Pt.: –7 C; Pour Pt.: –18 C
Toxicity/Handling: Not highly toxic by ingestion; avoid prolonged contact with skin and eyes
Std. Pkgs.: 55 gal (450 lb net) drums
Ref.: Tech. bulletin

A-Fax 500. Hercules Inc.
Chem. Descrip.: Amorphous polypropylene
Category/Applications: General purpose thermoplastic used as a base polymer in hot-melt adhesives formulations
Form: Waxy solid; Color: Off-white
Gen'l. Prop.: M.W.: 23,000; Density: 0.86 kg/l (22 C); Visc. (Brookfield): 2000 cps (160 C); Soften. Pt. (R&B): 150 C (E28); Fire Pt.: 280 C (D92)
Mech. Prop.: Tens. Str.: 6 kg/cm^2 (D-638); Tens. Elong.: 50% (@ break, D-638)
Therm. Prop.: Heat of Fusion: 7.6 cal/g (DuPont 990); Glass Trans. Temp.: –25 C (DuPont 990); Sp. Heat: 0.65 cal/g/C (Perkin-Elmer DSC-2)
Elec. Prop.: Dielec. Str.: 700 V/mil (short time, @ 60–70 mils, D149); Dielec. Const.: 2.2 (1 KHz, D150); Vol. Resist.: 6×10^{17} ohm-cm (D257); Dissip. Factor: 0.0004 (1 KHz, D150)
Std. Pkgs.: Avail. in bulk tank cars or in metric ton palletized units of 5-kg slabs
Ref.: Data bulletin PSW-101C, products catalogue

A-Fax 600. Hercules Inc.
Chem. Descrip.: Amorphous polypropylene
Category/Applications: High-quality, thermoplastic electrical-grade flooding compound for underground and marine communication and power cable; also for hot-melt and pressure sensitive adhesives formulations
Form: Waxy solid; Color: Off-white
Gen'l. Prop.: M.W.: 12,000; Density: 0.86 kg/l (22 C); Visc. (Brookfield): 240 cps (175 C); Soften. Pt. (R&B): 120 C (E28); Fire Pt.: 280 C (D92)
Therm. Prop.: Heat of Fusion: 7.6 cal/g (DuPont 990); Glass Trans. Temp.: –20 C (DuPont 990); Sp. Heat: 0.65 cal/g/C (Perkin-Elmer DSC-2)
Elec. Prop.: Dielec. Str.: 600 V/mil (short time, @ 60–70 mils, D149); Dielec. Const.: 2.2 (1 kHz, D150); Vol. Resist.: 3×10^{17} ohm-cm (D257); Dissip. Factor: 0.0001 (1 kHz, D150)
Std. Pkgs.: Avail. in bulk tank cars, 72-gal fiber or 55-gal metal drums
Ref.: Data bulletin PSW-102C, products catalogue

A-Fax 800. Hercules Inc.
Chem. Descrip.: Amorphous polypropylene
Category/Applications: Thermoplastic with much lower softening point and viscosity than A-Fax 500; can be used as a replacement or extender for some polybutenes in caulk and sealant compounds; greaselike consistency at R.T. with exceptional green tack
Gen'l. Prop.: Visc. (Brookfield): 500 cps (100 C); Soften. Pt. (R&B): 20 C
Std. Pkgs.: 55-gal metal drums
Ref.: Products catalogue

A-Fax 940. Hercules Inc.
Chem. Descrip.: Amorphous polypropylene
Category/Applications: Thermoplastic with higher softening point and viscosity than A-Fax 500 (improves cohesive strength of paper-to-paper bonds); used in wire and cable saturants, caulks and sealants, hot-melt adhesives, carpet backing, and modification of polyolefin, wax, and asphalt
Form: Waxy solid; Color: Off-white
Gen'l. Prop.: M.W.: 43,000; Density: 0.86 kg/l (22 C); Visc. (Brookfield): 2000 cps (190 C); Soften. Pt. (R&B): 155 C (E28); Fire Pt.: 280 C (D92)
Mech. Prop.: Tens. Str.: 8 kg/cm^2 (D-638); Tens. Elong.: 40% (@ break, D-638)
Therm. Prop.: Heat of Fusion: 7.6 cal/g (DuPont 990); Glass Trans. Temp.: −25 C (DuPont 990); Sp. Heat: 0.65 cal/g/C (Perkin-Elmer DSC-2)
Elec. Prop.: Dielec. Str.: 700 V/mil (short time, @ 60–70 mils, D149); Dielec. Const.: 2.2 (1 kHz, D150); Vol. Resist.: 6×10^{17} ohm-cm (D257); Dissip. Factor: 0.0004 (1 kHz, D150)
Std. Pkgs.: Avail. in bulk tank cars or metric ton palletized units of 5-kg slabs
Ref.: Data bulletin PSW-112B, products catalogue

Agent #5032. Polymer Research Corp. of America
Chem. Descrip.: Product is characterized by a large number of ether linkages and by two terminal hydroxyl groups
Category/Applications: Antitarnishing agent for pigment roller printing; solvent, lubricant, antistat, and conditioner for textile fibers
Form: Clear liquid; Color: Water white
Composition: 25% active sol'n. in water
Gen'l. Prop.: Density: 9.16 lb/gal (55 C); Stability: Heat stable and inert to many chemical agents; pH: 7
Toxicity/Handling: Prevent freezing
Std. Pkgs.: 30 and 55 gal fiber drums
Ref.: Product information bulletin

Alathon 7030. Du Pont Co.
Chem. Descrip.: High-density polyethylene resin
Category/Applications: Narrow m.w. distribution homopolymer recommended for injection molding applications where max. stiffness, dimensional stability, and warp resistance are prime requirements; offers exceptional toughness; FDA approved for food packaging applications
Gen'l. Prop.: Density: 0.96 g/cc (D-792); Soften. Pt. (Vicat): 260 F (D-1525); Melt Index: 2.8 dg/min (D1238)
Mech. Prop.: Tens. Str.: 4200 psi (yield, D638); Tens. Elong.: 900% (D638); Flex. Mod.: 220,000 psi (D790); Impact Str. (Izod): 1.8 ft lb/in. notch (D256); Hardness: 65 (D scale, D985)
Therm. Prop.: Distort. Temp.: 160 F (@ 66 psi, D648)
Ref.: Bulletin E-27266

Alathon 7040. Du Pont Co.
Chem. Descrip.: High-density polyethylene resin
Category/Applications: See Alathon 7030
Gen'l. Prop.: Density: 0.96 g/cc (D792);

Soften. Pt. (Vicat): 260 F (D1525); Melt Index: 6.0 dg/min (D1238)
Mech. Prop.: Tens. Str.: 4300 psi (yield, D638); Tens. Elong.: 900% (D638); Flex. Mod.: 220,000 psi (D790); Impact Str. (Izod): 1.5 ft lb/in. notch (D256); Hardness: 65 (D scale, D985)
Therm. Prop.: Distort. Temp.: 160 F (@ 66 psi, D648)
Ref.: Bulletin E-27266

Alathon 7050. Du Pont Co.
Chem. Descrip.: High-density polyethylene resin
Category/Applications: See Alathon 7030
Gen'l. Prop.: Density: 0.96 g/cc (D792); Soften. Pt. (Vicat): 260 F (D1525); Melt Index: 17.5 dg/min (D1238)
Mech. Prop.: Tens. Str.: 4300 psi (yield, D638); Tens. Elong.: 400% (D638); Flex. Mod.: 220,000 psi (D790); Impact Str. (Izod): 0.8 ft lb/in. notch (D256); Hardness: 65 (D scale, D985)
Therm. Prop.: Distort. Temp.: 160 F (@ 66 psi, D648)
Ref.: Bulletin E-27266

Alathon 7140. Du Pont Co.
Chem. Descrip.: High-density polyethylene resin
Category/Applications: Narrow m.w. distribution copolymer with lower density and offering increased toughness over the homopolymer without sacrificing dimensional stability and warp resistance; for injection molding applications; FDA approved for food packaging applications
Gen'l. Prop.: Density: 0.95 g/cc (D792); Soften. Pt. (Vicat): 255 F (D1525); Melt Index: 4.5 dg/min (D1238)
Mech. Prop.: Tens. Str.: 3700 psi (yield, D638); Tens. Elong.: 900% (D638); Flex. Mod.: 175,000 psi (D790); Impact Str. (Izod): 1.4 ft lb/in. notch (D256); Hardness: 62 (D scale, D985)
Therm. Prop.: Distort. Temp.: 150 F (@ 66 psi, D648)
Ref.: Bulletin E-27266

Alathon 7220. Du Pont Co.
Chem. Descrip.: High-density polyethylene resin
Category/Applications: Intermediate m.w. distribution copolymer for injection molding applications; offers enhanced stress crack resistance with only a slight sacrifice in stiffness, and significantly better processability than corresponding Alathon 7000 series resins; FDA approved for food packaging applications
Gen'l. Prop.: Density: 0.952 g/cc (D-792); Soften. Pt. (Vicat): 256 F (D-1525); Melt Index: 4.0 dg/min (D1238)
Mech. Prop.: Tens. Str.: 3800 psi (yield, D638); Tens. Elong.: 1000% (D638); Flex. Mod.: 185,000 psi (D790); Impact Str. (Izod): 1.3 ft lb/in. notch (D256); Hardness: 62 (D scale, D985)
Therm. Prop.: Distort. Temp.: 152 F (@ 66 psi, D648)
Ref.: Bulletin E-27266

Alathon 7230. Du Pont Co.
Chem. Descrip.: High-density polyethylene resin
Category/Applications: See Alathon 7220
Gen'l. Prop.: Density: 0.955 g/cc (D-792); Soften. Pt. (Vicat): 258 F (D-1525); Melt Index: 6.0 dg/min (D1238)
Mech. Prop.: Tens. Str.: 4000 psi (yield, D638); Tens. Elong.: 1300% (D638); Flex. Mod.: 205,000 psi (D790); Impact Str. (Izod): 1.0 ft lb/in. notch (D256); Hardness: 63 (D scale, D985)
Therm. Prop.: Distort. Temp.: 155 F (@ 66 psi, D648)
Ref.: Bulletin E-27266

Alathon 7240. Du Pont Co.
Chem. Descrip.: High-density polyethylene resin
Category/Applications: See Alathon 7220
Gen'l. Prop.: Density: 0.955 g/cc (D-792); Soften. Pt. (Vicat): 258 F (D-1525); Melt Index: 12.0 dg/min (D-1238)
Mech. Prop.: Tens. Str.: 4000 psi (yield, D638); Tens. Elong.: 850% (D638); Flex. Mod.: 205,000 psi (D790); Impact Str. (Izod): 0.5 ft lb/in. notch (D256); Hardness: 63 (D scale, D985)
Therm. Prop.: Distort. Temp.: 155 F (@ 66 psi, D648)
Ref.: Bulletin E-27266

Alathon 7245. Du Pont Co.
Chem. Descrip.: High-density polyethylene resin
Category/Applications: See Alathon 7220
Gen'l. Prop.: Density: 0.955 g/cc (D-792); Soften. Pt. (Vicat): 258 F (D-1525); Melt Index: 20.0 dg/min (D-1238)
Mech. Prop.: Tens. Str.: 4000 psi (yield, D638); Tens. Elong.: 100% (D638); Flex. Mod.: 205,000 psi (D790); Impact Str. (Izod): 0.5 ft lb/in. notch (D256); Hardness: 63 (D scale, D985)
Ref.: Bulletin E-27266

Alathon 7320. Du Pont Co.
Chem. Descrip.: High-density polyethylene resin
Category/Applications: Broad m.w. distribution copolymer recommended for injection molding applications where stress crack resistance and impact toughness are of paramount importance, and some sacrifice in stiffness can be tolerated; FDA approved for food packaging applications
Gen'l. Prop.: Density: 0.95 g/cc (D792); Soften. Pt. (Vicat): 255 F (D1525); Melt Index: 1.2 dg/min (D1238)
Mech. Prop.: Tens. Str.: 3700 psi (yield, D638); Tens. Elong.: 1100% (D638); Flex. Mod.: 175,000 psi (D790); Impact Str. (Izod): 2.0 ft lb/in. notch (D256); Hardness: 62 (D scale, D985)
Therm. Prop.: Distort. Temp.: 150 F (@ 66 psi, D648)
Ref.: Bulletin E-27266

Alathon 7340. Du Pont Co.
Chem. Descrip.: High-density polyethylene resin
Category/Applications: See Alathon 7320
Gen'l. Prop.: Density: 0.95 g/cc (D792); Soften. Pt. (Vicat): 255 F (D1525); Melt Index: 4.0 dg/min (D1238)
Mech. Prop.: Tens. Str.: 3700 psi (yield, D638); Tens. Elong.: 1100% (D638); Flex. Mod.: 175,000 psi (D790); Impact Str. (Izod): 1.2 ft lb/in. notch (D256); Hardness: 62 (D scale, D985)
Therm. Prop.: Distort. Temp.: 150 F (@ 66 psi, D648)
Ref.: Bulletin E-27266

Alathon 7440. Du Pont Co.
Chem. Descrip.: High-density polyethylene resin
Category/Applications: Injection molding resin offering a higher level of short chain branching (lower density) than the other grades; provides toughness and stress crack resistance superior to most high-density molding resins, while retaining the warp resistance characteristics of a narrow m.w. distribution resin; FDA approved for food packaging applications
Gen'l. Prop.: Density: 0.94 g/cc (D792); Soften. Pt. (Vicat): 239 F (D1525); Melt Index: 4.0 dg/min (D1238)
Mech. Prop.: Tens. Str.: 3300 psi (yield, D638); Tens. Elong.: 800% (D638); Flex. Mod.: 140,000 psi (D790); Impact

Str. (Izod): 1.5 ft lb/in. notch (D256); Hardness: 59 (D scale, D985)
Therm. Prop.: Distort. Temp.: 140 F (@ 66 psi, D648)
Ref.: Bulletin E-27266

Alathon 7810. Du Pont Co.
Chem. Descrip.: High-density polyethylene resin
Category/Applications: Copolymer resin designed for blown extrusion processing; can be processed as monofilm or in coextrusion processes with EVA or LDPE resins; as monofilms, produced in thin film gauges (15–35 μ) for use as sheets for interleavings of delicatessen wrap; in heavier gauges, converted into bags on standard machinery for carry-out packaging, e.g., merchandise bags; 7810 is recommended for very high toughness and slightly less stiffness; heavier gauge also suitable as an inner ply in multiwall bag construction
All properties–1.25 mil film:
Gen'l. Prop.: Density: 0.944 g/cc (D-792); Melt Index: 0.25 dg/min (D1238)
Mech. Prop.: Tens. Str.: 23.5 MPa (yield, TD, D882); Tens. Elong.: 800% (TD, D-882); Tens. Mod.: 690 MPa (secant, TD, D882); Tear Str. (Elmendorf): 116–193 mN/μm (D1922); Impact Str. (Spencer): 93.4–133 kJ/μm (D3420)
Toxicity/Handling: Low concs. of fumes can evolve at temps. above 260 C–use with adequate ventilation
Ref.: Bulletin E-20183

Alathon 7815. Du Pont Co.
Chem. Descrip.: High-density polyethylene resin
Category/Applications: See Alathon 7810; 7815 recommended where moderate toughness and stiffness is required
All properties–1.25 mil film:
Gen'l. Prop.: Density: 0.950 g/cc (D-792); Melt Index: 0.45 dg/min (D1238)
Mech. Prop.: Tens. Str.: 25.5 MPa (yield, TD, D882); Tens. Elong.: 800% (TD, D-882); Tens. Mod.: 793 MPa (secant, TD, D882); Tear Str. (Elmendorf): 58–116 mN/μm (TD, D1922); Impact Str. (Spencer): 58–85 kJ/μm (D3420)
Toxicity/Handling: See Alathon 7810
Ref.: Bulletin E-20183

Alathon 7820. Du Pont Co.
Chem. Descrip.: High-density polyethylene resin
Category/Applications: See Alathon 7810; 7820 recommended where max. stiffness and barrier properties are required; FDA approved for food contact applications
All properties–1.25 mil film:
Gen'l. Prop.: Density: 0.957 g/cc (D-792); Melt Index: 0.45 dg/min (D1238)
Mech. Prop.: Tens. Str.: 26.9 MPa (yield, TD, D882); Tens. Elong.: 850% (TD, D-882); Tens. Mod.: 966 MPa (secant, TD, D882); Tear Str. (Elmendorf): 96–170 mN/μm (TD, D1922); Impact Str. (Spencer): 53.4–85 kJ/μm (D3420)
Toxicity/Handling: See Alathon 7810
Ref.: Bulletin E-20183

Alathon 7835. Du Pont Co.
Chem. Descrip.: High-density polyethylene resin
Category/Applications: Film grade homopolymer resin designed for use in snack food applications; suitable for coextrusion or lamination to other substreates; FDA approved for contact with food; generally used in chill roll casting
All properties–1.25 mil film:
Gen'l. Prop.: Density: 0.960 g/cc (D-792); Melt Index: 3.0 dg/min (D1238)
Mech. Prop.: Tens. Str.: 20.0 MPa (yield, TD, D882); Tens. Elong.: 750% (TD, D-882); Tens. Mod.: 725 MPa (secant, TD, D882); Tear Str. (Elmendorf): 190 mN/μm (TD, D1922); Impact Str. (Spencer):

89 kJ/μm (D3420)
Toxicity/Handling: See Alathon 7810
Ref.: Bulletin E-20183

Alathon 7840. Du Pont Co.
Chem. Descrip.: High-density polyethylene resin
Category/Applications: Narrow m.w. distribution film grade homopolymer designed for the chill roll extrusion process; recommended for use in snack food applications where low MVTR is required; FDA approved
All properties–1.25 mil film:
Gen'l. Prop.: Density: 0.960 g/cc (D-792); Melt Index: 6.0 dg/min (D1238)
Mech. Prop.: Tens. Str.: 20.7 MPa (yield, TD, D882); Tens. Elong.: 850% (TD, D-882); Tens. Mod.: 897 MPa (secant, TD, D882); Tear Str. (Elmendorf): 135 mN/μm (TD, D1922); Impact Str. (Spencer): 53.4 kJ/μm (D3420)
Toxicity/Handling: See Alathon 7810
Ref.: Bulletin E-20183

Alathon 7860. Du Pont Co.
Chem. Descrip.: High-density polyethylene resin
Category/Applications: Moderately broad m.w. distribution, film grade copolymer recommended for water bath or chill roll film where relative ease of processing is desired; FDA approved for food contact applications
All properties–1.25 mil film:
Gen'l. Prop.: Density: 0.955 g/cc (D-792); Melt Index: 6.0 dg/min (D1238)
Mech. Prop.: Tens. Str.: 18.6 MPa (yield, TD, D882); Tens. Elong.: 850% (TD, D-882); Tens. Mod.: 725 MPa (secant, TD, D882); Tear Str. (Elmendorf): 115 mN/μm (TD, D1922); Impact Str. (Spencer): 53.4 kJ/μm (D3420)
Toxicity/Handling: See Alathon 7810
Ref.: Bulletin E-20183

Alathon 7910. Du Pont Co.
Chem. Descrip.: High-density polyethylene resin
Category/Applications: Copolymer for pipe extrusion for applications requiring PPI and NSF approvals; offers superior stress crack resistance and toughness
Gen'l. Prop.: Density: 0.944 g/cc (D-1505); Melt Index: 0.25 g/10 min (D-1238E)
Mech. Prop.: Tens. Str.: 22.9 MPa (yield, D638); Tens. Elong.: 890% (D638); Flex. Mod.: 931 MPa (D790); Tens. Impact: 514 kJ/m^2 (D1822); Stress Crack Resist.: > 1000 F_{50} h (D1693B)
Therm. Prop.: Brittleness Temp.: < 76 C (D746)
Toxicity/Handling: See Alathon 7810
Ref.: Bulletin E-21938

Alathon 7915. Du Pont Co.
Chem. Descrip.: High-density polyethylene resin
Category/Applications: See Alathon 7910
Gen'l. Prop.: Density: 0.950 g/cc (D-1505); Melt Index: 0.45 g/10 min (D-1238E)
Mech. Prop.: Tens. Str.: 25.2 MPa (yield, D638); Tens. Elong.: 800% (D638); Flex. Mod.: 1069 MPa (D790); Tens. Impact: 231 kJ/m^2 (D1822); Stress Crack Resist.: > 200 F_{50} h (D1693B)
Therm. Prop.: Brittleness Temp.: < –76 C
Toxicity/Handling: See Alathon 7810
Ref.: Bulletin E-21939

Alathon 7960. Du Pont Co.
Chem. Descrip.: High-density polyethylene resin
Category/Applications: Copolymer for pipe extrusion for applications not requiring PPI and NSF approvals; offers superior stress crack resistance and toughness

Gen'l. Prop.: Density: 0.944 g/cc (D-1505); Melt Index: 0.25 g/10 min (D-1238E)
Mech. Prop.: Tens. Str.: 22.9 MPa (yield, D638); Tens. Elong.: 890% (D638); Flex. Mod.: 931 MPa (D790); Tens. Impact: 514 kJ/m^2 (D1822); Stress Crack Resist.: > 1000 F_{50} h (D1693B)
Therm. Prop.: Brittleness Temp.: < 76 C (D746)
Toxicity/Handling: See Alathon 7810
Ref.: Bulletin E-21940

Alathon 7965. Du Pont Co.
Chem. Descrip.: High-density polyethylene resin
Category/Applications: See Alathon 7960
Gen'l. Prop.: Density: 0.950 g/cc (D-1505); Melt Index: 0.45 g/10 min (D-1238E)
Mech. Prop.: Tens. Str.: 25.2 MPa (yield, D638); Tens. Elong.: 800% (D638); Flex. Mod.: 1069 MPa (D790); Tens. Impact: 231 kJ/m^2 (D1822); Stress Crack Resist.: > 200 F_{50} h (D1693B)
Therm. Prop.: Brittleness Temp.: < −76 C (D746)
Toxicity/Handling: See Alathon 7810
Ref.: Bulletin E-21941

Alathon 7970. Du Pont Co.
Chem. Descrip.: High-density polyethylene resin
Category/Applications: Copolymer for pipe extrusion applications not requiring PPI and NSF approval; offers outstanding toughness, and good stress crack resistance and stiffness
Gen'l. Prop.: Density: 0.957 g/cc (D-1505); Melt Index: 0.45 g/10 min (D-1238E)
Mech. Prop.: Tens. Str.: 26.9 MPa (yield, D638); Tens. Elong.: 800% (D638); Flex. Mod.: 1276 MPa (D790); Tens. Impact: 158 kJ/m^2 (D1822); Stress Crack Resist.: 65 F_{50} h (D1693B)
Therm. Prop.: Brittleness Temp.: < −76 C (D746)
Toxicity/Handling: See Alathon 7810
Ref.: Bulletin E-21942

Alathon PE-5510. Du Pont Co.
Chem. Descrip.: High-density polyethylene resin
Category/Applications: Injection molding copolymer with outstanding stress crack resistance and good low temp. impact strength; contains an antistat acceptable to the FDA for the packaging of nonfatty foods; used for applications requiring intermediate stiffness, high impact strength, and stress crack resistance on exposure to chemical environments
Gen'l. Prop.: Density: 0.948 g/cc (D-792); Melt Index: 1.2 g/10 min (D1238); Soften. Pt. (Vicat): 123 C (D1525)
Mech. Prop.: Tens. Str.: 25.5 MPa (yield, D638); Tens. Elong.: 1100% (D638); Flex. Mod.: 1207 MPa (D790); Tens. Impact: 284 kJ/m^2 (D1822); Impact Str. (Izod): 107 J/m notch (D256); Stress Crack Resist.: 100 F_{50} h (D-1693B); Hardness: 62 (D scale, D2240)
Therm. Prop.: Distort. Temp.: 66 C (@ 0.455 MPa, D648); Brittleness Temp.: < −76 C (D746)
Toxicity/Handling: See Alathon 7810
Ref.: Bulletin E-20190

Alkapen EC-101. GAF Corp.
Chem. Descrip.: Modified polyether (anionic)
Category/Applications: Cleaner concentrate for textile industry; cleans dye residue from becks, jets, and package machines; low foaming above 150 F; does not present a pollution problem
Form: Liquid

Gen'l. Prop.: Stability: Stable to caustic and hydro
Ref.: Catalogue 2303-036

Allabond Twenty/twenty Adhesive. Bacon Industries Inc.
Chem. Descrip.: Epoxy adhesive
Category/Applications: Epoxy adhesive with > 20 min working life at room temp. and a 20 s cure at 250 F; suggested for production applications now requiring soldering, welding, or brazing; for use with Activator BA-66B; also available is Twenty/twenty CLEAR, containing no fillers and pigments, and Conducting Twenty/twenty, which is electrically conductive
Form: Paste; Color (after cure): Black
Gen'l. Prop.: Sp. Gr. (after cure): 1.50; Visc. (of mixed adhesive): 55 poise
Mech. Prop. (after cure): Flex. Str.: 6000 psi (24 h); Flex. Mod.: 0.6×10^6 psi (24 h); Shear Str.: 2800 psi (to aluminum, 24 h); Hardness (Shore D): 88
Therm. Prop. (after cure): Distort. Temp.: 163 F (264 psi)
Elec. Prop. (after cure): Dielec. Const.: 5.68 (@ 1 kHz); Vol. Resist.: 3×10^{15} ohm-cm
Toxicity/Handling: May cause injury to skin on prolonged or repeated contact; use with adequate ventilation; the activator can cause eye injury and skin burns
Std. Pkgs.: Qt (3 lb) and gal (12 lb) paint cans, 5 gal (60 lb) open head pails; Activator BA-66B avail. in pt (1 lb), qt (2 lb), and gal (8 lb) screw-top cans
Ref.: Tech. data sheet 2620

Allabond Twenty/twenty NM. Bacon Industries Inc.
Chem. Descrip.: Epoxy adhesive
Category/Applications: Nonmagnetic version of Allabond Twenty/twenty Adhesive
Form: Paste; Color (after cure): Black
Gen'l. Prop.: Sp. Gr. (after cure): 1.45; Visc. (of mixed adhesive): 90 poise
Mech. Prop. (after cure): Shear Str.: 2600 psi (24 h); Hardness (Shore D): 88
Toxicity/Handling: May cause injury to skin on prolonged or repeated contact; use with adequate ventilation; the activator can cause eye injury and skin burns
Std. Pkgs.: Qt (3 lb) and gal (12 lb) paint cans, 5 gal (60 lb) open head pails; Activator BA-66B avail. in pt (1 lb), qt (2 lb), and gal (8 lb) screw-top cans
Ref.: Tech. data sheet 2620

Amberol ST-137. Rohm & Haas Co.
Chem. Descrip.: Heat-hardening type phenolic resin
Category/Applications: Used in the production of tung oil varnishes (produces med. and long oil varnishes which are quick drying, pale, and very durable)
Form: Solid; also avail. in crushed form in ¾ in. chunks (Amberol ST-150 C); Color: WW min. (rosin standards)
Gen'l. Prop.: Sol.: Sol. in mineral thinner, aromatic petroleum hydrocarbons, turpentine, toluene, acetone, ether, ethyl acetate, ethylene dichloride; partly sol. in ethanol; Sp. Gr.: 1.04; Visc. (Gardner-Holdt): F-M (64% in toluene); M.P. (Capillary tube): 65 C; Acid No.: 10–30
Ref.: Data sheet C-364

Amberol ST-149. Rohm & Haas Co.
Chem. Descrip.: Nonheat-hardening phenolic resin
Category/Applications: Used in the production of tung oil varnishes (produces durable, chemical resistant finishes of various oil lengths); gasproofing of tung

oil varnishes; as a tackifier to provide bite to butyl rubber used in caulks and sealants
Form: Solid; also avail. in crushed form (Amberol ST-140C) or flaked form (Amberol ST-140F); Color: WG-X (rosin standards)
Gen'l. Prop.: Sol.: See Amberol ST-137; Sp. Gr.: 1.04; Visc. (Gardner-Holdt): F-L (64% in toluene); M.P. (R&B): 95 C; Acid No.: 20–40
Ref.: Data sheet C-364

Ameripol CB. BF Goodrich Co.
Chem. Description: Cis-polybutadiene
Category/Applications: Tough, resilient elastomer with good abrasion resistance; minimizes groove cracking in tires and improves low-temp. performance; good blendability and adhesion to metals; used in automobile, truck, and airplane tires, mechnical goods, adhesives, solid golf balls, heels and soles, hose and tubing, high-impact polystyrene
Ref.: Product information

Ameripol SBR. BF Goodrich Co.
Chem. Descrip.: Styrene-butadiene rubber
Category/Applications: Versatile elastomer with good flexibility and resilience; used for molded and extruded mechanical goods, tires, shoe soles and heels, adhesives, proofed goods, specialty products, milking inflations, wire and cable, valve diaphragms, rug underlay, automotive battery cases; has good heat aging characteristics
Gen'l. Prop.: Stability: Good resistance to wear, to weathering, to ozone, water, oil, chemicals, and permeation by gases
Ref.: Product information

Ameripol SN. BF Goodrich Co.
Chem. Descrip.: Cis-polyisoprene rubber
Category/Applications: Synthetic rubber with good resilience, tack, and heat aging of vulcanizates; used in tires, power transmission belting, motor mounts, milk inflations and tubing, gaskets, O-rings, diaphragms, hospital sheeting, sponge, drug sundries, wire insulation, adhesives, caulks, and sealants, flooring and matting, conveyor belts; low hysteresis (heat build-up); lower shrinkage in extruded and calendered goods
Ref.: Product information

ANE. Ciech Plastofarb Div.
Chem. Descrip.: Unsaturated polyester resin
Category/Applications: Engineering resin with good mechanical properties after cure and high chemical resistance; in styrene solution used for the manufacture of cast resins or glass fiber-reinforced laminates; used in the production of fuel storage tanks, tanks for chemicals (acids, solvents, aldehydes, ketones), fermentation and agricultural silos, chemical plant equipment components, machinery housings, motor-car engine casings, ventilation ducts for the chemical industry, industrial chimneys, etc.
Form: Powder, granulate, or beads; also avail. in a styrene solution
Gen'l. Prop.: Sp. Gr.: 1.09 g/cc (55% in styrene solution, 20 C); Visc.: 430 cps (55% in styrene solution, 20 C); Soften. Pt.: 90 C (solid); Acid No.: 37 mg KOH/g (solid); Stability: Resistant to alcohols, organic and inorganic acids, hydrocarbon solvents, fuels, aq. solutions of inorganic salts, elevated temps., etc.
Mech. Prop. (cured cast resin): Tens. Str.: 410 kG/cm^2; Tens. Elong.: 1.6% (@ break); Flex. Str.: 850 kG/cm^2; Flex. Mod.: 36,000 kG/cm^2; Impact Str.: 6.5; Hardness (Barcol): 47
Elec. Prop.: Vol. Resist.: $> 10^{15}$ ohm-

cm; Surf. Resist.: > 10^{15} ohm
Curing Charact.: Gel Time: 30–180 min (55% styrene solution, 20 C)
Ref.: Product data

ANE-D20. Ciech Plastofarb Div.
Chem. Descrip.: Unsaturated polyether-ester resin
Category/Applications: Engineering resin of high thermal stability and high mechanical strength; used in the manufacture of glass fiber-reinforced products, chemical plant equipment and fittings, tanks, pipelines, housings, pumps, connector pipes, manholes, chimneys and heating flues, agricultural silos, hulls of boats and yachts, machine elements, motor cars, engine casings, tank trucks, etc.
Form: Powder or granulate, or as a 40–50% solution in styrene
Gen'l. Prop.: Sp. Gr.: 1.09 g/cc (55% in styrene, 20 C); Visc.: 420 cps (55% in styrene, 20 C); Soften. Pt. (Vicat): 105 C (cured resin); Stability: Stable to chemically aggressive media (esp. acids) and to elevated temps.
Mech. Prop. (cured cast resin): Tens. Str.: 500 kG/cm^2; Tens. Elong.: 3.4% (@ break); Flex. Str.: 1200 kG/cm^2; Flex. Mod.: 32,000 kG/cm^2; Impact Str.: 6.0; Hardness (Barcol): 42
Elec. Prop.: See ANE
Curing Charact.: 30–150 min (55% in styrene solution, 20 C)
Ref.: Product data

ANE-D30. Ciech Plastofarb Div.
Chem. Descrip.: Unsaturated polyester resin
Category/Applications: See ANE-D20
Form: Powder, granules, or in a styrene solution
Gen'l. Prop.: Sp. Gr.: 1.08 g/cc (55% in styrene, 20 C); Visc.: 380 cps (55% in styrene, 20 C); Soften. Pt. (Vicat): 110 C (cured cast resin); Stability: See ANE-D20
Mech. Prop. (cured cast resin): Tens. Str.: 380 kG/cm^2; Tens. Elong.: 3.0% (@ break); Flex. Str.: 1000 kG/cm^2; Flex. Mod.: 34,000 kG/cm^2; Impact Str.: 6.5; Hardness (Barcol): 44
Curing Charact.: Gel Time: 30–150 min (55% in styrene, 20 C)
Ref.: Product data

ANE-K10. Ciech Plastofarb Div.
Chem. Descrip.: Unsaturated polyester resin
Category/Applications: Engineering resin of high flexibility and high resistance to water, acids, and dilute alkalis; can be used in the production of machine elements, car bodies and elements of car engines, hulls of vessels, chemical plant equipment, pipelines, etc.; with the addition of powdered fillers can be used for seamless floors, lutes, and for sealing purposes, and can replace cement in structural elements
Form: Powder, granules, or in a styrene solution
Gen'l. Prop.: Sp. Gr.: 1.08 g/cc (55% in styrene solution, 20 C); Visc.: 420 cps (55% in styrene solution, 20 C); Soften. Pt. (Vicat): 100 C (cured cast resin); Stability: Resistant to inorganic and organic acids, dilute alkalis, aq. solutions of inorganic salts
Mech. Prop. (cured cast resin): Tens. Str.: 450 kG/cm^2; Tens. Elong.: 4.4% (@ break); Flex. Str.: 1000 kG/cm^2; Flex. Mod.: 34,300 kG/cm^2; Impact Str.: 8.5; Hardness (Barcol): 38
Elec. Prop.: See ANE
Curing Charact.: Gel Time: 30–180 min (55% in styrene, 20 C)
Ref.: Product data

ANE-K20. Ciech Plastofarb Div.
Chem. Descrip.: Unsaturated polyester resin
Category/Applications: See ANE-K10
Form: Powder, granules, or in a styrene solution
Gen'l. Prop.: Sp. Gr.: 1.08 g/cc (55% in styrene, 20 C); Visc.: 340 cps (55% in styrene, 20 C); Soften. Pt. (Vicat): 108 C (cured cast resin); Stability: See ANE-K10
Mech. Prop. (cured cast resin): Tens. Str.: 520 kG/cm^2; Tens. Elong.: 3.5% (@ break); Flex. Str.: 1150 kG/cm^2; Flex. Mod.: 33,000 kG/cm^2; Impact Str.: 9.5; Hardness (Barcol): 45
Elec. Prop.: See ANE
Curing Charact.: Gel Time: 30–150 min (55% in styrene)
Ref.: Product data

ANE-P10. Ciech Plastofarb Div.
Chem. Descrip.: Unsaturated polyester resin
Category/Applications: Plastifying resin used for the modification of other ANE resins to improve their flexibility
All properties—ANE resin with 10% ANE-P10:
Gen'l. Prop.: Sp. Gr.: 1.09 g/cc (55% in styrene solution, 20 C); Visc.: 360 cps (55% in styrene solution, 20 C); Soften. Pt. (Vicat): 115 C (cast cured resin)
Mech. Prop. (cured cast resin): Tens. Str.: 350 kG/cm^2; Tens. Elong.: 2.4% (@ break); Flex. Str.: 850 kG/cm^2; Impact Str.: 6; Hardness (Barcol): 46
Curing Charact.: Gel Time: 30–120 min (55% in styrene, 20 C)
Ref.: Product data

ANE-P20. Ciech Plastofarb Div.
Chem. Descrip.: Unsaturated polyester resin
Category/Applications: See ANE-P10
All properties—ANE resin with 10% ANE-P20:
Gen'l. Prop.: Sp. Gr.: 1.08 g/cc (55% in styrene, 20 C); Visc.: 350 cps (55% in styrene, 20 C); Soften. Pt. (Vicat): 110 C (cured cast resin)
Mech. Prop. (cured cast resin): Tens. Str.: 470 kG/cm^2; Tens. Elong.: 2.3% (@ break); Flex. Str.: 1100 kG/cm^2; Flex. Mod.: 33,000 kG/cm^2; Impact Str.: 7.5; Hardness (Barcol): 40
Curing Charact.: Gel Time: 30–150 min (55% in styrene, 20 C)
Ref.: Product data

Aromatic Oil 745. Neville Chem. Co.
Chem. Descrip.: Aromatic plasticizer
Category/Applications: Aromatic, chemically inert, nonsaponifiable plasticizer exhibiting low reactivity; used in adhesives (mastic, pressure sensitive), rubber (cements, mechanical and molded goods, tires), and caulking compounds
Form: Liquid; Color: Dark brown
Gen'l. Prop.: M.W.: 200 (D2503); Sol.: Sol. in ethers and chlorinated, aromatic, naphthenic, and terpene hydrocarbons; partly sol. in esters, ketones (except acetone), and paraffinic hydrocarbons; Sp. Gr.: 1.050 (D71); Stability: Resistant to alkalis, acids, and moisture
Therm. Prop.: Flash Pt. (COC): 265 F (D92)
Ref.: Product pamphlet

Aropol 7020. Ashland Chem. Co.
Chem. Descrip.: High reactivity isophthalic polyester resin
Category/Applications: Resin which can be chemically thickened for use in low shrink and controlled shrink applications; suggested for low shrink SMC and BMC, and controlled shrink SMC and BMC

Form: Solid; Color (Gardner): 4 max.
Composition: 29–31% styrene
Gen'l. Prop.: Visc.: 6000–9000 cps (Brookfield, #3, 30 rpm); Acid No.: 26–30 (solids)
Mech. Prop. (after cure): Tens. Str.: 4800 psi (D638); Tens. Elong.: 0.9% (D-638); Flex. Str.: 13,200 psi (D790); Flex. Mod.: 470,000 psi (D790); Hardness (Barcol): 46
Therm. Prop. (after cure): Distort. Temp.: 111 C
Curing Charact. (std. 180 F bath–1% benzoyl peroxide): SPI Gel Time: 4–6 min (150–190 F); Total time to peak exotherm: 5.0–7.5 min (150 to peak F); Peak Exotherm: 430–460 F
Toxicity/Handling: Avoid exposure to heat and moisture in storage
Std. Pkgs.: 55 gal (500 lb net) drums
Ref.: Tech. data

Aropol 7021. Ashland Chem. Co.
Chem. Descrip.: Isophthalic polyester resin
Category/Applications: High-strength isophthalic, heat-resistant polyester resin with good electrical properties; suggested for glass reinforced electrical laminates
Form: Liquid; Color (Gardner): 2 max.
Composition: 33.5–36.5% styrene
Gen'l. Prop.: Density: 9.3 lb/gal; Visc.: 1300–1700 cps (Brookfield, #3, 30 rpm); Acid No.: 23–27 (solids)
Mech. Prop. (after cure): Tens. Str.: 5500 psi (D638); Tens. Elong.: 1.1% (D-638); Flex. Str.: 14,000 psi (D790); Flex. Mod.: 530,000 psi (D790); Hardness (Barcol): 48
Therm. Prop. (after cure): Distort. Temp.: 124 C
Elec. Prop. (after heat aging): Power Factor: 0.0107 ore megacycle/s (73 F); Dielec. Str.: 443 V/mil (73 F); Dielec. Const.: 4.29 ore megacycle/s (73 F)
Curing Charact. (std. 180 F bath–1% benzoyl peroxide): SPI Gel Time: 3.25 min (150–190 F); Total time to peak exotherm: 4.3 min (150–peak F); Peak Exotherm: 460 F
Toxicity/Handling: Avoid exposure to heat and moisture in storage
Std. Pkgs.: 55 gal (500 lb net) drums
Ref.: Tech. data

Aropol 7030. Ashland Chem. Co.
Chem. Descrip.: Isophthalic polyester resin
Category/Applications: High-reactivity, thickenable isophthalic, polyester resin for use in low shrink SMC and BMC, and controlled shrink and pigmentable SMC and BMC applications
Form: Liquid; Color (APHA): 4 max.
Composition: 37–39% styrene
Gen'l. Prop.: Visc.: 1000–1400 cps (Brookfield, #3, 30 rpm); Acid No.: 26–30 (solids)
Mech. Prop. (after cure): Tens. Str.: 4800 psi (D638); Tens. Elong.: 0.9% (D-638); Flex. Str.: 13,200 psi (D790); Flex. Mod.: 470,000 psi (D790); Hardness (Barcol): 46
Therm. Prop. (after cure): Distort. Temp.: 111 C
Curing Charact. (std. 180 F bath–1% benzoyl peroxide): SPI Gel Time: 5.0 min (150–190 F); Total time to peak exotherm: 6.0 min (150–peak F); Peak Exotherm: 445 F
Toxicity/Handling: Avoid exposure to heat and moisture in storage
Std. Pkgs.: 55 gal (500 lb net) drums
Ref.: Tech. data

Aropol 7131. Ashland Chem. Co.
Chem. Descrip.: Neopentyl glycol isophthalic polyester resin

Category/Applications: Resilient, med.-viscosity, unpromoted, water-resistant isophthalic polyester resin for gel coat manufacture, e.g., marine gel coats and applications requiring outstanding toughness with good water and heat resistance
Form: Liquid; Color (Gardner): 2 max.
Composition: 40% styrene
Gen'l. Prop.: Sp. Gr. (after cure): 1.15; Visc.: 900 cps (Brookfield, #3, 30 rpm); Acid No.: 22 (solids)
Mech. Prop. (after cure): Tens. Str.: 7900 psi (D638); Tens. Elong.: 1.95% (D638); Flex. Str.: 17,800 psi (D790); Flex. Mod.: 520,000 psi (D790); Hardness (Barcol): 45
Therm. Prop. (after cure): Distort. Temp.: 133 C; Flamm.: DOT rating 73–100 F (flammable liquid)
Curing Charact. (R.T. cure–1% methyl ethyl ketone peroxide (60%), 0.5% cobalt octoate (6%): Gel Time: 7 min; Total time to peak exotherm: 15 min; Peak Exotherm: 405 F
Toxicity/Handling: Store < 80 F; avoid exposure to heat and moisture
Std. Pkgs.: 55 gal (500 lb net) drums
Ref.: Tech. data

Aropol 7141. Ashland Chem. Co.
Chem. Descrip.: Isophthalic polyester resin
Category/Applications: Resilient, low-visc., unpromoted, water-resistant isophthalic polyester resin for gel coat manufacture, e.g., marine gel coats, flat and corrugated sheet, and applications requiring outstanding toughness with good water and heat resistance
Form: Liquid; Color (Gardner): 2 max.
Composition: 44–46% styrene
Gen'l. Prop.: Sp. Gr. (after cure): 1.15; Visc.: 300–450 cps (Brookfield, #3, 30 rpm); Acid No.: 20–24 (solids)
Mech. Prop. (after cure): Tens. Str.: 7900 psi (D638); Tens. Elong.: 1.95% (D638); Flex. Str.: 17,800 psi (D790); Flex. Mod.: 520,000 psi (D790); Hardness (Barcol): 45
Therm. Prop. (after cure): Distort. Temp.: 133 C
Curing Charact. (std. 180 F bath–1% benzoyl peroxide): SPI Gel Time: 6 min; Total time to peak exotherm: 7 min; Peak Exotherm: 455 F
Toxicity/Handling: Store < 80 F; avoid exposure to heat and moisture
Std. Pkgs.: 55 gal (500 lb net) drums
Ref.: Tech. data

Aropol 7221. Ashland Chem. Co.
Chem. Descrip.: Isophthalic polyester resin
Category/Applications: Med-visc., unpromoted, chemical-resistant isophthalic polyester resin; for press molding, filled molding compounds, filament winding, contact molding, sand grout compounds, and applications requiring chemical resistance
Form: Liquid; Color (Gardner): 4 max.
Composition: 35% styrene
Gen'l. Prop.: Visc.: 1700 cps (Brookfield, #3, 30 rpm); Acid No.: 18 (solids)
Mech. Prop. (after cure): Tens. Str.: 9075 psi (D638); Tens. Elong.: 1.5% (D-638); Flex. Str.: 23,400 psi (D790); Flex. Mod.: 560,000 psi (D790); Hardness (Barcol): 45
Therm. Prop. (after cure): Distort. Temp.: 126 C; Flamm.: DOT rating–flammable liquid
Curing Charact. (std. 180 F bath–1% benzoyl peroxide): SPI Gel Time: 3.5 min; Total Time: 4.75 min; Peak Exotherm: 420 F
Toxicity/Handling: Store < 80 F; avoid exposure to heat and moisture
Std. Pkgs.: 55 gal (500 lb net) drums
Ref.: Tech. data

Aropol 7229. Ashland Chem. Co.
Chem. Descrip.: Isophthalic polyester resin
Category/Applications: Unpromoted, med.-viscosity, resilient resin for press molding; used for high impact premix molding, and for preform and mat molding
Form: Liquid; Color (Gardner): 5 max.
Composition: 27–29% styrene
Gen'l. Prop.: Visc.: 1900–2400 cps (Brookfield, #3, 30 rpm); Acid No.: 17–23 (solids)
Mech. Prop. (after cure): Tens. Str.: 5500 psi (D638); Tens. Elong.: 1.5% (D638); Flex. Str.: 13,000 psi (D790); Flex. Mod.: 480,000 psi (D790); Hardness (Barcol): 40
Therm. Prop. (after cure): Distort. Temp.: 103 C
Curing Charact. (std. 180 F bath–1% benzoyl peroxide): SPI Gel Time: 10 min; Total time to peak exotherm: 13 min; Peak Exotherm: 426 F
Toxicity/Handling: Store $<$ 80 F; avoid exposure to heat and moisture
Std. Pkgs.: 55 gal (500 lb net) drums
Ref.: Tech. data

Aropol 7240. Ashland Chem. Co.
Chem. Descrip.: Isophthalic polyester resin
Category/Applications: Low-viscosity, unpromoted, chemical-resistant resin for press molding, filled molding compounds, contact molding, sand grout compounds, and applications requiring good chemical resistance
Form: Liquid; Color (Gardner): 2 max.
Gen'l. Prop.: Sp. gr. (after cure): 1.19; Density: 9.0 lb/gal; Visc.: 300 cps (Brookfield, #2, 30 rpm); Acid No.: 10 (solids)
Mech. Prop. (after cure): Tens. Str.: 8600 psi (D638); Tens. Elong.: 1.9% (D638); Flex. Str.: 20,500 psi (D790); Flex. Mod.: 550,000 psi (D790); Hardness (Barcol): 49
Therm. Prop.: Distort. Temp. (after cure): 121 C; Flamm. (DOT rating): 73–100 F–flamm. liquid
Curing Charact. (std. 180 F bath–1% benzoyl peroxide): SPI Gel Time: 5 min; Total time to peak exotherm: 7 min; Peak Exotherm: 445 F
Toxicity/Handling: Store $<$ 80 F; avoid exposure to heat and moisture
Std. Pkgs.: 55 gal (500 lb net) drums
Ref.: Tech. data

Aropol 7240T-15. Ashland Chem. Co.
Chem. Descrip.: Isophthalic polyester resin
Category/Applications: Thixotropic, prepromoted, high m.w., chemical-resistant resin containing no wax; for use in hand lay-up and spray-up, fume hoods and ducts, tanks, pipes, storage tank repairs and linings, and applications requiring good chemical resistance
Form: Liquid
Gen'l. Prop.: Visc.: 450 cps (Brookfield, #3, 60 rpm); Acid No.: 10 (solids)
Mech. Prop. (after cure): Tens. Str.: 9000 psi (D638); Tens. Elong.: 2.1% (D-638); Flex. Str.: 21,500 psi (D790); Flex. Mod.: 540,000 psi (D790); Hardness (Barcol): 49
Therm. Prop. (after cure): Distort. Temp.: 123 C; Flamm. (DOT rating): 73–160 F–flamm. liquid
Curing Charact. [R.T.–1% methyl ethyl ketone peroxide (60%)]: Gel Time: 15 min; Total time to peak exotherm: 28 min; Peak Exotherm: 383 F
Toxicity/Handling: Store $<$ 80 F; avoid exposure to heat and moisture; settling may occur
Std. Pkgs.: 55 gal (500 lb net) drums
Ref.: Tech. data

Aropol 7240WT-12. Ashland Chem. Co.
Chem. Descrip.: Isophthalic polyester resin
Category/Applications: Thixotropic, prepromoted, chemical-resistant resin containing wax; for use in hand lay-up and spray-up, fume hoods and ducts, tanks, storage tank bottom repair and linings, and applications requiring good chemical resistance
Form: Liquid
Gen'l. Prop.: Sp. gr. (after cure): 1.17; Density: 8.9 lb/gal; Visc.: 500 cps (Brookfield, #3, 60 rpm); Acid No.: 10 (solids)
Mech. Prop. (after cure): Tens. Str.: 9000 psi (D638); Tens. Elong.: 2.1% (D-638); Flex. Str.: 21,500 psi (D790); Flex. Mod.: 540,000 psi (D790); Hardness (Barcol): 49
Therm. Prop.: Distort. Temp. (after cure): 123 C; Flamm. (DOT rating): 73-100 F–flamm. liquid
Curing Charact. [R.T.–1% methyl ethyl ketone peroxide (60%)] : Gel Time: 12 min; Total time to peak exotherm: 21 min; Peak Exotherm: 392 F
Toxicity/Handling: Store < 80 F; avoid exposure to heat and moisture; settling may occur
Std. Pkgs.: 55 gal (500 lb net) drums
Ref.: Tech. data

Aropol 7241-12. Ashland Chem. Co.
Chem. Descrip.: Isophthalic polyester resin
Category/Applications: Low-viscosity, prepromoted, chemical-resistant resin for use in press molding, filled molding compounds, filament winding, contact molding by hand lay-up and spray-up, sand grout compounds, and applications requiring good chemical resistance
Form: Turbid liquid; Color: Blue
Composition: 42–44% styrene
Gen'l. Prop.: Visc.: 350–450 cps (Brookfield, #3, 30 rpm); Acid No.: 16–20
Mech. Prop. (after cure): Tens. Str.: 9075 psi (D638); Tens. Elong.: 1.5% (D-638); Flex. Str.: 23,400 psi (D790); Flex. Mod.: 560,000 psi (D790); Hardness (Barcol): 45
Therm. Prop. (after cure): Distort. Temp.: 126 C
Curing Charact. [R.T.–1% methyl ethyl ketone peroxide (60%)] : Gel Time: 11–13 min; Total time to peak exotherm: 17–23 min; Peak Exotherm: 360–390 F
Toxicity/Handling: Store < 80 F; avoid exposure to heat and moisture; 6-mo. shelf life
Std. Pkgs.: 55 gal (500 lb net) drums
Ref.: Tech. data

Aropol 7241T-15. Ashland Chem. Co.
Chem. Descrip.: Isophthalic polyester resin
Category/Applications: Thixotropic, prepromoted, chemical-resistant resin containing no wax; for use in hand lay-up and spray-up, fume hoods and ducts, tanks, pipe, storage tank repairs, and applications requiring good chemical resistance
Form: Turbid liquid; Color: Blue
Gen'l Prop.: Visc.: 450 cps (Brookfield, #3, 60 rpm); Acid No.: 18 (solids)
Mech. Prop. (after cure): Tens. Str.: 9100 psi (D638); Tens. Elong.: 1.5% (D638);; Flex. Str.: 23,400 psi (D790); Flex. Mod.: 560,000 psi (D790); Hardness (Barcol): 46
Therm. Prop. (after cure): Distort. Temp.: 126 C
Curing Charact. [R.T.–1% methyl ethyl ketone peroxide (60%)] : Gel Time: 15 min; Total time to peak exotherm: 28 min; Peak Exotherm: 383 F
Toxicity/Handling: Store < 80 F; avoid

exposure to heat and moisutre; settling may occur; 3-mo. shelf life
Std. Pkgs.: 55 gal (500 lb net) drums
Ref.: Tech. data

Aropol 7242T-15. Ashland Chem. Co.
Chem. Descrip.: Isophthalic polyester resin
Category/Applications: Thixotropic, prepromoted, chemical-resistant resin containing no wax designed for FDA-type applications (FDA Regulation CFR 21, ¶ 177.2420); used for hand lay-up and spray-up, tanks, pipes, ducts, and hoods
Form: Turbid liquid; Color: Blue
Gen'l. Prop.: Visc.: 450 cps (Brookfield, #3, 60 rpm); Acid No.: 20 (solids)
Mech. Prop. (after cure): Tens. Str.: 9100 psi (D638); Tens. Elong.: 1.5% (D638); Flex. Str.: 23,400 psi (D790); Flex. Mod.: 560,000 psi (D790); Hardness (Barcol): 46
Therm. Prop. (after cure): Distort. Temp.: 126 C; Flamm. (DOT rating): 73–100 F–flamm. liquid
Curing Charact. [R.T.–1% methyl ethyl ketone peroxide (60%)]: Gel Time: 15 min; Total time to peak exotherm: 28 min; Peak Exotherm: 380 F
Toxicity/Handling: Store < 80 F; avoid exposure to heat and moisture; settling may occur; 3-mo. shelf life
Std. Pkgs.: 55 gal (500 lb net) drums
Ref.: Tech. data

Aropol 7280. Ashland Chem. Co.
Chem. Descrip.: Monomer-free isophthalic polyester resin solution in acetone
Category/Applications: Low-visc., chemical-resistant resin for preimpregnation of fiberglass, paper, and asbestos
Form: Liquid; Color (Gardner): 2 max.
Composition: 28% acetone
Gen'l. Prop.: Sp. Gr. (after cure): 1.17; Visc.: 2000 cps (LVT, #3, 30 rpm); Acid No.: 10 (solids); Stability: Good chemical resistance; compatible with styrene, vinyl toluene, methyl methacrylate, diallyl phthalate, diacetone acrylamide, polyfunctional methacrylates
Mech. Prop. (after cure): Tens. Str.: 9000 psi (D638); Tens. Elong.: 2.1% (D-638); Flex. Str.: 21,500 psi (D790); Flex. Mod.: 540,000 psi (D790); Hardness (Barcol): 49
Therm. Prop.: Distort. Temp. (after cure): 123 C; Flamm.: Flamm. liquid
Curing Charact. (acetone-free, std. 180 F bath–1% benzoyl peroxide): SPI Gel Time: 3 min (150–190 F); Total time to peak exotherm: 4 min (150–peak exotherm); Peak Exotherm: 475 F
Toxicity/Handling: One-year shelf life @ < 120 F; use caution in opening drums due to possible pressure build-up from acetone
Std. Pkgs.: 55 gal (500 lb net) drums
Ref.: Tech. data

Aropol 7320. Ashland Chem. Co.
Chem. Descrip.: Isophthalic polyester resin
Category/Applications: Resilient, unpromoted, med.-visc. resin for mat, preform, and premix molding, and for pultrusion
Form: Liquid; Color (Gardner): 4 max.
Composition: 29–31% styrene
Gen'l. Prop.: Visc.: 2500–3000 cps (Brookfield, #3, 30 rpm); Acid No.: 28–32 (solids)
Mech. Prop. (after cure): Tens. Str.: 10,900 psi (D638); Tens. Elong.: 3.8% (D638); Flex. Str.: 21,000 psi (D790); Flex. Mod.: 550,000 psi (D790); Hardness (Barcol): 39
Therm. Prop. (after cure): Distort. Temp.: 87 C
Curing Charact. (std. 180 F bath–1%

benzoyl peroxide): SPI Gel Time: 5 min; Total time to peak exotherm: 6 min; Peak Exotherm: 430 F
Toxicity/Handling: Store < 80 F; avoid exposure to heat and moisutre; 6-mo. storage life
Std. Pkgs.: 55 gal (500 lb net) drums
Ref.: Tech. data

Aropol 7329. Ashland Chem. Co.
Chem. Descrip.: Isophthalic polyester resin
Category/Applications: Unpromoted, med.-visc., very resilient resin for mat, preform, and high impact premix molding
Form: Liquid; Color (Gardner): 5 max.
Composition: 28–30% styrene
Gen'l. Prop.: Visc.: 1700–2200 cps (Brookfield, #3, 30 rpm); Acid No.: 17–23 (solids)
Mech. Prop. (after cure): Tens. Str.: 5700 psi (D638); Tens. Elong.: 1.7% (D638); Flex. Str.: 13,300 psi (D790); Flex. Mod.: 420,000 psi (D790); Hardness (Barcol): 40
Therm. Prop. (after cure): Distort. Temp.: 80 C
Curing Charact. (std. 180 F bath–1% benzoyl peroxide): SPI Gel Time: 15 min; Total time to peak exotherm: 17.5 min; Peak Exotherm: 395 F
Toxicity/Handling: Store < 80 F; avoid exposure to heat and moisutre; 6-mo. storage life
Std. Pkgs.: 55 gal (500 lb net) drums
Ref.: Tech. data

Aropol 7362. Ashland Chem. Co.
Chem. Descrip.: Neopentyl glycol/isophthalic polyester resin containing vinyl toluene monomer
Category/Applications: Resilient, heat-resistant, electrical grade resin for high temperature electrical applications and gel coat mfg.; used for preform and mat molding, premix molding, pultrusion, and for applications requiring good corrosion resistance and toughness with good water and heat resistance
Form: Liquid; Color (Gardner): 6 max.
Composition: 33.5–37.5% vinyl toluene
Gen'l. Prop.: Sp. Gr.: 1.07–1.09; Visc.: 3250–4150 cps (Brookfield, #2, 30 rpm); Acid No.: 22–28 (solids)
Mech. Prop. (after cure): Tens. Str.: 5500 psi (D638); Tens. Elong.: 1.50% (D638); Flex. Str.: 11,000 psi (D790); Flex. Mod.: 1.5×10^6 psi (D790); Hardness (Barcol): 45
Therm. Prop. (after cure): Distort. Temp.: 185 C
Elec. Prop. (after cure): Dielec. Str.: 429 V/mil (S/T); Dielec. Const.: 2.75 (@ 1 MC); Dissip. Factor: 0.010 (@ 1 MC); Arc Resis.: 126 s (Method B)
Curing Charact. (std. 180 F bath–1% benzoyl peroxide): SPI Gel Time: 6.0 min; Total time to peak exotherm: 8.5 min; Peak Exotherm: 365 F
Toxicity/Handling: Store < 80 F; avoid exposure to heat and moisutre; 3-mo. storage life
Std. Pkgs.: 55 gal (500 lb net) drums
Ref.: Tech. data

Aropol 7420. Ashland Chem. Co.
Chem. Descrip.: Isophthalic polyester resin
Category/Applications: Unpromoted, med.-visc., resilient resin for preform, mat, and premix molding
Form: Liquid; Color (Gardner): 4 max.
Composition: 29–31% styrene
Gen'l. Prop.: Visc.: 2100–2700 cps (Brookfield, #3, 30 rpm); Acid No.: 23–27 (solids)
Mech. Prop. (after cure): Tens. Str.: 11,000 psi (D638); Tens. Elong.: 5.5% (D638); Flex. Str.: 22,000 psi (D790);

Flex. Mod.: 560,000 psi (D790); Hardness (Barcol): 40–42
Therm. Prop. (after cure): Distort. Temp.: 73 C
Curing Charact. (std. 180 F bath–1% benzoyl peroxide): SPI Gel Time: 3.0 min; Total time to peak exotherm: 4.0 min; Peak Exotherm: 425 F
Toxicity/Handling: Store < 80 F; avoid exposure to heat and moisutre; 6-mo. storage life
Std. Pkgs.: 55 gal (500 lb net) drums
Ref.: Tech. data

Aropol 7530T-16. Ashland Chem. Co.
Chem. Descrip.: Isophthalic polyester resin containing no wax
Category/Applications: Resilient, thixotropic, prepromoted resin for hand layup and spray-up applications, ducts and housing, marine and transportation equipment, appliance applications, and construction materials
Form: Turbid liquid; Color: Cobalt
Composition: 42% styrene
Gen'l. Prop.: Sp. Gr. (after cure): 1.19; Density: 9.1 lb/gal; Visc.: 450 cps (Brookfield, #3, 60 rpm); Acid No.: 18 (solids)
Mech. Prop. (after cure): Tens. Str.: 9300 psi (D638); Tens. Elong.: 2.1% (D638); Flex. Str.: 21,000 psi (D790); Flex. Mod.: 580,000 psi (D790); Hardness (Barcol): 49
Therm. Prop.: Distort. Temp.: 80 C; Flamm. (DOT rating) 73–100 F
Curing Charact. [R.T.–1% MEK peroxide (60%)]: Gel Time: 16 min; Total time to peak exotherm: 29 min; Peak Exotherm: 317 F
Toxicity/Handling: Store < 80 F; avoid exposure to heat and moisture; settling may occur; 3-mo. shelf life
Std. Pkgs.: 55 gal (500 lb net) drums
Ref.: Tech. data

Aropol 7532. Ashland Chem. Co.
Chem. Descrip.: Isophthalic polyester resin
Category/Applications: Resilient, low-visc., chemical-resistant resin for filament wound tanks, pipe, utility poles
Form: Liquid; Color (Gardner): 4
Composition: 60% nonvolatile
Gen'l. Prop.: Visc.: 450 cps (Brookfield, #2, 30 rpm); Acid No.: 19 (solids)
Mech. Prop. (after cure): Tens. Str.: 11,300 psi (D638); Tens. Elong.: 4.1% (D638); Tens. Mod.: 540,000 psi (D-638); Flex. Str.: 19,000 psi (D790); Flex. Mod.: 510,000 psi (D790); Hardness (Barcol): 42
Therm. Prop.: Distort. Temp. (after cure): 68 C; Flamm. (DOT rating): 73–100 F
Curing Charact. [R.T.–1% Lupersol DDM (11% active), 0.5% of 6% cobalt octoate]: Gel Time: 8 min; Total time to peak exotherm: 18 min; Peak Exotherm: 317 F
Toxicity/Handling: Store < 80 F; avoid exposure to heat and moisture; 3-mo. shelf life
Std. Pkgs.: 55 gal (500 lb net) drums
Ref.: Tech. data

Aropol 8101. Ashland Chem. Co.
Chem. Descrip.: Polyester resin
Category/Applications: Low-visc., rapid wetting vehicle for grinding pigment dispersions (concentrates and pastes)
Form: Liquid; Color (Gardner): 2 max.
Composition: 100% solids
Gen'l. Prop.: Density: 9.4 lb/gal; Visc.: 1200–1700 cps (Brookfield, #3, 30 rpm); Acid No.: 8–12 (solids)
Therm. Prop.: Flamm. (P-M flash pt.): 275 F
Curing Charact. (std. 180 F bath–1% benzoyl peroxide, 20% 12T styrene): SPI Gel Time: 10 min (150–190 F); Total time to peak exotherm: 19 min

(150–peak exotherm); Peak Exotherm: 210 F
Toxicity/Handling: Store < 100 F; avoid exposure to heat and moisture
Std. Pkgs.: 55 gal (500 lb net) drums
Ref.: Tech. data

Aropol 8110. Ashland Chem. Co.
Chem. Descrip.: Polyester resin
Category/Applications: Rigid, med.-visc. resin for general purpose matched-die molding; used for preform molded chairs, tote bins, trays and boxes, outboard motor cowlings, pultrusion applications
Form: Liquid; Color (Gardner): 1 max.
Composition: 25–27% styrene
Gen'l. Prop.: Density: 9.6 lb/gal; Visc.: 3300–3900 cps (Brookfield, #3, 30 rpm); Acid No.: 24–30 (solids)
Mech. Prop. (after cure): Tens. Str.: 5300 psi (D638); Tens. Elong.: 0.8% (D638); Flex. Str.: 14,300 psi (D790); Flex. Mod.: 690,000 psi (D790); Hardness (Barcol): 50
Therm. Prop. (after cure): Distort. Temp.: 73 C
Curing Charact. (std. 180 F bath–1% benzoyl peroxide): SPI Gel Time: 5.5 min (150–190 F); Total time to peak exotherm: 7.5 min (150–peak F); Peak Exotherm: 375 F
Toxicity/Handling: Store < 80 F; avoid exposure to heat and moisture; 6-mo. storage life
Std. Pkgs.: 55 gal (500 lb net) drums
Ref.: Tech. data

Aropol 8310. Ashland Chem. Co.
Chem. Descrip.: Polyester resin
Category/Applications: Med.-visc. resin for general purpose matched-die molding; used for preform molded chairs, tote bins, trays and boxes, outboard motor cowlings, pultrusion applications
Form: Liquid; Color (Gardner): 1 max.
Composition: 28–30% styrene
Gen'l. Prop.: Density: 9.7 lb/gal; Visc.: 2000–2400 cps (Brookfield, #3, 30 rpm); Acid No.: 18–24 (solids)
Mech. Prop. (after cure): Tens. Str.: 10,000 psi (D638); Tens. Elong.: 1.9% (D638); Flex. Str.: 17,000 psi (D790); Flex. Mod.: 610,000 psi (D790); Hardness (Barcol): 50
Therm. Prop. (after cure): Distort. Temp.: 73 C
Curing Charact. (std. 180 F bath–1% benzoyl peroxide): SPI Gel Time: 5 min (150–190 F); Total time to peak exotherm: 7 min (150–peak F); Peak Exotherm: 400 F
Toxicity/Handling: Store < 80 F; avoid exposure to heat and moisture; 6-mo. storage life
Std. Pkgs.: 55 gal (500 lb net) drums
Ref.: Tech. data

Aropol 8319. Ashland Chem. Co.
Chem. Descrip.: Polyester resin
Category/Applications: Med.-visc., resilient resin for general purpose matched-die molding; used for preform molded chairs, tote bins, trays and boxes, outboard motor cowlings, pultrusion applications
Form: Liquid; Color (Gardner): 6 max.
Composition: 25–27% styrene
Gen'l. Prop.: Sp. Gr.: 1.10–1.15; Density: 9.35–9.55 lb/gal; Visc.: 1800–2400 cps (Brookfield, #3, 30 rpm); Acid No.: 17–23 (solids)
Mech. Prop. (after cure): Tens. Str.: 9400 psi (D638); Tens. Elong.: 3.5% (D638); Flex. Str.: 17,800 psi (D790); Flex. Mod.: 520,000 psi (D790); Hardness (Barcol): 46
Therm. Prop. (after cure): Distort. Temp.: 60 C

Curing Charact. (std. 180 F bath–1% benzoyl peroxide): SPI Gel Time: 5 min (150–190 F); Total time to peak exotherm: 7.5 min (150–peak F); Peak Exotherm: 335 F
Toxicity/Handling: Store < 80 F; avoid exposure to heat and moisture; 6-mo. storage life
Std. Pkgs.: 55 gal (500 lb net) drums
Ref.: Tech. data

Aropol 8321. Ashland Chem. Co.
Chem. Descrip.: Polyester resin
Category/Applications: Med.-visc., resilient, light colored resin for general purpose matched-die molding; used for preform molded chairs, tote bins, trays and boxes, outboard motor cowlings, pultrusion applications, button casting
Form: Liquid; Color (APHA): 75 max.
Composition: 29–31% styrene
Gen'l. Prop.: Density: 9.7 lb/gal; Visc.: 2100–2700 cps (Brookfield, #3, 30 rpm); Acid No.: 31–37 (solids)
Mech. Prop. (after cure): Tens. Str.: 8700 psi (D638); Tens. Elong.: 1.6% (D638); Flex. Str.: 14,000 psi (D790); Flex. Mod.: 640,000 psi (D790); Hardness (Barcol): 50
Therm. Prop. (after cure): Distort. Temp.: 89 C
Curing Charact. (std. 180 F bath–1% benzoyl peroxide): SPI Gel Time: 4 min (150–190 F); Total time to peak exotherm: 6 min (150–peak F); Peak Exotherm: 390 F
Toxicity/Handling: Store < 80 F; avoid exposure to heat and moisture; 6-mo. storage life
Std. Pkgs.: 55 gal (500 lb net) drums
Ref.: Tech. data

Aropol 8329. Ashland Chem. Co.
Chem. Descrip.: Polyester resin
Category/Applications: Low-visc., resilient resin for high filler loading; used for preform molded chairs, electrical potting, trays and boxes, bowling ball cores, pultrusion applications
Form: Liquid; Color (Gardner): 6 max.
Composition: 34–36% styrene
Gen'l. Prop.: Sp. Gr. (after cure): 1.25; Density: 9.2 lb/gal; Visc.: 350–450 cps (Brookfield, #2, 30 rpm); Acid No.: 17–23 (solids)
Mech. Prop. (after cure): Tens. Str.: 9400 psi (D638); Tens. Elong.: 3.5% (D638); Flex. Str.: 17,800 psi (D790); Flex. Mod.: 520,000 psi (D790); Hardness (Barcol): 46
Therm. Prop. (after cure): Distort. Temp.: 60 C
Curing Charact. (std. 180 F bath–1% benzoyl peroxide): SPI Gel Time: 5 min (150–190 F); Total time to peak exotherm: 7.5 min (150–peak F); Peak Exotherm: 385 F
Toxicity/Handling: Store < 80 F; avoid exposure to heat and moisture; 6-mo. storage life
Std. Pkgs.: 55 gal (500 lb net) drums
Ref.: Tech. data

Aropol 8329-25. Ashland Chem. Co.
Chem. Descrip.: Polyester resin
Category/Applications: Prepromoted, resilient, high-visc. resin for vacuum bag molding; used for hand lay-up and spray-up of parts not requiring thixotropy
Form: Liquid; Color: Blue
Composition: 30–32% styrene
Gen'l. Prop.: Sp. Gr. (after cure): 1.23; Visc.: 1000–1100 cps (Brookfield, #3, 30 rpm); Acid No.: 17–23 (solids)
Mech. Prop. (after cure): Tens. Str.: 10,000 psi (D638); Tens. Elong.: 3.0% (D638); Flex. Str.: 17,000 psi (D790); Flex. Mod.: 500,000 psi (D790); Hardness (Barcol): 50
Therm. Prop. (after cure): Distort. Temp.: 65 C

Curing Charact. [R.T.–1% MEK peroxide (60%)]: Gel Time: 25 min; Total time to peak exotherm: 35 min; Peak Exotherm: 356 F
Toxicity/Handling: Store < 80 F; avoid exposure to heat and moisture; 6-mo. storage life
Std. Pkgs.: 55 gal (500 lb net) drums
Ref.: Tech. data

Aropol 8339T-12. Ashland Chem. Co.
Chem. Descrip.: Polyester resin
Category/Applications: Resilient, thixotropic, prepromoted resin for hand layup and spray-up, appliance uses, transportation equipment, marine applications
Form: Turbid liquid; Color: Cobalt
Composition: 42% styrene
Gen'l. Prop.: Sp. Gr. (after cure): 1.23; Visc.: 350 cps (Brookfield, #2, 60 rpm); Acid No.: 20 (solids)
Mech. Prop. (after cure): Tens. Str.: 10,000 psi (D638); Tens. Elong.: 3.0% (D638); Flex. Str.: 17,000 psi (D790); Flex. Mod.: 500,000 psi (D790); Hardness (Barcol): 50
Therm. Prop.: Distort. Temp. (after cure): 65 C; Flamm. (DOT rating): 73–100 F
Curing Charact. [R.T.–1% MEK peroxide (60%)]: Gel Time: 12 min; Total time to peak exotherm: 21 min; Peak Exotherm: 332 F
Toxicity/Handling: Store < 80 F; avoid exposure to heat and moisture; settling may occur; 3-mo. shelf life
Std. Pkgs.: 55 gal (500 lb net) drums
Ref.: Tech. data

Aropol 8343T-11. Ashland Chem. Co.
Chem. Descrip.: Polyester resin
Category/Applications: Thixotropic, low-visc., prepromoted, resilient resin formulated for extended trim time; used for spray-up with catalyst injection for general applications, e.g., silo tops, boats, shower enclosures
Form: Liquid; Color (Gardner): Cobalt
Composition: 44% styrene
Gen'l. Prop.: Sp. Gr. (after cure): 1.23; Visc.: 325 cps (Brookfield, #2, 60 rpm); Acid No.: 20 (plastic)
Mech. Prop. (after cure): Tens. Str.: 10,000 psi (D638); Tens. Elong.: 3.0% (D638); Flex. Str.: 17,000 psi (D790); Flex. Mod.: 500,000 psi (D790); Hardness (Barcol): 50
Therm. Prop.: Distort. Temp. (after cure): 65 C; Flamm. (DOT rating): 73–100 F
Curing Charact. [R.T.–1% MEK peroxide (60%)]: Gel Time: 11 min; Total time to peak exotherm: 23 min; Peak Exotherm: 335 F
Toxicity/Handling: Store < 80 F; avoid exposure to heat and moisture; settling may occur; 3-mo. shelf life
Std. Pkgs.: 55 gal (500 lb net) drums
Ref.: Tech. data

Aropol 8343T-12. Ashland Chem. Co.
Chem. Descrip.: Polyester resin
Category/Applications: Thixotropic, low-visc., prepromoted, resilient resin formulated for rapid cure; used for spray-up with catalyst injection for general applications, e.g., silo tops, boats, shower enclosures
Form: Liquid; Color (Gardner): Cobalt
Composition: 44% styrene
Gen'l. Prop.: Sp. Gr. (after cure): 1.23; Visc.: 325 cps (Brookfield, #2, 60 rpm); Acid No.: 20 (plastic)
Mech. Prop. (after cure): Tens. Str.: 10,000 psi (D638); Tens. Elong.: 3.0% (D638); Flex. Str.: 17,000 psi (D790); Flex. Mod.: 500,000 psi (D790); Hardness (Barcol): 50

Therm. Prop.: Distort. Temp. (after cure): 65 C; Flamm. (DOT rating): 73–100 F

Curing Charact. [R.T.–1% MEK peroxide (60%)]: Gel Time: 12 min; Total time to peak exotherm: 21 min; Peak Exotherm: 332 F

Toxicity/Handling: Store < 80 F; avoid exposure to heat and moisture; settling may occur; 3-mo. shelf life

Std. Pkgs.: 55 gal (500 lb net) drums

Ref.: Tech. data

Aropol 8343T-21. Ashland Chem. Co.
Chem. Descrip.: Polyester resin

Category/Applications: Thixotropic, low-visc., prepromoted, resilient resin formulated for extended trim time; used for spray-up with catalyst injection for general applications, e.g., silo tops, boats, shower enclosures

Form: Liquid; Color (Gardner): Cobalt-milky

Gen'l. Prop.: Sp. Gr. (after cure): 1.23; Visc.: 325 cps (Brookfield, #2, 60 rpm); Acid No.: 20 (plastic)

Mech. Prop. (after cure): Tens. Str.: 10,000 psi (D638); Tens. Elong.: 3.0% (D638); Flex. Str.: 17,000 psi (D790); Flex. Mod.: 500,000 psi (D790); Hardness (Barcol): 50

Therm. Prop.: Distort. Temp. (after cure): 65 C; Flamm. (DOT rating): 73–100 F

Curing Charact. [R.T.–1% MEK peroxide (60%)]: Gel Time: 21 min; Total time to peak exotherm: 33 min; Peak Exotherm: 332 F

Toxicity/Handling: Store < 80 F; avoid exposure to heat and moisture; settling may occur; 3-mo. shelf life

Std. Pkgs.: 55 gal (500 lb net) drums

Ref.: Tech. data

Aropol 8343T-22. Ashland Chem. Co.
Chem. Descrip.: Polyester resin

Category/Applications: Thixotropic, low-visc., prepromoted, resilient resin for rapid cure; used for spray-up with catalyst injection for general applications, e.g., silo tops, boats, shower enclosures

Form: Liquid; Color (Gardner): Cobalt

Gen'l. Prop.: Sp. Gr. (after cure): 1.23; Visc.: 325 cps (Brookfield, #2, 60 rpm); Acid No.: 20 (plastic)

Mech. Prop. (after cure): Tens. Str.: 10,000 psi (D638); Tens. Elong.: 3.0% (D638); Flex. Str.: 17,000 psi (D790); Flex. Mod.: 500,000 psi (D790); Hardness (Barcol): 50

Therm. Prop.: Distort. Temp. (after cure): 65 C; Flamm. (DOT rating): 73–100 F

Curing Charact. [R.T.–1% MEK peroxide (60%)]: Gel Time: 22 min; Total time to peak exotherm: 31 min; Peak Exotherm: 330 F

Toxicity/Handling: Store < 80 F; avoid exposure to heat and moisture; settling may occur; 3-mo. shelf life

Std. Pkgs.: 55 gal (500 lb net) drums

Ref.: Tech. data

Aropol 8349-15. Ashland Chem. Co.
Chem. Descrip.: Polyester resin

Category/Applications: Low-visc., prepromoted, resilient resin for hand lay-up and spray-up using hydrated alumina filler, rigidizing vacuum formed acrylic, transportation equipment, and marine applications

Form: Turbid liquid; Color: Blue

Composition: 45–47% styrene

Gen'l. Prop: Sp. Gr. (after cure): 1.23; Visc.: 45–85 cps (Brookfield, #2, 60 rpm); Acid No.: 17–23 (solids)

Mech. Prop. (after cure): Tens. Str.: 10,000 psi (D638); Tens. Elong.: 3.0% (D638); Flex. Str.: 17,000 psi (D790); Flex. Mod.: 500,000 psi (D790); Hardness (Barcol): 50
Therm. Prop. (after cure): Distort. Temp.: 65 C
Curing Charact. [R.T.–1% MEK peroxide (60%)]: Gel Time: 15 min; Total time to peak exotherm: 23 min; Peak Exotherm: 348 F
Toxicity/Handling: Store < 80 F; avoid exposure to heat and moisture; 6-mo. storage life
Std. Pkgs.: 55 gal (500 lb net) drums
Ref.: Tech. data

Aropol 8349T-09. Ashland Chem. Co.
Chem. Descrip.: Polyester resin
Category/Applications: Low-visc., thixotropic, prepromoted, resilient resin for filled spray-up systems; used for hand lay-up and spray-up using hydrated alumina filler, rigidizing vacuum formed acrylic, transportation equipment, and marine applications
Form: Liquid; Color: Cobalt
Composition: 46% styrene
Gen'l. Prop.: Sp. Gr. (after cure): 1.23; Visc.: 150 cps (Brookfield, #2, 60 rpm); Acid No.: 20 (solids)
Mech. Prop. (after cure): Tens. Str.: 10,000 psi (D638); Tens. Elong.: 3.0% (D638); Flex. Str.: 17,000 psi (D790); Flex. Mod.: 500,000 psi (D790); Hardness (Barcol): 50
Therm. Prop.: Distort. Temp. (after cure): 65 C; Flamm. (DOT rating): 73–100 F
Curing Charact. [filled, 50/50 resin/hydrated alumina, R.T.–1% MEK peroxide (60%)]: Gel Time: 9 min; Total time to peak exotherm: 24 min; Peak Exotherm: 220 F
Toxicity/Handling: Store < 80 F; avoid exposure to heat and moisture; settling may occur; 3-mo. shelf life
Std. Pkgs.: 55 gal (500 lb net) drums
Ref.: Tech. data

Aropol 8349T-21. Ashland Chem. Co.
Chem. Descrip.: Polyester resin
Category/Applications: Thixotropic, low-visc., prepromoted, resilient resin for hand lay-up and spray-up with high filler loading, rigidizing acrylic sheet, transportation equipment, marine applications
Form: Turbid liquid; Color: Blue
Composition: 45–47% styrene
Gen'l. Prop.: Sp. Gr. (after cure): 1.23; Visc.: 180–280 cps (Brookfield, #2, 60 rpm); Acid No.: 17–23 (solids)
Mech. Prop. (after cure): Tens. Str.: 10,000 psi (D638); Tens. Elong.: 3.0% (D638); Flex. Str.: 17,000 psi (D790); Flex. Mod.: 500,000 psi (D790); Hardness (Barcol): 50
Therm. Prop. (after cure): Distort. Temp.: 65 C
Curing Charact. [R.T.–1% MEK peroxide (60%): Gel Time: 21 min; Total time to peak exotherm: 32 min; Peak Exotherm: 348 F
Toxicity/Handling: Store < 80 F; avoid exposure to heat and moisture; settling may occur; 6-mo. storage life
Std. Pkgs.: 55 gal (500 lb net) drums
Ref.: Tech. data

Aropol 8410-12. Ashland Chem. Co.
Chem. Descrip.: Polyester resin
Category/Applications: Prepromoted, high-visc., resilient resin for use where high-visc. mix is desired; suggested for synthetic marble vanity tops with integral basin and back splash
Form: Liquid; Color: Purple-red
Composition: 28% styrene
Gen'l. Prop.: Density: 9.7 lb/gal; Visc.:

3250 cps (Brookfield, #3, 30 rpm); Acid No.: 18–24 (solids)
Mech. Prop. (after cure): Tens. Str.: 8200 psi (D638); Tens. Elong.: 1.8% (D638); Flex. Str.: 16,000 psi (D790); Flex. Mod.: 620,000 psi (D790); Hardness (Barcol): 50
Therm. Prop.: Distort. Temp. (after cure): 56 C; Flamm. (DOT rating): 73–100 F–flamm. liquid
Curing Charact. [R.T.–1% MEK peroxide (60%)]: Gel Time: 12 min; Total time to peak exotherm: 23 min; Peak Exotherm: 302 F
Toxicity/Handling: Store < 80 F; avoid exposure to heat and moisture; 6-mo. storage life
Std. Pkgs.: 55 gal (500 lb net) drums
Ref.: Tech. data

Aropol 8420. Ashland Chem. Co.
Chem. Descrip.: Polyester resin
Category/Applications: Resilient, nonpromoted, light-colored resin containing styrene monomer for use in matched-die molding, pultrusion, architectural sheeting, gel coats, casting
Form: Liquid; Color (Gardner): 1 max.
Composition: 30 ± 1% styrene
Gen'l. Prop.: Density: 9.6 lb/gal; Visc.: 1900 cps (Brookfield, #3, 30 rpm); Acid No.: 21 (plastic)
Mech. Prop. (after cure): Tens. Str.: 9000 psi (D638); Tens. Elong.: 2.1% (D638); Flex. Str.: 13,000 psi (D790); Flex. Mod.: 600,000 psi (D790); Compr. Str. (of glass fiber mat laminate–4 ply, 2 oz mat): 23,000 psi; Impact Str. (of glass fiber mat laminate–4 ply, 2 oz mat, Izod): 22 ft lb/in. notched; Hardness (Barcol): 51
Therm. Prop.: Flamm. (DOT rating): 73-100 F–flamm. liquid
Curing Charact. (std. 180 F bath–1% benzoyl peroxide): SPI Gel Time: 5 min; Total time to peak exotherm: 7.5 min; Peak Exotherm: 355 F
Std. Pkgs.: 55 gal (500 lb net) drums
Ref.: Tech. data

Aropol 8420-09. Ashland Chem. Co.
Chem. Descrip.: Polyester resin
Category/Applications: Prepromoted, fast-cure, resilient resin for use where high filler loadings are needed; used for synthetic marble counter tops and sinks
Form: Liquid; Color: Purple-red
Composition: 33% styrene
Gen'l. Prop.: Density: 9.6 lb/gal; Visc.: 1100 cps (Brookfield, #3, 60 rpm); Acid No.: 21 (solids)
Mech. Prop. (after cure): Tens. Str.: 8200 psi (D638); Tens. Elong.: 1.8% (D638); Flex. Str.: 16,000 psi (D790); Flex. Mod.: 620,000 psi (D790); Hardness (Barcol): 50
Therm. Prop.: Distort. Temp. (after cure): 56 C; Flamm. (DOT rating): 73–100 F
Curing Charact. [R.T.–1% MEK peroxide (60%)]: Gel Time: 9 min; Total time to peak exotherm: 20 min; Peak Exotherm: 318 F
Toxicity/Handling: Store < 80 F; avoid exposure to heat and moisture; 3-mo. shelf life
Std. Pkgs.: 55 gal (500 lb net) drums
Ref.: Tech. data

Aropol 8449T-15. Ashland Chem. Co.
Chem. Descrip.: Polyester resin
Category/Applications: Wax-free thixotropic, low-visc., prepromoted, resilient resin formulated for rapid cure; used for spray-up with catalyst injection for general purpose applications
Form: Liquid; Color: Blue
Composition: 45.5% styrene
Gen'l. Prop.: Density: 8.9 lb/gal; Visc.: 325 cps (Brookfield, #2, 60 rpm); Acid No.: 14 (plastic)

Mech. Prop. (after cure): Tens. Str.: 7000 psi (D638); Tens. Elong.: 2.1% (D638); Flex. Str.: 13,000 psi (D790); Flex. Mod.: 300,000 psi (D790); Hardness (Barcol): 40
Therm. Prop.: Flamm. (DOT rating): 73–100 F
Curing Charact. [R.T.–1% MEK peroxide (60%)]: Gel Time: 15 min; Total time to peak exotherm: 25 min; Peak Exotherm: 305 F
Toxicity/Handling: Store < 80 F; avoid exposure to heat and moisture; settling may occur; 3-mo. shelf life
Std. Pkgs.: 55 gal (500 lb net) drum
Ref.: Tech. data

Aropol 8520-14. Ashland Chem. Co.
Chem. Descrip.: Polyester resin
Category/Applications: Light-colored, prepromoted, resilient, crack and craze resistant resin for transparent castings; used for casting objects into clear, transparent pieces, solid spheres, jewelry, imbedding specimens, paper weights
Form: Liquid; Color: Faint blue-green
Composition: 38% styrene
Gen'l. Prop.: Density: 9.31 lb/gal; Visc.: 460 cps (Brookfield, #3, 30 rpm)
Mech. Prop. (after cure): Tens. Str.: 7500 psi (D638); Tens. Elong.: 1.4% (D638); Flex. Str.: 13,500 psi (D790); Flex. Mod.: 580,000 psi (D790); Hardness (Barcol): 50
Therm. Prop.: Distort. Temp. (after cure): 61 C; Flamm. (DOT rating): 73–100 F–flamm. liquid
Curing Charact. [R.T.–1% MEK peroxide (60%)]: Gel Time: 14 min; Total time to peak exotherm: 38 min; Peak Exotherm: 292 F
Toxicity/Handling: Store < 80 F; avoid exposure to heat and moisture; 3-mo. shelf life
Std. Pkgs.: 55 gal (500 lb net) drums
Ref.: Tech. data

Aropol 8621-16. Ashland Chem. Co.
Chem. Descrip.: Polyester resin
Category/Applications: Prepromoted, low-visc., low-exotherm, resilient resin offering improved hot/cold cycling performance and reduced exotherm cracking; used for synthetic marble vanity tops with integral basin and back splash, synthetic marble bathtubs
Form: Clear liquid; Color: Blue
Composition: 30% styrene
Gen'l. Prop.: Sp. Gr. (liq., after cure): 1.111; Visc.: 725 cps (Brookfield, #3, 30 rpm); Acid No.: 34 (solids)
Mech. Prop. (after cure): Tens. Str.: 3300 psi (D638); Tens. Elong.: 23% (D638); Flex. Str.: 8500 psi (D790); Flex. Mod.: 270,000 psi (D790); Hardness (Barcol): 20
Therm. Prop.: Distort. Temp. (after cure): 36 C; Flamm. (DOT rating): 73–100 F
Curing Charact. [R.T.–1% MEK peroxide (60%)]: Gel Time: 16 min; Total time to peak exotherm: 36 min; Peak Exotherm: 250 F
Toxicity/Handling: Store < 80 F; avoid exposure to heat and moisture; 3-mo. shelf life
Std. Pkgs.: 55 gal (500 lb net) drums
Ref.: Tech. data

Aropol 8800. Ashland Chem. Co.
Chem. Descrip.: Polyester resin
Category/Applications: Flexible, monomer-free, unpromoted resin to provide toughness when blended with rigid and/or resilient resins; used for button casting, electrical laminates, press molding applications
Form: Liquid; Color (APHA): 80 max.
Composition: Contains no styrene
Gen'l. Prop.: Visc.: 200–300 cps (Brookfield, #2, 60 rpm, 30% styrene); Acid No.: 18–22 (solids)

Curing Charact. [R.T.–1% MEK peroxide (60%), 0.5% cobalt octoate (6%)]: Gel Time: 23 min; Total time to peak exotherm: 56 min; Peak Exotherm: 248 F
Toxicity/Handling: Avoid exposure to heat; may be stored several years ⩽ 80 F without danger of gellation

Std. Pkgs.: 55 gal (500 lb net) drums

Ref.: Tech. data

Arylef U 100. Solvay
Chem. Descrip.: Polyarylate

Category/Applications: High-performance thermoplastic engineering polymer with good dielectric and elastic properties, and high impact strength; processed by injection, extrusion, blow molding, and other methods at high temps.; used in electronics and electrical engineering (equipment housings, switchgear, connectors, insulator plates, etc.), mechanical (pulleys, gearwheels, threaded rods, rings, pads, sleeves, etc.), lighting and optical uses (furnace inspection ports, tinted glazing, spectacles, lenses, etc.), medical applications (dentures, bottles for eye lotion, etc.), household electrical goods (hair-dryers, microwave ovens, electrical tools, office machinery, etc.)
Form: Solid
Gen'l. Prop.: Density: 1.21 mg/m^3 (D-1505); Stability: Good dimensional stability; good resistance to heat, aging, UV radiation, and to combustion; Ref. Index: 1.61 (D542)

Mech. Prop.: Tens. Str.: 71 MPa (D638); Tens. Elong.: 50% (D638), Tens. Mod.: 2000 MPa (D638); Flex. Str.: 81 MPa (D790); Flex. Mod.: 2000 Mpa (D790); Compr. Str.: 94 MPa (D695); Compr. Mod.: 2050 MPa (D695); Impact Str. (Izod): 220 J/m notched (3.2 mm bar, D256); Hardness (Rockwell): 125 R, M 95 (D785)

Therm. Prop.: Distort. Temp.: 175 C (1.8 MPa, D648); Coeff. of Linear Exp.: 6.2 × 10^{-5} K^{-1} (–30 to +30 C, D696); Sp. Heat: 1.3 kJ kg^{-1} K^{-1} (20–300 C, D-696); Flamm.: SE (D635), V-O (UL 94); Self-Ignit. Temp.: 545 C (D1929)

Elec. Prop.: Dielec. Str.: 30 kV/mm (D-149); Dielec. Const.: 2.73 (60 Hz, D-150); Vol. Resist.: 3 × 10^{16} ohm-cm (23 C, D257); Surf. Resist.: > 2 × 10^{17} ohm (0.03 mm film)

Ref.: Product information

AS-10GF. Compounding Technology, Inc.

Chem. Descrip.: Glass-reinforced ABS resin

Category/Applications: Thermoplastic resin noted for good dimensional stability and toughness; used in business machine housings, cabinetry, tool housings, appliances, automotive and construction items

Composition: 10% glass fiber

Gen'l. Prop.: Sp. Gr.: 1.10 (D792); M.P.: 460–500 F; Soften. Pt. (Vicat): 221 F (D1525)

Mech. Prop.: Tens. Str.: 9800 psi (D-638); Tens. Elong.: 3–5% (D638); Tens. Mod.: 675,000 psi (D638); Flex. Str.: 14,900 psi (D790); Flex. Mod.: 680,000 psi (D790); Compr. Str.: 12,300 psi (D-695); Impact Str. (Izod): 1.30 ft lb/in. notch (¼ in., D256)

Therm. Prop.: Distort. Temp.: 205 F (@ 264 psi, D648); Coeff. of Linear Exp.: 2.4 × 10^{-5} in./in./F (D696)

Elec. Prop.: Dielec. Str.: 400 V/mil (D-149); Vol. Resist.: 10^{15} ohm-cm (D257)

Ref.: Data sheet

AS-20GF. Compounding Technology, Inc.
Chem. Descrip.: Glass-reinforced ABS resin
Category/Applications: See AS-10GF
Composition: 20% glass fiber
Gen'l. Prop.: Sp. Gr.: 1.20 (D792)
Mech. Prop.: Tens. Str.: 12,500 psi (D-638); Tens. Elong.: 2–4% (D638); Tens. Mod.: 950,000 psi (D638); Flex. Str.: 16,500 psi (D790); Flex. Mod.: 900,000 psi (D790); Compr. Str.: 14,000 psi (D-695); Shear Str.: 6500 psi (D732); Impact Str. (Izod): 1.25 ft lb/in. notch (¼ in., D256); Hardness (Rockwell): R 122 (D785)
Therm. Prop.: Conduct.: 1.4 Btu/h/ft^2/F/in. (Cenco); Distort. Temp.: 210 F (@ 264 psi, D648); Coeff. of Linear Exp.: 2.1×10^{-5} in./in./F (D696); Flamm.: 94 HB (UL94)
Elec. Prop.: Dissip. Factor: 0.009 (10^6 Hz, D150); Dielec. Str.: 400 V/mil (D-149); Dielec. Const.: 3.3 (10^6 Hz, D-150); Vol. Resist.: 10^{15} ohm-cm (D257)
Ref.: Data sheet

AS-30GF. Compounding Technology, Inc.
Chem. Descrip.: Glass-reinforced ABS resin
Category/Applications: See AS-10GF
Composition: 30% glass fiber
Gen'l. Prop.: Sp. Gr.: 1.29 (D792)
Mech. Prop.: Tens. Str.: 13,800 psi (D-638); Tens. Elong.: 2–4% (D638); Tens. Mod.: 1.1×10^6 psi (D638); Flex. Str.: 18,000 psi (D790); Flex. Mod.: 1.15×10^6 psi (D790); Compr. Str.: 15,000 psi (D695); Shear Str.: 7000 psi (D732); Impact Str. (Izod): 1.1 ft lb/in. notch (¼ in., D256); Hardness (Rockwell): R-123 (D785)
Therm. Prop.: Conduct.: 1.5 Btu/h/ft^2/F/in. (Cenco); Distort. Temp.: 215 F (@ 264 psi, D648); Coeff. of Linear Exp.: 1.8×10^{-5} in./in./F (D696); Flamm.: 94HB (UL94)
Elec. Prop.: Dissip. Factor: 0.009 (10^6 Hz, D150); Dielec. Str.: 400 V/mil (D-149); Dielec. Const.: 3.4 (10^6 Hz, D-150); Vol. Resist.: 10^{15} ohm-cm (D257)
Ref.: Data sheet

AS-40GF. Compounding Technology Inc.
Chem. Descrip.: Glass-reinforced ABS resin
Category/Applications: See AS-10GF
Composition: 40% glass fiber
Gen'l. Prop.: Sp. Gr.: 1.38 (D792)
Mech. Prop.: Tens. Str.: 15,500 psi (D-638); Tens. Elong.: 1–3% (D638); Tens. Mod.: 1.2×10^6 psi (D638); Flex. Str.: 20,500 psi (D790); Flex. Mod.: 1.25×10^6 psi (D790); Compr. Str.: 16,000 psi (D695); Shear Str.: 7300 psi (D732); Impact Str. (Izod): 1.0 ft lb/in. notch (¼ in., D256); Hardness (Rockwell): R-123 (D785)
Therm. Prop.: Conduct.: 1.6 Btu/h/ft^2/F/in. (Cenco); Distort. Temp.: 220 F (@ 264 psi, D648); Coeff. of Linear Exp.: 1.4×10^{-5} in./in./F (D696); Flamm.: 94HB (UL94)
Elec. Prop.: Dissip. Factor: 0.005 (10^6 Hz, D150); Dielec. Str.: 400 V/mil (D-149); Dielec. Const.: 3.5 (10^6 Hz, D-150); Vol. Resist.: 10^{15} ohm-cm (D257)
Ref.: Data sheet

AT-20GF. Compounding Technology, Inc.
Chem. Descrip.: Glass-reinforced acetal resin
Category/Applications: Reinforced resin which excels in applications requiring good lubricity, and chemical and hot water resistance; used in automotive, hardware, and plumbing applications
Composition: 20% glass fiber

Gen'l. Prop.: Sp. Gr.: 1.55 (D792); M.P.: 390 F
Mech. Prop.: Tens. Str.: 13,500 psi (D-638); Tens. Elong.: 2% (D638); Tens. Mod.: 1.3 × 10^6 psi (D638); Flex. Str.: 16,500 psi (D790); Flex. Mod.: 900,000 psi (D790); Compr. Str.: 12,000 psi (D695); Impact Str. (Izod): 1.0 ft lb/in. notch (¼ in., D256)
Therm. Prop.: Conduct.: 2.1 Btu/h/ft^2/F/in. (Cenco); Distort. Temp.: 325 F (@ 264 psi, D648); Coeff. of Linear Exp.: 2.6 × 10^{-5} in./in./F (D696)
Elec. Prop.: Dissip. Factor: 0.006 (10^6 Hz, D150); Dielec. Str.: 500 V/mil (D-149); Dielec. Const.: 3.9 (50–100 Hz, D150); Vol. Resist.: 10^{14} ohm-cm (D-257)
Ref.: Data sheet

AT-30GF. Compounding Technology, Inc.
Chem. Descrip.: Glass-reinforced acetal resin
Category/Applications: See AT-20GF
Composition: 30% glass fiber
Gen'l. Prop.: Sp. Gr.: 1.63 (D792); M.P.: 390 F
Mech. Prop.: Tens. Str.: 15,500 psi (D-638); Tens. Elong.: 2% (D638); Tens. Mod.: 1.4 × 10^6 psi (D638); Flex. Str.: 17,500 psi (D790); Flex. Mod.: 1.1 × 10^6 psi (D790); Compr. Str.: 11,500 psi (D695); Impact Str. (Izod): 0.8 ft lb/in. notch (¼ in., D256)
Therm. Prop.: Conduct.: 2.2 Btu/h/ft^2/F/in. (Cenco); Distort. Temp.: 325 F (@ 264 psi, D648); Coeff. of Linear Exp.: 2.4 × 10^{-5} in./in./F (D696)
Elec. Prop.: See AT-20GF
Ref.: Data sheet

AT-40GF. Compounding Technology, Inc.
Chem. Descrip.: Glass-reinforced acetal resin
Category/Applications: See AT-20GF
Composition: 40% glass fiber
Gen'l. Prop.: Sp. Gr.: 1.74 (D792); M.P.: 390 F
Mech. Prop.: Tens. Str.: 14,000 psi (D-638); Tens. Elong.: 2% (D638); Tens. Mod.: 1.6 × 10^6 psi (D638); Flex. Str.: 18,000 psi (D790); Flex. Mod.: 1.3 × 10^6 psi (D790); Compr. Str.: 11,000 psi (D695); Impact Str. (Izod): 0.9 ft lb/in. notch (¼ in., D256)
Therm. Prop.: Conduct.: 2.3 Btu/h/ft^2/F/in. (Cenco); Distort. Temp.: 330 F (@ 66 psi, D648); Coeff. of Linear Exp.: 1.9 × 10^{-5} in./in./F (D696)
Elec. Prop.: Dissip. Factor: 0.006 (10^6 Hz, D150); Dielec. Str.: 500 V/mil (D-149); Dielec. Const.: 3.9 (50–100 Hz, D150)
Ref.: Data sheet

B

Barco Bond MB-14X. Astro Chem. Co., Inc.

Chem. Descrip.: Epoxy adhesive

Category/Applications: General purpose epoxy adhesive with good flexibility and peel strength; bonding agent for factory use in bonding metals and other nonporous surfaces where precise control of cleaning methods is not possible

Form: Paste; Color: Clear after mixing the green tube (resin) and red tube (curing agent)

Toxicity/Handling: Minimum toxicity ingredients; avoid skin contact and use with good ventilation

Std. Pkgs.: Tubes

Barco Bond MB-100X. Astro Chem. Co., Inc.

Chem. Descrip.: Adhesive resin

Category/Applications: General purpose adhesive, electrical insulating material, and surface filling and sealing resin; used for: electrical insulation (repair and reinsulation of electrical equipment, motors, generators, transformers); bonding of metal, glass, ceramics, crockery, concrete, and some plastics; patching water and chemical piping, metal castings, pumps, valves, tanks, porcelain insulators; sealing caulk wood, concrete, metal surfaces, seal porous castings, pipes bushings, and wire terminations

Form: Paste; Color: Neutral

Gen'l. Prop.: Stability: Resistant to weather, water, oil, gasoline, and other chemicals

Mech. Prop.: Tens. Str.: 9000 psi; Tens. Shear: 4000 psi; Bond Str.: 3000–4000 psi (½ in., 78 F, Al/Al bond)

Therm. Prop.: Conduct.: 35×10^{-4} Cal/(s)(cm^2)(C)(cm); Distort. Temp.: 100 C (1 wk @ 77 F); Coeff. of Linear Exp.: 52×10^{-6} (in./in./C)

Elec. Prop.: Dielec. Str.: 200 V/mil (@ 10 mil)

Curing Charact.: Pot Life: 30 min (70 F); Cure Time: 4 days (70 F), or 1½ h (175 F)

Toxicity/Handling: Avoid skin contact; use with good ventilation

Std. Pkgs.: Tubes

Ref.: Product information

Barco Bond MB-127. Astro Chem. Co., Inc.

Chem. Descrip.: Epoxy adhesive

Category/Applications: General purpose thixotropic resilient epoxy adhesive with good flexibility and peel strength; bonding agent for factory use in bonding metals and other nonporous surfaces where precise control of cleaning methods is not possible

Form: Part A–resin, Part B–curing agent; Color: Translucent after cure
Gen'l. Prop.: Sp. Gr.: 1.15–1.25 (Part A), 0.98–0.99 (Part B); Visc.: 300,000 cps (Part A)
Mech. Prop.: Tens. Shear: 3000 psi (A1/A1)
Therm. Prop.: Conduct.: 3×10^{-4} Cal (s)(cm^2)(C)(cm); Distort. Temp.: 185 F
Elec. Prop.: Dielec. Str.: 1500 V/mil (@ 10 mil); Resist.: 1×10^{12} ohm/cm^2
Curing Charact.: Pot Life: 1 h (1 lb, 70 F); Cure Time: 16 h (good cure, 77 F); 3 days (full cure, 77 F); 10 min (full cure, 300 F)
Toxicity/Handling: Avoid contact with skin; use with good ventilation
Std. Pkgs.: Tubes
Ref.: Product information

Barco Bond MB-165. Astro Chem. Co., Inc.
Chem. Descrip.: Adhesive resins
Category/Applications: Epoxy adhesive with good physical, electrical, and chemical properties; used for stanking electrical or electronic parts for assembly, mounting instruction of name plates, as sealing insulation for screws, nuts, or terminals, sealing cracks or holes in tanks or pipes used for gases, water, certain chemicals, petroleum, and other liquids, general purpose repair material for marine and household use
Form: Resin; Color: Clear after cure
Mech. Prop.: Shear Str. (after 24 h bond): 350 psi (acrylic surfaces), 300 psi (epoxy surfaces), 350 psi (aluminum surfaces), 1000 psi (copper surfaces), 2200 psi (steel surfaces)
Curing Charact. (R.T.): Initial Set: 60 s; 80% Bond Set: 5 min (slight heat will hasten cure)
Toxicity/Handling: Avoid skin contact; use with good ventilation
Std. Pkgs.: Self-metered tubes
Ref.: Product information

Barco Bond MB-175. Astro Chem. Co., Inc.
Chem. Descrip.: Adhesive resins
Category/Applications: See Barco Bond MB-165
Form: Paste; Color: Clear after cure
Mech. Prop.: Shear Str. (after 24 h bond): 350 psi (acrylic surfaces), 350 psi (epoxy surfaces), 750 psi (aluminum surfaces), 3000 psi (copper surfaces), 2500 psi (steel surfaces)
Curing Charact. (R.T.): Initial Set: 3 min; 80% Bond Set: 30 min (slight heat will hasten cure)
Toxicity/Handling: Avoid skin contact; use with good ventilation
Std. Pkgs.: Self-metered tubes
Ref.: Product information

Barco Bond MB-185. Astro Chem. Co., Inc.
Chem. Descrip.: Adhesive resins
Category/Applications: See Barco Bond MB-165
Form: Paste; Color: Clear after cure
Mech. Prop.: Shear Str. (after 24 h bond): 350 psi (acrylic surfaces), 370 psi (epoxy surfaces), 1000 psi (aluminum surfaces), 4000 psi (copper surfaces), 3000 psi (steel surfaces)
Curing Charact. (R.T.): Initial Set: 5 min; 80% Bond Set: 40 min (slight heat will hasten cure)
Toxicity/Handling: Avoid skin contact; use with good ventilation
Std. Pkgs.: Self-metered tubes
Ref.: Product information

Bayblend MC-2500. Mobay Chem. Corp.
Chem. Descrip.: ABS/PC alloy resin
Category/Applications: General purpose injection molding and extrusion resin offering med. flow, high impact strength, high deflection temp., cosmetic surface appearance, and moderate processing temps.; used for appliance, automotive, and consumer product components;

continuous use temps. to 60 C (without impact, UL746B)
Form: Opaque, colorable
Gen'l. Prop.: Density: 1.12 mg/m^3; Stability: Good dimensional stability and chemical resistance
Mech. Prop.: Tens. Str.: 50 MPa (break); Tens. Elong.: 15% (break); Flex. Str.: 90 MPa (break, 3.2 mm); Flex. Mod.: 2.43 GPa (3.2 mm); Impact Str. (Izod): 560 J/m notch (23 C, 3.2 mm); Hardness (Rockwell): R 117
Therm. Prop.: Distort. Temp.: 116 C (@ 1.82 MPa, unannealed); Coeff. of Linear Exp.: 63 × 10^{-6} mm/mm/C; Flamm.: HB (@ 1.47 mm, UL94)
Elec. Prop.: Dissip. Factor: 0.0038 (60 Hz); Dielec. Str.: > 17 kV/mm; Dielec. Const.: 2.90 (60 Hz); Arc Resist.: 91 s
Ref.: Product information

Bayblend MD-6500. Mobay Chem. Corp.
Chem. Descrip.: ABS/PC alloy resin
Category/Applications: Flame-retardant resin with high impact strength, high deflection temp., cosmetic surface appearance, moderate processing temps.; used in appliance, electrical/electronic, communications, and business machine applications; continuous use temps. to 95 C (without impact, UL746B)
Form: Opaque, colorable
Gen'l. Prop.: Density: 1.20 mg/m^3; Stability: See Bayblend MC-2500
Mech. Prop.: Tens. Str.: 48 MPa (break); Tens. Elong.: 10% (break); Flex. Str.: 94 MPa (break, 3.2 mm); Flex. Mod.: 2.60 GPa (3.2 mm); Compr. Str.: 78 MPa (yield); Impact Str. (Izod): 560 J/m notch (23 C, 3.2 mm); Hardness (Rockwell): R 117
Therm. Prop.: Distort. Temp.: 104 C (@ 1.82 MPa, unannealed); Coeff. of Linear Exp.: 67 × 10^{-6} mm/mm/C; Flamm.: V-O (@1.47 mm, UL94)
Elec. Prop.: Dissip. Factor: 0.0041 (60 Hz); Dielec. Str.: 18 kV/mm; Dielec. Const.: 2.95 (60 Hz); Arc Resist.: 23 s
Ref.: Product information

Bayblend MH-6500. Mobay Chem. Corp.
Chem. Descrip.: ABS/PC alloy resin
Category/Applications: Flame-retardant injection molding resin with improved color stability under indoor lighting exposure; offers high impact strength, high deflection temp., cosmetic surface appearance, and moderate processing temps.; used for electrical/electronic equipment housings, and lighting fixtures
Form: Opaque, colorable
Gen'l. Prop.: Density: 1.20 mg/m^3; Stability: See Bayblend MC-2500
Mech. Prop.: Tens. Str.: 47 MPa (break); Tens. Elong.: 26% (break); Flex. Str.: 93 MPa (break, 3.2 mm); Flex. Mod.: 2.60 GPa (3.2 mm); Impact Str. (Izod): 267–427 J/m notch (23 C, 3.2 mm)
Therm. Prop.: Distort. Temp.: 107 C (@ 1.82 MPa, unannealed); Flamm.: V-O @ 1.6 mm; 5V @ 3.2 mm (UL94)
Ref.: Product information

Bayblend MJ-2500. Mobay Chem. Corp.
Chem. Descrip.: ABS/PC alloy resin
Category/Applications: Injection molding and extrusion resin comparable to Bayblend MC-2500 in properties; complies with FDA regulations for use in certain food contact applications
Form: Opaque, colorable
Gen'l. Prop.: Density: 1.12 mg/m^3; Stability: See Bayblend MC-2500
Mech. Prop.: Tens. Str.: 50 MPa (break); Tens. Elong.: 65% (break); Flex. Str.: 84 MPa (break, 3.2 mm); Flex. Mod.: 2.20 GPa (3.2 mm); Impact Str. (Izod): 538 J/m notch (23 C, 3.2 mm); Hardness (Rockwell): R 120
Therm. Prop.: Distort. Temp.: 104 C (@ 1.82 MPa, unannealed); Coeff. of Linear Exp.: 67 × 10^{-6} mm/mm/C; Flamm.: HB (@ 1.6 mm, UL94)

Elec. Prop.: Dissip. Factor: 0.0040 (60 Hz); Dielec. Str.: > 17 kV/mm; Dielec. Const.: 2.90 (60 Hz); Arc Resist.: > 50 s
Ref.: Product information

Bayblend T 45 MN. Bayer AG
Chem. Descrip.: Blend of Makrolon (PC) and Novodur (ABS)
Category/Applications: Thermoplastic blend with dimensional stability, surface gloss, impact resistance, high heat deflection temp., and good stiffness and strength; may be used in injection molding, extrusion, blow molding, thermoforming, machining, welding, bonding, screwing, coating, printing, vacuum metallizing, electroplating, and other processing techniques for automotive, electrical, domestic appliances, camera bodies, electroplated components, and other applications
Form: Cylindrical granules; Color: Natural opaque
Gen'l. Prop.: Density: 1.12 g/cc (DIN 53479, ISO/R 1183); Soften. Pt. (Vicat): 120 C (A, DIN 53460, ISO 306); Stability: Resistant to dilute mineral acids, neutral and acid salt solutions, aliphatic hydrocarbons, and many oils and fats
Mech. Prop.: Tens. Str.: 35 MPa (@ break, DIN 53455, ISO/R 527); Tens. Elong.: 40% (@ break, DIN 53455, ISO/R 527); Tens. Mod.: 1950 MPa (DIN 53457); Impact Str.: 25 kJ/m^2 notch (23 C, DIN 53453, ISO/R 179); Hardness (Ball Indentation): 80 N/mm^2 (DIN 53456, ISO 2039)
Therm. Prop.: Conduct.: 0.18 W/K m (based on VDE 0304/1 and DIN 52612); Distort. Temp.: 95 C (A, DIN 53461, ISO 75); Coeff. of Linear Exp.: 80–85 10^{-6} K^{-1} (–40 to 100 C, VDE 0304/1, DIN 53752); Sp. Heat: 1.34 kJ/kg K (35 C); Flamm.: HB (UL94)
Elec. Prop.: Dissip. Factor: 4×10^{-3} (dry, 50 Hz, VDE 0304, DIN 53482); Dielec. Str.: 24 kV/mm (dry, DIN 53481, VDE 0303/2, in transformer oil); Dielec. Const.: 2.9 (dry, 50 Hz, VDE 0304, DIN 53482); Vol. Resist.: 10^{16} ohm-cm (dry, DIN 53482, VDE 0303); Surf. Resist.: > 10^{14} ohm (dry, DIN 53482, VDE 0303/3)
Std. Pkgs.: 25-kg sacks
Ref.: Pamphlet KL 46150e

Bayblend T 65 MN. Bayer AG
Chem. Descrip.: Blend of Makrolon (PC) and Novodur (ABS)
Category/Applications: See Bayblend T 45 MN
Form: Cylindrical granules; Color: Natural opaque
Gen'l. Prop.: Density: 1.15 g/cc (DIN 53479, ISO/R 1183); Soften. Pt. (Vicat): 130 C (A, DIN 53460, ISO 306); Stability: See Bayblend T 45 MN
Mech. Prop.: Tens. Str.: 40 MPa (@ break, DIN 53455, ISO/R 527); Tens. Elong.: 45% (@ break, DIN 53455, ISO/R 527); Tens. Mod.: 2100 MPa (DIN 53457); Impact Str.: 30 kJ/m^2 notch (23 C, DIN 53453, ISO/R 179); Hardness (Ball Indentation): 90 N/mm^2 (DIN 53456, ISO 2039)
Therm. Prop.: Conduct.: 0.19 W/k m (based on VDE 0304/1, DIN 52612); Distort. Temp.: 100 C (A, DIN 53461, ISO 75); Coeff. of Linear Exp.: 75-80 10^{-6} K^{-1} (–40 to 100 C, VDE 0304/1, DIN 53752); Sp. Heat: 1.28 kJ/kg K (35 C); Flamm.: HB (UL94)
Elec. Prop.: Dissip. Factor: 4×10^{-3} (dry, 50 Hz, VDE 0304, DIN 53482); Dielec. Str.: 24 kV/mm (dry, DIN 53481, VDE 0303/2, in transformer oil); Dielec. Const.: 2.9 (dry, 50 Hz, VDE 0304, DIN 53482); Vol. Resist.: > 10^{16} ohm-cm (dry, DIN 53482, VDE 0303); Surf. Resist.: > 10^{14} ohm (dry, DIN 53482, VDE 0303/3)
Std. Pkgs.: 25-kg sacks
Ref.: Pamphlet KL 46150e

Bayblend T 85 MN. Bayer AG
Chem. Descrip.: Blend of Makrolon (PC) and Novodur (ABS)
Category/Applications: See Bayblend T 45 MN
Form: Cylindrical granules; Color: Natural opaque
Gen'l. Prop.: Density: 1.16 g/cc (DIN 53479, ISO/R 1183); Soften. Pt. (Vicat): 137 C (A, DIN 53460, ISO 306); Stability: See Bayblend T 45 MN
Mech. Prop.: Tens. Str.: 45 MPa (@ break, DIN 53455, ISO/R 527); Tens. Elong.: 50% (@ break, DIN 53455, ISO/R 527); Tens. Mod.: 2200 MPa (DIN 53457); Impact Str.: 30 kJ/m^2 notch (23 C, DIN 53453, ISO/R 179); Hardness (Ball Indentation): 90 N/mm^2 (DIN 53456, ISO 2039)
Therm. Prop.: Conduct.: 0.20 W/k m (based on VDE 0304/1 and DIN 52612); Distort. Temp.: 105 C (A, DIN 53461, ISO 75); Coeff. of Linear Exp.: 70-75 10^{-6} K^{-1} (−40 to 100 C, VDE 0304/1, DIN 53752); Sp. Heat: 1.26 kJ/kg K (35 C); Flamm.: HB (UL94)
Elec. Prop.: Dissip. Factor: 4 $\times 10^{-3}$ (dry, 50 Hz, VDE 0304, DIN 53482); Dielec. Str.: 24 kV/mm (dry, DIN 53481, VDE 0303/2, in transformer oil); Dielec. Const.: 3.0 (dry, 50 Hz, VDE 0304, DIN 53482); Vol. Resist.: $> 10^{16}$ ohm-cm (dry, DIN 53482, VDE 0303); Surf. Resist.: $> 10^{14}$ ohm (dry,, DIN 53482, VDE 0303/3)
Std. Pkgs.: 25-kg sacks
Ref.: Pamphlet KL 46150e

Baycoll 12. Bayer AG
Chem. Descrip.: Polyester polyol
Category/Applications: Used with Desmodur crosslinking agents for solventless or low-solvent reaction adhesives; such adhesives have outstanding wetting action on smooth substrates and penetrate porous materials well; the bonds are very tough and stable; good adhesion to polar substrates
Composition: 5.0 ± 0.3% OH content
Gen'l. Prop.: Visc. (Brookfield): ≈ 20 Pa·s (23 C, spindle 4, 6 rpm)
Ref.: Bulletin KA 30536 e

Baycoll 14. Bayer AG
Chem. Descrip.: Polyester polyol
Category/Applications: See Baycoll 12
Composition: 3.3 ± 0.2% OH content
Gen'l. Prop.: Visc. (Brookfield): ≈ 15 Pa·s (23 C, spindle 4, 6 rpm)
Ref.: Bulletin KA 30536 e

Baycoll 22. Bayer AG
Chem. Descrip.: Polyester polyol
Category/Applications: See Baycoll 12
Composition: 1.8 ± 0.2% OH content
Gen'l. Prop.: Visc. (Brookfield): ≈ 25 Pa·s (23 C, spindle 4, 6 rpm)
Ref.: Bulletin KA 30536 e

Baycoll 31. Bayer AG
Chem. Descrip.: Polyether polyol
Category/Applications: Used with Desmodur crosslinking agents for solventless or low-solvent reaction adhesives; such adhesives have outstanding wetting action on smooth substrates and penetrate porous materials well; the bonds are very tough and stable; the polyether polyols are low viscosity and give films of improved stability
Composition: 1.7 ± 0.2% OH content
Gen'l. Prop.: Visc. (Brookfield): ≈ 0.35 Pa·s (23 C, spindle 3, 30 rpm)
Ref.: Bulletin KA 30536 e

Baycoll 81. Bayer AG
Chem. Descrip.: Polyether polyol
Category/Applications: See Baycoll 31
Composition: 11.0 ± 0.2% OH content
Gen'l. Prop.: Visc. (Brookfield): ≈ 0.6 Pa·s (23 C, spindle 3, 30 rpm)
Ref.: Bulletin KA 30536 e

Bayer CM 2552. Bayer AG
Chem. Descrip.: Chlorinated polyethylene
Category/Applications: Polymer with good mechanical properties, high wear resistance, and dynamic strength; used for the rubber and cable industries in cable and wire sheaths and insulations for low voltage cables, hoses (exp. hydraulic and chemical hoses), technical moldings, roll covers, conveyor belting, rubberized fabrics, cellular rubber
Form: Fine powder; Color: Light
Composition: 25% chlorine content
Gen'l. Prop.: Density: 1.10 g/cc; Visc. (Mooney): ≈ 130 (ML 5 + 4, 120 C); Stability: Good resistance to ageing in oils and hot air, to ozone, weathering, color change, to oils and liquid fuels; very good resistance to chemicals
Std. Pkgs.: 20-kg paper pags, polyethylene-lined, on 600-kg pallets
Ref.: Pamphlet KA 32125 e

Bayer CM 3610. Bayer AG
Chem. Descrip.: Chlorinated polyethylene
Category/Applications: See Bayer CM 2552
Form: Fine powder; Color: Light
Composition: 36% chlorine content
Gen'l. Prop.: Density: 1.19 g/cc; Visc. (Mooney): ≈ 35 (ML 1 + 4, 120 C); Stability: See Bayer CM 2552
Std. Pkgs.: See Bayer CM 3630
Ref.: Pamphlet KA 32125 e

Bayer CM 3630. Bayer AG
Chem. Descrip.: Chlorinated polyethylene
Category/Applications: See Bayer CM 2552; 3630 is the main and general purpose grade
Form: Fine powder; Color: Light
Composition: 36% chlorine content
Gen'l. Prop.: Density: 1.19 g/cc; Visc. (Mooney): ≈ 85 (ML 1 + 4, 120 C); Stability: See Bayer CM 2552
Std. Pkgs.: 25-kg paper bags, polyethylene-lined, on 750-kg pallets
Ref.: Pamphlet KA 32125 e

Bayer CM 3631. Bayer AG
Chem. Descrip.: Chlorinated polyethylene
Category/Applications: See Bayer CM 2552; 3631 swells less than 3630 in water
Form: Fine powder; Color: Light
Composition: 36% chlorine content
Gen'l. Prop.: Density: 1.19 g/cc; Visc. (Mooney): ≈ 85 (ML 1 + 4, 120 C); Stability: See Bayer CM 2552
Std. Pkgs.: See Bayer CM 3630
Ref.: Pamphlet KA 32125 e

Bayer CM 3632. Bayer AG
Chem. Descrip.: Chlorinated polyethylene
Category/Applications: See Bayer CM 2552
Form: Fine powder; Color: Light
Composition: 36% chlorine content
Gen'l. Prop.: Density: 1.19 g/cc; Visc. (Mooney): ≈ 110 (ML 1 + 4, 120 C); Stability: See Bayer CM 2552
Std. Pkgs.: See Bayer CM 3630
Ref.: Pamphlet KA 32125 e

Bayer CM 4230. Bayer AG
Chem. Descrip.: Chlorinated polyethylene
Category/Applications: See Bayer CM 2552
Form: Fine powder; Color: Light
Composition: 42% chlorine content
Gen'l. Prop.: Density: 1.25 g/cc; Visc. (Mooney): ≈ 85 (ML 1 + 4, 120 C); Stability: See Bayer CM 2552
Std. Pkgs.: See Bayer CM 3630
Ref.: Pamphlet KA 32125 e

Bayer SBR Latex 200 C. Bayer AG
Chem. Descrip.: Styrene-butadiene latex
Category/Applications: Nonstaining, hard component for blended latex goods
Composition: ≈ 80% styrene content; 50 ± 0.5% solids
Gen'l. Prop.: Density: 1.00 g/cc; pH: ≈ 9.5
Std. Pkgs.: 120-kg plastic drums, 600-kg plastic containers, 16–23-ton tank wagons
Ref.: Pamphlet KA 32125 e

Bayer SBR Latex 210 C. Bayer AG
Chem. Descrip.: Styrene-butadiene latex
Category/Applications: Nonstaining latex used for hard, heat-formable toe caps; paper and cardboard impregnation
Composition: ≈ 80% styrene content; 50 ± 0.5% solids
Gen'l. Prop.: Density: 1.03 g/cc; pH: 8.5 ± 0.5
Std. Pkgs.: See Bayer SBR Latex 200 C
Ref.: Pamphlet KA 32125 e

Bayer SBR Latex 310 C. Bayer AG
Chem. Descrip.: Styrene-butadiene latex
Category/Applications: Nonstaining, hard binder for paper and needle felt reinforcement
Composition: ≈ 65% styrene content; 50 ± 0.5% solids
Gen'l. Prop.: Density: 1.03 g/cc; pH: 8.5 ± 0.5
Std. Pkgs.: See Bayer SBR Latex 200 C
Ref.: Pamphlet KA 32125 e

Bayer SBR Latex 320 C. Bayer AG
Chem. Descrip.: Styrene-butadiene latex
Category/Applications: Nonstaining, hard binder for carpet backs with heavy filler loadings
Composition: ≈ 65% styrene content; 50 ± 1.0% solids
Gen'l. Prop.: Density: 1.03 g/cc; pH: 8.5 ± 0.5
Std. Pkgs.: See Bayer SBR Latex 200 C
Ref.: Pamphlet KA 32125 e

Bayer SBR Latex 410 C. Bayer AG
Chem. Descrip.: Styrene-butadiene latex
Category/Applications: Nonstaining, soft binder for needle felt reinforcement
Composition: ≈ 55% styrene content; 50 ± 0.5% solids
Gen'l. Prop.: Density: 1.01 g/cc; pH: 8.5 ± 0.5
Std. Pkgs.: See Bayer SBR Latex 200 C
Ref.: Pamphlet KA 32125 e

Bayer SBR Latex 420 C. Bayer AG
Chem. Descrip.: Styrene-butadiene latex
Category/Applications: Nonstaining soft binder for carpet backs with heavy filler loadings; binder for lamination of textile secondary backings on tufted floor coverings
Composition: ≈ 55% styrene content; 50 ± 1.0% solids
Gen'l. Prop.: Density: 1.01 g/cc; pH: 8.5 ± 0.5
Std. Pkgs.: See Bayer SBR Latex 200 C
Ref.: Pamphlet KA 32125 e

Bayer SBR Latex 603 C. Bayer AG
Chem. Descrip.: Styrene-butadiene latex
Category/Applications: Nonstaining, soft component for blended latex goods; binder for cork scrap and ground rubber scrap
Composition: ≈ 34% styrene content; 54 ± 1.0% solids
Gen'l. Prop.: Density: 1.00 g/cc; pH: 8.5 ± 0.5
Std. Pkgs.: See Bayer SBR Latex 200 C
Ref.: Pamphlet AK 32125 e

Baypren 110. Bayer AG
Chem. Descrip.: Chloroprene rubber
Category/Applications: Synthetic polymer with good mechanical properties,

incl. high resistance to abrasion; grades with slight to med. crystallization tendencies used for moldings and extrudates of all types, reinforced hoses, roll covers, belting, cable sheathings and insulations for low voltage wires, sponge rubber, corrosion-resistant linings, sheeting (esp. for roofing), fabric proofings, footwear (boots), food-contacting goods; grades with very strong crystallization tendencies used in adhesives for footwear, furniture, and applications in the building industry
Form: Flat chips; Color: Whitish to pale brown
Gen'l. Prop.: Density: 1.23 g/cc; Visc. (Mooney): 40–45 and 45–50 (ML 4, 100 C); Stability: Good resistance to weathering, ozone, and chemicals; very good resistance to ageing and hot air; med. resistance to oil and petrol; Crystallization: Slight
Std. Pkgs.: 25-kg polyethylene-lined paper bags on 1000-kg pallets
Ref.: Pamphlet KA 32125 e

Baypren 112. Bayer AG
Chem. Descrip.: Chloroprene rubber
Category/Applications: See Baypren 110
Form: Flat chips; Color: Whitish to pale brown
Gen'l. Prop.: Density: 1.23 g/cc; Visc. (Mooney): 40–45 and 45–50 (ML 4, 100 C); Stability: See Baypren 110; Crystallization: Slight to med.
Std. Pkgs.: See Baypren 110
Ref.: Pamphlet KA 32125 e

Baypren 115. Bayer AG
Chem. Descrip.: Chloroprene rubber, precrosslinked grade
Category/Applications: See Baypren 110; very easy to process
Form: Flat chips; Color: Whitish to pale brown
Gen'l. Prop.: Density: 1.23 g/cc; Visc. (Mooney): 45–55 (ML 4, 100 C); Stability: See Baypren 110; Crystallization: Slight
Std. Pkgs.: See Baypren 110
Ref.: Pamphlet KA 32125 e

Baypren 124. Bayer AG
Chem. Descrip.: Chloroprene rubber, precrosslinked grade
Category/Applications: See Baypren 110; very easy to process
Form: Flat chips; Color: Whitish to pale brown
Gen'l. Prop.: Density: 1.23 g/cc; Visc. (Mooney): 65–75 (ML 4, 100 C); Stability: See Baypren 110; Crystallization: Slight
Std. Pkgs.: See Baypren 110
Ref.: Pamphlet KA 32125 e

Baypren 130. Bayer AG
Chem. Descrip.: Chloroprene rubber
Category/Applications: See Baypren 110
Form: Flat chips; Color: Whitish to pale brown
Gen'l. Prop.: Density: 1.23 g/cc; Visc. (Mooney): 100–120 (ML 4, 100 C); Stability: See Baypren 110; Crystallization: Slight
Std. Pkgs.: See Baypren 110
Ref.: Pamphlet KA 32125 e

Baypren 210. Bayer AG
Chem. Descrip.: Chloroprene rubber
Category/Applications: See Baypren 110
Form: Flat chips; Color: Whitish to pale brown
Gen'l. Prop.: Density: 1.23 g/cc; Visc. (Mooney): 40–45 and 45–50 (ML 4, 100 C); Stability: See Baypren 110; Crystallization: Med.
Std. Pkgs.: See Baypren 110
Ref.: Pamphlet KA 32125 e

Baypren 211. Bayer AG
Chem. Descrip.: Chloroprene rubber
Category/Applications: See Baypren 110
Form: Flat chips; Color: Whitish to pale brown
Gen'l. Prop.: Density: 1.23 g/cc; Visc. (Mooney): 35–40 (ML 4, 100 C); Stability: See Baypren 110; Crystallization: Med.
Std. Pkgs.: See Baypren 110
Ref.: Pamphlet KA 32125 e

Baypren 213. Bayer AG
Chem. Descrip.: Chloroprene rubber
Category/Applications: See Baypren 110; used in adhesives
Form: Chips
Gen'l. Prop.: Visc. (Mooney): ≈ 45; Crystallization: Med.
Ref.: Bulletin KA 30536 e

Baypren 214. Bayer AG
Chem. Descrip.: Chloroprene rubber, precrosslinked grade
Category/Applications: See Baypren 110; very easy to process
Form: Flat chips; Color: Whitish to pale brown
Gen'l. Prop.: Density: 1.23 g/cc; Visc. (Mooney): 50–60 (ML 4, 100 C); Stability: See Baypren 110; Crystallization: Med., < Baypren 210
Std. Pkgs.: See Baypren 110
Ref.: Pamphlet KA 32125 e

Baypren 215. Bayer AG
Chem. Descrip.: Chloroprene rubber, precrosslinked grade
Category/Applications: See Baypren 110; very easy to process
Form: Flat chips; Color: Whitish to pale brown
Gen'l. Prop.: Density: 1.23 g/cc; Visc. (Mooney): 45–55 (ML 4, 100 C); Stability: See Baypren 110; Crystallization: Med.
Std. Pkgs.: See Baypren 110
Ref.: Pamphlet KA 32125 e

Baypren 220. Bayer AG
Chem. Descrip.: Chloroprene rubber
Category/Applications: See Baypren 110
Form: Flat chips; Color: Whitish to pale brown
Gen'l. Prop.: Density: 1.23 g/cc; Visc. (Mooney): 70–80 (ML 4, 100 C); Stability: See Baypren 110; Crystallization: Med.
Std. Pkgs.: See Baypren 110
Ref.: Pamphlet KA 32125 e

Baypren 230. Bayer AG
Chem. Descrip.: Chloroprene rubber
Category/Applications: See Baypren 110
Form: Flat chips; Color: Whitish to pale brown
Gen'l. Prop.: Density: 1.23 g/cc; Visc. (Mooney): 95–110 and 110–125 (ML 4, 100 C); Stability: See Baypren 110; Crystallization: Med.
Std. Pkgs.: See Baypren 110
Ref.: Pamphlet KA 32125 e

Baypren 233. Bayer AG
Chem. Descrip.: Chloroprene rubber
Category/Applications: See Baypren 110; used in adhesives
Form: Flat chips; Color: Whitish to pale brown
Gen'l. Prop.: Density: 1.23 g/cc; Visc. (Mooney): 95–110 and 110–125 (ML 4, 100 C); Stability: See Baypren 110; Crystallization: Med.
Std. Pkgs.: See Baypren 110
Ref.: Pamphlet KA 32125 e

Baypren 235. Bayer AG
Chem. Descrip.: Chloroprene rubber, precrosslinked grade
Category/Applications: See Baypren 110; very easy to process
Form: Flat chips; Color: Whitish to pale brown
Gen'l. Prop.: Density: 1.23 g/cc; Visc.

(Mooney): 90-110 (ML 4, 100 C); Stability: See Baypren 110; Crystallization: Med.
Std. Pkgs.: See Baypren 110
Ref.: Pamphlet KA 32125 e

Baypren 236. Bayer AG
Chem. Descrip.: Chloroprene rubber
Category/Applications: See Baypren 110
Form: Flat chips; Color: Whitish to pale brown
Gen'l. Prop.: Density: 1.23 g/cc; Visc. (Mooney): 90-105 (ML 4, 100 C); Stability: See Baypren 110; Crystallization: Med.
Std. Pkgs.: See Baypren 110
Ref.: Pamphlet KA 32125 e

Baypren 243. Bayer AG
Chem. Descrip.: Chloroprene rubber
Category/Applications: See Baypren 110; used in adhesives
Form: Chips
Gen'l. Prop.: Visc. (Mooney): ≈ 120; Crystallization: Med.
Ref.: Bulletin KA 30536 e

Baypren 320. Bayer AG
Chem. Descrip.: Chloroprene rubber
Category/Applications: See Baypren 110
Form: Flat chips; Color: Whitish to pale brown
Gen'l. Prop.: Density: 1.23 g/cc; Visc. (Mooney): 80-95 (ML 4, 100 C); Stability: See Baypren 110; Crystallization: Strong
Std. Pkgs.: See Baypren 110
Ref.: Pamphlet KA 32125 e

Baypren 320 GR. Bayer AG
Chem. Descrip.: Chloroprene rubber
Category/Applications: See Baypren 110
Form: Granules
Gen'l. Prop.: Density: 1.23 g/cc; Visc. (Mooney): 75-85 (ML 4, 100 C); Stability: See Baypren 110; Crystallization: Strong
Std. Pkgs.: See Baypren 110
Ref.: Pamphlet KA 32125 e

Baypren 321, 331. Bayer AG
Chem. Descrip.: Chloroprene rubber, stabilized with thiuram disulfide
Category/Applications: See Baypren 110
Form: Flat chips; Color: Whitish to pale brown
Gen'l. Prop.: Density: 1.23 g/cc; Visc. (Mooney): 70-90, 90-105 (ML 4, 100 C) respectively; Stability: See Baypren 110; Crystallization: Strong
Std. Pkgs.: See Baypren 110
Ref.: Pamphlet KA 32125 e

Baypren 321 GR. Bayer AG
Chem. Descrip.: Chloroprene rubber, stabilized with thiuram disulfide
Category/Applications: See Baypren 110
Form: Granules
Gen'l. Prop.: Density: 1.23 g/cc; Visc. (Mooney): 70-75 (ML 4, 100 C); Stability: See Baypren 110; Crystallization: Strong
Std. Pkgs.: See Baypren 110
Ref.: Pamphlet KA 32125 e

Baypren 330. Bayer AG
Chem. Descrip.: Chloroprene rubber
Category/Applications: See Baypren 110
Form: Flat chips; Color: Whitish to pale brown
Gen'l. Prop.: Density: 1.23 g/cc; Visc. (Mooney): 95-105 (ML 4, 100 C); Stability: See Baypren 110; Crystallization: Strong
Std. Pkgs.: See Baypren 110
Ref.: Pamphlet KA 32125 e

Baypren 610, 710. Bayer AG
Chem. Descrip.: Chloroprene rubber, sulfur-modified grades
Category/Applications: See Baypren 110

Form: Flat chips; Color: Whitish to pale brown
Gen'l. Prop.: Density: 1.23 g/cc; Visc. (Mooney): 35–55 (ML 4, 100 C); Stability: See Baypren 110; Crystallization: Slight and med., respectively (610, 710)
Std. Pkgs.: See Baypren 110
Ref.: Pamphlet KA 32125 e

Baypren Latex Series. Bayer AG
Chem. Descrip.: Polychloroprene latices
Category/Applications: Similar to Baypren solid rubber grades with corresponding crystallization tendencies; used in dipped goods, proofings, coatings, lamination of fabrics and paper; binders for animal, vegetable, and synthetic fibers; modification of bituminous compositions for use in road construction, waterproofing, and corrosion protection; adhesives, sealing compositions; mine safety applications
Ref.: Pamphlet KA 32125 e

Baypren Latex B. Bayer AG
Chem. Descrip.: Polychloroprene latex
Category/Applications: See Baypren Latex Series; this grade used mainly for improvement of bituminous compositions
Composition: 58 ± 0.5% solids
Gen'l. Prop.: Density: 1.13 g/cc; pH: 12.5–13.0; Crystallization: Slight to med.
Std. Pkgs.: See Baypren Latex 4 R
Ref.: Pamphlet KA 32125 e

Baypren Latex GK. Bayer AG
Chem. Descrip.: Polychloroprene latex
Category/Applications: See Baypren Latex Series; this grade used as binder for various fibers; asbestos paper with good flame retardant properties
Composition: 32 ± 1.0% solids
Gen'l. Prop.: Density: ≈ 1.08 g/cc; pH: 12.5–13.0; Crystallization: Slight
Std. Pkgs.: See Baypren Latex 4 R
Ref.: Pamphlet KA 32125 e

Baypren Latex MKB. Bayer AG
Chem. Descrip.: Polychloroprene latex
Category/Applications: See Baypren Latex Series; this grade is a general purpose latex used mainly as a binder for various fibers
Composition: 58 ± 0.5% solids
Gen'l. Prop.: Density: 1.13 g/cc; pH: 12.5–13.0; Crystallization: Slight to med.
Std. Pkgs.: See Baypren Latex 4 R
Ref.: Pamphlet KA 32125 e

Baypren Latex SK. Bayer AG
Chem. Descrip.: Polychloroprene latex
Category/Applications: See Baypren Latex Series; this grade used in production of stiff toe caps and counters for footwear; dispersion adhesives
Composition: 55 ± 0.5% solids
Gen'l. Prop.: Density: 1.12 g/cc; pH: 12.5–13.0; Crystallization: Very strong
Std. Pkgs.: See Baypren Latex 4 R
Ref.: Pamphlet KA 32125 e

Baypren Latex T. Bayer AG
Chem. Descrip.: Polychloroprene latex
Category/Applications: See Baypren Latex Series; this grade used mainly for dipped goods
Composition: 58 ± 0.5% solids
Gen'l. Prop.: Density: 1.13 g/cc; pH: 12.5–13.0; Crystallization: Very slight
Std. Pkgs.: See Baypren Latex 4 R
Ref.: Pamphlet KA 32125 e

Baypren Latex 4 R. Bayer AG
Chem. Descrip.: Polychloroprene latex with 4% methacrylic acid
Category/Applications: See Baypren Latex Series; this grade suitable for R.T. curing; used in can sealing compositions and the binding of fibers of all types

Composition: 50 ± 0.5% solids
Gen'l. Prop.: Density: 1.11 g/cc; pH: ≈ 6.5; Crystallization: Very slight
Std. Pkgs.: 125-kg or 625-kg plastic drums; 16–23-ton tank wagons
Ref.: Pamphlet KA 32125 e

BCI Nylon #808–809. Belding Chem. Industries
Chem. Descrip.: Modified polyamide; alkoxy alkyl substituted 6:6 nylon
Category/Applications: Resin for use in textile, wood, metals, rubber industries, etc.; for protective coating and bonding, sewing thread, fabric finishes, barriers in military and commercial aircraft fuel tanks, etc.
Form: Granular to be converted to solvent solution
Gen'l. Prop.: Sol.: Sol. in the lower aliphatic alcohols, in phenols, and in solvent systems containing these with various cosolvents; Sp. Gr. (of films cast from 70:30 methanol-water solvent, air-dried @ R.T., uncured): 1.09 (D792); M.P. (of films, see above): 145–153 C; Stability: Excellent resistance (noncross-linked resin) to aqueous alkalis, oxygen, ozone, carbon dioxide, hydrogen, nitrogen, aliphatic hydrocarbons, aromatic hydrocarbons, ethers, mineral oils; good resistance to organic sulfides, chlorinated hydrocarbons, esters, amines, amides, and vegetable oils
Mech. Prop. (of films, see above): Tens. Str.: 3900 psi (D882); Tens. Elong.: 550% (D882)
Therm. Prop. (of films, see above): Sp. Heat: 0.40
Ref.: Tech. bulletin VIII-A

BCI Nylon #818–819. Belding Chem. Industries
Chem. Descrip.: Modified polyamide; alkoxy alkyl substituted 6:6 nylon
Category/Applications: See BCI Nylon #808–809
Form: Granular to be converted to solvent solution
Gen'l. Prop.: Sol.: Sol. in the lower aliphatic alcohols, in phenols, and in solvent systems containing these with various cosolvents; Sp. Gr. (of films cast from 70:30 methanol-water solvent, air-dried @ R.T., uncured): 1.09 (D792); M.P. (of films, see above): 140–148 C; Stability: Excellent resistance (non-cross-linked resin) to aq. alkalis, oxygen, ozone, carbon dioxide, hydrogen, nitrogen, aliphatic hydrocarbons, aromatic hydrocarbons, ethers, mineral oils; good resistance to organic sulfides, chlorinated hydrocarbons, esters, amines, amides, and vegetable oils
Mech. Prop. (of films, see above): Tens. Str.: 3500 psi (D882); Tens. Elong.: 550 (D882)
Therm. Prop. (of films, see above): Sp. Heat: 0.40
Ref.: Tech. bulletin VIII-A

BCI Nylon #819-S-10%. Belding Chem. Industries
Chem. Descrip.: Modified polyamide; alkoxy alkyl substituted 6:6 nylon
Category/Applications: See BCI Nylon #808–809
Form: 10% nylon resin solution in alcohol/water vehicle
Gen'l. Prop.: Sol.: Sol. in the lower aliphatic alcohols, in phenols, and in solvent systems containing these with various cosolvents; Stability: Excellent resistance (noncross-linked resin) to aq. alkalis, oxygen, ozone, carbon dioxide, hydrogen, nitrogen, aliphatic hydrocarbons, aromatic hydrocarbons, ethers, mineral oils, good resistance to organic sulfides, chlorinated hydrocarbons, esters, amines, amides, and vegetable oils
Ref.: Tech. bulletin VIII-A

BCI Nylon #829. Belding Chem. Industries
Chem. Descrip.: Modified polyamide; alkoxy alkyl substituted 6:6 nylon
Category/Applications. See BCI Nylon #808–809
Form: Granular to be converted to solvent solution
Gen'l. Prop.: Sol.: Sol. in the lower aliphatic alcohols, in phenols, and in solvent systems containing these with various cosolvents; Sp. Gr. (of films cast from 70:30 methanol-water solvent, air-dried @ R.T., uncured): 1.09 (D792); M.P. (of films, see above): 140–148 C; Stability: Excellent resistance (non-cross-linked resin) to aq. alkalis, oxygen, ozone, carbon dioxide, hydrogen, nitrogen, aliphatic hydrocarbons, aromatic hydrocarbons, ethers, mineral oils; good resistance to organic sulfides, chlorinated hydrocarbons, esters, amines, amides, and vegetable oils
Mech. Prop. (of films, see above): Tens. Str.: 2800 psi (D882); Tens. Elong.: 700 (D882)
Therm. Prop. (of films, see above): Sp. Heat: 0.40
Ref.: Tech. bulletin VIII-A

BCI Nylon LX-3249. Belding Chem. Industries
Chem. Descrip.: Linear polyamides (terpolymers)
Category/Applications: See BCI Nylon No. 651 (LX-3249 is structurally stronger and of somewhat higher melting pt.)
Form: Pellet
Gen'l. Prop. (of films cast from an ethanol-water (80:20) solvent system, 10% solids, air-dried at R.T.): Sp. Gr.: 1.12–1.13 (D792); M.P.: 158–170 C (D789); Stability: Generally unaffected by organic reagents (except those which act as solvents, i.e., the lower alcohols, and the phenols); little affected by inorganic bases and by neutral salts; inorganic acids and certain strong organic acids dissolve it with degradation
Mech. Prop. (of films, see above): Tens. Str.: 6500 psi (D882); Tens. Elong.: 200% (D882)
Therm. Prop. (of films, see above): Sp. Heat: 0.4
Ref.: Product bulletin BCI 600 Series

BCI Nylon LX-3250. Belding Chem. Industries
Chem. Descrip.: Linear polyamides (terpolymers)
Category/Applications: See BCI Nylon No. 653 (LX-3250 is structurally stronger and of somewhat higher melting pt.)
Form: Pellet
Gen'l. Prop. (of molded specimens): Sp. Gr.: 1.13 (D792); M.P.: 319 F (D-789)
Mech. Prop. (of molded specimens): Tens. Str.: 6800 psi (73 F, D638); Tens. Elong.: 275% (73 F, D638); Tens. Mod.: 45,000 psi (73 F, D638); Flex. Mod.: 53,000 psi (73 F, D790); Impact Str. (Izod): 1.75 ft lb/in. (73 F, D256); Hardness (Shore): D-70 (D1706)
Therm. Prop. (of molded specimens): Distort. Temp.: 133 F (66 psi, D648); Coeff. of Linear Exp.: 820,000 F (D-696); Sp. Heat: 0.40 Cal/g/F
Elec. Prop. (of molded specimens): Power Factor: 0.20 (60 cycles, D150); Dielec. Str.: 400 V/mil (short-time, D-149); Dielec. Const.: 11.0 (60 cycles, D-150); Vol. Resist.: 0.6×10^{13} ohm-cm (D257)
Ref.: Product bulletin BCI 600 Series

BCI Nylon No. 651. Belding Chem. Industries
Chem. Descrip.: Linear polyamide (terpolymer)
Category/Applications: Resin designed for solution work, with high strength

and elongation potential, good chemical resistance, relatively high temp. tolerance; used for thread bonding, wire coatint, fuel cell barrier, textile gummed edging, component of structural metal to metal adhesives, abrasive resistant finishes for woven glass, gasoline and oil barrier coatings, phenolic resin modifier, etc.
Form: Pellets;
Gen'l. Prop. (of films cast from an ethanol-water (80:20) solvent system, 10% solids, air-dried at R.T.): Sp. Gr.: 1.12–1.13 (D792); M.P.: 148–160 C (D789); Stability: Generally unaffected by organic reagents (except those which act as solvents); little affected by inorganic bases and by neutral salts; inorganic acids and certain strong acids dissolve it with degradation
Mech. Prop. (of films, see above): Tens. Str.: 5500–5800 psi (D882); Tens. Elong.: 250–350% (D882)
Therm. Prop. (of films, see above): Sp. Heat: 0.4
Ref.: Product bulletin BCI 600 Series

BCI Nylon No. 653. Belding Chem. Industries
Chem. Descrip.: Linear polyamide (terpolymer)
Category/Applications: Resin designed for thermal processing, with high strength and elongation potential, good chemical resistance, and relatively high temp. tolerance; used for injection (gaskets, high resiliency mechanical components), extrusion (mechanical wire jacketing, heat sealable strip stock), etc.
Form: Pellet; Color (of molded specimens): Translucent
Gen'l. Prop. (of molded specimens): Sp. Gr.: 1.13 (D792); M.P.: 309 F (D789)
Mech. Prop. (of molded specimens): Tens. Str.: 6500 psi (73 F, D638); Tens. Elong.: 300% (73 F, D638); Tens. Mod.: 35,000 psi (73 F, D638); Flex. Str.: 3900 psi (73 F, D790); Flex. Mod.: 28,000 psi (73 F, D790); Impact Str. (Izod): 2.0 ft lb/in. (73 F, D256); Hardness (Shore): D-65 (D1706)
Therm. Prop. (of molded specimens): Distort. Temp.: 131 F (@ 66 psi, D648); Coeff. of Linear Exp.: 820,000 F (D-696); Sp. Heat: 0.40 Cal/g/F
Elec. Prop. (of molded specimens): Power Factor: 0.20 (60 cycles, D150); Dielec. Str.: 400 V/mil (short-time, D-149); Dielec. Const.: 11.0 (60 cycles, D150); Vol. Resist.: 0.6 × 10^{13} ohm-cm (D257)
Ref.: Product bulletin BCI 600 Series

Bead Cellulose. Chemapol
Chem. Descrip.: Functional biopolymer; regenerated cellulose
Category/Applications: High-porosity bead form cellulose for use as construction materials, in textiles, paints, stationery products, soluble additives, insoluble filter aids, adsorbents, ion exchangers, polymeric carriers in immobilization of enzymes, affinity chromatography, controlled release of biologically active substances, etc.
Form: Beads
Ref.: Product information

Binder #5008. Polymer Research Corp. of America
Category/Applications: Metallic screen print nonionic binder for use with natural or synthetic blended fabrics; exhibits a good soft hand; cure @ 325 F, 3 min
Form: Paste; Color: Amber
Composition: 30% latex activity; 14.4% latex solid
Gen'l. Prop.: pH: 8.0
Toxicity/Handling: Store at R.T.
Std. Pkgs.: Fiber drums
Ref.: Tech. bulletin

Biozan. Hercules Inc.
Chem. Descrip.: Xanthan gum
Category/Applications: Water-sol. polysaccharide produced by pure culture fermentation, with exceptional suspending power, thickening efficiency, and tolerance for extremes of temp., pH, and dissolved salt content; used as a thickener, emulsifier, and suspension aid in slurry explosives, foundry coatings, acid and caustic cleaning compounds, cosmetics, pharmaceuticals, in oil field chemicals, and other aq. systems; food-grade xanthan used as a thickener, emulsifier, and suspension aid for pourable salad dressings, sauces, puddings, and pie fillings
Ref.: Products catalogue

Biozan SPX 5423. Hercules Inc.
Chem. Descrip.: Xanthan gum
Category/Applications: Industrial-grade xanthan gum for applications requiring thickening, suspending power, excellent pH and temp. stability, and compatibility with metal salts
Form: Free-flowing powder; Color: Tan
Gen'l. Prop.: Visc. (Brookfield): 1300–1600 (1% sol'n., spindle no. 3, 30 rpm); Stability: Excellent pH and temp. stability; stable in presence of a wide range of acids, bases, salts, surfactants, and other thickeners
Std. Pkgs.: 50-kg cartons with polyethylene liners
Ref.: Bulletin BD-101

Bisphenol A. Dow Chem. Co.
Category/Applications: Resin intermediate for phenolic resins, epoxy resins, Bisphenol A-formaldehyde resins, and high performance polyester resins which are used for coatings, adhesives, reinforced plastics, filament wound pipe and tanks, fixtures, forming tools, electrical laminates, part encapsulation, construction materials, barn paints, enamels, high-quality printing inks, varnishes; also used for specialty products, e.g., antioxidants for rubber, soap, and brake fluids, and in fungicides, stabilizers, and dyeability enhancers
Form: Flakes; Color: White
Gen'l. Prop.: M.W.: 228; Sol. (g/100 g solvent): 85 g acetone, 33 g ether, 210 g methanol; Sp. Gr.: 1.195; Density (bulk): 37–41 lb/ft^3; B.P.: 220 C (4 mm Hg); F.P.: 154 C min; Stability: Incompatible with oxidizing materials
Therm. Prop.: Flamm.: Suspended dust in air can be flammable; Flash Pt.: 415 F (COC)
Toxicity/Handling: Moderate eye irritation; may cause corneal injury; repeated prolonged skin contact may cause mild to moderate irritation; dust irritating to upper respiratory passages; 50% sol'ns. may cause skin burns and severe eye injury
Std. Pkgs.: 50 lb multiwall paper bags; carloads and truckloads
Ref.: Product booklet

Blackhawk 500. Blackhawk Resin & Chem. Co.
Chem. Descrip.: Natural hydrocarbon resin of mineral origin
Category/Applications: Resin with good hardness, stability, solubility in hydrocarbon solvents, wettability, gloss, and solvent-release properties; used in the production of printing inks and flushed printing-ink colors; as a tackifier resin in synthetic and natural rubber-based pressure sensitive and hot-melt adhesives, construction adhesives, hot-melt-applied protective coatings, and rubber compounding; modifier or extender resin in various cold-cut and cooled oleoresinous-based varnish vehicles for primer- and finished-coat systems
Form: Solid, flakes, Color: Dark

Gen'l. Prop.: M.W.: ≈ 1000; Sol. (@ 20% conc.): Sol. in most aliphatic, aromatic, and chlorinated hydrocarbons, incl. heptane, cyclohexane, min. spirits, VM&P naphtha, kerosene, benzene, toluene, xylene, carbon tetrachloride, ethylene dichloride, and trichloroethylene; also sol. in methylamyl and methyl isobutyl ketones; insol. in esters, acetone, alcohols, and water; Density: 1.02 kg/l; Visc. (Gardner-Holdt): Z-Z1 (60% solids in min. spirits); Soften. Pt. (R&B): 180 C; Acid No.: 3; Sapon. No.: 3; Stability: Excellent resistance to heat, light, oxidation, moisture, and alkalis; Ref. Index: 1.544 (20 C); Flash Pt.: 271 C (COC)
Std. Pkgs.: 205-kg net drums (solid); 22.7-kg net multiwall paper bags (flakes)
Ref.: Product information BR-101D

Bonding Agent 2001, 2005. Bayer AG
Chem. Descrip.: Reactive aromatic polyisocyanates
Category/Applications: One-part bonding agents giving high adhesion between PVC paste coatings and textile substrates of synthetic fibers; used where the material is exposed to severe conditions, e.g., tarpaulins for trucks, inflatable tents, flexible containers, tent roofs, awnings, protective clothing, conveyor belting, etc.
Form: Liquid
Composition: 2001 is solventless; 2005 is solvent-containing
Ref.: Pamphlet KL 42911 e

Bonding Agent TN. Bayer AG
Chem. Descrip.: Polyester with hydroxyl groups
Category/Applications: Two-part bonding agent used with an isocyanate (Desmodur L 75, TT, N) for the coating of synthetic fiber fabrics with PVC pastes
Form: Liquid (solution)
Ref.: Pamphlet KL 42911 e

Bresin 2 Resin. Hercules Inc.
Chem. Descrip.: Natural resin derived from pinewood
Category/Applications: Thermoplastic resin used in hot-melt adhesives, mastics, and construction adhesives
Form: Solid; Color: Dark
Gen'l. Prop.: Soften. Pt. (Drop): 90 C; Acid No.: 62
Ref.: Products catalogue

Buna CB 10. Bayer AG
Chem. Descrip.: Butadiene rubber
Category/Applications: Synthetic rubber with good low temp. flexibility and high elasticity; used in tires, conveyor belting, caterpillar tread blocks, footwear soles, transmission belting; blending with NR for goods having thick sections and needing high resistance to reversion, e.g., buffers; roll covers needing high abrasion resistance; seals, strip and other goods, esp. when produced by injection molding; free from rubber poisons, e.g., Mn and Cu compounds; compatible with other diene rubbers
Form: Bales; Color: Cream
Composition: ≈ 98% cis-1,4 content
Gen'l. Prop.: Density: 0.91 g/cc; Visc. (Mooney): ≈ 48 (ML 4, 100 C); Stability: Good resistance to ageing, reversion, abrasion, and flexcracking
Std. Pkgs.: 30-kg polyethylene bags on 900-kg pallets
Ref.: Pamphlet KA 32125 e

Buna CB 11. Bayer AG
Chem. Descrip.: Butadiene rubber
Category/Applications: See Buna CB 10
Form: Bales; Color: Light brown
Composition: ≈ 95% cis-1,4 content
Gen'l. Prop.: Density: 0.91 g/cc; Visc. (Mooney): ≈ 46 (ML 4, 100 C); Stability: See Buna CB 10
Std. Pkgs.: See Buna CB 10; also avail. on 750-kg pallets
Ref.: Pamphlet KA 32125 e

Buna CB 30. Bayer AG
Chem. Descrip.: Butadiene rubber
Category/Applications: See Buna CB 10
Form: Bales; Color: Dark brown
Composition: ≈ 95% cis-1,4 content
Gen'l. Prop.: Density: 0.92 g/cc; Visc. (Mooney): ≈ 34 and 40 (ML 4, 100 C); Stability: See Buna CB 10
Std. Pkgs.: See Buna CB 10
Ref.: Pamphlet KA 32125 e

Butacite. Du Pont Co.
Chem. Descrip.: Polyvinyl butyral resin
Category/Applications: Sheeting containing an ultraviolet absorbent and used in the production of safety glass laminates for automotive windshields and architurectural applications; avail. in dusted form or refrigerated form; Butacite 125, 106, and 140 are formulated to give a controlled adhesion to glass and optimum impact performance in laminated high-penetration resistance windshields; Butacite 10 and 14 are high-adhesion products used primarily in architectural glazing; solid-colored Butacite 10 and 14 are used in architectural and transportation glazing where reduced light transmission is desired
Form: Avail. in 15- and 30-mil sheeting in various widths; Color: Clear, standard solid colors, and with gradient bands on clear or colors
Gen'l. Prop.: Sol.: Sol. in the common oxygen-containing solvents (alcohols, ketones, esters, etc.); limited sol. in higher esters such as plasticizers, and in most chlorinated hydrocarbons; swells when subjected to aromatic solvents; Sp. Gr.: 1.07–1.08; Stability: Relatively stable under alkaline conditions; unstable in the presence of acids; Ref. Index: 1.48
Mech. Prop.: Tens. Str.: 3500 psi (@ 0.3% water)
Therm. Prop.: Conduct.: 0.10 PCU/h-ft^2-C/ft; Sp. Heat: 0.45 PCU/lb-C
Ref.: Bulletins A-79309, A-79310, E-40602

Butaprene PL-3116, PL-3133, PL-3147. Firestone Plastics Co.
Chem. Descrip.: Styrene-butadiene copolymer latex
Category/Applications: Latex designed for the formulation of interior latex paints and specialty coatings, with improved flow and leveling properties
Form: Liquid; Odor: Low
Composition: 48% solids
Gen'l. Prop.: Sp. Gr.: 1.01; Visc. (Brookfield): 100 cps max. (20 rpm); Surface Tension: 31 dynes/cm; Stability: Good mechanical and freeze/thaw stability; pH: 9.5
Ref.: Data sheet

Butaprene PL-3131. Firestone Plastics Co.
Chem. Descrip.: Styrene-butadiene copolymer
Category/Applications: Latex for concrete and mortar mix applications with excellent film-forming properties; used in preparing rubber modified cement patching and ceramic tile adhesive compounds; as a binder in floor leveling cement compounds where thin section, good adhesion, flexibility, and feathering are required
Form: Liquid
Composition: 48 ± 1% solids
Gen'l. Prop.: Sp. Gr.: 1.000–1.015; Visc. (Brookfield): 100 cps max. (20 rpm); Surface Tension: 27–35 dynes/cm; Stability: Freeze-thaw stability; stable to mechanical and chemical shock from compounding or handling; excellent shelf stability; pH: 9.5 ± 1
Ref.: Data sheet

BY-59-18. Neville Chem. Co.
Chem. Descrip.: Modified hydrocarbon resin
Category/Applications: Resin possessing reactive unsaturation characteristics; used as replacement resins for rosin derivatives, and in inks (gelled varnishes, gravure, heat set, letterpress, and lithographic)
Form: Flakes; Color (Gardner): 12 (50% in toluene, D1544)
Gen'l. Prop.: M.W.: 1120 (D2503); Sol.: Sol. in esters, ethers, ketones (except acetone), and chlorinated, aromatic, and terpene hydrocarbons; partly sol. in naphthenic hydrocarbons; Sp. Gr.: 1.080 (D71); Soften. Pt. (Ring & Ball): 144 C (E28); Acid No.: 12; Iodine No. (Wijs): 65 (D1959)
Therm. Prop.: Flash Pt. (COC): 520 F (D92)
Ref.: Product pamphlet

C

C-84. Bacon Industries Inc.
Chem. Descrip.: Epoxy comp.
Category/Applications: General purpose potting comp. used with Activator BA-63 for potting and casting applications of up to 1 lb mass; offers an inherently high degree of flame resistance without the use of noxious additives; mix ratio: 7.3 phr of Activator BA-63 by wt.
Form: Color: Black
Gen'l. Prop.: Sp. Gr.: 1.68 (cured); Visc.: 600 poise (comp.), 60 poise (activated comp.)
All properties—after recommended cure:
Mech. Prop.: Flex. Str.: 10,000 psi; Flex. Mod. (Young's): 1.1 × 10^6 psi; Hardness (Shore D): 86
Therm. Prop.: Distort. Temp.: 154 F (@ 264 psi); Coeff. of Linear Exp.: 34 × 10^{-6}/F (75–200 F); Flamm.: 35 s and 5 mm (time and extent of burning, D-635)
Elec. Prop.: Dissip. Factor: 0.008 (1 kHz); Dielec. Const.: 4.3 (1 kHz); Vol. Resist.: 10^{16} ohm-cm
Curing Charact.: Pot Life: 30 min @ 77 F (300 g) or 120 min @ 77 F (100 g); Recommended Cure: 3 h @ 77 F plus 2 h @ 250 F
Toxicity/Handling: May cause injury to skin on prolonged or repeated contact; Activator BA-63 may cause eye injury or skin burns; use with adequate ventilation; > 1 yr shelf life when stored in unopened containers below 85 F
Std. Pkgs.: Qt (3 lb net) and gal (12 lb net) paint cans; 5 gal (60 lb net) open-head pails; BA-63 avail. in pt, qt, and gal screw-top cans containing 1, 2, and 8 lb net respectively
Ref.: Data sheet 1384-63

CA-394-60. Eastman Chem. Products, Inc.
Chem. Descrip.: Cellulose acetate ester
Category/Applications: Very high-melting ester with limited solubility and compatibility; generally used where high strength and good resistance to heat, uv light, oils, and greases are required
Form: Fine, dry powder; Color: White; Odor: Low
Composition: 39.4% acetyl content
Gen'l. Prop.: Density: 1.32 kg/l; Visc.: 228.0 poise (D871/A, D1343); M.P.: 240–260 C
Mech. Prop.: Hardness (Tukon): 20 (Knoops, D1474)
Ref.: Bulletin E-146F

CA-398-3, CA-398-6, CA-398-10, CA-398-30. Eastman Chem. Products, Inc.
Chem. Descrip.: Cellulose acetate ester
Category/Applications & Form: See CA-394-60
Composition: 39.8% acetyl content
Gen'l. Prop.: Density: 1.31 kg/l; Visc.: 11.4, 22.8, 38.0, 114.0 poise respectively (D871/A, D1343); M.P.: 230–250 C
Mech. Prop.: Hardness (Tukon): 19 (Knoops, D1474)
Ref.: Bulletin E-146F

CAB-171-15S. Eastman Chem. Products, Inc.
Chem. Descrip.: Cellulose acetate butyrate ester
Category/Applications: Unfiltered ester with limited solubility and compatibility; used in cloth coatings for airplanes, wire coatings, leather coatings, and plastic coatings
Form: Flakes
Composition: 29.5% acetyl, 17% butyryl
Gen'l. Prop.: Density: 1.26 kg/l; Visc.: 57.0 poise (D817/A, D1343); M.P.: 230–240 C
Mech. Prop.: Hardness (Tukon): 16 (Knoops, D1474)
Ref.: Bulletin E-146F

CAB-381-0.1, 381-0.5, 381-2, 381-20. Eastman Chem. Products, Inc.
Chem. Descrip.: Cellulose acetate butyrate esters
Category/Applications: Used in lacquers of all types; higher viscosity esters are toughest
Form: Fine, dry powder; Color: White
Composition: 13.0% acetyl, 37% butyryl
Gen'l. Prop.: Sol. (CAB-381-01): Sol. in diethylene glycol diacrylate, 1,3-butylene glycol diacrylate, glycidyl methacrylate; sol. with haze in styrene/diacetone acrylamide (90/10), styrene/methyl methacrylate (80/20), and butyl acrylate; Density: 1.20 kg/l (all); Visc.: 0.38 poise, 1.9 poise, 7.6 poise, 76.0 poise respectively (D817/A, D1343); M.P.: 155–165 C, 155–165 C, 171–184 C, 195–205 C respectively
Mech. Prop.: Hardness (Tukon): 12 (Knoops, D1474)
Ref.: Bulletin E-146F

CAB-500-1, 500-5. Eastman Chem. Products, Inc.
Chem. Descrip.: Cellulose acetate butyrate esters
Category/Applications: Hot melts and additives for polyurethanes
Form: Fine, dry powder; Color: White
Composition: 5.0% acetyl, 49% butyryl
Gen'l. Prop.: Density: 1.18 kg/l; Visc.: 3.8 and 19.0 poise respectively (D817/A, D1343); M.P.: 165–175 C
Mech. Prop.: Hardness (Tukon): 10 (Knoops, D1474)
Ref.: Bulletin E-146F

CAB-531-1. Eastman Chem. Products, Inc.
Chem. Descrip.: Cellulose acetate butyrate ester
Category/Applications: Very compatible with many modifying resins; may be used with CAB-551-0.2 for coating wood, plastic, or metal
Form: Fine, dry powder; Color: White
Composition: 2.8% acetyl, 50% butyryl
Gen'l. Prop.: Sol.: Sol. in glycidyl methacrylate, diethylene glycol diacrylate, 1,3-butylene glycol diacrylate; sol. with haze in styrene/diacetone acrylamide (90/10), styrene/methyl methacrylate (80/20), and butyl acrylate; Density: 1.17 kg/l; Visc.: 3.8 poise (D817/A, D-1343); M.P.: 135–150 C
Mech. Prop.: Hardness (Tukon): 9 (Knoops, D1474)
Ref.: Bulletin E-146F

CAB-551-0.01. Eastman Chem. Products, Inc.

Chem. Descrip.: Cellulose acetate butyrate ester

Category/Applications: Used in high-solids coatings and thermosetting powder coatings

Form: Fine, dry powder; Color: White

Composition: 2.0% acetyl, 53% butyryl

Gen'l. Prop.: Sol.: Sol. in styrene/diacetone acrylamide (90/10), styrene/methyl methacrylate (80/20), glycidyl methacrylate, butyl acrylate, diethylene glycol diacrylate, 1,3-butylene glycol diacrylate; sol. with haze in styrene, 2-ethylhexyl acrylate; Density: 1.16 kg/l; Visc.: 0.038 poise (D817/A, D1343); M.P.: 110–125 C

Mech. Prop.: Hardness (Tukon): 8 (Knoops, D1474)

Ref.: Bulletin E-146F

CAB-551-0.2. Eastman Chem. Products, Inc.

Chem. Descrip.: Cellulose acetate butyrate ester

Category/Applications: Most compatible of the CAB's with modifying resins for coating wood, plastic, or metal

Form: Fine, dry powder; Color: White

Composition: 2.0% acetyl, 52% butyryl

Gen'l. Prop.: Sol.: Sol. in styrene/unsaturated polyester (45/55), glycidyl methacrylate, diethylene glycol diacrylate, 1,3-butylene glycol diacrylate; sol. with haze in styrene/diacetone acrylamide (90/10), styrene/methyl methacrylate (80/20), and butyl acrylate; Density: 1.16 kg/l; Visc.: 0.76 poise (D817/A, D1343); M.P.: 130–140 C

Mech. Prop.: Hardness (Tukon): 8 (Knoops, D1474)

Ref.: Bulletin E-146F

CAB-553-0.4. Eastman Chem. Products, Inc.

Chem. Descrip.: Cellulose acetate butyrate ester

Category/Applications: Used in printing inks, coatings for wood, paper, and plastics

Form: Fine, dry powder; Color: White

Composition: 2.0% acetyl, 47% butyryl

Gen'l. Prop.: Density: 1.20 kg/l; Visc.: 1.5 poise (D817/A, D1343); M.P.: 150–160 C

Mech. Prop.: Hardness (Tukon): 10 (Knoops, D1474)

Ref.: Bulletin E-146F

Cadon 112. Monsanto Plastics & Resins Co.

Chem. Descrip.: Impact-modified styrene-maleic anhydride terpolymer

Category/Applications: Engineering thermoplastic with superior impact resistance and moderately high heat distortion; used in injection molding applications

Gen'l. Prop.: Sp. Gr.: 1.07 (73 F, D-792); Soften. Pt. (Vicat): 125 C (D-1525)

Mech. Prop. (@ 73 F): Tens. Str.: 35.8 MPa (yield, D638); Tens. Elong.: 1.8% @ yield, 40% @ fail (D638); Tens. Mod.: 2.07 GPa (D538); Flex. Str.: 55.2 MPa (yield, D790); Flex. Mod.: 2.21 GPa (D-790); Impact Str. (Izod): 250 J/m (D-256/A); Hardness (Rockwell): R 109 (D785)

Therm. Prop.: Distort. Temp.: 106 C (unannealed, @ 264 psi, D648); Coeff. of Linear Exp.: 8.4 × 10^{-5} cm/cm/C (–30 to 30 C, D696)

Ref.: Data sheet no. 6505 A

Cadon 127. Monsanto Plastics & Resins Co.

Chem. Descrip.: Impact-modified styrene-maleic anhydride terpolymer

Category/Applications: Engineering thermoplastic offering superior heat distortion properties and high impact resistance; recommended when the heat resistance of a modified PPO is needed but better flow and consistency are desired; used in injection molding applications
Gen'l. Prop.: Sp. Gr.: 1.07 (73 F, D-792); Soften. Pt. (Vicat): 127 C (1 kg, D1525)
Mech. Prop. (@ 73 F): Tens. Str.: 34.5 MPa (yield, D638); Tens. Elong.: 1.7% @ yield, 40% @ fail (D638); Tens. Mod.: 2.14 GPa (D638); Flex. Str.: 57.9 MPa (D790); Flex. Mod.: 2.35 GPa (D790); Impact Str. (Izod): 225 J/m (D256/A); Hardness (Rockwell): R 109 (D785)
Therm. Prop.: Distort. Temp.: 112 C (unannealed, @ 264 psi, D648); Coeff. of Linear Exp.: 9.4 X 10^{-5} cm/cm/C (–30 to 30 C, D696)
Ref.: Data sheet no. 6504 A

Cadon 135-PG. Monsanto Plastics & Resins Co.
Chem. Descrip.: Impact-modified styrene-maleic anhydride terpolymer
Category/Applications: Engineering thermoplastic plating grade resin formulated for high impact but with higher heat distortion than a plating grade ABS; can be electroplated in standard plating systems with no special etches or longer cycles; provides parts with high plate adhesion, excellent surface appearance, and high resistance to plate cracking and blistering; for injection molding applications
All properties–unplated:
Gen'l. Prop.: Sp. Gr.: 1.07 (D792)
Mech. Prop.: Tens. Str.: 33.1 MPa @ yield, 30.4 MPa @ fail (D638); Tens. Elong.: 1.8% @ yield, 20.0% @ fail (D-638); Tens. Mod.: 2.26 GPa (D638); Flex. Str.: 55.2 MPa (@ yield, D790); Flex. Mod.: 2.14 GPa (D790); Impact Str. (Izod): 151 J/m (D256/A); Hardness (Rockwell): R 102 (D785)
Therm. Prop.: Distort. Temp.: 106 C (unannealed, @ 264 psi, D648); Coeff. of Linear Exp.: 8.4 X 10^{-5} cm/cm/C (D696)
Ref.: Data sheet no. 6513

Calthane 1400. Cal Polymers, Inc.
Chem. Descrip.: Two-component urethane system
Category/Applications: Urethane molding compound for making male-master molds and in electrical encapsulation and potting; cures at R.T., does not require degassing, and has good dielectrical properties; mixing ratio: 1A:4B, by weight; also available in thixotropic grade for adhesive applications
Form: Liquid
Gen'l. Prop.: Sp. Gr.: 1.56; Density: 13 lb/gal; Visc. (mixed): 20,000 cps (20 C), Stability: Good dimensional stability and chemical resistance
Mech. Prop.: Hardness (Shore D): 80
Curing Charact.: Demold at R.T. in 3-5 h; ultimate cure 2–3 days at R.T.
Std. Pkgs.: 5 gal (60 lb net) pails
Ref.: Product information

Calthane 1700. Cal Polymers, Inc.
Chem. Descrip.: Two-part, filled urethane elastomer based on MDI; MOCA-free
Category/Applications: Urethane elastomer designed to meet OSHA requirements; used as a mold material for high density rigid urethane foam, molding decorative polyester parts, etc.; Part A is a low-visc., MDI-based prepolymer; Part B is a med.-visc., resin filler system; mixing ratio: 1A:7B, by weight; may be cured at R.T.
Form: Liquid (Parts A and B)
Gen'l. Prop.: Sp. Gr.: 1.20 (Part A),

1.40 (Part B); Stability: Good chemical and temperature resistance
Mech. Prop. (mixed): Tens. Str.: 990 psi; Tens. Elong.: 180%; Hardness (Shore A): 70 ± 3
Std. Pkgs.: 5 gal (58 lb net) unit containing 8 lb of Part A, 50 lb Part B; drum quantities shipped in 5 gal pails
Ref.: Product information

Calthane 1900. Cal Polymers, Inc.
Chem. Descrip.: Two-component, filled, urethane elastomer system
Category/Applications: Urethane elastomer system for making flexible molds; cures at R.T.; does not require degassing; Part A is a TDI-based prepolymer; Part B is a resin filler; mixing ratio: 1A:9B, by weight
Form: Liquid (Parts A and B); Color: Green (Part A); tan (Part B); tan (mixed)
Gen'l. Prop.: Sp. Gr.: 1.14 (Part A), 1.40 (Part B), 1.30 (20 C, mixed); Density: 11.3 lb/gal (mixed); Visc.: 2000 cps (20 C, Part A), 3000 cps (20 C, Part B), $<$ 3000 cps (20 C, mixed)
Mech. Prop.: Hardness (Shore A): 40–60 depending on mixing ratios
Ref.: Product information

Calthane ND 1100. Cal Polymers, Inc.
Chem. Descrip.: Two-part urethane elastomer system; does not contain TDI, MOCA, or solvents
Category/Applications: Urethane elastomer system with excellent nondiscoloring characteristics for use in flexible and rigid castings, prosthetics, underwater sports equipment, optical lenses, transparent or pigmented clear art objects, mechanical devices, flexible windshields, industrial tubing or sheeting, toys, novelties, transparent bicycle tires, adhesives; mixing ratio: 1A:1B, by weight or volume
Form: Liquid; Color: Clear
Mech. Prop. (after R.T. cure): Tens. Str.: 2300 psi; Tens. Elong.: 130%; Hardness (Shore A): 93 ± 5
Std. Pkgs.: 4 gal (32 lb net) unit containing 16 lb A and 16 lb B; 5 gal (40 lb net) pail; 50 gal (400 lb net) drum
Ref.: Product information

Calthane ND 2300. Cal Polymers, Inc.
Chem. Descrip.: Two-part urethane elastomer system; does not contain TDI, MOCA, or solvents
Category/Applications: Urethane elastomer system with excellent nondiscoloring characteristics for use in flexible and rigid castings, prosthetics, underwater sports equipment, optical lenses, transparent or pigmented clear art objects, mechanical devices, flexible windshields, industrial tubing or sheeting, toys, novelties, transparent bicycle tires, adhesives; mixing ratio: 2A: 3B by weight or volume; alternate mixing ratio of 1A:2B by weight or volume produces a very flexible adhesive for laminating clear plastics, e.g., polycarbonate, and acrylic to glass, in transparent ballistic composites
Form: Liquid; Color: Clear
Mech. Prop. (after R.T. cure): Tens. Str.: 800 psi; Tens. Elong.: 180%; Hardness (Shore A); 65 ± 5
Std. Pkgs.: 5 gal (40 lb net) unit containing 16 lb A, 24 lb B; 50 gal (400 lb net) drums
Ref.: Product information

Calthane ND 3200. Cal Polymers, Inc.
Chem. Descrip.: Two-part urethane elastomer system; does not contain MOCA, TDI, or solvents
Category/Applications: Urethane elastomer system with excellent nondiscoloring characteristics for use in flexible and rigid castings, prosthetics, underwater sports equipment, optical lenses, transparent or pigmented clear art objects,

mechanical devices, flexible windshields, industrial tubing or sheeting, toys, novelties, transparent bicycle tires, adhesives; lends itself to applications that previously called for thermoforming cast sheets of polycarbonate or acrylic; due to low viscosities of Parts A and B, and extensive pot life after mixing, large castings of very hard, clear, abrasive resistant urethanes are possible; mixing ratio: 3A:2B by weight or volume
Form: Liquid; Color: Clear
Gen'l. Prop. (after R.T. cure): Sp. Gr.: 1.04
Mech. Prop. (after R.T. cure): Tens. Str.: 6300 psi; Tens. Elong.: 10%; Hardness (Shore D): 80
Std. Pkgs.: 5 gal (40 lb net) unit containing 24 lb A and 16 lb B; 50 gal (400 lb net) drums
Ref.: Product information

Calthane NF 310 W. Cal Polymers, Inc.
Chem. Descrip.: Two-part nonfilled urethane elastomer system; contains no MOCA or TDI
Category/Applications: Urethane elastomer system for casting industrial and print rollers, and for bumper pads, diaphragms, dust boots, engine mounts, gaskets, liners, flexible and rigid molds, and wear pads; mixing ratio: 3A:10B by weight
Form: Low-visc., clear liquid (Part A); white solid (Part B); Color (after mixing): Pale orange
Gen'l. Prop.: M.P. (Part B): 150 F
Mech. Prop. (after cure): Tens. Str.: 5300 psi min.; Tens. Elong.: 500%; Hardness (Shore A): 60 ± 3
Curing Charact.: Cure @ 150–175 F for ½ h or @ R.T. overnight
Std. Pkgs.: 5 gal (65 lb net) unit containing 15 lb A and 50 lb B; 650 lb net drums containing 150 lb A and 500 lb B
Ref.: Product information

Calthane NF 0810. Cal Polymers, Inc.
Chem. Descrip.: Two-part urethane elastomer system; both MOCA and TDI-free
Category/Applications: Urethane elastomer casting system which duplicates the properties of MOCA cured Adiprene urethanes but contains no MOCA or TDI; mixing ratio: 8A:10B by weight
Form: Low-visc., amber liquid (Part A); clear semisolid that remains liquid ⩾ 75 F (Part B)
Gen'l. Prop.: Sp. Gr.: 1.20 (Part A); 1.01 (Part B)
Mech. Prop. (after mixing): Tens. Str.: 4500 psi; Tens. Elong.: 245%; Hardness (Shore A): 90 ± 2
Curing Charact.: May be cured for 5 min into preheated metal molds, to 3–5 h @ R.T.
Std. Pkgs.: 5 gal (44 lb net) unit containing 20 lb A and 24 lb B; drum quantity (570 lb net) consisting of 250 lb A and 320 lb B in 5 gal pails
Ref.: Product information

Calthane NF 1300. Cal Polymers, Inc.
Chem. Descrip.: Two-part nonfilled urethane elastomer; MOCA-free
Category/Applications: Urethane elastomer system for casting industrial and print rollers, and for bumper pads, diaphragms, dust boots, engine mounts, gaskets, liners, flexible and rigid molds, and wear pads; mixing ratio: 1A:3B (B should be premixed); also used as an adhesive and potting compound
Form: Low-visc. liquid (Part A); med.-visc. liquid (Part B); Color (after R.T. cure): Opaque
Gen'l. Prop. (after cure): Sp. Gr.: 1.0; Density: 8.3 lb/gal; Stability: Excellent hydrolytic stability and low-temp. flexibility
Mech. Prop. (after R.T. cure): Tens. Str.: 2000 psi; Tens. Elong.: 300%; Hardness (Shore A): 83 ± 3

Elec. Prop. (after cure): Dielec. Str.: 450 V/mil (75 F); Dielec. Const.: 3.1 (@ 1K Hz, 75 F); Vol. Resist.: 6 $\times$ 10^{15} ohm-cm; Dissip. Factor: 0.030 (@ 1K Hz, 75 F)
Std. Pkgs.: 5 gal (54 lb net) unit containing 14 lb A and 40 lb B; 540 lb net drum quantity supplied in 5 gal pails
Ref.: Product information

Calthane NF 1300-60A. Cal Polymers, Inc.
Chem. Descrip.: Two-part nonfilled urethane elastomer; MOCA and TDI free; Part A is a MDI-based prepolymer; Part B is a polyester
Category/Applications: Urethane elastomer formulated to resist gasoline and oil; mixing ratio: 1A:3B by weight or volume
Form: Low-visc. liquid (Part A); Liquid (Part B)
Gen'l. Prop.: Sp. Gr.: 1.20 (Part A); 1.20 (Part B); Visc.: 1850 cps (Part B)
Mech. Prop. (after R.T. cure): Tens. Str.: 480 psi; Tens. Elong.: 240%; Hardness (Shore A): 60 ± 3
Std. Pkgs.: 5 gal (67 lb net) unit containing 17 lb A and 50 lb B; 670 lb drum quantity containing 170 lb A and 500 lb B
Ref.: Product information

Calthane NF 1500. Cal Polymers, Inc.
Chem. Descrip.: Two-part urethane elastomer; MOCA-free; Part A is a TDI-based prepolymer; Part B is a polyether resin
Category/Applications: Urethane elastomer for potting and encapsulating electronic components; excellent dielectric properties (can be used at 500 V and 10,000 megohms without electrical leakage); mixing ratio: 1A:5B by weight
Form: Green liquid (Part A); Clear liquid (Part B)
Gen'l. Prop.: Sp. Gr.: 1.14 (Part A); 1.00 (Part B); 1.03 (mixed); Visc.: 2000 cps (Part A); 600 cps (Part B)
Mech. Prop. (after R.T. cure): Hardness (Shore A): 50 ± 3
Std. Pkgs.: 5 gal (48 lb net) unit containing 8 lb A and 40 lb B; 560 lb net drum quantity containing 100 lb A and 460 lb B
Ref.: Product information

Calthane NF 2300. Cal Polymers, Inc.
Chem. Descrip.: Two-part nonfilled urethane elastomer; Part A is a MDI-based prepolymer; Part B is a resin system; TDI and MOCA-free
Category/Applications: Urethane elastomer designed to meet OSHA requirements; low viscosity system for use for machine dispensed urethane sheets and coated fabrics, and mechanical parts; mixing ratio: 2A:3B by weight
Form: Liquid; Color (after cure): Opaque
Gen'l. Prop.: Sp. Gr.: 1.20 (Part A); 1.02 (Part B); Visc.: Low-visc. (Part A); med.-visc. (Part B)
Mech. Prop. (after R.T. cure): Tens. Str.: 2620 psi; Tens. Elong.: 140%; Hardness (Shore A): 88 ± 3
Std. Pkgs.: 5 gal (40 lb net) unit containing 16 lb A and 24 lb B; 450 lb net drum quantities
Ref.: Product information

CAP-482-0.5. Eastman Chem. Products, Inc.
Chem. Descrip.: Cellulose acetate propionate ester
Category/Applications: Used in printing inks, paper coatings, and cloth coatings; excellent grease barrier properties
Form: Fine, dry powder; Color: White; Odor: Low
Composition: 2.5% acetyl; 44% propionyl

Gen'l. Prop.: Sol.: Sol. in 1,6-hexandiol diacrylate, trimethylolpropane triacrylate, tetraethylene glycol diacrylate; Density: 1.22 kg/l; Visc.: 1.9 poise (D817/A, D1343); M.P.: 188–210 C
Mech. Prop.: Hardness (Tukon): 13 (Knoops, D1474)
Ref.: Bulletin E-146F

CAP-482-20. Eastman Chem. Products, Inc.
Chem. Descrip.: Cellulose acetate propionate ester
Category/Applications: High viscosity ester for blending with CAP-482-0.5
Form: Fine, dry powder; Color: White
Composition: 2.5% acetyl, 46% propionyl
Gen'l. Prop.: Density: 1.22 kg/l; Visc.: 76.0 poise (D817/A, D1343); M.P.: 188–210 C
Mech. Prop.: Hardness (Tukon): 13 (Knoops, D1474)
Ref.: Bulletin E-146F

CAP-504-0.2. Eastman Chem. Products, Inc.
Chem. Descrip.: Cellulose acetate propionate ester
Category/Applications: Performs well in flexographic and rotogravure inks; sol. in alcohol/water blends; forms stable one-package thermoset systems; excellent grease barrier properties
Form: Fine, dry powder; Color: White; Odor: Low
Composition: 2.5% acetyl, 40% propionyl
Gen'l. Prop.: Sol.: Sol. in tetraethylene glycol diacrylate; partly sol. in 1,6-hexanediol diacrylate; Density: 1.26 kg/l; Visc.: 0.76 poise (D817/A, D1343); M.P.: 188–210 C
Mech. Prop.: Hardness (Tukon): 12 (Knoops, D1474)
Ref.: Bulletin E-146F

CAPA 200. Solvay
Chem. Descrip.: Linear polycaprolactone
Category/Applications: Thermoplastic polyurethane for use for polyurethane elastomers and 2-part surface coatings, and as low-visc. vehicles for dispersions of plastic additives, e.g., "liquid colors"
Form: Liquid
Gen'l. Prop.: M.W.: 550; M.P.: 15 C; Hydroxyl Value: 204
Ref.: Interox Chem. bulletin P20-130

CAPA 205. Solvay
Chem. Descrip.: Linear polycaprolactone
Category/Applications: Thermoplastic polyurethane used for polyurethane elastomers and fabric coatings; confers good handle and weathering resistance
Gen'l. Prop.: M.W.: 830; M.P.: 25 C; Hydroxyl Value: 135
Ref.: Interox Chem. bulletin P20-130

CAPA 210. Solvay
Chem. Descrip.: Linear polycaprolactone
Category/Applications: Thermoplastic polyurethane used for polyurethane elastomers and fabric coatings; confers good handle and weathering resistance
Form: Paste
Gen'l. Prop.: M.W.: 1000; M.P.: 37 C; Hydroxyl Value: 112
Ref.: Interox Chem. bulletin P20-130

CAPA 212. Solvay
Chem. Descrip.: Linear polycaprolactone
Category/Applications: Thermoplastic polyurethane used for polyurethane elastomers and fabric coatings
Form: Paste
Gen'l. Prop.: M.W.: 1000; M.P.: 37 C; Hydroxyl Value: 112
Ref.: Interox Chem. bulletin P20-130

CAPA 215. Solvay
Chem. Descrip.: Linear polycaprolactone
Category/Applications: Thermoplastic polyurethane used for fabric coatings; confers good handle and weathering resistance
Form: Wax-like
Gen'l. Prop.: M.W.: 1250; M.P.: 40 C; Hydroxyl Value: 90
Ref.: Interox Chem. bulletin P20-130

CAPA 220. Solvay
Chem. Descrip.: Linear polycaprolactone
Category/Applications: Thermoplastic polyurethane used for polyurethane elastomers, and microcellular foams with good processability and low-temp. performance
Form: Wax-like
Gen'l. Prop.: M.W.: 2000; M.P.: 47 C; Hydroxyl Value: 56
Ref.: Interox Chem. bulletin P20-130

CAPA 222. Solvay
Chem. Descrip.: Linear polycaprolactone
Category/Applications: Thermoplastic polyurethane used for polyurethane elastomers, and microcellular foams with good processability and low-temp. performance
Form: Wax-like
Gen'l. Prop.: M.W.: 2000; M.P.: 47 C; Hydroxyl Value: 56
Ref.: Interox Chem. bulletin P20-130

CAPA 223. Solvay
Chem. Descrip.: Linear polycaprolactone
Category/Applications: Thermoplastic polyurethane used for polyurethane elastomers and microcellular foams with good processability and low-temp. performance; esp. suitable for extrusion grades and where long pot-lives are required
Form: Wax-like
Gen'l. Prop.: M.W.: 2000; M.P.: 47 C; Hydroxyl Value: 56
Ref.: Interox Chem. bulletin P20-130

CAPA 231. Solvay
Chem. Descrip.: Linear polycaprolactone
Category/Applications: Thermoplastic polyurethane used for polyurethane adhesives; produces rapid crystallizing polyurethanes with excellent retention of bond-strength at high temps.
Form: Wax-like
Gen'l. Prop.: M.W.: 3000; M.P.: 52 C; Hydroxyl Value: 36
Ref.: Interox Chem. bulletin P20-130

CAPA 240. Solvay
Chem. Descrip.: Linear polycaprolactone
Category/Applications: Thermoplastic polyurethane used for adhesives; produces rapid crystallizing polyurethanes with excellent retention of bond-strength at high temps.
Form: Wax-like
Gen'l. Prop.: M.W.: 4000; M.P.: 55 C; Hydroxyl Value: 28
Ref.: Interox Chem. bulletin P20-130

CAPA 304. Solvay
Chem. Descrip.: Branched (f = 2.4) polycaprolactone
Category/Applications: Thermoplastic polyurethane used for rigid polyurethane foams where its low viscosity and high reactivity confer easy processing with outstanding mold reception; also as a low-viscosity vehicle for dispersions of plastics additives, e.g., "liquid colors"
Form: Liquid
Gen'l. Prop.: M.W.: 250; Hydroxyl Value: 540
Ref.: Interox Chem. bulletin P20-130

CAPA 305. Solvay
Chem. Descrip.: Branched (f = 3) polycaprolactone
Category/Applications: Thermoplastic polyurethane used for 2-part polyurethane surface coatings and as a low-visc. vehicle for dispersions of plastics additives, e.g., "liquid colors"
Form: Liquid
Gen'l. Prop.: M.W.: 540; Hydroxyl Value: 310
Ref.: Interox Chem. bulletin P20-130

CAPA 520. Solvay
Chem. Descrip.: Linear polycaprolactone
Category/Applications: Thermoplastic polyurethane used for PU microcellular foams; particularly recommended for low-visc. systems; good compatibility with the common chain extenders
Form: Paste
Gen'l. Prop.: M.W.: 2000; M.P.: 26 C; Hydroxyl Value: 56
Ref.: Interox Chem. bulletin P20-130

CAPA 600. Solvay
Chem. Descrip.: Linear polycaprolactone
Category/Applications: Thermoplastic polyurethane
Form: Tough polymer
Gen'l. Prop.: M.W.: 20,000; M.P.: 58 C; Hydroxyl Value: ≈ 5
Ref.: Interox Chem. bulletin P20-130

CAPA 600M. Solvay
Chem. Descrip.: Linear polycaprolactone
Category/Applications: Thermoplastic polyurethane used as dispersion medium (recommended for universal masterbatches because of their low melt viscosity and wide compatibility with ABS, nylon, and polycarbonates); also used for hot melt adhesives
Form: Tough polymer
Gen'l. Prop.: M.W.: 20,000; M.P.: 58 C; Hydroxyl Value: ≈ 5
Ref.: Interox Chem. bulletin P20-130

CAPA 601. Solvay
Chem. Descrip.: Linear polycaprolactone
Category/Applications: Thermoplastic polyurethane
Form: Tough polymer
Gen'l. Prop.: M.W.: 45,000; M.P.: 58 C; Hydroxyl Value: ≈ 2
Ref.: Interox Chem. bulletin P20-130

CAPA 601M. Solvay
Chem. Descrip.: Linear polycaprolactone
Category/Applications: Thermoplastic polyurethane with a sharp melting pt. and a very low melt viscosity used in golf-ball covers, hot melt adhesives, and for orthopedic applications
Form: Tough polymer
Gen'l. Prop.: M.W.: 45,000; M.P.: 58 C; Hydroxyl Value: ≈ 2
Ref.: Interox Chem. bulletin P20-130

CAPA Monomer. Solvay
Chem. Descrip.: Caprolactone monomer
Category/Applications: Thermoplastic polyurethane used as a reactive diluent in epoxy resins, and for modifying synthetic resins
Form: Liquid
Gen'l. Prop.: M.W.: 114; M.P.: 1 C
Ref.: Interox Chem. bulletin P20-130

Capran Films. Allied Chem., Fibers & Plastics Co.
Chem. Descrip.: A polyamide thermoplastic film, nylon 6 type
Category/Applications: Thermoplastic film having high elongation and exceptional tensile properties; general purpose packaging film; readily thermoformed;

low permeability to gases; high tensile properties; good machinability; various types available for specific applications; excellent sterilization packaging film
Form: Extruded films in various widths and thicknesses from 0.6 mil; Color: Clear (transparency varies with thickness); Odor: Little or no odor
Gen'l. Prop.: Stability: Resistant to oils, fats, waxes, greases, and most solvents; affected by mineral acids and aq. sol'ns.
Mech. Prop.: Good resistance to puncture, abrasion, flex, pinholing, and stress cracking; high tensile properties
Therm. Prop.: Good thermal resistance
Toxicity/Handling: Generally nontoxic
Std. Pkgs.: Available on 6-in. cores
Ref.: Tech. data bulletin

Capron Nylon (Glass-Reinforced). Allied Chem., Fibers & Plastics Co.
Chem. Descrip.: Nylon molding resin, broad range of grades
Category/Applications: Glass-reinforced injecting molding nylon with excellent stiffness and dimensional stability
Form: Pellets
Mech. Prop.: Flex. Mod.: 365,000–1,350,000 psi; Impact Str. (Izod): 0.7–2.1 ft lb/in. notched; Drop Wt. Impact: 11–25 in. lb
Therm. Prop.: Distort. Temp.: 300–420 F (@ 264 psi)
Ref.: Tech. data bulletin

Capron Nylon (Grafted Copolymers). Allied Chem., Fibers & Plastics Co.
Chem. Descrip.: Nylon molding resin, broad range of grades
Category/Applications: Grafted copolymers; extrusion and injection molding nylon with excellent flexibility and impact resistance
Form: Pellets
Mech. Prop.: Flex. Mod.: 55,000–350,000 psi; Impact Str. (Izod): 2.0–2.3 ft lb/in. notched; Drop Wt. Impact: 105–150 ft lb
Therm. Prop.: Distort. Temp.:115–130 F (@ 264 psi)
Ref.: Tech. data bulletins

Capron Nylon (Homolymers). Allied Chem., Fibers & Plastics Co.
Chem. Descrip.: Nylon molding resin, broad range of grades
Category/Applications: Homolymers; conventional injection molding nylon with excellent abrasion and chemical resistance, excellent rigidity and stiffness, and high speed processing capability
Form: Pellets
Mech. Prop.: Flex. Mod.: 170,000–450,000 psi; Impact Str. (Izod): 1.1–4.0 ft lb/in. notched; Drop Wt. Impact: 75–125 ft lb
Therm. Prop.: Distort. Temp.: 150–165 F (@ 264 psi)
Ref.: Tech. data bulletins

Capron Nylon (Mineral Filler Reinforced). Allied Chem., Fibers & Plastics Co.
Chem. Descrip.: Mineral filler reinforced injection molding nylon with excellent sink mark resistance and warp-free characteristics; specialty grade platable using standard procedures
Form: Pellets
Mech. Prop.: Flex. Mod.: 950,000–1,300,000 psi; Impact Str. (Izod): 0.5–0.9 ft lb/in. notched; Drop Wt. Impact: 10–30 in. lb
Therm. Prop.: Distort. Temp.: 250–390 F (@ 264 psi)
Ref.: Tech. data bulletins

Carbopol Resins. BF Goodrich Co.
Chem. Descrip.: Modified polyacrylic acids of high molecular weight
Category/Applications: Thickening agent for aqueous and solvent systems for cosmetics, pharmaceuticals, agricultural, industrial, chemical specialties,

and textile printing formulations; provide crystal clarity for cosmetic gels, sustained release for pharmaceutical tablets, controlled flow properties and body for water-based textile print pastes and chemical specialties; suspends solids, stabilizes emulsions, etc.
Form: Dry, fluffy powder; Color: White
Gen'l. Prop.: Stability: Unlimited stability; unaffected by aging; resistant to bacteria; wide compatibility
Ref.: Product information

Carboset 514A. BF Goodrich Co.
Chem. Descrip.: Acrylic resin
Category/Applications: Thermoplastic resin
Form: Clear solution; Color: Colorless
Composition: 70% in isopropyl alcohol
Gen'l. Prop.: M.W.: 30,000; Density: 1030 kg/m^3; Visc.: 10,000–15,000 MPa·s; Acid No.: 60–70
Mech. Prop.: Tens. Str.: 11.0 MPa; Tens. Elong.: 50%; Hardness (Sward): 24
Ref.: Product data

Carboset 514H. BF Goodrich Co.
Chem. Descrip.: Acrylic resin
Category/Applications: Thermoplastic resin
Form: Dispersion; Color: Milky white
Composition: 40% in ammonia water
Gen'l. Prop.: M.W.: 30,000; Density: 1052 kg/m^3; Visc.: 350 MPa·s; Acid No.: 60–70; pH: 6.7–7.3
Mech. Prop.: Tens. Str.: 11.0 MPa; Tens. Elong.: 50%; Hardness (Sward): 24
Ref.: Product data

Carboset 515. BF Goodrich Co.
Chem. Descrip.: Acrylic resin
Category/Applications: Thermoplastic resin
Form: Clear liquid; Color: Colorless
Composition: 100% liquid resin
Gen'l. Prop.: M.W.: 7000; Density: 1100 kg/m^3; Visc.: 2.0 × 10^6 MPa·s; Acid No.: 60–65
Ref.: Product data

Carboset 525. BF Goodrich Co.
Chem. Descrip.: Acrylic resin
Category/Applications: Thermoplastic resin
Form: Granular solid; Color: White
Composition: 100% solid resin
Gen'l. Prop.: M.W.: 260,000; Acid No.: 76–85
Mech. Prop.: Tens. Str.: 26.2 MPa; Tens. Elong.: 165%; Hardness (Sward): 24
Ref.: Product data

Carboset 531. BF Goodrich Co.
Chem. Descrip.: Acrylic resin
Category/Applications: Thermoset resin
Form: Translucent solution; Color: White
Composition: 25% in ammonia water
Gen'l. Prop.: M.W.: 10^6 (cured); Density: 1040 kg/m^3; Visc.: 2000 MPa·s; Acid No.: 60–70; pH: 7.4–8.2
Mech. Prop.: Hardness (Sward): 32
Ref.: Product data

Carboset Resins. BF Goodrich Co.
Chem. Descrip.: Acrylic film-forming resins
Category/Applications: Thermoset and thermoplastic type resins; used in permanent and temporary coatings that can be removed with alkali sol'ns.; excellent dispersant, leveling, film-forming, and binding characteristics for use in protective metal coatings, paints, ceramics, adhesives, textiles, paper, leather, cosmetics, floor polishes, and chemical specialties; the thermoplastic resins can be crosslinked with epoxy or melamine formaldehyde resins to yield tough, permanent protective coatings with excellent chemical resistance

Gen'l. Prop.: Sol.: Water-sol. resins air-dry to water-insolubility
Ref.: Product information

Carboset XL-11. BF Goodrich Co.
Chem. Descrip.: Acrylic resin
Category/Applications: Thermoplastic resin
Form: Dispersion; Color: Milky white
Composition: 30% in ammonia water
Gen'l. Prop.: M.W.: 40,000; Density: 1040 kg/m^3; Visc.: 50 MPa·s; Acid No.: 70–80; pH: 6.4–7.0
Mech. Prop.: Hardness (Sward): 35
Ref.: Product data

Carboset XL-19. BF Goodrich Co.
Chem. Descrip.: Acrylic resin
Category/Applications: Thermoplastic resin
Form: Dispersion; Color: Milky white
Composition: 40% in ammonia water
Gen'l. Prop.: M.W.: 30,000; Density: 1052 kg/m^3; Visc.: 35 MPa·s; Acid No.: 29–35; pH: 7.2–7.8
Mech. Prop.: Hardness (Sward): 21
Ref.: Product data

Carpet Guard. Fabri-Coate Co.
Chem. Descrip.: Copolymer resin in petroleum naphtha
Category/Applications: Resin for use as carpet finish to impart soil resistance and protect against spots, spills, and stains
Form: Clear liquid; Odor: Slight oily odor
Gen'l. Prop.: Sol.: Negligible sol. in water; Sp. Gr.: 0.74; B.P.: 360–400 F
Therm. Prop.: Flash Pt.: 132 F (TCC)
Toxicity/Handling: May cause eye irritation, drying of skin; excessive inhalaton causes headache, dizziness, and nausea; avoid contact with moisture and strong oxidizing agents
Std. Pkgs.: 14 oz aerosol can, 12 cans/case; bulk: 4 X 1 case; 5-, 30-, or 55-gal drums
Ref.: Product pamphlet, safety data sheet

Carstab 700. Cincinnati Milacron Chem.
Chem. Descrip.: 2-Hydroxy-4-n-octoxybenzophenone
Category/Applications: Ultraviolet light stabilizer (absorbing ultraviolet radiaton in the 230–360 nm range) for polypropylene, high-density polyethylene, low-density polyethylene, ethylene vinyl acetate, polyvinyl chloride (flexible and rigid), polycarbonates, acetals, epoxies, and some cellulosic materials
Form: Crystalline powder; Color: Off-white
Composition: 98.5% pure
Gen'l. Prop.: M.W.: 326; Sol. (g/100 ml solvent): 74.3 g in acetone, 72.7 g in benzene, 180 g in ethyl acetate, 65 g in MEK; insol. in water; Sp. Gr.: 1.16 g/cc; F.P.: 47.4 C; Stability: Photo and heat stable; relatively inert compound, unaffected by most acidic or alkaline media
Toxicity/Handling: Low dermal and oral toxicity
Std. Pkgs.: 51 gal (200 lb net) polyethylene-lined fiber drums
Ref.: Tech. bulletin

Carstab 701. Cincinnati Milacron Chem.
Chem. Descrip.: 2-Hydroxy-4-isooctoxybenzophenone
Category/Applications: Ultraviolet light stabilizer (absorbs ultraviolet radiation in the 230–360 nm region) for polypropylene, high-density polyethylene, low-density polyethylene, ethylene vinyl acetate, polyvinyl chloride (flexible and rigid), polycarbonates, acetals, epoxies, and some cellulosic materials
Form: Liquid; Color: Pale yellow
Gen'l. Prop.: M.W.: 326; Sol. (g/100 g

solvent): 14 g in methanol, 20 g in ethanol (95%); miscible with acetone, benzene, hexane, ethyl acetate, MEK, perchloroethylene; insol. in water; Sp. Gr.: 1.066 (24 C); Pour Pt.: −15 C; Stability: Heat and photo stable; relatively inert; unaffected by most acidic or alkaline media
Toxicity/Handling: Low oral and dermal toxicity
Ref.: Tech. bulletin

Carstab 702. Cincinnati Milacron Chem.
Chem. Descrip.: 2-Hydroxy-4-isooctoxy-benzophenone
Category/Applications: Ultraviolet light stabilizer (absorbs ultraviolet radiation in the 230–360 nm range) for polypropylene, high-density polyethylene, low-density polyethylene, ethylene vinyl acetate, polyvinyl chloride (flexible and rigid), polycarbonate, acetals, epoxies, and some cellulosic materials
Form: Liquid; Color: Pale yellow
Gen'l. Prop.: M.W.: 326; Sol. (g/100 g solvent): 20 g in ethanol (95%); miscible with acetone, heptane, MEK, perchloroethylene, toluene; insol. in water; Sp. Gr.: 1.504 (24 C); Pour Pt.: −20 C; Stability: Photo and heat stable; relatively inert; unaffected by most acidic or alkaline media
Toxicity/Handling: Low oral and dermal toxicity; store indoors in a dry location < 35 C; keep drums tightly closed during storage
Ref.: Tech. bulletin

Carstab 705. Cincinnati Milacron Chem.
Chem. Descrip.: 2-Hydroxy-4-n-octoxy-benzophenone
Category/Applications: Completely free-flowing and free from lumping and caking modified form of Carstab 700; UV light stabilizer for polypropylene, polyethylene, ethylene vinyl acetate, cellulosics, and epoxies; effective in cosmetics and protective coatings; imparts long-term protection to plastics due to excellent compatibility, low volatility, good heat stability, and low water extractability
Form: Noncaking crystalline powder; Color: Off white
Gen'l. Prop.: Sol.: Readily sol. in most common organic solvents incl. aliphatic and aromatic hydrocarbons, ketones, and esters; insol. in water; Density (bulk): 5.9 lb/gal; M.P.: 47.0 C; Stability: Heat and photo stable
Std. Pkgs.: 51 gal (200 lb net) polyethylene-lined fiber drums
Ref.: Tech. bulletin

Carstab DLTDP. Cincinatti Milacron Chem.
Chem. Descrip.: Dilauryl thiodipropionate
Category/Applications: Thioester synergist for use as an effective long-term heat aging stabilizer when used in conjunction with primary antioxidants; highly effective in polypropylene and high-density polyethylene
Form: Flake; Color: White
Gen'l. Prop.: M.W.: 514; Sol. (g/100 g): 30 g in benzene; insol. in water; Sp. Gr.: 0.896 (80 C); F.P.: 41 C; Acid No.: 0.5; Stability: Good heat stability; resistant to discoloration and volatilization; relatively inert; does not react with most commonly-used plastic additives
Toxicity/Handling: Low oral and dermal toxicity
Std. Pkgs.: 55 gal (200 lb net) polyethylene-lined fiber drums
Ref.: Tech. bulletin

Carstab DMTDP. Cincinnati Milacron Chem.
Chem. Descrip.: Dimyristyl thiodipropionate

Category/Applications: Antioxidant for polyolefins; most effective used in synergistic combination with hindered phenolic antioxidants (retards embrittlement and color formation due to oxidative degradation during processing and long term aging)

Form: Small flake; Color: White; Odor: Good

Composition: 99% purity

Gen'l. Prop.: M.W.: 570; Sol.: Sol. in benzene, chloroform, toluene; insol. in water, alcohol, ethyl acetate; Sp. Gr.: 0.885 (80 C); Density (bulk): 3.99 lb/gal; M.P.: 51 C; Acid Value: 0.4

Std. Pkgs.: 200 lb net polyethylene-lined fiber drums

Ref.: Tech. bulletin

Carstab DSTDP. Cincinnati Milacron Chem.

Chem. Descrip.: Distearyl thiodipropionate

Category/Applications: Thioester synergist for use as an effective long-term heat aging stabilizer when used in conjunction with primary antioxidants; highly effective in polypropylene and high-density polyethylene

Form: Flake or powder; Color: White

Composition: 98% pure

Gen'l. Prop.: M.W.: 682; Sol. (g/100 g solvent): 29 g in benzene, 2 g in acetone and hexane; insol. in water; Sp. Gr.: 0.858 (80 C); F.P.: 65 C; Acid No.: 1.0; Stability: Good thermal stability; nonvolatile; relatively inert; does not react with most commonly used plastic additives

Toxicity/Handling: Low oral and dermal toxicity

Std. Pkgs.: 55 gal (200 lb net) polyethylene-lined fiber drums

Ref.: Tech. bulletin

Cast Cellulose Acetate Film 904. Georgia-Pacific Corp.

Chem. Descrip.: Cast cellulose acetate film

Form: 1.2 mil gauge film

Mech. Prop.: Tens. Str.: 9000–12,000 psi; Tens. Elong.: 15–25%

Elec. Prop.: Dielec. Str.: 2300–3000 V/mil (50% r.h., 25 C); Dielec. Const.: 4.8 ohm-cm (@ 1 kilocycle, 50% r.h., 25 C)

Ref.: Product report CPR 76-132

Cast Cellulose Acetate Sheet 867. XCEL Corp.

Chem. Descrip.: Cast cellulose acetate sheet

Form: 5 mil gauge sheet

Mech. Prop.: Tens. Str.: 11,500 psi; Tens. Elong.: 40%

Therm. Prop.: Flamm.: 0.45 $in.^2/s$

Ref.: Product report JJF-73-104 (A)

Cast Cellulose Di Acetate Film 912. Georgia-Pacific Corp.

Chem. Descrip.: Cast cellulose diacetate film

Form: 0.0017 in. gauge; Finish: Clear and matte

Gen'l. Prop.: M.W.: 912

Mech. Prop.: Tens. Str.: 10,100 psi; Tens. Elong.: 35%

Ref.: Product report JJF75-212

Cast Cellulose Triacetate HP-556. XCEL Corp.

Chem. Descrip.: Cast cellulose triacetate sheet

Mech. Prop. (4.7 mil gauge): Tens. Str.: 16,500 psi; Tens. Elong.: 24%

Ref.: Product report JJF-77-53

Cast Cellulose Triacetate Sheet HP-550. XCEL Corp.

Chem. Descrip.: Cast cellulose triacetate sheet

Mech. Prop. (3.0 mil gauge): Tens. Str.: 15,000 psi; Tens. Elong.: 22%

Therm. Prop.: Flamm.: Self-extinguishing, no burning rate (D568-61)
Ref.: Product report JJF-76-20

Catalyst R # 5024. Polymer Research Corp. of America

Category/Applications: Catalyst recommended for pigment screen and roller printing; accelerates the curing cycle and enables curing to proceed at lower temps. while maintaining the AATCC fastness properties; use 1 oz/gal of the print paste

Form: Opaque liquid; Color: Amber
Composition: 28% sol'n. in water

Gen'l. Prop.: Sol.: Disp. in water producing stable, off-white emulsion; Density: 9.6 lb/gal; pH: 7.0
Toxicity/Handling: Store @ R.T.
Std. Pkgs.: 30 and 55 gal fiber drums
Ref.: Tech. bulletin

CB-4-34. Neville Chem. Co.
Chem. Descrip.: Aromatic plasticizer

Category/Applications: Aromatic, chemically inert, nonsaponifiable plasticizer exhibiting low reactivity; used in adhesives (mastic, pressure sensitive), rubber (cements, mechanical and molded goods, tires), and caulking compounds

Form: Liquid; Color (Gardner): 14 (50% in toluene, D1544)

Gen'l. Prop.: M.W.: 515 (D2503); Sol.: Sol. in ethers and chlorinated, aromatic, naphthenic, and terpene hydrocarbons; partly sol. in esters, ketones (except acetone), and paraffinic hydrocarbons; Sp. Gr.: 1.003 (D71); Soften. Pt. (Ring & Ball): 28 C (D1544); Iodine No. (Wijs): 36 (D1959); Stability: Resistant to alkalis, acids, and moisture

Therm. Prop.: Flash Pt. (COC): 390 F (D92)

Ref.: Product pamphlet

Celanese Nylon 1000-1. Celanese Corp.
Chem. Descrip.: 6/6 type polyamide nylon resin

Category/Applications: General purpose nylon resin for injection molding and extrusion; used for mechanical parts, gears, bearings, hardware, etc.

Form: Cylindrical, free-flowing pellets; Color: Natural and black

Gen'l. Prop. (as molded): Density: 0.0412 lb/in.^{3} (D792); M.P.: 495 F (D-789); Stability: Excellent resistance to aromatic, chlorinated, and ketone solvents, gasolines, oils, greases, animal and vegetable oils; satisfactory resistance to most mild or moderate base sol'ns.; limited resistance to most acids and poor resistance to strong acids, e.g., sulfuric, hydrochloric, formic, nitric, etc.; attacked by phenols, sodium hypochlorite, chlorine water, potassium permanganate

Mech. Prop. (as molded, @ 73 F): Tens. Str.: 12,000 psi (D638); Tens. Elong.: 40–80% (D638); Flex. Str.: 17,000 psi (D790); Flex. Mod.: 420,000 psi (D-790); Shear Str.: 9500 psi (D732); Impact Str. (Izod): 1.0 ft lb/in. notched (1/8 in. bar, D256); Hardness (Rockwell): M79-M86 (D785)

Therm. Prop. (as molded): Distort. Temp.: 170 F (@ 264 psi, D648); Coeff. of Linear Exp.: 5×10^{-5} in./in./F (D-696); Sp. Heat: 0.4 (Cenco-Fitch); Flamm.: UL-94 (V2)

Elec. Prop. (as molded, 73 F): Power Factor: 0.020 (60 Hz, D150); Dielec. Str.: 600 V/mil (short time, D149); Dielec. Const.: 5.5 (100 Hz, D150); Vol. Resist.: 10^{14} ohm-cm

Toxicity/Handling: May cause burns or ignite other flamm. materials @ normal processing temps. ≥ 500 F

Ref.: Bulletin N1A

Celanese Nylon 1000-2. Celanese Corp.
Chem. Descrip.: 6/6 type polyamide nylon resin
Category/Applications: Surface-lubricated form of Celanese Nylon 1000-1; used for mechanical parts, gears, bearings, hardware, etc.
Form: Cylindrical, free-flowing pellets; Color: Natural and black
Gen'l., Mech., Therm., and Elec. Prop.: See Celanese Nylon 1000-1
Toxicity/Handling: See Celanese Nylon 1000-1
Ref.: Bulletin N1A

Celanese Nylon 1000-4. Celanese Corp.
Chem. Descrip.: 6/6 type polyamide nylon resin
Category/Applications: General purpose resin for easy mold release; used for mechanical parts, gears, bearings, hardware, etc.
Form: Cylindrical, free-flowing pellets; Color: Natural and black
Gen'l., Mech., Therm., and Elec. Prop.: See Celanese Nylon 1000-1
Toxicity/Handling: See Celanese Nylon 1000-1
Ref.: Bulletin N1A

Celanese Nylon 1003-1. Celanese Corp.
Chem. Descrip.: 6/6 type polyamide nylon resin
Category/Applications: Heat-stabilized form of Celanese Nylon 1000-1; used for high-temp. resistance applications, e.g., under-the-hood automotive parts
Form: Cylindrical, free-flowing pellets; Color: Natural and black
Gen'l., Mech., Therm., and Elec. Prop.: See Celanese Nylon 1000-1
Toxicity/Handling: See Celanese Nylon 1000-1
Ref.: Bulletin N1A

Celanese Nylon 1003-2. Celanese Corp.
Chem. Descrip.: 6/6 type polyamide nylon resin
Category/Applications: Surface-lubricated form of Celanese Nylon 1003-1; used for high-temp. resistant applications, e.g., under-the-hood automotive parts
Form: Cylindrical, free-flowing pellets; Color: Natural and black
Gen'l., Mech., Therm., and Elec. Prop.: See Celanse Nylon 1000-1
Toxicity/Handling: See Celanese Nylon 1000-1
Ref.: Bulletin N1A

Celanese Nylon 1200-1. Celanese Corp.
Chem. Descrip.: Nylon resin
Category/Applications: High-viscosity extrusion grade nylon resin; used for extruded profiles, film, sheet, and injection molded parts with improved toughness
Ref.: Bulletin N3A

Celanese Nylon 1300 Series. Celanese Corp.
Chem. Descrip.: 6/6 nylon resin
Category/Applications: Controlled crystallization grade nylon resin for faster cycling, improved processability (flow and mold release), toughness, and increased surface hardness
Form: Cylindrical, free-flowing pellets; Color: Natural and black
Gen'l. Prop. (as molded): Density: 0.0412 lb/in.^{3} (D792); M.P.: 495 F (D-789); Stability: Excellent resistance to aromatic, chlorinated, and ketone solvents, gasoline, oils, greases, animal and vegetable oils; satisfactory resistance to most mild or moderate base sol'ns.; limited resistance to most acids and poor resistance to strong acids, e.g., sulfuric, hydrochloric, formic, nitric, etc.; attacked by phenols, sodium hypochlorite, chlorine water, potassium permanganate

Mech. Prop. (as molded, @ 73 F): Tens. Str.: 14,000 psi (D638); Tens. Elong.: 25–40% (D638); Flex. Str.: 19,000 psi (D790); Flex. Mod.: 500,000 psi (D-790); Shear Str.: 9400 psi (D732); Impact Str. (Izod): 0.8–0.9 ft lb/in. notched (1/8 in. bar, D256); Hardness (Rockwell): M86–M89 (D785)
Therm. Prop. (as molded): Distort. Temp.: 174 F (@ 264 psi, D648); Sp. Heat: 0.4 (Cenco-Fitch); Flamm.: UL-94 (V2)
Elec. Prop. (as molded, 73 F): Power Factor: 0.037 (100 Hz, D150); Dielec. Str.: 600 V/mil (short time, D149); Dielec. Const.: 5.4 (100 Hz, D150); Vol. Resist.: 10^{14} ohm-cm
Toxicity/Handling: May cause burns or ignite other flamm. materials @ normal processing temps. ⩾ 500 F
Ref.: Bulletin N1A

Celanese Nylon 1310-1. Celanese Corp.
Chem. Descrip.: 6/6 type nylon resin
Category/Applications: Controlled crystallization resin for fast cycling in injection molding with general purpose product toughness; used for parts similar to 1000 series applications which can benefit from higher production rates due to faster cycles, easy mold release, etc.
Form: Cylindrical, free-flowing pellets; Color: Natural and black
Gen'l. Prop. (as molded): Density: 0.0412 lb/in.^{3} (D792); M.P.: 495 F (D-789); Stability: Excellent resistance to aromatic, chlorinated, and ketone solvents, gasoline, oils, greases, animal and vegetable oils; satisfactory resistance to most mild or moderate base sol.ns; limited resistance to most acids and poor resistance to strong acids, e.g., sulfuric, hydrochloric, formic, nitric, etc.; attacked by phenols, sodium hypochlorite, chlorine water, potassium permanganate
Mech. Prop. (as molded, @ 73 F): Tens. Str.: 13,600 psi (D638); Tens. Elong.: 35–55% (D638); Flex. Str.: 17,900 psi (D790); Flex. Mod.: 470,000 psi (D-790); Shear Str.: 9400 psi (D732); Impact Str. (Izod): 1.0 ft lb/in. notched (1/8 in. bar, D256); Hardness (Rockwell): M84–M88 (D785)
Therm. Prop. (as molded): Distort. Temp.: 171 F (@ 264 psi, D648); Sp. Heat: 0.4 (Cenco-Fitch); Flamm. UL-94 (V2)
Elec. Prop. (as molded, 73 F): Power Factor: 0.037 (100 Hz, D150); Dielec. Str.: 600 V/mil (short time, D149); Dielec. Const.: 5.4 (100 Hz, D150); Vol. Resist.: 10^{14} ohm-cm
Toxicity/Handling: May cause burns or ignite other flamm. materials @ normal processing temps. ⩾ 500 F
Ref.: Bulletin N1A

Celanese Nylon 1310-2. Celanese Corp.
Chem. Descrip.: 6/6 type nylon resin
Category/Applications: Surface-lubricated form of Celanese Nylon 1310-1 for easier processing and mold release
Form: Cylindrical, free-flowing pellets; Color: Natural and black
Gen'l., Mech., Therm., and Elec. Prop.: See Celanese Nylon 1310-1
Toxicity/Handling: See Celanese Nylon 1310-1
Ref.: Bulletin N1A

Celanese Nylon 1310-4. Celanese Corp.
Chem. Descrip.: 6/6 type nylon resin
Category/Applications: Controlled crystallization resin for easy mold release
Form: Cylindrical, free-flowing pellets; Color: Natural and black
Gen'l., Mech., Therm., and Elec. Prop.: See Celanese Nylon 1310-1
Toxicity/Handling: See Celanese Nylon 1310-1
Ref.: Bulletin N1A

Celanese Nylon 1500-1. Celanese Corp.

Chem. Descrip.: Glass fiber (33%) reinforced nylon 6/6

Cateory/Applications: Injection molding grade nylon resin for use in screw injection molding machines; used for injected molded parts with excellent impact, high tensile strength, and greatest stiffness

Form: Solid

Gen'l. Prop. (as molded): Density: 0.0498 lb/in.^{3}; M.P.: 495–507 F (D789); Stability: Excellent resistance to aromatic, chlorinated, and ketone solvents, gasoline, oils, greases, ethers, animal and vegetable oils; satisfactory resistance to most mild or moderate base sol'ns.; limited resistance to most acids and poor resistance to strong acids, e.g., sulfuric, hydrochloric, formic, nitric, etc.; attacked by phenols, sodium hypochlorite, chlorine water, potassium permanganate

Mech. Prop. (as molded): Tens. Str.: 28,000 psi (D638); Tens. Elong.: 4% (D-638); Flex. Str.: 41,000 psi (D790); Flex. Mod.: 1.3 $\times$ 10^6 psi (D790); Compr. Str.: 29,400 psi (D695); Shear Str.: 13,000 psi (D732); Impact Str. (Izod): 2.2 ft lb/in. (D256); Hardness (Rockwell): M100 (D785)

Therm. Prop. (as molded): Conduct.: 1.5 Btu/h/ft^2/F/in.; Distort. Temp.: 495 F (@ 264 psi, D648); Coeff. of Linear Exp.: 1.6 $\times$ 10^{-5} in./in./F (D-696); Sp. Heat: 0.3 (Cenco-Fitch); Flamm.: 1.1 in./min (D635), HB (UL-94)

Elec. Prop. (as molded): Power Factor: 0.006 (@ 60 Hz, D150); Dielec. Str.: 400 V/mil (D149); Dielec. Const.: 4.1 (60 Hz, D150); Vol. Resist.: 10^{14} ohm-cm

Toxicity/Handling: May cause burns or ignite other flamm. materials @ normal processing temps. ≥ 500 F
Ref.: Bulletin N1A, N1B

Celanese Nylon 1500-2. Celanese Corp.
Chem. Descrip.: Glass-fiber (33%) reinforced nylon 6/6, surface-lubricated
Category/Applications: Surface-lubricated form of Celanese Nylon 1500-1 for use in screw or ram type injection molding machine; used for injection molded parts with excellent impact, high tens. strength, and greatest stiffness
Form: Solid
Gen'l., Mech., Therm., and Elec. Prop.: See Celanese Nylon 1500-1
Toxicity/Handling: See Celanese Nylon 1500-1
Ref.: Bulletin N1A, N1B

Celanese Nylon 1503-1. Celanese Corp.
Chem. Descrip.: Glass fiber (33%) reinforced nylon 6/6
Category/Applications: Heat-stabilized form of Celanese Nylon 1500-1 for use in screw injection molding machines; used for strong, highest strength parts requiring temp. resistance up to 350 F
Form: Solid
Gen'l., Mech., Therm., and Elec. Prop.: See Celanese Nylon 1500-1
Toxicity/Handling: See Celanese Nylon 1500-1
Ref.: Bulletin N1A, N1B

Celanese Nylon 1503-2. Celanese Corp.
Chem. Descrip.: Glass fiber (33%) reinforced nylon 6/6, surface-lubricated and heat-stabilized
Category/Applications: Surface-lubricated form of Celanese Nylon 1503-1; used as general purpose glass fiber-reinforced grade in both screw and ram type injection molding machines; used for strong,

highest strength parts requiring temp. resistance up to 350 F
Form: Solid
Gen'l., Mech., Therm., and Elec. Prop.: See Celanese Nylon 1500-1
Toxicity/Handling: May cause burns or ignite other flamm. materials @ normal processing temps. ⩾ 500 F
Ref.: Bulletin N1A, N1B

Celanese Nylon 1600-1. Celanese Corp.
Chem. Descrip.: Glass fiber (40%) reinforced nylon 6/6
Category/Applications: Nylon resin injection molding grade for use in screw machines; used for injection molded parts with exceptionally high impact, high tens. strength, and superior stiffness
Form: Solid
Gen'l. Prop. (as molded): Density: 0.0574 lb/in.^{3}; M.P.: 495-507 F (D789); Stability: Excellent resistance to aromatic, chlorinated, and ketone solvents, gasoline, oils, greases, ethers, animal and vegetable oils; satisfactory resistance to most mild or moderate base sol'ns.; limited resistance to most acids and poor resistance to strong acids, e.g., sulfuric, hydrochloric, formic, nitric, etc.; attacked by phenols, sodium hypochlorite, chlorine water, potassium permanganate
Mech. Prop. (as molded): Tens. Str.: 34,000 psi (D638); Tens. Elong.: 3% (D638); Flex. Str.: 48,000 psi (D790); Flex. Mod.: 1.6×10^6 psi (D790); Shear Str.: 13,000 psi (D732); Impact Str. (Izod): 2.6 ft lb/in. (D256); Hardness (Rockwell): M100 (D785)
Therm. Prop. (as molded): Distort. Temp.: 488 F (@ 264 psi, D648); Coeff. of Linear Exp.: 1.4×10^{-5} in./in./F (D-696); Flamm. HB (UL94)
Elec. Prop. (as molded): Power Factor: 0.009 (100 Hz, D150); Dielec. Str.: 400 V/mil (D149); Dielec. Const.: 4.4 (100 Hz, D150); Vol. Resist.: 10^{14} ohm-cm
Toxicity/Handling: May cause burns or ignite other flamm. materials @ normal processing temps. ⩾ 500 F
Ref.: Bulletin N1A, N1B

Celanese Nylon 1600-2. Celanese Corp.
Chem. Descrip.: Glass fiber (40%) reinforced nylon 6/6
Category/Applications: Surface-lubricated form of Celanese Nylon 1600-1; used for injection molded parts with exceptionally high impact, high tens. strength, and superior stiffness
Form: Solid
Gen'l., Mech., Therm., and Elec. Prop.: See Celanese Nylon 1600-1
Toxicity/Handling: See Celanese Nylon 1600-1
Ref.: Bulletin N1A, N1B

Celanese Nylon 1603-1. Celanese Corp.
Chem. Descrip.: Glass fiber (40%) reinforced nylon 6/6, heat-stabilized
Category/Applications: Heat stabilized form of Celanese Nylon 1600-1 for use in screw injection molding machines; used for parts requiring extra high strength and rigidity and temp. resistance up to 350 F
Form: Solid
Gen'l., Mech., Therm., and Elec. Prop.: See Celanese Nylon 1600-1
Toxicity/Handling: See Celanese Nylon 1600-1
Ref.: Bulletin N1A, N1B

Celanese Nylon 1603-2. Celanese Corp.
Chem. Descrip.: Glass fiber (40%) reinforced nylon 6/6, surface-lubricated and heat stabilized
Category/Applications: Surface lubricated form of Celanese Nylon 1603-1 for use in screw injection molding machines;

used for parts requiring extra high strength and rigidity and temp. resistance up to 350 F
Form: Solid
Gen'l., Mech., Therm., and Elec. Prop.: See Celanese Nylon 1600-1
Toxicity/Handling: See Celanese Nylon 1600-1
Ref.: Bulletin N1A, N1B

Celanex 2000. Celanese Corp.
Chem. Descrip.: Unreinforced thermoplastic polyester resin
Category/Applications: Thermoplastic resin with exceptional processing characteristics combining product consistency with good surface gloss and high flow to achieve fast cycle times; also available with an internal lubricant system to facilitate mold release (identified by suffix -2)
Form: Solid
Gen'l. Prop.: Sp. Gr.: 1.31 (D792); M.P.: 442 F; Stability: Good dimensional stability; excellent R.T. resistance to water, weak acids and bases, detergents, aliphatic hydrocarbons, chlorinated aliphatic hydrocarbons, oils, fats, ethers, alcohols, and glycols
Mech. Prop. (73 F): Tens. Str.: 8300 psi (D638); Tens. Elong.: 5% (D638); Flex. Stress: 12,400 psi (@ 5% strain, D790); Flex. Mod.: 360,000 psi (D790); Impact Str. (Izod): 0.7 ft lb/in. notched (D256); Hardness (Rockwell): M 75 (D785)
Therm. Prop.: Distort. Temp.: 123 F (@ 264 psi, D648); Flamm.: 94HB (UL94)
Elec. Prop. (73 F): Dielec. Str.: 420 V/mil (0.125 in., short term, D149); Dielec. Const.: 3.2 (100 Hz, D150); Vol. Resist.: 10^{15} ohm-cm (D257); Arc Resist.: 75 s (D495); Dissip. Factor: 0.02 (100 Hz, D150)
Ref.: Product information

Celanex 2012. Celanese Corp.
Chem. Descrip.: Unreinforced thermoplastic polyester resin
Category/Applications: Thermoplastic resin with exceptional product consistency, and mechanical and electrical properties; also available in lubricated grade identified by the suffix -2
Form: Solid
Gen'l. Prop.: Sp. Gr.: 1.43 (D792); M.P.: 442 F; Stability: Excellent R.T. resistance to water, weak acids and bases, detergents, aliphatic hydrocarbons, chlorinated aliphatic hydrocarbons, oils, fats, ethers, alcohols, glycols
Mech. Prop. (73 F): Tens. Str.: 8000 psi (D638); Tens. Elong.: 15% (D638); Flex. Stress: 14,000 psi (@ 5% strain, D790); Flex. Mod.: 400,000 psi (D790); Impact Str. (Izod): 0.5 ft lb/in. notched (D256); Hardness (Rockwell): M 80 (D785)
Therm. Prop.: Distort. Temp.: 135 F (@ 264 psi, D648); Flamm.: 94V-O (@ 1/32 in., UL94)
Elec. Prop. (73 F): Dielec. Str.: 450 V/mil (0.125 in., short term, D149); Dielec. Const.: 3.2 (100 Hz, D150); Vol. Resist.: 10^{15} ohm-cm (D257); Arc Resist.: 110 s (D495); Dissip. Factor: 0.0020 (100 Hz, D150)
Ref.: Product information

Celanex 3200. Celanese Corp.
Chem. Descrip.: 15% glass reinforced thermoplastic polyester resin
Category/Applications: General purpose thermoplastic resin suited for applications requiring greater flexibility; highest flow glass-reinforced grade for easy processing with excellent surface gloss; also available in lubricated grade identified by suffix -2
Form: Solid
Gen'l. Prop.: Sp. Gr.: 1.41 (D792); M.P.: 442 F (D618); Stability: See Celanex 2012

Mech. Prop. (73 F): Tens. Str.: 13,000 psi (D638); Tens. Elong.: 2% (D638); Flex. Stress: 21,000 psi (@ break, D-790); Flex. Mod.: 700,000 psi (D790); Impact Str. (Izod): 0.9 ft lb/in. notched (D256)
Therm. Prop.: Distort. Temp.: 378 F (@ 264 psi, D648); Flamm.: 94HB (UL-94)
Elec. Prop. (73 F): Dielec. Str.: 460 V/mil (0.125 in., short term, D149); Arc Resist.: 130 s (D495)
Ref.: Product information

Celanex 3210. Celanese Corp.
Chem. Descrip.: 18% glass reinforced thermoplastic polyester resin
Category/Applications: Thermoplastic resin with excellent thermal and electrical properties of Celanex 3310, but with greater flexibility; highest flow glass-reinforced flame retarded grade for easy processing; also available in lubricated grade identified by suffix -2
Form: Solid
Gen'l. Prop.: Sp. Gr.: 1.62 (D792); M.P.: 442 F; Stability: See Celanex 2012
Mech. Prop. (73 F): Tens. Str.: 16,500 psi (D638); Tens. Elong.: 2–3% (D638); Flex. Stress: 24,000 psi (@ break, D-790); Flex. Mod.: 1.1 × 10^6 psi (D790); Impact Str. (Izod): 1.0 ft lb/in. notched (D256)
Therm. Prop.: Conduct.: 1.63 Btu/h/ft^2/F/in. (Cenco-Fitch); Distort. Temp.: 395 F (D@ 264 psi, D648); Coeff. of Linear Exp.: 1.3 × 10^{-5} in./in./F (75–100 F, flow direction, D696); Flamm.: 94V-O (at 1/32 in., UL94)
Elec. Prop. (73 F): Dielec. Str.: 480 V/mil (0.0125 in., short term, D149); Dielec. Const.: 3.8 (100 Hz, D150); Vol. Resist.: 5 × 10^{15} ohm-cm (D257); Arc. Resist.: 125 s (D495); Dissip. Factor: 0.0020 (100 Hz, D150)
Ref.: Product information

Celanex 3211. Celanese Corp.
Chem. Descrip.: 18% glass reinforced thermoplastic polyester resin
Category/Applications: Thermoplastic resin which complies with all U.L. requirements for direct support of current carrying parts, incl. a high volt track rate of < 1.0 in./min; thermal and elec. prop. similar to Celanex 3311 but with greater flexibility; also available in lubricated grade identified by suffix -2
Form: Solid
Gen'l. Prop.: Sp. Gr.: 1.60 (D792); M.P.: 417 F; Stability: See Celanex 2012
Mech. Prop. (73 F): Tens. Str.: 17,000 psi (D638); Tens. Elong.: 2–3% (D638); Flex. Stress: 26,500 psi (@ break, D-790); Flex. Mod.: 1.2 × 10^6 psi (D790); Impact Str. (Izod): 1.2 ft lb/in. notched (D256); Hardness (Rockwell): M 93 (D-785)
Therm. Prop.: Distort. Temp.: 365 F (@ 264 psi, D648); Sp. Heat: 0.22 Btu/lb/F (C351); Flamm.: 94V-O (1/32 in., UL-94)
Elec. Prop. (73 F): Dielec. Str.: 570 V/mil (0.125 in., short term, D149); Dielec. Const.: 3.7 (100 Hz, D150); Vol. Resist.: 4 × 10^{15} ohm-cm (D257); Arc Resist.: 130 s (D495); Dissip. Factor: 0.0041 (100 Hz, D150)
Ref.: Product information

Celanex 3300. Celanese Corp.
Chem. Descrip.: 30% glass reinforced thermoplastic polyester resin
Category/Applications: General purpose

thermoplastic resin with excellent thermal, elec., and mech. prop. and outstanding processability; also available in lubricated grade identified by the suffix -2
Form: Solid
Gen'l. Prop.: Sp. Gr.: 1.54 (D792); M.P.: 442 F; Stability: See Celanex 2012
Mech. Prop. (73 F): Tens. Str.: 19,500 psi (D638); Tens. Elong.: 2% (D638); Flex. Stress: 28,000 psi (@ break, D-790); Flex. Mod.: 1.2 X 10^6 psi (D790); Compr. Str.: 18,000 psi (D695); Shear Str.: 8100 psi (D732); Impact Str. (Izod): 1.7 ft lb/in. notched (D256); Hardness (Rockwell): M 90 (D785)
Therm. Prop.: Conduct.: 1.54 Btu/h/ft^2/F/in. (Cenco-Fitch); Distort. Temp.: 403 F (@ 264 psi, D648); Coeff. of Linear Exp.: 1.3 X 10^{-5} in./in./F (75–150 F, flow direction, D696); Sp. Heat: 0.30 Btu/lb/F (C351); Flamm.: 94HB (UL94)
Elec. Prop. (73 F): Dielec. Str.: 560 V/mil (0.125 in., short term, D149); Dielec. Const.: 3.7 (100 Hz, D150); Vol. Resist.: 10^{16} ohm-cm (D257); Arc Resist.: 130 s (D495); Dissip. Factor: 0.0015 (100 Hz, D150)
Ref.: Product information

Celanex 3310. Celanese Corp.
Chem. Descrip.: 30% glass reinforced thermoplastic polyester resin
Category/Applications: Thermoplastic resin with outstanding mech. and elec. prop. throughout a wide range of aggressive end-use conditions, superior thermal stability, and good processing characteristics; also available in lubricated grade identified by suffix -2
Form: Solid
Gen'l. Prop.: Sp. Gr.: 1.69 (D792); M.P.: 442 F; Stability: See Celanex 2012
Mech. Prop. (73 F): Tens. Str.: 19,500 psi (D638); Tens. Elong.: 1–2% (D638); Flex. Stress: 28,000 psi (@ break, D-790); Flex. Mod.: 1.5 X 10^6 psi (D790); Compr. Str.: 18,000 psi (D695); Shear Str.: 8900 psi (D732); Impact Str. (Izod): 1.3 ft lb/in. notched (D256); Hardness (Rockwell): M 90 (D785)
Therm. Prop.: Conduct.: 1.71 Btu/h/ft^2/F/in. (Cenco-Fitch); Distort. Temp.: 406 F (@ 264 psi, D648); Coeff. of Linear Exp.: 1.3 X 10^{-5} in./in./F (75–150 F, flow direction, D696); Sp. Heat: 0.27 Btu/lb/F (C351); Flamm.: 94V-O (1/32 in., UL94)
Elec. Prop. (73 F): Dielec. Str.: 490 V/mil (0.125 in., short term, D149); Dielec. Const.: 3.9 (100 Hz, D150); Vol. Resist.: 5 X 10^{15} ohm-cm (D257); Arc Resist.: 130 s (D495); Dissip. Factor: 0.0064 (100 Hz, D150)
Ref.: Product information

Celanex 3311. Celanese Corp.
Chem. Descrip.: 30% glass reinforced thermoplastic polyester resin
Category/Applications: Thermoplastic resin which complies with U.L. requirements for direct support of live parts, incl. a high volt track rate of < 1.0 in./min; exceptional elec. and mech. properties throughout a wide range of aggressive environments; more ductile, but with a lower heat deflection temp. than Celanex 3310; also available in lubricated grade identified by the suffix -2
Form: Solid
Gen'l. Prop.: Sp. Gr.: 1.65 (D792); M.P.: 417 F; Stability: See Celanex 2012
Mech. Prop. (73 F): Tens. Str.: 19,500 psi (D638); Tens. Elong.: 1–2% (D638), Flex. Stress: 29,000 psi (@ break, D-790); Flex. Mod.: 1.5 X 10^6 psi (D790); Compr. Str.: 22,000 psi (D695); Shear Str.: 9700 psi (D732); Impact Str. (Izod): 1.4 ft lb/in. notched (D256); Hardness (Rockwell): M 94 (D785)

Therm. Prop.: Distort. Temp.: 378 F (@ 264 psi, D648); Sp. Heat: 0.23 Btu/lb/F (C351); Flamm.: 94 V-O (1/32 in., UL94)
Elec. Prop. (73 F): Dielec. Str.: 530 V/mil (0.125 in., short term, D149); Dielec. Const.: 3.8 (100 Hz, D150); Vol. Resist.: 4 × 10^{15} ohm-cm (D257); Arc Resist.: 128 s (D495); Dissip. Factor: 0.0012 (100 Hz, D150)
Ref.: Product information

Celcon GC-25A. Celanese Corp.
Chem. Descrip.: Crystalline acetal copolymer based on trioxane
Category/Applications: Thermoplastic resin reinforced with 25% by wt. of glass fibers together with a coupling agent; provides increased stiffness, tens. strength, and creep resistance; used for windshield wiper pivots, gears, pulleys, and other parts requiring the maximum in stiffness and strength
Form: Free-flowing pellets; Color: Translucent white
Gen'l. Prop.: Sp. Gr.: 1.59 (D792); Sp. Vol.: 17.54 in.^{3}/lb; Density: 0.057 lb/in.^{3}; M.P.: 331 F; Soften. Pt. (Vicat): 324 F (D1525); Stability: Excellent chem. resistance; unaffected by common solvents (acetone, ethyl alcohol, and lacquer solvents @ R.T.); little affected by oils (mineral oil, motor oil, and brake fluids); exceptionally resistant to strong alkalis; attacked by strong mineral acids (sulfuric, nitric, and hydrochloric)
Mech. Prop. (50% R.H., 73 F): Tens. Str.: 16,000 psi (D638); Tens. Elong.: 2–3% (D638); Tens. Mod.: 1.2 × 10^6 psi (D638); Flex. Mod.: 1.05 × 10^6 psi (D790); Shear Str.: 13,000 psi (D732); Impact Str. (Izod): 1.1 ft lb/in. notched (D256)
Therm. Prop.: Distort. Temp.: 322 F (@ 264 psi, D648); Coeff. of Linear Exp.: 2.2 × 10^{-5} (−30 to +30 C, flow direction, D696)
Elec. Prop. (50% R.H., 73 F): Power Factor: 0.003 (0.04 in. sheet, 10^2 Hz, D150); Dielec. Str.: 600 V/.001 in. (0.09 in. sheet, short time, D149); Dielec. Const.: 4.12 (0.04 in. sheet, 100 Hz, D150); Vol. Resist.: 8 ohm-cm (1/8 in. thick, D257); Surface Resist.: 2 × 10^{16} ohm (1/8 in. thick, D257); Arc Resist.: 142 s (1/8 in. thick, D495)
Toxicity/Handling: Releases formaldehyde when heated > melt temp.; use adequate ventilation for heating and processing; keep dry
Std. Pkgs.: 50 lb multiwall bags; 1200 lb boxes
Ref.: Bulletin C1A

Celcon M25-01. Celanese Corp.
Chem. Descrip.: Crystalline acetal copolymer based on trioxane
Category/Applications: Thermoplastic resin with excellent processability in extrusion; used for rod, tube, sheet, slab, wire coating
Form: Free-flowing pellets; Color: Translucent white
Gen'l. Prop.: Sp. Gr.: 1.41 (D792); Sp. Vol.: 19.7 in.^{3}/lb; Density: 0.0507 lb/in.^{3}; M.P.: 329 F; Soften. Pt. (Vicat): 324 F (D1525); Stability: See Celcon GC-25A
Mech. Prop. (50% R.H., 73 F): Tens. Str.: 8800 psi (D638); Tens. Elong.: 75% (D638); Tens. Mod.: 410,000 psi (D638); Flex. Stress: 13,000 psi (@ 5% deformation, D790); Compr. Stress: 4500 psi (@ 1% deflection, D695); Shear Str.: 7700 psi (D732); Impact Str. (Izod): 1.5 ft lb/in. notched (D256); Hardness (Rockwell): M 80 (D785)
Therm. Prop.: Conduct.: 1.6 Btu/ft^2/F/in.; Distort. Temp.: 230 F (@ 264 psi, D648); Coeff. of Linear Exp.: 4.7 × 10^{-5} (−30 to +30 C, flow direction, D696);

Sp. Heat: 0.35 Btu/lb/F; Flamm.: 1.1 in./min (D635)
Elec. Prop. (50% R.H., 73 F): Power Factor: 0.001 (0.04 in. sheet, 100 Hz, D150); Dielec. Str.: 2100 V/.001 in. 0.005 in. film, short time, D149); Dielec. Const.: 3.7 (0.04 in. sheet, 100 Hz, D-150); Vol. Resist.: 1 × 10^{14} ohm-cm (1/8 in. thick, D257); Surf. Resist.: 1.3 × 10^{16} ohm (1/8 in., D257); Arc Resist.: 240 s (burns) (1/8 in., D495)
Toxicity/Handling: See Celcon GC-25A adequate ventilation for heating and processing; keep dry
Std. Pkgs.: 50 lb multiwall bags; 1200 lb boxes
Ref.: Bulletin C1A

Celcon M25-04. Celanese Corp.
Chem. Descrip.: Crystalline acetal copolymer based on trioxane
Category/Applications: Internally lubricated form of Celcon M25-01; good processability for injection molding in easy-to-fill molds; possesses greater toughness and elongation than M90 grades; used for injection molded parts requiring optimum toughness and elongation
Form: Free-flowing pellets; Color: Translucent white
Gen'l., Mech., Therm., and Elec. Prop.: See Celcon M25-01
Toxicity/Handling: See Celcon GC-25A
Std. Pkgs.: 50 lb multiwall bags; 1200 lb boxes
Ref.: Bulletin C1A

Celcon M90-04. Celanese Corp.
Chem. Descrip.: Crystalline acetal copolymer based on trioxane
Category/Applications: Thermoplastic resin with excellent moldability due to high flow characteristics and stability in processing; high surface gloss and good dimensional stability; internally lubricated; used for general injection molding
Form: Free-flowing pellets; Color: Translucent white
Gen'l. Prop.: See Celcon M25-01
Mech. Prop. (50% R.H., 73 F): Tens. Str.: 8800 psi (D638); Tens. Elong.: 20% (D638); Tens. Mod.: 410,000 psi psi (D638); Flex. Stress: 13,000 psi (@ 5% deformation, D790); Flex. Mod.: 375,000 psi (D790); Compr. Stress: 4500 psi (@ 1% deflection, D695); Shear Str.: 7700 psi (D732); Impact Str. (Izod): 1.3 ft lb/in. notched (D256); Hardness (Rockwell): M 80 (D785)
Therm. and Elec. Prop.: See Celcon M25-01
Toxicity/Handling: See Celcon GC-25A
Std. Pkgs.: 50 lb multiwall bags; 1200 lb boxes
Ref.: Bulletin C1A

Celcon M90-07. Celanese Corp.
Category/Applications: Celcon color concentrates are provided in a wide range of standard colors for letdown into natural Celcon via extrusion or screw injection molding machines; used to obtain body colored injection molded or extruded products
Ref.: Bulletin C1A

Celcon M90-08. Celanese Corp.
Chem. Descrip.: Crystalline acetal copolymer based on trioxane
Category/Applications: Ultraviolet-stabilized thermoplastic resin; has good maintenance of physical properties and resistance to chalking in outdoor exposures; used for injection molded seasonal outdoor applications
Form: Free-flowing pellets; Color: Translucent white
Gen'l., Mech., Therm., and Elec. Prop.: See Celcon M90-04
Toxicity/Handling: See Celcon GC-25A
Std. Pkgs.: 50 lb multiwall bags; 1200 lb boxes
Ref.: Bulletin C1A

Celcon M270-04. Celanese Corp.
Chem. Descrip.: Crystalline acetal copolymer based on trioxane
Category/Applications: Internally-lubricated thermoplastic resin with highest flow in Celcon series; superior moldability for hard-to-fill molds with somewhat less toughness than M90; used for high speed injection molding for multicavity parts
Form: Free-flowing pellets; Color: Translucent white
Gen'l. Prop.: See Celcon M25-01
Mech. Prop. (50% R.H., 73 F): Tens. Str.: 8800 psi (D638); Tens. Elong.: 15% (D638); Tens. Mod.: 410,000 psi (D638); Flex. Stress: 13,000 psi (@ 5% deformation, D790); Flex. Mod.: 375,000 psi (D790); Compr. Stress: 4500 psi (@1% deflection, D695); Shear Str.: 7700 psi (D732); Impact Str. (Izod): 1.0 ft lb/in. notched (D256); Hardness (Rockwell): M80 (D785)
Therm. & Elec. Prop.: See Celcon M25-01
Toxicity/Handling: See Celcon GC-25A
Std. Pkgs.: 50 lb multiwall bags; 1200 lb boxes
Ref.: Bulletin C1A

Celcon U10-11. Celanese Corp.
Chem. Descrip.: Crystalline acetal copolymer based on trioxane
Category/Applications: Thermoplastic resin with excellent processability in extrusion blow molding injection blow molding, and extrusion; high melt strength; low odor grade for aerosols; used for aerosols, containers, industrial articles, rod, tube, slab, profiles
Form: Free-flowing pellets; Color: Translucent white
Gen'l. Prop.: Stability: Excellent chem. resistance; unaffected by common solvents (acetone, ethyl alcohol, and lacquer solvents @ R.T.); little affected by oils (mineral oils, motor oil, and brake fluids); exceptionally resistant to strong alkalis; attacked by strong mineral acids (sulfuric, nitric, and hydrochloric)
Toxicity/Handling: See Celcon GC-25A
Std. Pkgs.: 50 lb multiwall bags; 1200 lb boxes
Ref.: Bulletin C1A

Cellidor A. Bayer AG
Chem. Descrip.: Organic cellulose acetate ester
Category/Applications: Thermoplastic material for use in automotive industry, tool industry, building and furniture industries, domestic appliances, lighting, electrical, radio and TV industries, optical and photographic industries, stationery, toy industry, toiletries, packaging
Form: Cylindrical granules, diced granules; Color: Avail. in wide range of transparent, translucent, and opaque colors
Gen'l. Prop.: Sol.: Sol. in ethyl acetate, methylene chloride, acetone, orthochlorphenol; Density: 1.16–1.32 mg/m^3 (unfilled); Stability: Resistant to water, carbon tetrachloride, trichloroethylene, perchloroethylene, benzene, xylene, petrol, motor fuel mixture, paraffin, linseed oil, turpentine, ether
Ref.: Pamphlet KL 40001 e

Cellidor B 500-05 (formerly BH). Bayer AG
Chem. Descrip.: Organic cellulose acetate butyrate ester
Category/Applications: See Cellidor A
Form: See Cellidor A
Gen'l. Prop.: Sol.: See Cellidor A; Density: 1.21 g/cc (DIN53479/ISO/R1183); Soften. Pt. (Vicat): 95 C (B, DIN53460/ISO306); Stability: Resistant to water, petrol, paraffin, linseed oil; resistant with swelling to carbon tetrachloride,

perchloroethylene, motor fuel mixture, turpentine, ether, formalin; Ref. Index: 1.48 (DIN53491)
All properties @ 23 C, 50% r.h. after 10 days storage:
Mech. Prop.: Tens. Mod.: 1.9 GPa (DIN-53457); Flex. Mod.: 56 MPa (DIN-53452); Impact Str. (Izod): 95 J/m (D-256/ISO/R180; Hardness (Rockwell): R 108 (D785)
Therm. Prop.: Conduct.: 0.20 W/Km (VDE0304); Distort. Temp.: 85 C (@ 1.81 MPa, DIN53461/ISO/R75); Coeff. of Linear Exp.: 119 10^{-6} K^{-1} (VDE-0304); Sp. Heat: 1.3–1.7 kJ/kgK; Flamm.: 30 mm/min (D635-72)
Elec. Prop.: Dissip. Factor: 0.007 (dry, 50 Hz, VDE0303/4, DIN53483); Dielec. Str.: 35 kV/mm (dry, 50 Hz, 0.5 kV/s, VDE0303/2, DIN53481); Dielec. Const.: 4.0 (dry, 50 Hz, VDE0303/4, DIN-53483); Vol. Resist.: 10^{16} ohm-cm (dry, VDE0303/3, DIN 53482); Surf. Resist.: 10^{14} ohm (dry, VDE0303/3, DIN-53482)
Ref.: Pamphlet KL 40001 e

Cellidor B 500-10 (formerly BM). Bayer AG
Chem. Descrip.: Organic cellulose acetate butyrate ester
Category/Applications & Form: See Cellidor A
Gen'l. Prop.: Sol.: See Cellidor A; Density: 1.19 g/cc (DIN53479/ISO/R1183); Soften. Pt. (Vicat): 80 C (B, DIN53460/ ISO306); Stability: See Cellidor B 500-05; Ref. Index: 1.48 (DIN53491)
All properties @ 23 C, 50% r.h. after 10 days storage:
Mech. Prop.: Tens. Mod.: 1.7 GPa (DIN-53457); Flex. Mod.: 45 MPa (DIN-53452); Impact Str. (Izod): 370 J/m (D-256/ISO/R180); Hardness (Rockwell): R 97 (D785)
Therm. Prop.: Conduct.: 0.21 W/Km (VDE0304); Distort. Temp.: 69 C (@ 1.81 MPa, DIN53461/ISO/R75); Coeff. of Linear Exp.: 129 10^{-6} K^{-1} (VDE-0304); Sp. Heat: 1.3–1.7 kJ/kgK; Flamm.: 30 mm/min (D635-72)
Elec. Prop.: Dissip. Factor: 0.006 (dry, 50 Hz, VDE0303/4, DIN53483); Dielec. Str.: 35 kV/mm (dry, 50 Hz, 0.5 kV/s, VDE0303/2, DIN53481); Dielec. Const.: 4.0 (dry, 50 Hz, VDE0303/4, DIN-53483); Vol. Resist.: 10^{15} ohm-cm (dry, VDE0303/3, DIN 53482); Surf. Resist.: 10^{14} ohm (dry, VDE0303/3, DIN-53482)
Ref.: Pamphlet KL 40001 e

Cellidor B 500-15 (formerly BW). Bayer AG
Chem. Descrip.: Organic cellulose acetate butyrate ester
Category/Applications & Form: See Cellidor A
Gen'l. Prop.: Sol.: See Cellidor A; Density: 1.18 g/cc (DIN53479/ISO/R1183); Soften. Pt. (Vicat): 65 C (B, DIN53460/ ISO306); Stability: See Cellidor B 500-05; Ref. Index: 1.48 (DIN53491)
All properties @ 23 C, 50% r.h. after 10 days storage:
Mech. Prop.: Tens. Mod.: 1.2 GPa (DIN-53457); Flex. Mod.: 33 MPa (DIN-53452); Impact Str. (Izod): 540 J/m (D-256/ISO/R180); Hardness (Rockwell): R 72 (D785)
Therm. Prop.: Conduct.: 0.21 W/Km (VDE0304); Distort. Temp.: 54 C (@ 1.81 MPa, DIN53461/ISO/R75); Coeff. of Linear Exp.: 136 10^{-6} K^{-1} (VDE-0304); Sp. Heat: 1.3–1.7 kJ/kgK; Flamm.: 30 mm/min (D635-72)
Elec. Prop.: Dissip. Factor: 0.006 (dry, 50 Hz, VDE0303/4, DIN53483); Dielec. Str.: 35 kV/mm (dry, 50 Hz, 0.5 kV/s, VDE0303/2, DIN53481); Dielec. Const.: 4.1 (dry, 50 Hz, VDE0303/4, DIN-53483); Vol. Resist.: 10^{15} ohm-cm (dry,

VDE0303/3, DIN53482); Surf. Resist.: 10^{13} ohm (dry, VDE0303/3, DIN53482)
Ref.: Pamphlet KL 40001 e

Cellidor B 500-20 (formerly BWW). Bayer AG
Chem. Descrip.: Organic cellulose acetate butyrate ester
Category/Applications & Form: See Cellidor A
Gen'l. Prop.: Sol.: See Cellidor A; Density: 1.17 g/cc (DIN53479/ISO/R1183); Soften. Pt. (Vicat): 50 C (B, DIN53460/ ISO306); Stability: See Cellidor B 500-05; Ref. Index: 1.47 (DIN53491)
All properties @ 23 C, 50% r.h. after 10 days storage:
Mech. Prop.: Tens. Mod.: 0.8 GPa (DIN-53457); Flex. Mod.: 24 MPa (DIN-53452); Impact Str. (Izod): 620 J/m (D256/ISO/R180); Hardness (Rockwell): R 42 (D785)
Therm. Prop.: Conduct.: 0.22 W/Km (VDE0304); Distort. Temp.: 37 C (@ 1.81 MPa, DIN53461/ISO/R75); Coeff. of Linear Exp.: 148 10^{-6} K^{-1} (VDE-0304); Sp. Heat: 1.3–1.7 kJ/kgK; Flamm.: 30 mm/min (D635-72)
Elec. Prop.: Dissip. Factor: 0.005 (dry, 50 Hz, VDE0303/4, DIN53483); Dielec. Str.: 34.5 kV/mm (dry, 50 Hz, 0.5 kV/s, VDE0303/2, DIN53481); Dielec. Const.: 4.1 (dry, 50 Hz, VDE0303/4, DIN-53483); Vol. Resist.: 10^{14} ohm-cm (dry, VDE0303/3, DIN53482); Surf. Resist.: 10^{13} ohm (dry, VDE0303/3, DIN53482)
Ref.: Pamphlet KL 40001 e

Cellidor B 700-04 (formerly BHH). Bayer AG
Chem. Descrip.: Organic cellulose acetate butyrate ester
Category/Applications & Form: See Cellidor A
Gen'l. Prop.: Sol.: See Cellidor A; Density: 1.22 g/cc (DIN53479/ISO/R1183); Soften. Pt. (Vicat): 104 C (B, DIN-53460/ISO306); Stability: See Cellidor B 500-05; Ref. Index: 1.48 (DIN53491)
All properties @ 23 C, 50% r.h. after 10 days storage:
Mech. Prop.: Tens. Mod.: 2.1 GPa (DIN-53457); Flex. Mod.: 65 MPa (DIN-53452); Impact Str. (Izod): 50 J/m (D-256/ISO/R180); Hardness (Rockwell): R 112 (D785)
Therm. Prop.: Conduct.: 0.20 W/Km (VDE0304); Distort. Temp.: 92 C (@ 1.81 MPa, DIN53461/ISO/R75); Coeff. of Linear Exp.: 115 10^{-6} K^{-1} (VDE-0304); Sp. Heat: 1.3–1.7 kJ/kgK; Flamm.: 30 mm/min (D635-72)
Elec. Prop.: Dissip. Factor: 0.007 (dry, 50 Hz, VDE0303/4, DIN53483); Dielec. Str.: 38 kV/mm (dry, 50 Hz, 0.5 kV/s, VDE0303/2, DIN53481); Dielec. Const.: 3.6 (dry, 50 Hz, VDE0303/4, DIN-53483); Vol. Resist.: 10^{16} ohm-cm (dry, VDE0303/3, DIN53482); Surf. Resist.: 10^{15} ohm (dry, VDE0303/3, DIN53482)
Ref.: Pamphlet KL 40001 e

Cellidor CP 300-08 (formerly CPHH). Bayer AG
Chem. Descrip.: Organic thermoplastic cellulose propionate ester
Category/Applications & Form: See Cellidor A
Gen'l. Prop.: Sol.: Sol. in ethyl acetate, methylene chloride, acetone, orthochlorphenol; Density: 1.22 g/cc (DIN-53479/ISO/R1183); Soften. Pt. (Vicat): 105 C (B, DIN53460/ISO306); Stability: Resistant to water, petrol, mineral oil (paraffin), linseed oil; resistant with swelling to carbon tetrachloride, perchloroethylene, motor fuel mixture (high octane), turpentine, ether, formalin; Ref. Index: 1.47 (DIN53491)
All properties @ 23 C, 50% r.h. after 10 days storage:
Mech. Prop.: Tens. Mod.: 2.4 GPa (DIN-

53457); Flex. Mod.: 70 MPa (DIN-53452); Impact Str. (Izod): 60 J/m (D-256/ISO/R180); Hardness (Rockwell): R 117 (D785)
Therm. Prop.: Conduct.: 0.21 W/Km (VDE0304); Distort. Temp.: 96 C (@ 1.81 MPa, DIN53461/ISO/R75); Coeff. of Linear Exp.: 120 10^{-6} K^{-1} (VDE-0304); Sp. Heat: 1.3–1.7 kJ/kgK; Flamm.: 30 mm/min (D635-72)
Elec. Prop.: Dissip. Factor: 0.005 (dry, 50 Hz, VDE0303/4, DIN53483); Dielec. Str.: 35.5 kV/mm (dry, 50 Hz, 0.5 kV/s, VDE0303/2, DIN53481); Dielec. Const.: 4.1 (dry, 50 Hz, VDE0303/4, DIN-53483); Vol. Resist.: 10^{16} ohm-cm (dry, VDE0303/3, DIN53482); Surf. Resist.: 10^{13} ohm (dry, VDE0303/3, DIN-53482)
Ref.: Pamphlet KL 4001 e

Cellidor CP 400-05 (formerly CPH). Bayer AG
Chem. Descrip.: Organic cellulose propionate ester
Category/Applications & Form: See Cellidor A
Gen'l. Prop.: Sol. & Stability: See Cellidor CP 300-08; Density: 1.23 g/cc (DIN-53479/ISO/R1183); Soften. Pt. (Vicat): 100 C (B, DIN53460/ISO306); Ref. Index: 1.48 (DIN53491)
All properties @ 23 C, 50% r.h. after 10 days storage:
Mech. Prop.: Tens. Mod.: 2.1 GPa (DIN-53457); Flex. Mod.: 65 MPa (DIN-53452); Impact Str. (Izod): 70 J/m (D-256/ISO/R180); Hardness (Rockwell): R 112 (D785)
Therm. Prop.: Conduct.: 0.20 W/Km (VDE0304); Distort. Temp.: 93 C (@ 1.81 MPa, DIN53461/ISO/R75); Coeff. of Linear Exp.: 114 10^{-6} K^{-1} (VDE-0304); Sp. Heat: 1.7 kJ/kgK; Flamm.: 30 mm/min (D635-72)
Elec. Prop.: Dissip. Factor: 0.005 (dry, 50 Hz, VDE0303/4, DIN53483); Dielec. Str.: 35.5 kV/mm (dry, 50 Hz, 0.5 kV/s, VDE0303/2, DIN53481); Dielec. Const.: 4.0 (dry, 50 Hz, VDE0303/4, DIN-53483); Vol. Resist.: 10^{16} ohm-cm (dry, VDE0303/3, DIN53482); Surf. Resist.: 10^{14} ohm (dry, VDE0303/3, DIN53482)
Ref.: Pamphlet KL 40001 e

Cellidor CP 400-10 (formerly CPM). Bayer AG
Chem. Descrip.: Organic cellulose propionate ester
Category/Applications & Form: See Cellidor A
Gen'l. Prop.: Sol. & Stability: See Cellidor CP 300-08; Density: 1.22 g/cc (DIN-53479/ISO/R1183); Soften. Pt. (Vicat): 90 C (B, DIN53460/ISO306); Ref. Index: 1.48 (DIN53491)
All properties @ 23 C, 50% r.h. after 10 days storage:
Mech. Prop.: Tens. Mod.: 1.9 GPa (DIN-53457); Flex. Mod.: 53 MPa (DIN-53452); Impact Str. (Izod): 350 J/m (D-256/ISO/R180); Hardness (Rockwell): R 105 (D785)
Therm. Prop.: Conduct.: 0.21 W/Km (VDE0304); Distort. Temp.: 77 C (@ 1.81 MPa, DIN53461/ISO/R75); Coeff. of Linear Exp.: 123 10^{-6} K^{-1} (VDE-0304); Sp. Heat: 1.7 kJ/kgK; Flamm.: 30 mm/min (D635-72)
Elec. Prop.: Dissip. Factor: 0.005 (dry, 50 Hz, VDE0303/4, DIN53483); Dielec. Str.: 35.5 kV/mm (dry, 50 Hz, 0.5 kV/s, VDE0303/2, DIN53481); Dielec. Const.: 4.1 (dry, 50 Hz, VDE0303/4, DIN-53483); Vol. Resist.: 10^{16} ohm-cm (dry, VDE0303/3, DIN53482); Surf. Resist.: 10^{14} ohm (dry, VDE0303/3, DIN53482)
Ref.: Pamphlet KL 40001 e

Cellidor CP 400-15 (formerly CPW). Bayer AG

Chem. Descrip.: Organic cellulose propionate ester

Category/Applications & Form: See Cellidor A

Gen'l. Prop.: Sol. & Stability: See Cellidor CP 300-08; Density: 1.21 g/cc (DIN-53479/ISO/R1183); Soften. Pt. (Vicat): 72 C (B, DIN53460/ISO306); Ref. Index: 1.47 (DIN53491)
All properties @ 23 C, 50% r.h. after 10 days storage:

Mech. Prop.: Tens. Mod.: 1.4 GPa (DIN-53457); Flex. Mod.: 40 MPa (DIN-53452); Impact Str. (Izod): 500 J/m (D-256/ISO/R180); Hardness (Rockwell): R 85 (D785)

Therm. Prop.: Conduct.: 0.21 W/Km (VDE0304); Distort. Temp.: 59 C (@ 1.81 MPa, DIN53461/ISO/R75); Coeff. of Linear Exp.: 131 10^{-6} K^{-1} (VDE-0304); Sp. Heat: 1.7 kJ/kgK; Flamm.: 30 mm/min (D635-72)

Elec. Prop.: Dissip. Factor: 0.005 (dry, 50 Hz, VDE0303/4, DIN53483); Dielec. Str.: 35 kV/mm (dry, 50 Hz, 0.5 kV/s, VDE0303/2, DIN53481); Dielec. Const.: 4.2 (dry, 50 Hz, VDE0303/4, DIN-53483); Vol. Resist.: 10^{15} ohm-cm (dry, VDE0303/3, DIN53482); Surf. Resist.: 10^{14} ohm (dry, VDE0303/3, DIN53482)
Ref.: Pamphlet KL 40001 e

Cellidor CP 400-20 (formerly CPWW). Bayer AG
Chem. Descrip.: Organic cellulose propionate ester
Category/Applications & Form: See Cellidor A
Gen'l. Prop.: Sol. & Stability: See Cellidor CP 300-08; Density: 1.19 g/cc (DIN-53479/ISO/R1183); Soften. Pt. (Vicat): 55 C (B, DIN53460/ISO306); Ref. Index: 1.47 (DIN53491)
All properties @ 23 C, 50% r.h. after 10 days storage:

Mech. Prop.: Tens. Mod.: 1 GPa (DIN-53457); Flex. Mod.: 30 MPa (DIN-53452); Impact Str. (Izod): 560 J/m (D-256/ISO/R180); Hardness (Rockwell): R 55 (D785)

Therm. Prop.: Conduct.: 0.22 W/Km (VDE0304); Distort. Temp.: 42 C (@ 1.81 MPa, DIN53461/ISO/R75); Coeff. of Linear Exp.: 145 10^{-6} K^{-1} (VDE-0304); Sp. Heat: 1.7 kJ/kgK; Flamm.: 30 mm/min (D635-72)

Elec. Prop.: Dissip. Factor: 0.005 (dry, 50 Hz, VDE0303/4, DIN53483); Dielec. Str.: 34.5 kV/mm (dry, 50 Hz, 0.5 kV/s, VDE0303/2, DIN53481); Dielec. Const.: 4.2 (dry, 50 Hz, VDE0303/4, DIN-53483); Vol. Resist.: 10^{15} ohm-cm (dry, VDE0303/3, DIN53482); Surf. Resist.: 10^{14} ohm (dry, VDE0303/3, DIN53482)
Ref.: Pamphlet KL 40001 e

Cellidor CP 790-19 (formerly Cp-Ro). Bayer AG
Chem. Descrip.: Organic cellulose propionate ester
Category/Applications: Rotational molding grade
Form: Granules
Gen'l. Prop.: Sol. & Stability: See Cellidor CP 300-08; Density: 1.20 g/cc (DIN-53479/ISO/R1183); Soften. Pt. (Vicat): 52 C (B, DIN53460/ISO306)
All properties @ 23 C, 50% r.h. after 10 days storage:
Mech. Prop.: Flex. Mod.: 26 MPa (DIN-53452); Hardness (Ball Indentation): 26 N/mm^2 (DIN53456 tentative)
Therm. Prop.: Distort. Temp.: 48 C (@ 1.81 MPa, DIN53461/ISO/R75)
Ref.: Pamphlet KL 40001 e

Cellidor S 100-25 (formerly AM). Bayer AG

Chem. Descrip.: Organic cellulose acetate ester

Category/Applications & Form: See Cellidor A

Gen'l. Prop.: Sol. & Stability: See Cellidor A; Density: 1.28 g/cc (DIN53479/ISO/R1183); Soften. Pt. (Vicat): 80 C (B, DIN53460/ISO306); Ref. Index: 1.50 (DIN53491)

All properties @ 23 C, 50% r.h. after 10 days storage:

Mech. Prop.: Tens. Mod.: 2.2 GPa (DIN-53457); Flex. Mod.: 56 MPa (DIN-53452); Impact Str. (Izod): 110 J/m (D-256/ISO/R180); Hardness (Rockwell): R 105 (D785)

Therm. Prop.: Conduct.: 0.21 W/Km (VDE0304); Distort. Temp.: 64 C (@ 1.81 MPa, DIN53461/ISO/R75); Coeff. of Linear Exp.: 96 10^{-6} K^{-1} (VDE-0304); Sp. Heat: 1.3-1.7 kJ/kgK; Flamm.: 30 mm/min (D635-72)

Elec. Prop.: Dissip. Factor: 0.009 (dry, 50 Hz, VDE0303/4, DIN53483); Dielec. Str.: 31.5 kV/mm (dry, 50 Hz, 0.5 kV/s, VDE0303/2, DIN53481); Dielec. Const.: 5.1 (dry, 50 Hz, VDE0303/4, DIN-53483); Vol. Resist.: 10^{15} ohm-cm (dry, VDE0303/3, DIN53482); Surf. Resist.: 10^{14} ohm (dry, VDE0303/3, DIN53482)

Ref.: Pamphlet KL 40001 e

Cellidor S 100-30 (formerly AW). Bayer AG

Chem. Descrip.: Organic cellulose acetate ester

Category/Applications & Form: See Cellidor A

Gen'l. Prop.: Sol. & Stability: See Cellidor A; Density: 1.27 g/cc (DIN53479/ISO/R1183); Soften. Pt. (Vicat): 67 C (B, DIN53460/ISO306); Ref. Index: 1.50 (DIN53491)

All properties @ 23 C, 50% r.h. after 10 days storage:

Mech. Prop.: Tens. Mod.: 1.8 GPa (DIN-53457); Flex. Mod.: 42 MPa (DIN-53452); Impact Str. (Izod): 250 J/m (D-256/ISO/R180); Hardness (Rockwell): R 90 (D785)

Therm. Prop.: Conduct.: 0.22 W/Km (VDE0304); Distort. Temp.: 52 C (@ 181 MPa, DIN53461/ISO/R75); Coeff. of Linear Exp.: 102 10^{-6} K^{-1} (VDE-0304); Sp. Heat: 1.3-1.7 kJ/kgK; Flamm.: 30 mm/min (D635-72)

Elec. Prop.: Dissip. Factor: 0.008 (dry, 50 Hz, VDE0303/4, DIN53483); Dielec. Str.: 30.5 kV/mm (dry, 50 Hz, 0.5 kV/s, VDE0303/2, DIN53481); Dielec. Const.: 5.5 (dry, 50 Hz, VDE0303/4, DIN-53483); Vol. Resist.: 10^{15} ohm-cm (dry, VDE0303/3, DIN53482); Surf. Resist.: 10^{13} ohm (dry, VDE0303/3, DIN53482)

Ref.: Pamphlet KL 40001 e

Cellidor S 100-33 (formerly AWW). Bayer AG

Chem. Descrip.: Organic cellulose acetate ester

Category/Applications & Form: See Cellidor A

Gen'l. Prop.: Sol. & Stability: See Cellidor A; Density: 1.26 g/cc (DIN53479/ISO/R1183); Soften. Pt. (Vicat): 60 C (B, DIN53460/ISO306); Ref. Index: 1.51 (DIN53491)

All properties @ 23 C, 50% r.h. after 10 days storage:

Mech. Prop.: Tens. Mod.: 1.4 GPa (DIN-53457); Flex. Mod.: 35 MPa (DIN-53452); Impact Str. (Izod): 300 J/m (D-256/ISO/R180); Hardness (Rockwell): R 73 (D785)

Therm. Prop.: Conduct.: 0.22 W/Km (VDE0304); Distort. Temp.: 43 C (@ 1.81 MPa, DIN53461/ISO/R75); Coeff. of Linear Exp.: 108 10^{-6} K^{-1} (VDE-

0304); Sp. Heat: 1.3–1.7 kJ/kgK; Flamm.: 30 mm/min (D635-72)
Elec. Prop.: Dissip. Factor: 0.007 (dry, 50 Hz, VDE0303/4, DIN53483); Dielec. Str.: 28.5 kV/mm (dry, 50 Hz, 0.5 kV/s, VDE0303/2, DIN53481); Dielec. Const.: 5.6 (dry, 50 Hz, VDE0303/4, DIN-53483); Vol. Resist.: 10^{14} ohm-cm (dry, VDE0303/3, DIN53482); Surf. Resist.: 10^{13} ohm (dry, VDE0303/3, DIN53482)
Ref.: Pamphlet KL 40001 e

Cellidor S 200-17 (formerly SH-17). Bayer AG
Chem. Descrip.: Organic cellulose acetate ester
Category/Applications & Form: See Cellidor A
Gen'l. Prop.: Sol. & Stability: See Cellidor A; Density: 1.29 g/cc (DIN53479/ISO/R1183); Soften. Pt. (Vicat): 100 C (B, DIN53460/ISO306); Ref. Index: 1.49 (DIN53491)
All properties @ 23 C, 50% r.h. after 10 days storage:
Mech. Prop.: Tens. Mod.: 3.0 GPa (DIN-53457); Flex. Mod.: 75 MPa (DIN-53452); Impact Str. (Izod): 50 J/m (D-256/ISO/R180); Hardness (Rockwell): R 120 (D785)
Therm. Prop.: Conduct.: 0.21 W/Km (VDE0304); Distort. Temp.: 82 C (@ 1.81 MPa, DIN53461/ISO/R75); Coeff. of Linear Exp.: 100 10^{-6} K^{-1} (VDE-0304); Sp. Heat: 1.3–1.7 kJ/kgK; Flamm.: 30 mm/min (D635-72)
Elec. Prop.: Dissip. Factor: 0.013 (dry, 50 Hz, VDE0303/4, DIN53483); Dielec. Str.: 33 kV/mm (dry, 50 Hz, 0.5 kV/s, VDE0303/2, DIN53481); Dielec. Const.: 5.1 (dry, 50 Hz, VDE0303/4, DIN-53483); Vol. Resist.: 10^{15} ohm-cm (dry, VDE0303/3, DIN53482); Surf. Resist.: 10^{14} ohm (dry, VDE0303/3, DIN53482)
Ref.: Pamphlet KL 40001 e

Cellidor S 200-22 (formerly SM). Bayer AG
Chem. Descrip.: Organic cellulose acetate ester
Category/Applications & Form: See Cellidor A
Gen'l. Prop.: Sol. & Stability: See Cellidor A; Density: 1.28 g/cc (DIN53479/ISO/R1183); Soften. Pt. (Vicat): 88 C (B, DIN53460/ISO306); Ref. Index: 1.49 (DIN53491)
All properties @ 23 C, 50% r.h. after 10 days storage:
Mech. Prop.: Tens. Mod.: 2.7 GPa (DIN-53457); Flex. Mod.: 65 MPa (DIN-53452); Impact Str. (Izod): 90 J/m (D-256/ISO/R180); Hardness (Rockwell): R 110 (D785)
Therm. Prop.: Conduct.: 0.21 W/Km (VDE0304); Distort. Temp.: 73 C (@ 1.81 MPa, DIN53461/ISO/R75); Coeff. of Linear Exp.: 105 10^{-6} K^{-1} (VDE-0304); Sp. Heat: 1.3–1.7 kJ/kgK; Flamm.: 30 mm/min (D635-72)
Elec. Prop.: Dissip. Factor: 0.009 (dry, 50 Hz, VDE0303/4, DIN53483); Dielec. Str.: 31.5 kV/mm (dry, 50 Hz, 0.5 kV/s, VDE0303/2, DIN53481); Dielec. Const.: 5.1 (dry, 50 Hz, VDE0303/4, DIN-53483); Vol. Resist.: 10^{15} ohm-cm (dry, VDE0303/3, DIN53482); Surf. Resist.: 10^{14} ohm (dry, VDE0303/3, DIN53482)
Ref.: Pamphlet KL 40001 e

Cellidor S 200-27 (formerly SW). Bayer AG
Chem. Descrip.: Organic cellulose acetate ester
Category/Applications & Form: See Cellidor A
Gen'l. Prop.: Sol. & Stability: See Cellidor A; Density: 1.27 g/cc (DIN53479/ISO/R1183); Soften. Pt. (Vicat): 75 C (B, DIN53460/ISO306); Ref. Index: 1.49 (DIN53491)

All properties @ 23 C, 50% r.h. after 10 days storage:
Mech. Prop.: Tens. Mod.: 2.0 GPa (DIN-53457); Flex. Mod.: 50 MPa (DIN-53452); Impact Str. (Izod): 200 J/m (D-256/ISO/R180); Hardness (Rockwell): R 100 (D785)
Therm. Prop.: Conduct.: 0.22 W/Km (VDE0304); Distort. Temp.: 59 C (@ 1.81 MPa, DIN53461/ISO/R75); Coeff. of Linear Exp.: 115 10^{-6} K^{-1} (VDE-0304); Sp. Heat: 1.3–1.7 kJ/kgK; Flamm.: 30 mm/min (D635-72)
Elec. Prop.: Dissip. Factor: 0.008 (dry, 50 Hz, VDE0303/4, DIN53483); Dielec. Str.: 30.5 kV/mm (dry, 50 Hz, 0.5 kV/s, VDE0303/2, DIN53481); Dielec. Const.: 5.5 (dry, 50 Hz, VDE0303/4, DIN-53483); Vol. Resist.: 10^{15} ohm-cm (dry, VDE0303/3, DIN53482); Surf. Resist.: 10^{13} ohm (dry, VDE0303/3, DIN53482)
Ref.: Pamphlet KL 40001 e

Cellidor S 200-32 (formerly SWW). Bayer AG
Chem. Descrip.: Organic cellulose acetate ester
Category/Applications & Form: See Cellidor A
Gen'l. Prop.: Sol. & Stability: See Cellidor A; Density: 1.26 g/cc (DIN53479/ISO/R1183); Soften. Pt. (Vicat): 63 C (B, DIN53460/ISO306); Ref. Index: 1.49 (DIN53491)
All properties @ 23 C, 50% r.h. after 10 days storage:
Mech. Prop.: Tens. Mod.: 1.5 GPa (DIN-53457); Flex. Mod.: 37 MPa (DIN-53452); Impact Str. (Izod): 260 J/m (D-256/ISO/R180); Hardness (Rockwell): R 77 (D785)
Therm. Prop.: Conduct.: 0.22 W/Km (VDE0304); Distort. Temp.: 47 C (@ 1.81 MPa, DIN53461/ISO/R75); Coeff. of Linear Exp.: 120 10^{-6} K^{-1} (VDE-0304); Sp. Heat: 1.3–1.7 kJ/kgK; Flamm.: 30 mm/min (D635-72)
Elec. Prop.: Dissip. Factor: 0.007 (dry, 50 Hz, VDE0303/4, DIN53483); Dielec. Str.: 28.5 kV/mm (dry, 50 Hz, 0.5 kV/s, VDE0303/2, DIN53481); Dielec. Const.: 5.6 (dry, 50 Hz, VDE0303/4, DIN-53483); Vol. Resist.: 10^{14} ohm-cm (dry, VDE0303/3, DIN53482); Surf. Resist.: 10^{13} ohm (dry, VDE0303/3, DIN53482)
Ref.: Pamphlet KL 40001 e

Cellolyn 21. Hercules Inc.
Chem. Descrip.: Phthalate ester of technical hydroabietyl alcohol
Category/Applications: Low-softening point resin with good solubility, pigment-wetting characteristics, and wide compatibility; used in hot-melt and pressure sensitive adhesives, specialty nitrocellulose lacquers, printing inks
Form: Solid; Color: Pale
Gen'l. Prop.: Sol.: Sol. in esters, ketones, higher alcohols, coal tar, and petroleum hydrocarbons; insol. in water, methanol, ethanol, and ethylene glycol; Density: 1.05 kg/l; Soften. Pt. (Hercules Drop): 63 C; Acid No.: 8; Sapon. No.: 129; Ref. Index: 1.527 (20 C); Stability: Good color stability
Std. Pkgs.: 200-kg net light-gauge metal drums
Ref.: Data sheet 717-7

Cellolyn 21-80H. Hercules Inc.
Chem. Descrip.: 80% solution of Cellolyn 21 in hexane
Category/Applications: See Cellolyn 21
Form: Liquid; Color (Gardner): 5 max (75% solids in toluene)
Gen'l. Prop.: Sol.: See Cellolyn 21; Visc. (Gardner-Holdt): G-M; Soften. Pt. (Hercules Drop): 60–70 C (solids only); Acid No.: 10 max. (solids only)

Std. Pkgs.: Heavy-gauge, ICC 17E-type steel drums
Ref.: Data sheet 7275-1

Cellolyn 95-80T. Hercules Inc.
Chem. Descrip.: Balsamic, oil-free alkyd resin derived from technical hydroabietyl alcohol
Category/Applications: Resin solution used in the manufacture of pale, flexible, highly color-stable wood lacquers; in alcohol-resistant wood finishes; as an extender for nondrying oil alkyds in pigmented lacquers, paper lacquers, and printing inks; and in grinding vehicles for lacquers
Form: Liquid; Color (Gardner): 2
Composition: 80% solids in toluene/isopropanol (16.2/3.8)
Gen'l. Prop.: Sol.: Sol. in all ordinary lacquer solvent and diluent combinations; Sp. Gr.: 1.02; Density: 8.5 lb/gal; Visc. (Gardner-Holdt): S; Acid No.: 9; Flash Pt.: 27 C (COC)
Std. Pkgs.: Tank cars, tank trucks, and 208-l (200-kg net) steel drums
Ref.: Data sheet 718-5

Cellolyn 98-80T. Hercules Inc.
Chem. Descrip.: Oil-free plasticizing alkyd resin solution
Category/Applications: Used in the manufacture of printing inks for flexible substrates; as the sole resin component in both sealer and topcoat lacquer finishes; general applications where nondrying alkyds are employed
Form: Liquid; Color (Gardner): 2
Composition: 80% solids in toluene/isopropanol (15.8/4.2)
Gen'l. Prop.: Sol.: Sol. in all ordinary lacquer solvent and diluent combinations; Sp. Gr.: 1.06; Density: 8.8 lb/gal; Visc. (Gardner-Holdt): Z; Acid No.: 18; Flash Pt.: 27 C (COC)
Ref.: Data sheet 7113-3

Cellolyn 102. Hercules Inc.
Chem. Descrip.: Modified pentaerythritol ester of rosin
Category/Applications: Thermoplastic resin with wide solubility, quick solvent release, and fully compatible with nitrocellulose, ethyl hydroxyethyl cellulose, and N-type ethyl cellulose; used in lacquers, gravure and screen-process ink vehicles, short oil blending varnishes, and cold-cut varnishes for modifying architectural enamels
Form: Flakes; Color (USDA Rosin Scale): WG
Gen'l. Prop.: Sol.: Sol. in aromatic hydrocarbons, isopropanol, high-KB aliphatic hydrocarbons, esters, and ketones; Density: 1.123 kg/l; Visc. (Gardner-Holdt): J (60% in toluene); Soften. Pt. (Hercules Drop): 134 C; Acid No.: 36; Stability: Prone to gradual oxidation resulting in darkening or having an adverse effect on solubility in organic solvents; Ref. Index: 1.543 (20 C)
Std. Pkgs.: 22.7-kg net multiwall paper bags
Ref.: Data sheet 719-6

Cellolyn 104. Hercules Inc.
Chem. Descrip.: Highly modified pentaerythritol ester of rosin
Category/Applications: Internally plasticized synthetic resin for coatings
Gen'l. Prop.: Soften. Pt. (Drop): 101 C; Acid No.: 30
Ref.: Products catalogue

Cellulose Gum. Hercules Inc.
Chem. Descrip.: Sodium salt of carboxymethylcellulose
Category/Applications: Anionic water-soluble high-purity polymer avail. in a number of grades and viscosities; used as a thickener, binder, stabilizer, suspending agent, and film-former in textile processing, emulsion-type coatings,

water-based paints, lithography, paper and paperboard manufacture, adhesives, insecticides, cosmetics, foods, pharmaceutical, laundry, petroleum, and detergent applications
Form: Powder; Color: White to light cream
Composition: 99.5% min. purity
Gen'l. Prop.: Sol.: Sol. in hot or cold water; has good tolerance for ethanol and acetone; Sp.Gr.: 1.0068 (2% sol'n.); Visc.: Range from < 20 cps in 2% aq. sol'ns. to 5000 cps in 1% aq. sol'ns.; Density (Bulk): 0.75 g/ml; Stability: Films of cellulose gum are unaffected by oils, greases, and organic solvents incl. alcohols, ketones, and hydrocarbons; pH: 7.5 (2% sol'n.)
Toxicity/Handling: Store in a cool, dry place
Std. Pkgs.: 50-lb multiwall bags
Ref.: Bulletin No. 405-6

Cereclor 42. ICI United States Inc.
Chem. Descrip.: Chlorinated paraffin
Category/Applications: Used as a plasticizer extender in polyvinyl chloride; as an extreme pressure additive in gear oils and metal working lubricants; as a plasticizer for chemical resistant paints; and as a flame retardant additive for plastics, rubbers, and textiles; Cereclor 42 is esp. suited for adhesives, sealants, and paints plasticizers, flame retardant applications, and plasticizer in Neoprene compounding
Form: Clear, viscous liquid; Color (Gardner): 1
Composition: 42% chlorine content
Gen'l. Prop.: Sol.: Miscible with benzene, esters, ketones, ethers, cyclohexanol, petroleum ether, trichloroethylene, and similar solvents, and with veg. oils such as linseed and castor oils; Density: 1.17 g/ml; Visc.: 24 poise; Stability: Resistant to acids and alkalis; stable at ordinary temps.; unaffected by sunlight except on prolonged exposure when slight darkening takes place
Ref.: Leaflet 761-25R

Cereclor 42P. ICI United States Inc.
Chem. Descrip.: Chlorinated paraffin
Category/Applications: See Cereclor 42; 42B is used esp. as PVC plasticizer and in flame retardant applications
Form: Clear, viscous liquid; Color (Gardner): 4
Composition: 42% chlorine content
Gen'l. Prop.: Sol.: See Cereclor 42; Density: 1.16 g/ml; Visc.: 28 poise; Stability: See Cereclor 42
Ref.: Leaflet 761-25R

Cereclor 50LV. ICI United States Inc.
Chem. Descrip.: Chlorinated paraffin
Category/Applications: See Cereclor 42; 50LV is used esp. as a lubricant additive
Form: Clear, viscous liquid; Color (Gardner): 1
Composition: 50% chlorine content
Gen'l. Prop.: Sol.: See Cereclor 42; Density: 1.19 g/ml; Visc.: 1 poise; Stability: See Creclor 42
Ref.: Leaflet 761-25R

Cereclor 51-L. ICI United States Inc.
Chem. Descrip.: Chlorinated paraffin
Category/Applications: See Creclor 42; 51-L used esp. as lubricant additive
Form: Clear, viscous liquid; Color (Gardner): 6
Composition: 51% chlorine content
Gen'l. Prop.: Sol.: See Cereclor 42; Density: 1.25 g/ml; Visc.: 13 poise; Stability: See Cereclor 42
Ref.: Leaflet 761-25R

Cereclor 52P. ICI United States Inc.
Chem. Descrip.: Chlorinated paraffin
Category/Applications: See Cereclor 42; 52P is used esp. as adhesives, sealants,

and paints plasticizer, and in flame retardant applications
Form: Clear, viscous liquid; Color (Gardner): 4
Composition: 52% chlorine content
Gen'l. Prop.: Sol.: See Cereclor 42; Density: 1.25 g/ml; Visc.: 12 poise; Stability: See Cereclor 42
Ref.: Leaflet 761-25R

Cereclor 70L. ICI United States Inc.
Chem. Descrip.: Chlorinated paraffin
Category/Applications: See Cereclor 42; 70L is used esp. as lubricant additive and in flame retardant applications
Form: Clear, viscous liquid; Color (Gardner): 11
Composition: 70% chlorine content
Gen'l. Prop.: Sol.: See Cereclor 42; Density: 1.55 g/ml; Visc.: 20,000 poise; Stability: See Cereclor 42
Ref.: Leaflet 761-25R

Cereclor AP45. ICI United States Inc.
Chem. Descrip.: Chlorinated paraffin
Category/Applications: See Cereclor 42; AP45 is used esp. in flame retardant and specialty applications
Form: Clear, viscous liquid; Color (Gardner): < 1
Composition: 45% chlorine content
Gen'l. Prop.: Sol.: See Cereclor 42; Density: 1.16 g/ml; Visc.: 1 poise; Stability: See Cereclor 42
Ref.: Leaflet 761-25R

Cereclor AP52. ICI United States Inc.
Chem. Descrip.: Chlorinated paraffin
Category/Applications: See Cereclor 42; AP52 is used esp. for flame retardant and specialty applications
Form: Clear, viscous liquid; Color (Gardner): < 1
Composition: 52% chlorine content
Gen'l. Prop.: Sol.: See Cereclor 42; Density: 1.25 g/ml; Visc.: 11 poise; Stability: See Cereclor 42
Ref.: Leaflet 761-25R

Cereclor LP4446. ICI United States Inc.
Chem. Descrip.: Chlorinated paraffin
Category/Applications: See Cereclor 42; LP4446 is used esp. as lubricant additive and for flame retardant applications
Form: Clear, viscous liquid; Color (Gardner): 4
Composition: 44% chlorine content
Gen'l. Prop.: Sol.: See Cereclor 42; Density: 1.19 g/ml; Visc.: 45 poise; Stability: See Cereclor 42
Ref.: Leaflet 761-25R

Cereclor LP4985. ICI United States Inc.
Chem. Descrip.: Chlorinated paraffin
Category/Applications: See Cereclor 42; LP4985 is used esp. as adhesives, sealants, and paints plasticizers, lubricant additives, and in flame retardant applications
Form: Clear, viscous liquid; Color (Gardner): 8
Composition: 50% chlorine content
Gen'l. Prop.: Sol.: See Cereclor 42; Density: 1.28 g/ml; Visc.: 169 poise; Stability: See Cereclor 42
Ref.: Leaflet 761-25R

Cereclor S45. ICI United States
Chem. Descrip.: Chlorinated paraffin
Category/Applications: See Cereclor 42; S45 is used esp. as adhesives, sealants, paints, PVC, and rubber plasticizers, and in flame retardant applications
Form: Clear, viscous liquid; Color (Gardner): 1
Composition: 45% chlorine content
Gen'l. Prop.: Sol.: See Cereclor 42; Density: 1.17 g/ml; Visc.: 2 poise; Stability: See Cereclor 42
Ref.: Leaflet 761-25R

Cereclor S-52. ICI United States Inc.
Chem. Descrip.: Chlorinated paraffin
Category/Applications: See Cereclor 42; S-52 is used esp. as adhesives, sealants, paints, PVC, and rubber plasticizers, and in flame retardant applications
Form: Clear, viscous liquid; Color (Gardner): 1
Composition: 52% chlorine content
Gen'l. Prop.: Sol.: See Cereclor 42; Density: 1.25 g/ml; Visc.: 12 poise; Stability: See Cereclor 42
Ref.: Leaflet 761-25R

Chemigum N3. Goodyear Chem.
Chem. Descrip.: Hot acrylonitrile polymer
Category/Applications: Polymer recommended for high quality stocks requiring exceptional resistance to solvents and oils, high physical strength, and the best heat resistance obtainable from nitrile polymers
Form: Liquid
Composition: 39% acrylonitrile content
Gen'l. Prop.: Sp. Gr.: 1.00; Visc. (Mooney): 75–93; Sol.: 44% min. in MEK
Mech. Prop. (compounded in a 50 part HAF black formulation, ≥6 min, 350 F); Tens. Str.: 4250 psi; Tens. Elong.: 370%; 300% Mod.: 3650 psi; Hardness (Shore A): 79
Ref.: Tech. bulletin CG-39

Chemigum N5. Goodyear Chem.
Chem. Descrip.: Hot acrylonitrile polymer
Category/Applications: Soluble high acrylonitrile polymer for use in cements, grouts, and adhesives
Form: Liquid
Composition: 39% acrylonitrile content
Gen'l. Prop.: Sp. Gr.: 1.00; Visc. (Mooney): 91–109; Sol.: 70% min. in MEK
Ref.: Tech. bulletin CG-39

Chemigum N6B. Goodyear Chem.
Chem. Descrip.: Hot acrylonitrile polymer
Category/Applications: Medium acrylonitrile polymer used in adhesives and light colored molded goods; has the best tack of all the Chemigums
Form: Liquid; Color: Light
Composition: 33% acrylonitrile content
Gen'l. Prop.: Sp. Gr.: 0.98; Visc. (Mooney): 45–63; Sol.: 72% min. in MEK
Mech. Prop. (compounded in a 50 part HAF black formulation, ≥6 min, 350 F); Tens. Str.: 3500 psi; Tens. Elong.: 410%; 300% Mod.: 2450 psi; Hardness (Shore A): 73
Ref.: Tech. bulletin CG-39

Chemigum N7. Goodyear Chem.
Chem. Descrip.: Hot acrylonitrile polymer
Category/Applications: Medium acrylonitrile high viscosity polymer for use in compounds which must withstand rugged or high-temp. service, as in hose tubes and belts
Form: Liquid
Composition: 33% acrylonitrile content
Gen'l. Prop.: Sp. Gr.: 0.98; Visc. (Mooney): 81–92; Sol.: 55% min. in MEK
Mech. Prop. (compounded in a 50 part HAF black formulation, ≥6 min, 350 F); Tens. Str.: 3800 psi; Tens. Elong.: 340%; 300% Mod.: 3500 psi; Hardness (Shore A): 76
Ref.: Tech. bulletin CG-39

Chemigum N8. Goodyear Chem.
Chem. Descrip.: Hot acrylonitrile polymer
Category/Applications: Medium acrylonitrile polymer characterized by its low shrinkage as compared to other nitrile

polymers; can be used as the sole polymer in a compound, but recommended for blending with other Chemigums to reduce calender shrinkage or die swell; also used to modify polyvinyl chloride and ABS resins; imparts flexibility and the hot tens. strength needed in vacuum forming
Form: Crumbs in various particle sizes
Composition: 33% acrylonitrile content
Gen'l. Prop.: Sp. Gr.: 0.98; Visc. (Mooney): 72–86; Sol.: 20% min. in MEK
Mech. Prop. (compounded in a 50 part HAF black formulation, ⩾6 min, 350 F); Tens. Str.: 3000 psi; Tens. Elong.: 240%; Hardness (Shore A): 79
Ref.: Tech. bulletin CG-39

Chemigum N206. Goodyear Chem.
Chem. Descrip.: Cold acrylonitrile polymer
Category/Applications: Ultra-high acrylonitrile polymer with the best oil and solvent resistances of all Chemigums; most resistant of Chemigums to Freon refrigerants; recommended for automotive air conditioning hose
Form: Liquid
Composition: 45% acrylonitrile content
Gen'l. Prop.: Sp. Gr.: 0.98; Visc. (Mooney): 50–70; Sol.: 98% min. in MEK
Mech. Prop. (compounded in a 50 part HAF black formulation, ⩾6 min, 350 F); Tens. Str.: 4375 psi; Tens. Elong.: 420%; 300% Mod.: 3200 psi; Hardness (Shore A): 81
Ref.: Tech. bulletin CG-39

Chemigum N300. Goodyear Chem.
Chem. Descrip.: Cold acrylonitrile polymer
Category/Applications: High acrylonitrile, med. viscosity, noncorrosive polymer for general compounding use where better oil and solvent resistance is required; the lower viscosity makes highly loaded stocks easier to extrude and mold
Form: Liquid
Composition: 39% acrylonitrile content
Gen'l. Prop.: Sp. Gr.: 1.00; Visc. (Mooney): 50–63; Sol.: 98% min. in MEK
Mech. Prop. (compounded in a 50 part HAF black formulation, ⩾6 min, 350 F); Tens. Str.: 4050 psi; Tens. Elong.: 500%; 300% Mod.: 2100 psi; Hardness (Shore A): 77
Ref.: Tech. bulletin CG-39

Chemigum N318B. Goodyear Chem.
Chem. Descrip.: Cold acrylonitrile polymer
Category/Applications: High acrylonitrile, high visc., noncorrosive polymer possessing slightly higher tensile than Chemigum N300; the higher viscosity facilitates good extrusions and gives higher back pressure for compression molding
Form: Liquid
Composition: 39% acrylonitrile content
Gen'l. Prop.: Sp. Gr.: 1.00; Visc. (Mooney): 70–80; Sol.: 98% min. in MEK
Mech. Prop. (compounded in a 50 part HAF black formulation, ⩾6 min, 350 F); Tens. Str.: 4400 psi; Tens. Elong.: 520%; 300% Mod.: 2475 psi; Hardness (Shore A): 77
Ref.: Tech. bulletin CG-39

Chemigum N328B. Goodyear Chem.
Chem. Descrip.: Cold acrylonitrile polymer
Category/Applications: High acrylonitrile, high viscosity, noncorrosive polymer similar to Chemigum N318B in physical properties, but with better mold flow and extrusion characteristics
Form: Liquid

Composition: 39% acrylonitrile content
Gen'l. Prop.: Sp. Gr.: 1.00; Visc. (Mooney): 70–80; Sol.: 98% min. in MEK
Mech. Prop. (compounded in a 50 part HAF black formulation, ≥ 6 min, 350 F); Tens. Str.: 3725 psi; Tens. Elong.: 450%; 300% Mod.: 2350 psi; Hardness (Shore A): 77
Ref.: Tech. bulletin CG-39

Chemigum N600. Goodyear Chem.
Chem. Descrip.: Cold acrylonitrile polymer
Category/Applications: Med. acrylonitrile polymer with a nonstaining antioxidant which makes it the choice for use in adhesives and light colored molded goods
Form: Liquid
Composition: 33% acrylonitrile content
Gen'l. Prop.: Sp. Gr.: 0.98; Visc. (Mooney): 40–58; Sol.: 90% min. in MEK
Mech. Prop. (compounded in a 50 part HAF black formulation, ≥ 6 min, 350 F); Tens. Str.: 3450 psi; Tens. Elong.: 545%; 300% Mod.: 1575 psi; Hardness (Shore A): 73
Ref.: Tech. bulletin CG-39

Chemigum N608. Goodyear Chem.
Chem. Descrip.: Cold acrylonitrile polymer
Category/Applications: Med. acrylonitrile, high viscosity, noncorrosive polymer; the higher viscosity makes it the choice for extrusions where dimensional stability is required, as in hose tubes
Composition: 33% acrylonitrile content
Gen'l. Prop.: Sp. Gr.: 0.99; Visc. (Mooney): 68–85; Sol.: 98% min. in MEK
Mech. Prop. (compounded in a 50 part HAF black formulation, ≥ 6 min, 350 F); Tens. Str.: 4200 psi; Tens. Elong.: 500%; 300% Mod.: 2325 psi; Hardness (Shore A): 73
Ref.: Tech. bulletin CG-39

Chemigum N612, N612B. Goodyear Chem.
Chem. Descrip.: Cold acrylonitrile polymer
Category/Applications: Med. acrylonitrile, low viscosity, noncorrosive polymers differing only in antioxidants (B designating a nonstaining type); both are outstanding for easy mixing and good mold flow; the low viscosities make highly loaded stocks easier to extrude and mold
Form: Liquid
Composition: 33% acrylonitrile content
Gen'l. Prop.: Sp. Gr.: 0.98; Visc. (Mooney): 20–30; Sol.: 98% min. in MEK
Mech. Prop. (compounded in a 50 part HAF black formulation, ≥ 6 min, 350 F); Tens. Str.: 3250 psi; Tens. Elong.: 600%; 300% Mod.: 1425 psi; Hardness (Shore A): 72
Ref.: Tech. bulletin CG-39

Chemigum N615, N615B. Goodyear Chem.
Chem. Descrip.: Cold acrylonitrile polymers
Category/Applications: Med. acrylonitrile, med. viscosity, noncorrosive polymers with easy mixing and processing characteristics and excellent physical properties; compression set resistance is excellent; these differ only in antioxidants (B designating the nonstaining type)
Form: Liquid
Composition: 33% acrylonitrile content
Gen'l. Prop.: Sp. Gr.: 0.98; Visc. (Mooney): 45–63; Sol.: 98% min. in MEK

Mech. Prop. (compounded in a 50 part HAF black formulation, ⩾ 6 min, 350 F); Tens. Str.: 4000 psi; Tens. Elong.: 535%; 300% Mod.: 1950 psi; Hardness (Shore A): 74
Ref.: Tech. bulletin CG-39

Chemigum N634. Goodyear Chem.
Chem. Descrip.: Cold acrylonitrile polymer containing 50 parts of dioctyl phthalate/100 parts of rubber hydrocarbon
Category/Applications: Noncorrosive polymer designed to aid in the mill mixing of low durometer stocks
Ref.: Tech. bulletin CG-39

Chemigum N715B. Goodyear Chem.
Chem. Descrip.: Cold acrylonitrile polymer
Category/Applications: Medium low acrylonitrile, med. viscosity polymer developed to meet specific needs in rubber applications, e.g., better low temp. flex properties while still providing adequate resistance to fuels and oils
Form: Liquid
Composition: 28% acrylonitrile content
Gen'l. Prop.: Sp. Gr.: 0.98; Visc. (Mooney): 45–63; Sol.: 98% min. in MEK
Mech. Prop. (compounded in a 50 part HAF black formulation, ⩾ 6 min, 350 F); Tens. Str.: 3675 psi; Tens. Elong.: 550%; 300% Mod.: 1650 psi; Hardness (Shore A): 70
Ref.: Tech. bulletin CG-39

Chemigum N917. Goodyear Chem.
Chem. Descrip.: Cold acrylonitrile polymer
Category/Applications: Low acrylonitrile polymer designed to meet the need for gaskets, seals, and O-rings for the aircraft industry and other polymer product applications where the best low temp. flexibility and resistance to fuel and oil is required
Form: Liquid
Composition: 22% acrylonitrile content
Gen'l. Prop.: Sp. Gr.: 0.97; Visc. (Mooney): 60–75; Sol.: ⩾ 70% in MEK
Mech. Prop. (compounded in a 50 part HAF black formulation, ⩾ 6 min, 350 F); Tens. Str.: 3650 psi; Tens. Elong.: 400%; 300% Mod.: 2475 psi; Hardness (Shore A): 68
Ref.: Tech. bulletin CG-39

Chemigum NBR Polymers. Goodyear Chem.
Chem. Descrip.: Family of nitrile polymers produced by catalytic emulsion copolymerization of acrylonitrile and butadiene
Category/Applications: Rubbers with the characteristic of low volume change in petroleum-based oils, greases, and solvents compared to general purpose rubber; available in a range of acrylonitrile contents in both "hot" and "cold" polymerization formulas; compatible with all Pliovic vinyl resins, forming wide range of compounds, e.g., chemically blown sponge for insulation, abrasion resistant shoe soles, ink and solvent resistant printer's rolls, automotive door panels and safety pads, cable jackets, and hose covers
Form: Liquid
Ref.: Tech. booklet CG-39

Chemplex 1005. Chemplex Co.
Chem. Descrip.: Polyethylene
Category/Applications: Homopolymer for general purpose film liner applications
Gen'l. Prop.: Density: 0.919 g/cc (D-1505); Melt Index: 3.5 g/10 min (D-1238)
Mech. Prop. (of 1.5-mil film): Tens. Str.:

1400 psi (TD, yield, D882), 3200 psi (TD, break, D882); Tens. Elong.: 450% (TD, D882); Tens. Mod.: 28,000 psi (TD, 1% secant, D822); Impact Str. (Dart Drop): 125 g (D1709/A)
Ref.: Data sheet

Chemplex 1007. Chemplex Co.
Chem. Descrip.: Polyethylene
Category/Applications: Soft and tough polyethylene resin for specialty film applications, e.g., diaper liners
Gen'l. Prop.: Density: 0.919 g/cc (D-1505); Melt Index: 2.5 g/10 min (D-1238)
Mech. Prop. (of 1.5-mil film): Tens. Str.: 1400 psi (TD, yield, D882), 3500 psi (TD, break, D882); Tens. Elong.: 450% (TD, D882); Tens. Mod.: 28,000 psi (TD, 1% secant, D822); Impact Str. (Dart Drop): 135 g (D1709/A)
Ref.: Data sheet

Chemplex 1008. Chemplex Co.
Chem. Descrip.: Polyethylene
Category/Applications: Polyethylene resin with good processability, for multiwall bag liners
Gen'l. Prop.: Density: 0.919 g/cc (D-1505); Melt Index: 5.0 g/10 min (D-1238)
Mech. Prop. (of 1.5-mil film): Tens. Str.: 1400 psi (TD, yield, D882), 3000 psi (TD, break, D882); Tens. Elong.: 450% (TD, D882); Tens. Mod.: 27,000 psi (TD, 1% secant, D822); Impact Str. (Dart Drop): 120 g (D1709/A)
Ref.: Data sheet

Chemplex 1013. Chemplex Co.
Chem. Descrip.: Polyethylene
Category/Applications: High speed, lightweight coating resin for locker wraps, sugar pouches, industrial and multiwall bags, and laminations
Gen'l. Prop.: Density: 0.918; Melt Index: 13
Ref.: Data sheet

Chemplex 1014. Chemplex Co.
Chem. Descrip.: Polyethylene
Category/Applications: General-purpose, heavy, med., and light coating weight resin; avail. with low, med., and high slip levels; used in general-purpose flexible packaging, foil mounting, and laminating
Gen'l. Prop.: Density: 0.918; Melt Index: 7.0
Ref.: Data sheet

Chemplex 1015. Chemplex Co.
Chem. Descrip.: Polyethylene
Category/Applications: High speed, med. weight coating resin for flexible packaging, milk cartons, food boards, set-up cartons, industrial papers, and foil mounting
Gen'l. Prop.: Density: 0.923; Melt Index: 4.5
Ref.: Data sheet

Chemplex 1016. Chemplex Co.
Chem. Descrip.: Polyethylene
Category/Applications: General-purpose, heavy and med. coating weight resin for flexible packaging, milk cartons, set-up cartons, and food boards; avail. with antistat or chill roll release
Gen'l. Prop.: Density: 0.923; Melt Index: 3.7
Ref.: Data sheet

Chemplex 1017. Chemplex Co.
Chem. Descrip.: Polyethylene
Category/Applications: High speed, lightweight coating version of Chemplex 1016; used for flexible packaging, foil mounting, milk cartons, food boards, and industrial papers

Gen'l. Prop.: Density: 0.923; Melt Index: 5.5
Ref.: Data sheet

Chemplex 1040. Chemplex Co.
Chem. Descrip.: Polyethylene
Category/Applications: Polyethylene resin for stretch and shrink applications, e.g., light pallet shrink wrapping, case wrap; offers high impact strength, tear resistance, and low shrink temp.
Gen'l. Prop.: Density: 0.923 g/cc (D-1505); Melt Index: 0.5 g/10 min (D-1238)
Mech. Prop. (of 2.0-mil film): Tens. Str.: 1700 psi (TD, yield, D882), 3300 psi (TD, break, D882); Tens. Elong.: 480% (TD, D882); Tens. Mod.: 36,000 psi (TD, 1% secant, D822); Impact Str. (Dart Drop): 220 g (D1709/A)
Ref.: Data sheet

Chemplex 1044. Chemplex Co.
Chem. Descrip.: Polyethylene
Category/Applications: Polyethylene resin for industrial film applications, with excellent stability for large, heavy gauge bubbles; used in shipping sacks, skin packaging
Composition: 3.5% vinyl acetate
Gen'l. Prop.: Density: 0.929 g/cc (D-1505); Melt Index: 0.5 g/10 min (D-1238)
Mech. Prop. (of 2.0-mil film): Tens. Str.: 1350 psi (TD, yield, D882), 3000 psi (TD, break, D882); Tens. Elong.: 500% (TD, D882); Tens. Mod.: 25,000 psi (TD, 1% secant, D822); Impact. Str. (Dart Drop): 235 g (D1709/A)
Ref.: Data sheet

Chemplex 1045. Chemplex Co.
Chem. Descrip.: Polyethylene
Category/Applications: Polyethylene resin for stretch and shrink applications, e.g., pallet shrink wrapping; offers high impact strength, tear resistance, and low shrink temp.
Composition: 2.0% vinyl acetate
Gen'l. Prop.: Density: 0.926 g/cc (D-1505); Melt Index: 0.5 g/10 min (D-1238)
Mech. Prop. (of 2.0-mil film): Tens. Str.: 1600 psi (TD, yield, D882), 3300 psi (TD, break, D882); Tens. Elong.: 490% (TD, D882); Tens. Mod.: 31,000 psi (TD, 1% secant, D822); Impact Str. (Dart Drop): 260 g (D1709/A)
Ref.: Data sheet

Chemplex 1050. Chemplex Co.
Chem. Descrip.: Polyethylene
Category/Applications: Copolymer for general-purpose liner film applications
Composition: 2.0% vinyl acetate
Gen'l. Prop.: Density: 0.926 g/cc (D-1505); Melt Index: 3.5 g/10 min (D-1238)
Mech. Prop. (of 1.5-mil film): Tens. Str.: 1500 psi (TD, yield, D882), 3200 psi (TD, break, D882); Tens. Elong.: 530% (TD, D882); Tens. Mod.: 30,000 psi (TD, 1% secant, D822); Impact Str. (Dart Drop): 130 g (D1709/A)
Ref.: Data sheet

Chemplex 1054. Chemplex Co.
Chem. Descrip.: Polyethylene
Category/Applications: Polyethylene resin with high impact strength and toughness, for tough liner films and shrink bundling
Composition: 3.5% vinyl acetate
Gen'l. Prop.: Density: 0.927 g/cc (D-1505); Melt Index: 2.0 g/10 min (D-1238)
Mech. Prop. (of 1.5-mil film): Tens. Str.: 1400 psi (TD, yield, D882), 3100 psi (TD, break, D882); Tens. Elong.: 490% (TD, D882); Tens. Mod.: 28,000 psi

(TD, 1% secant, D822); Impact Str. (Dart Drop): 150 g (D1709/A)
Ref.: Data sheet

Chemplex 1057. Chemplex Co.
Chem. Descrip.: Polyethylene
Category/Applications: Polyethylene resin for industrial film applications, esp. bag-in-box; offers high impact, high flex life, and superior ESCR
Composition: 7.5% vinyl acetate
Gen'l. Prop.: Density: 0.928 g/cc (D-1505); Melt Index: 1.2 g/10 min (D-1238)
All properties—of 2.0 mil film:
Mech. Prop. (of 2.0-mil film): Tens. Str.: 1200 psi (TD, yield, D882), 3500 psi (TD, break, D882); Tens. Elong.: 500% (TD, D882); Tens. Mod.: 16,000 psi (TD, 1% secant, D822); Impact Str. (Dart Drop): 340 g (D1709/A)
Ref.: Data sheet

Chemplex 1060. Chemplex Co.
Chem. Descrip.: Polyethylene
Category/Applications: Polyethylene resin for industrial film applications, e.g., ice bags, shipping sacks, bag-in-box, heavy duty skin packaging; offers high impact and low temp. toughness; also avail. with a 2.5 melt index (Chemplex 1061)
Composition: 7.5% vinyl acetate
Gen'l. Prop.: Density: 0.928 g/cc (D-1505); Melt Index: 0.5 g/10 min (D-1238)
Mech. Prop. (of 1.25-mil film): Tens. Str.: 1200 psi (TD, yield, D882), 3700 psi (TD, break, D882); Tens. Elong.: 570% (TD, D882); Tens. Mod.: 16,000 psi (TD, 1% secant, D822); Impact Str. (Dart Drop): 400 g (D1709/A)
Ref.: Data sheet

Chemplex 1101. Chemplex Co.
Chem. Descrip.: Polyethylene
Category/Applications: Med.-density resin for low coating weights at high machine speeds, offering better barrier properties and higher temp. resistance; used for boil-in-the bags, oil cans, milk and juice cartons, and multiwall bags
Gen'l. Prop.: Density: 0.931; Melt Index: 6.0
Ref.: Data sheet

Chemplex 3015. Chemplex Co.
Chem. Descrip.: Polyethylene
Category/Applications: Polyethylene resin for specialty applications, e.g., blown overwrap; offers good clarity, stiffness, and strength
Gen'l. Prop.: Density: 0.928 g/cc (D-1505); Melt Index: 1.7 g/10 min (D-1238)
Mech. Prop. (of 1.25-mil film): Tens. Str.: 2200 psi (TD, yield, D882), 2500 psi (TD, break, D882); Tens. Elong.: 460% (TD, D882); Tens. Mod.: 41,000 psi (TD, 1% secant, D822); Impact Str. (Dart Drop): 70 g (D1709/A)
Ref.: Data sheet

Chemplex 3024. Chemplex Co.
Chem. Descrip.: Polyethylene
Category/Applications: Polyethylene resin for specialty film applications; cast film resin with high clarity and gloss, and stiffness; used in textiles, bakery goods, paper products overwrap
Gen'l. Prop.: Density: 0.932 g/cc (D-1505); Melt Index: 3.0 g/10 min (D-1238)
Mech. Prop. (of 1.25-mil film): Tens. Str.: 2200 spi (TD, yield, D882), 2400 psi (TD, break, D882); Tens. Elong.: 540% (TD, D882); Tens. Mod.: 43,000 psi (TD, 1% secant, D822); Impact Str. (Dart Drop): 80 g (D1709/A)
Ref.: Data sheet

Chemplex 3040. Chemplex Co.
Chem. Descrip.: Polyethylene
Category/Applications: Polyethylene resin for specialty film applications, e.g., heavy produce and textiles, frozen foods, skin packaging; offers good clarity, high toughness, and good melt strength
Composition: 4.5% vinyl acetate
Gen'l. Prop.: Density: 0.928 g/cc (D-1505); Melt Index: 1.0 g/10 min (D-1238)
Mech. Prop. (of 1.25-mil film): Tens. Str.: 1400 psi (TD, yield, D882), 3200 psi (TD, break, D882); Tens. Elong.: 540% (TD, D882); Tens. Mod.: 24,000 psi (TD, 1% secant, D822); Impact Str. (Dart Drop): 160 g (D1709/A)
Ref.: Data sheet

Chemplex 3043. Chemplex Co.
Chem. Descrip.: Polyethylene
Category/Applications: Polyethylene resin for stretch and shrink applications, e.g., pallet stretch wrap, produce, laminating; offers clarity, toughness, and good melt strength
Composition: 4.5% vinyl acetate
Gen'l. Prop.: Density: 0.928 g/cc (D-1505); Melt Index: 2.0 g/10 min (D-1238)
Mech. Prop. (of 1.25-mil film): Tens. Str.: 1350 psi (TD, yield, D882), 3000 psi (TD, break, D882); Tens. Elong.: 540% (TD, D882); Tens. Mod.: 24,000 psi (TD, 1% secant, D822); Impact Str. (Dart Drop): 140 g (D1709/A)
Ref.: Data sheet

Chemplex 3044. Chemplex Co.
Chem. Descrip.: Polyethylene
Category/Applications: Polyethylene resin for specialty film applications, e.g., heavy produce and textiles, frozen foods, skin packaging; offers high clarity, toughness, and good melt strength
Composition: 3.0% vinyl acetate
Gen'l. Prop.: Density: 0.927 g/cc (D-1505); Melt Index: 1.5 g/10 min (D-1238)
Mech. Prop. (of 1.25-mil film): Tens. Str.: 1400 psi (TD, yield, D882), 3200 psi (TD, break, D882); Tens. Elong.: 550% (TD, D882); Tens. Mod.: 25,000 psi (TD, 1% secant, D822); Impact Str. (Dart Drop): 130 g (D1709/A)
Ref.: Data sheet

Chemplex 3050. Chemplex Co.
Chem. Descrip.: Vinyl acetate copolymer
Category/Applications: Copolymer formulated with an additive for chill roll release and slip to minimize, and in most cases eliminate, the need for starch; used in coated cellophane, Mylar, and polypropylene, for cheese wrap, and other critical applications
Composition: 18% vinyl acetate
Gen'l. Prop.: Density: 0.93+; Melt Index: 30
Ref.: Data sheet

Chemplex 3052. Chemplex Co.
Chem. Descrip.: Polyethylene
Category/Applications: Polyethylene resin for specialty film applications, e.g., household wrap; offers super clarity; cast film resin
Gen'l. Prop.: Density: 0.926 g/cc (D-1505); Melt Index: 5.2 g/10 min (D-1238)
Mech. Prop. (of 1.25-mil film): Tens. Str.: 2000 psi (TD, yield, D882), 1800 psi (TD, break, D882); Tens. Elong.: 600% (TD, D882); Tens. Mod.: 21,000 psi (TD, 1% secant, D822)
Ref.: Data sheet

Chemplex 3104. Chemplex Co.
Chem. Descrip.: Polyethylene
Category/Applications: Copolymer for general-purpose film applications, e.g., light produce, textile packaging, thin films, bags, and pouches; offers high clarity, good impact strength
Composition: 2.0% vinyl acetate
Gen'l. Prop.: Density: 0.926 g/cc (D-1505); Melt Index: 2.0 g/10 min (D-1238)
Mech. Prop. (of 1.25-mil film): Tens. Str.: 1600 psi (TD, yield, D882), 2900 psi (TD, break, D882); Tens. Elong.: 550% (TD, D882); Tens. Mod.: 28,000 psi (TD, 1% secant, D822); Impact Str. (Dart Drop): 115 g (D1709/A)
Ref.: Data sheet

Chemplex 3105. Chemplex Co.
Chem. Descrip.: Polyethylene
Category/Applications: Copolymer for general-purpose, general clarity film applications; same as Chemplex 3404 with vinyl acetate
Composition: 1.5% vinyl acetate
Gen'l. Prop.: Density: 0.925 g/cc (D-1505); Melt Index: 2.0 g/10 min (D-1238)
Mech. Prop. (of 1.25-mil film): Tens. Str.: 1500 psi (TD, yield, D882), 3100 psi (TD, break, D882); Tens. Elong.: 560% (TD, D882); Tens. Mod.: 28,000 psi (TD, 1% secant, D822); Impact Str. (Dart Drop): 120 g (D1709/A)
Ref.: Data sheet

Chemplex 3306. Chemplex Co.
Chem. Descrip.: Polyethylene
Category/Applications: Polyethylene resin with high VA for coextrusion; offers excellent heat sealability and toughness
Composition: 18% vinyl acetate
Gen'l. Prop.: Density: 0.940 g/cc (D-1505); Melt Index: 12.5 g/10 min (D 1238)
Mech. Prop. (of 2.0-mil film): Tens. Mod.: 8400 psi (TD, 1% secant, D822)
Ref.: Data sheet

Chemplex 3311. Chemplex Co.
Chem. Descrip.: Polyethylene
Category/Applications: Polyethylene resin with high VA for coextrusion and blends; offers excellent heat sealability
Composition: 18% vinyl acetate
Gen'l. Prop.: Density: 0.939 g/cc (D-1505); Melt Index: 0.3 g/10 min (D-1238)
Mech. Prop. (of 2.5-mil film): Tens. Str.: 3700 psi (TD, break, D882); Tens. Elong.: 600% (TD, D882); Tens. Mod.: 7200 psi (TD, 1% secant, D822)
Ref.: Data sheet

Chemplex 3401. Chemplex Co.
Chem. Descrip.: Polyethylene
Category/Applications: Polyethylene resin for stretch and shrink applications, e.g., clarity bundling; offers good blend of clarity and melt strength
Gen'l. Prop.: Density: 0.923 g/cc (D-1505); Melt Index: 1.0 g/10 min (D-1238)
Mech. Prop. (of 1.25-mil film): Tens. Str.: 1660 psi (TD, yield, D882), 3200 psi (TD, break, D882); Tens. Elong.: 550% (TD, D882); Tens. Mod.: 29,000 psi (TD, 1% secant, D822); Impact Str. (Dart Drop): 170 g (D1709/A)
Ref.: Data sheet

Chemplex 3402. Chemplex Co.
Chem. Descrip.: Polyethylene
Category/Applications: Cast film resin for specialty applications, e.g., bakery bags

Gen'l. Prop.: Density: 0.925 g/cc (D-1505); Melt Index: 2.0 g/10 min (D-1238)
Mech. Prop. (of 1.25-mil film): Tens. Str.: 1600 psi (TD, yield, D882), 2100 psi (TD, break, D882); Tens. Elong.: 530% (TD, D882); Tens. Mod.: 27,000 psi (TD, 1% secant, D822)
Ref.: Data sheet

Chemplex 3404. Chemplex Co.
Chem. Descrip.: Polyethylene
Category/Applications: Homopolymer for general-purpose film applications, e.g., light produce, textile packaging, thin films, bags, and pouches; offers premium clarity, good impact and stiffness
Gen'l. Prop.: Density: 0.922 g/cc (D-1505); Melt Index: 1.8 g/10 min (D-1238)
Mech. Prop. (of 1.25-mil film): Tens. Str.: 1660 psi (TD, yield, D882), 3100 psi (TD, break, D882); Tens. Elong.: 550% (TD, D882); Tens. Mod.: 29,000 psi (TD, 1% secant, D822); Impact Str. (Dart Drop): 135 g (D1709/A)
Ref.: Data sheet

Chemplex 3405. Chemplex Co.
Chem. Descrip.: Polyethylene
Category/Applications: Polyethylene resin for specialty film applications, e.g., thin, soft wraps, lettuce wraps; offers good clarity and cling
Gen'l. Prop.: Density: 0.923 g/cc (D-1505); Melt Index: 3.5 g/10 min (D-1238)
Mech. Prop. (of 1.25-mil film): Tens. Str.: 1600 psi (TD, yield, D882), 2700 psi (TD, break, D882); Tens. Elong.: 550% (TD, D882); Tens. Mod.: 29,000 psi (TD, 1% secant, D822); Impact Str. (Dart Drop): 105 g (D1709/A)
Ref.: Data sheet

Chemplex 5003. Chemplex Co.
Chem. Descrip.: High-density polyethylene
Category/Applictions: Copolymer for blow molding and extrusion, with excellent stress crack resistance; used in heavy duty detergents and solvents; avail. with or without antistat; melt index 0.3 g/10 min (D1238)
Gen'l. Prop.: Density: 0.952 g/cc (D-1505); Soften. Pt. (Vicat): 125 C (D-1525)
Mech. Prop.: Tens. Str.: 3900 psi (yield, D638); Tens. Elong.: > 600% (D638); Flex. Mod.: 160,000 psi (D790); Impact Str. (Izod): 2.5 ft lb/in. notch (D256); Hardness (Shore D): 65 (D1706)
Therm. Prop.: Distort. Temp.: 75 C (@ 66 psi, D648)
Ref.: Data sheet

Chemplex 5402. Chemplex Co.
Chem. Descrip.: High-density polyethylene
Category/Applications: Copolymer for blow molding and extrusion with max. stress crack resistance and high toughness; used in industrial shipping containers, pipe; avail. with or without antistat; melt index 0.17 g/10 min (D1238)
Gen'l. Prop.: Density: 0.942 g/cc (D-1505); Soften. Pt. (Vicat): 120 C (D-1525)
Mech. Prop.: Tens. Str.: 3200 psi (yield, D638); Tens. Elong.: > 600% (D638); Flex. Mod.: 120,000 psi (D790); Impact Str. (Izod): 5 ft lb/in. notch (D256); Hardness (Shore D): 62 (D1706)
Therm. Prop.: Distort. Temp.: 65 C (@ 66 psi, D648)
Ref.: Data sheet

Chemplex 5602. Chemplex Co.
Chem. Descrip.: High-density polyethylene
Category/Applications: Copolymer for blow molding and extrusion with good

rigidity and stress crack resistance; used in detergent and bleach bottles; avail. with or without antistat; melt index 0.22 g/10 min (D1238)
Gen'l. Prop.: Density: 0.955 g/cc (D-1505); Soften. Pt. (Vicat): 125 C (D-1525)
Mech. Prop.: Tens. Str.: 4200 psi (yield, D638); Tens. Elong.: > 600% (D638); Flex. Mod.: 180,000 psi (D790); Impact Str. (Izod): 6 ft lb/in. notch (D256); Hardness (Shore D): 65
Therm. Prop.: Distort. Temp.: 76 C (@ 66 psi, D648)
Ref.: Data sheet

Chemplex 5604. Chemplex Co.
Chem. Descrip.: High-density polyethylene
Category/Applications: See Chemplex 5602; melt index 0.35 g/10 min (D1238)
Gen'l. Prop.: Density: 0.954 g/cc (D-1505); Soften. Pt. (Vicat): 124 C (D-1525)
Mech. Prop.: Tens. Str.: 4200 psi (yield, D638); Tens. Elong.: > 600% (D638); Flex. Mod.: 170,000 psi (D790); Impact Str. (Izod): 6 ft lb/in. notch (D256); Hardness (Shore D): 65 (D1706)
Therm. Prop.: Distort. Temp.: 76 C (@ 66 psi, D648)
Ref.: Data sheet

Chemplex 5701. Chemplex Co.
Chem. Descrip.: High-density polyethylene
Category/Applications: High m.w. copolymer with superior impact strength, high environmental stress crack resistance, and ease of processing; used in 55-gal drums, gas tanks; melt index 11 g/10 min (D1238)
Gen'l. Prop.: Density: 0.953 g/cc (D-1505); Soften. Pt. (Vicat): 124 C (D-1525)
Mech. Prop.: Tens. Str.: 3900 psi (yield, D638); Tens. Elong.: > 600% (D638); Flex. Mod.: 160,000 psi (D790); Impact Str. (Izod): 4.5 ft lb/in. notch (D256); Hardness (Shore D): 64 (D1706)
Therm. Prop.: Distort. Temp.: 70 C (@ 66 psi, D648)
Ref.: Data sheet

Chemplex 5704. Chemplex Co.
Chem. Descrip.: High-density polyethylene
Category/Applications: High m.w. copolymer with high toughness and stress crack resistance, high melt strength, and ease of processing; used in large industrial tanks, DOT containers up to 10 gal, stadium seats; melt index 20 g/10 min (D1238)
Gen'l. Prop.: Density: 0.952 g/cc (D-1505); Soften. Pt. (Vicat): 124 C (D-1525)
Mech. Prop.: Tens. Str.: 3900 psi (yield, D638); Tens. Elong.: > 600% (D638); Flex. Mod.: 160,000 psi (D790); Impact Str. (Izod): 3.2 ft lb/in. notch (D256); Hardness (Shore D): 64 (D1706)
Therm. Prop.: Distort. Temp.: 69 C (@ 66 psi, D648)
Ref.: Data sheet

Chemplex 5705. Chemplex Co.
Chem. Descrip.: High-density polyethylene
Category/Applications: High m.w. copolymer with high stress crack resistance, high melt strength, and ease of processing; used in large industrial parts, thermoformed sheet; melt index 20 g/10 min (D1238)
Gen'l. Prop.: Density: 0.948 g/cc (D-1505); Soften. Pt. (Vicat): 124 C (D-1525)
Mech. Prop.: Tens. Str.: 3800 psi (yield, D638); Tens. Elong.: > 600% (D638); Flex. Mod.: 150,000 psi (D790); Impact

Str. (Izod): 3.2 ft lb/in. notch (D256); Hardness (Shore D): 63 (D1706)
Therm. Prop.: Distort. Temp.: 68 C (@ 66 psi, D648)
Ref.: Data sheet

Chemplex 6001. Chemplex Co.
Chem. Descrip.: High-density polyethylene
Category/Applications: Homopolymer for blowmolding and extrusion, with high melt strength for low sheet sag; used in sheet for multicavity thermoforming, large blow molded parts; melt index 0.25 g/10 min (D1238)
Gen'l. Prop.: Density: 0.96 g/cc (D-1505); Soften. Pt. (Vicat): 127 C (D-1525)
Mech. Prop.: Tens. Str.: 4500 psi (yield, D638); Tens. Elong.: > 600% (D638); Flex. Mod.: 220,000 psi (D790); Impact Str. (Izod): 8 ft lb/in. notch (D256); Hardness (Shore D): 68 (D1706)
Therm. Prop.: Distort. Temp.: 80 C (@ 66 psi, D648)
Ref.: Data sheet

Chemplex 6004. Chemplex Co.
Chem. Descrip.: High-density polyethylene
Category/Applications: Homopolymer for blow molding and extrusion with good processability and excellent rigidity; used for continuous extrusion blow molding of milk, juice, and water bottles; melt index 0.32 g/10 min (D1238)
Gen'l. Prop.: Density: 0.96 g/cc (D-1505); Soften. Pt. (Vicat): 127 C (D-1525)
Mech. Prop.: Tens. Str.: 4400 psi (yield, D638); Tens. Elong.: > 600% (D638); Flex. Mod.: 220,000 psi (D790); Impact Str. (Izod): 6 ft lb/in. notch (D256); Hardness (Shore D): 68 (D1706)
Therm. Prop.: Distort. Temp.: 80 C (@ 66 psi, D648)
Ref.: Data sheet

Chemplex 6006. Chemplex Co.
Chem. Descrip.: High-density polyethylene
Category/Applications: Homopolymer for blow molding and extrusion; high melt strength version of Chemplex 6009; used for ultra-light milk bottles; melt index 0.6 g/10 min (D1238)
Gen'l. Prop.: Density: 0.96 g/cc (D-1505); Soften. Pt. (Vicat): 127 C (D-1525)
Mech. Prop.: Tens. Str.: 4400 psi (yield, D638); Tens. Elong.: > 600% (D638); Flex. Mod.: 220,000 psi (D790); Impact Str. (Izod): 3 ft lb/in. notch (D256); Hardness (Shore D): 68 (D1706)
Therm. Prop.: Distort. Temp.: 80 C (@ 66 psi, D648)
Ref.: Data sheet

Chemplex 6008. Chemplex Co.
Chem. Descrip.: High-density polyethylene
Category/Applications: Homopolymer for blow molding and extrusion with good processability and rigidity; used in milk, juice, and water bottles; melt index 0.8 g/10 min (D1238)
Gen'l. Prop.: Density: 0.96 g/cc (D-1505); Soften. Pt. (Vicat): 127 C (D-1525)
Mech. Prop.: Tens. Str.: 4400 psi (yield, D638); Tens. Elong.: 500% (D648); Flex. Mod.: 220,000 psi (D790); Impact Str. (Izod): 4 ft lb/in. notch (D256); Hardness (Shore D): 68 (D1706)
Therm. Prop.: Distort. Temp.: 80 C (@ 66 psi, D648)
Ref.: Data sheet

Chemplex 6009. Chemplex Co.
Chem. Descrip.: High-density polyethylene
Category/Applications: Homopolymer for blow molding and extrusion; high swell version of Chemplex 6008; used in

milk, juice, and water bottles; melt index 0.8 g/10 min (D1238)
Gen'l. Prop.: Density: 0.96 g/cc (D-1505); Soften. Pt. (Vicat): 127 C (D-1525)
Mech. Prop.: Tens. Str.: 4400 psi (D-638); Tens. Elong.: > 600% (D638); Flex. Mod.: 220,000 psi (D790); Impact Str. (Izod): 3 ft lb/in. notch (D256); Hardness (Shore D): 68 (D1706)
Therm. Prop.: Distort. Temp.: 80 C (@ 66 psi, D648)
Ref.: Data sheet

Chemplex 6109. Chemplex Co.
Chem. Descrip.: Polyethylene
Category/Applications: High-density polyethylene with high stiffness for film applications, e.g., tissue, laminations coextrusion
Gen'l. Prop.: Density: 0.960+ g/cc (D-1505); Melt Index: 0.85 g/10 min (D-1238)
Mech. Prop. (of 1.25-mil film): Tens. Str.: 3600 psi (TD, yield, D882), 3200 psi (TD, break, D882); Tens. Elong.: 250% (TD, D882); Tens. Mod.: 175,000 psi (TD, 1% secant, D822)
Ref.: Data sheet

Chempol 10-1300. Freeman Chem. Corp.
Chem. Descrip.: Isophthalic acrylic/alkyd copolymer
Category/Applications: Used in water borne industrial finishes, machinery and implement enamels, primers, and maintenance paints where fast dry-to-handle properties are required; also in force-cured coatings with drier catalysts, or may be crosslinked with compatible urea or melamine resins in bake applications
Form: Liquid; Color: 3–8
Composition: 70% solids in isobutanol/butyl Cellosolve 67/33
Gen'l. Prop.: Sol.: Water-reducible with triethyl amine and/or ammonium hydroxide as solubilizers; Density: 8.15–8.35 lb/gal; Visc.: Z_4–Z_6; Acid No.: 32–42; Flash Pt. (TCC): 29 C
Ref.: Tech. data sheet

Chempol 11-3071. Freeman Chem. Corp.
Chem. Descrip.: Short soybean oil chain-stopped alkyd
Category/Applications: Used in air dried or baked coatings requiring reduced solvent emission during cure; high solids coatings formulated with Chempol 11-3071 have good exterior durability, high initial and retained gloss, and resistance to humid and corrosive environments; used for coatings for machinery and farm implements, railcars and other transport equipment, industrial parts and equipment, etc.
Form: Liquid; Color: 3–7
Composition: 75 ± 1% solids in methyl propylketone/toluene (88/12)
Gen'l. Prop.: Sol.: Can be reduced in xylol, butyl acetate, VM&P naphtha; Density: 8.85 ± 0.1 lb/gal; Visc.: Z_1–Z_3; Acid No.: 1–6; Flash Pt. (TCC): 6.6 C
Curing Charact. (air dry–3.0 mil wet film on glass): Set-to-touch: 20–30 min; Dry through: 3–5 h
Ref.: Tech. data sheet

Chempol 11-3363. Freeman Chem. Corp.
Chem. Descrip.: High solids oil-free alkyd
Category/Applications: Used in spray-applied coatings which cure at low temps., e.g., appliance finishes, coatings for metal furniture, metal extrusion coatings; also avail. at 80% solids in exempt solvent as Chempol 11-3362, and in slower evaporating solvent for coil coating and electrostatic spray applications as Chempol 11-3360
Form: Liquid; Color: 2–6

Composition: 75 ± 1% solids in xylol/butanol
Gen'l. Prop.: Density: 8.95 ± 0.1 lb/gal; Visc.: W–Z; Acid No.: 3–5; Flash Pt. (TCC): 27.2 C
Ref.: Tech. data sheet

Chempol 13-1905. Freeman Chem. Corp.
Chem. Descrip.: Vinyltoluene-modified alkyd copolymer
Category/Applications: Copolymer providing very fast drying, high gloss coatings, with high pigment loads; used in coatings for polystyrene, toy enamels, paper coatings, finishes for trailer interiors, aerosol coatings
Form: Liquid; Color: 3–6
Composition: 50 ± 1% solids in VM&P Naphtha
Gen'l. Prop.: Sol.: Sol. without haze—16–18% in mineral spirits, 22–24% in butanol, < 5% in xylol; Density: 7.25–7.45 lb/gal; Visc.: W–Z; Acid No.: 2–6; Flash Pt. (TCC): 35 F
Curing Charact. (1.5-mil wet film on glass at 50% solids by wt.): Set: 5 min; Dry Hard: 10 min; Dry Through: 15 min
Ref.: Tech. data sheet

Chempol 13-2211. Freeman Chem. Corp.
Chem. Descrip.: Vinyltoluene alkyd copolymer resin
Category/Applications: Very fast drying resin for coatings exhibiting high initial gloss plus above average toughness and flexibility; used in low-temp. cure coatings, aerosol systems, fast-dry industrial finishes; compatible with other film-forming resins and may be used as a modifier of slower drying products
Form: Liquid; Color: 3–6
Composition: 50 ± 1% solids in toluol/VM&P Naphtha 5/95 by wt.
Gen'l. Prop.: Sol.: Sol. without haze—< 5% in xylol, 16–18% in mineral spirits, 22–25% in n-butanol; Density: 7.3–7.5 lb/gal; Visc.: Y–Z; Acid No.: 4–7; Flash Pt. (TCC): 35 F
Curing Charact. (Air dry—1.5-mil wet film applied to cold rolled steel): Set-to-Touch: 5–10 min; Dry Hard: 10–15 min; Dry Through: 1–2 h
Ref.: Tech. data sheet

Chempol 19-4807. Freeman Chem. Corp.
Chem. Descrip.: High m.w. polyester resin
Category/Applications: Resin designed to provide tough, high-build coatings which cure by exposure to uv radiation with addition of a suitable photoinitiator (e.g., alkyl benzoin ethers); used for surface filler for particle board, ground coat for printed wood-graining
Form: Liquid; Color: 3–7
Composition: 55 ± 1% solids in styrene
Gen'l. Prop.: Density: 9.2–9.4 lb/gal; Visc.: 800–1200 cps; Acid No.: 20–28; Flash Pt. (TCC): 31 C
Toxicity/Handling: Store in closed containers ⩽ 38 C; avoid exposure to direct sunlight
Ref.: Tech. data sheet

Chempol 19-4815. Freeman Chem. Corp.
Chem. Descrip.: Urethane-acrylate resin
Category/Applications: Rapid radiation-curing, high viscosity resin for use in coatings; cured films have good mar resistance and adhesion to the substrate; used in lithographic inks, metal decorating; recommended uv photoinitiators include benzophenone/tertiary amine combinations, alkyl benzoin ethers, etc.
Form: Liquid; Color: 1–3
Composition: 52 ± 1% solids in trimethylolpropane triacrylate
Gen'l. Prop.: Density: 10.9–11.1 lb/gal; Visc.: 110,000–150,000 cps; Acid No.: 0–2; Hydroxyl No.: 10 max.
Ref.: Tech. data sheet

Chempol 19-4824. Freeman Chem. Corp.
Chem. Descrip.: Epoxy-acrylate resin
Category/Applications: Radiation-curing resin for coatings having rapid cure response, high flexibility, toughness, and stain resistance; used in metal decorating, topcoats for wood and plastic, inks, adhesives, etc.; suggested photoinitiators include benzophenone/tertiary amine combinations, alkyl benzoin ethers, 2-chlorothioxanthone; also avail. at 70% solids in tetra ethylene glycol diacrylate (Chempol 19-4824)
Form: Color: 5–9
Composition: 100% solids
Gen'l. Prop.: Density: 8.95–9.15 lb/gal; Visc.: 3000–5000 cps (180 F); Acid No.: 6–12
Ref.: Tech. data sheet

Chempol 19-4825. Freeman Chem. Corp.
Chem. Descrip.: Epoxy-acrylate resin
Category/Applications: Solvent-free resin for coatings characterized by hard, tough films having good stain and solvent resistance; used in laminating adhesives, wood graining laminates, printing inks, overprint varnishes; suggested uv photoinitiators include benzophenone/tertiary amine combinations and alkyl benzoin ethers
Form: Color: 1–4
Composition: 100% solids
Gen'l. Prop.: Density: 9.8–10.0 lb/gal; Visc.: 4000–6000 cps (140 F); Acid No.: 3–10
Ref.: Tech. data sheet

Chempol 19-4827. Freeman Chem. Corp.
Chem. Descrip.: Urethane-acrylate resin containing active acrylic unsaturation on the polymer molecule
Category/Applications: Base resin for clear and pigmented coatings, e.g., wire coatings, floor tile coatings, inks, overprint varnishes; suggested uv photoinitiators for radiation cure include benzophenone/tertiary amine combinations and alkyl benzoin ethers
Form: Liquid; Color: 1–4
Composition: 100% polymer solids
Gen'l. Prop.: Density: 9.2–9.4 lb/gal; Visc.: 3500–4500 cps (60 C); Acid No.: 0–3; Hydroxyl No.: 2–8
Ref.: Tech. data sheet

Chempol 19-4833. Freeman Chem. Corp.
Chem. Descrip.: Urethane-acrylate resin
Category/Applications: Nonyellowing resin for coatings having high abrasion and scuff resistance; used in clear coatings for floor tile, prefinished wall panels, barrier coatings for paper wall coverings; suggested uv photoinitiators for radiation cure include benzophenone/tertiary amine combinations and alkyl benzoin ethers
Form: Liquid; Color: 1–3
Composition: 90 ± 1% solids in N-vinyl-2-pyrrolidone
Gen'l. Prop.: Density: 9.2–9.4 lb/gal; Visc.: 1670–2670 cps (140 F); Acid No.: 0–3
Ref.: Tech. data sheet

Chempol 19-4836. Freeman Chem. Corp.
Chem. Descrip.: Epoxy-acrylate resin containing active unsaturation on the polymer molecule
Category/Applications: Low viscosity resin for coatings having low odor and good "slip" characteristics, e.g., packaging inks, overprint varnishes for wood, paper, and vinyl film; also used as a viscosity reducer for other radiation cure resins; suggested uv photoinitiators include benzophenone/tertiary amine combinations and alkyl benzoin ethers
Form: Liquid; Color: 3–12

Composition: 100% polymer solids
Gen'l. Prop.: Density: 9.6–9.8 lb/gal; Visc.: 800–1000 cps; Acid No.: 6–12
Ref.: Tech. data sheet

Chempol 19-4840. Freeman Chem. Corp.
Chem. Descrip.: Urethane acrylic resin
Category/Applications: Resin which requires no added amine for uv photoinitiation; forms highly flexible film or can be blended with more brittle radiation cure resin systems to provide a higher degree of toughness and formability; used as modifier for other energy cure resins, in printing inks, and electrical wire coatings; suggested uv photoinitiators include benzophenone and alkyl benzoin ethers
Form: Liquid; Color: 0–3
Composition: 100% polymer solids
Gen'l. Prop.: Density: 9.2–9.4 lb/gal; Visc.: 5000–6000 cps (140 F); Acid No.: 0–3
Ref.: Tech. data sheet

Chempol 19-4842. Freeman Chem. Corp.
Chem. Descrip.: Silicone-modified urethane-acrylate resin
Category/Applications: Nonyellowing resin providing a lower coefficient of friction; used in release coatings for pressure sensitive labels and tapes, shatter-resistant coatings for glass containers, and for overprint varnishes on surfaces where good "slip" and a slick feel are required; suggested uv photoinitiators for radiation cure include benzophenone/tertiary amine combinations and alkyl benzoin ethers
Form: Liquid; Color: 1–3
Composition: 100% polymer solids
Gen'l. Prop.: Density: 8.75–8.95 lb/gal; Visc.: 1300–1600 cps (60 C); Acid No.: 0–3; Flash Pt. (TCC): > 300 F
Ref.: Tech. data sheet

Chempol 30-1947/30-2032. Freeman Chem. Corp.
Chem. Descrip.: Polyether, quasi-prepolymer flexible urethane foam system
Category/Applications: Resin system designed for pour-in-place, vinyl-skinned, med. load bearing, high resiliency seating applications, featuring fast demold, rapid R.T. cure, low compression sets, and variable firmness; mixing ratio: 100/50 parts by wt. (30-1947/30-2032)
Gen'l. Prop.: Density: 8.53 lb/gal, 9.95 lb/gal respectively; Visc.: 700 cps, 100 cps respectively
Mech. Prop.: Tens. Str.: 16 psi; Tens. Elong.: 100%; Tear Str.: 1.0 pli
Toxicity/Handling: Avoid skin and eye contact and vapor inhalation; store components in tightly sealed containers and keep as dry as possible; may present a serious fire hazard
Ref.: Tech. data sheet

Chempol 32-1761/32-1601. Freeman Chem. Corp.
Chem. Descrip.: Freon blown foam system comprised of resin (32-1761) and isocyanate (32-1601)
Category/Applications: Freon blown foam system for applications where a lightweight part with excellent skin toughness is desired, e.g., in the point-of-purchase display industry, ceiling beam manufacture, and nonstructural decorative accessories; also used in marine applications where a high strength-to-weight ratio is desired; the system features excellent mold reproduction, low component viscosity, fast cycle times, long working time; mixing ratio: 100 pbw 32-1761 @ 70 F/100 pbw 32-1601 @ 90 F
Form: Color: Amber and dark brown respectively
Gen'l. Prop.: Density (Free-Rise): 2.0

pcf; Visc.: 600 cps (70 F) and 200 cps (77 F) respectively
Curing Charact.: Gel Time: 115 s
Toxicity/Handling: Avoid breathing fumes and direct contact with skin; may present a serious fire hazard
Ref.: Tech. data sheet

Chempol 33-3101/33-3201. Freeman Chem. Corp.
Chem. Descrip.: Urethane foam system comprised of resin (33-3101) and isocyanate (33-3201)
Category/Applications: Foam designed for pour-in-place packaging; mixing ratio: 100 pbw 33-3101/100 pbw 33-3201
Gen'l. Prop.: Density: 10.1 and 10.2 lb/gal respectively; Free-Rise Density: 0.9 pcf; Visc.: 150 cps and 200 cps respectively (70 F)
Mech. Prop.: Compr. Str.: 10 psi (parallel to rise), 5 psi (perpendicular to rise)
Curing Charact.: Gel Time: 40 s (75 F)
Toxicity/Handling: May present a serious fire hazard; material should be used @ 70–75 F–31-3101 contains fluorocarbon blowing agent which boils @ 75 F
Ref.: Tech. data sheet

Chempol 34-4199/34-5075. Freeman Chem. Corp.
Chem. Descrip.: Rigid polyurethane foam froth system with Fluorocarbon 12
Category/Applications: Rigid polyurethane foam froth pressure bottle system; mixing ratio: 95 pbw 34-4199/100 pbw 34-5075
Gen'l. Prop.: Density: 9.8 and 10.3 lb/gal respectively; Visc.: 300 cps (72 F) and 200 cps (77 F) respectively
Mech. Prop. (2.0 pcf packed core specimen): Tens. Str.: 43 psi (D1623-64); Compr. Str.: 35 psi (to rise @ yield, D-1621-64)
Curing Charact.: Gel Time: 95 s (75–80 F)
Toxicity/Handling: Avoid skin contact and inhalation of vapors; may present a serious fire hazard; avoid moisture contamination of components; maintain component temps. @ 75–80 F
Ref.: Tech. data sheet

Chempol 35-0017. Freeman Chem. Corp.
Chem. Descrip.: Polyester-TDI prepolymer
Category/Applications: Prepolymer yielding a strong, abrasion- and solvent-resistant urethane rubber when cured with MOCA
Form: Color (Gardner): 3
Gen'l. Prop.: Density: 9.6 lb/gal; Visc.: 6000 cps max. (158 F); F.P.: 125 F; Stability: 3 mo min. shelf life when stored in tightly sealed containers @ 60–70 F
Mech. Prop. (after cure): Tens. Str.: 6300 psi (D412); Tens. Elong.: 700% (D412); Tens. Mod.: 400 psi (100% mod., D412); Hardness (Durometer): 82 (D2240)
Toxicity/Handling: See Chempol 30-1947/30-2032
Ref.: Tech. data sheet

Chemprene 50, 75. Chemfax Inc.
Chem. Descrip.: Isoprenoidal polymer
Category/Applications: Polymer designed to meet the requirements of a neutral, mineral spirits soluble resin with excellent water, acid, and alkali resistance and excellent oxidative stability
Form: Viscous liquids; Color (Gardner): 10
Gen'l. Prop.: Sol.: Sol. in mineral spirits; Sp. Gr.: 1.1; Visc.: Z1 (G/H tube method, 70% mineral spirits); Soften. Pt.: 50 F and 75 F (Chemprene 50 and 75 respectively); Acid No.: 0; Sapon. No.: 0; Iodine Value: 150 (R-100); Stability:

Excellent water, acid, and alkali resistance; excellent oxidative stability
Std. Pkgs.: Heavy duty, open head drums with clamp type heads
Ref.: Product data

Chemprene 125, 160, 210, 235. Chemfax Inc.
Chem. Descrip.: Isoprenoidal polymer
Category/Applications: See Chemprene 50, 75
Form: Solid; Color (Gardner): 10
Gen'l. Prop.: Sol.: Sol. in mineral spirits; Sp. Gr.: 1.1; Visc.: Z1 (G/H tube method, 70% mineral spirits); Soften. Pt.: 125, 160, 210, and 235 F (Chemprene 125, 160, 210, 235 respectively); Acid No.: 0; Sapon. No.: 0; Iodine Value: 150 (R-100); Stability: Excellent water, acid, and alkali resistance; excellent oxidative stability
Std. Pkgs.: 29 gauge nonreturnable strippable drums; Chemprene 210 packaged in 4-ply natural kraft bags in flaked form but will remass on extended storage in hot climate; Chemprene 235 is packaged in 4-ply natural kraft bags in flaked form, but will not remass on storage
Ref.: Product data

Chemprene R-10. Chemfax Inc.
Chem. Descrip.: Thermoplastic isoprenoidal polymer
Category/Applications: Polymer for use in the compounding of synthetic, natural, and reclaim rubber; promotes good aging, flexing, and tear resistance; improves processing; reduces scorching and premature set-up in uncured compounds; improves heat resistance; acts as softeners, tackifiers, and reinforcing agents; due to excellent solvency, retards bloom, improves stability by reducing bleeding of organic colors; used in calendering and extruding, hard rubber, molded goods, in camelback, tubes, and tire stocks, in rubber cements, in wire and cable insulation (exhibits high electrical resistance), and in floor tile, shoe compounds, caulks, and curtain wall sealants; R series is specially inhibited to provide a remarkably stable resin for high temp. processing with no loss of solubility due to oxidation
Form: Viscous liquid to semisolid; Color (Barrett): 2
Gen'l. Prop.: Sol.: Wide range of sol. from min. spirits to ketones and cyclic alcohols; Sp. Gr.: 1.02; Soften. Pt.: 10 C; Stability: Exceptionally stable to oxidation, acids, alkalis, and water; excellent high temp. stability
Std. Pkgs.: Heavy duty, open head drums with clamp lid heads
Ref.: Tech. bulletin

Chemprene R-25. Chemfax Inc.
Chem. Descrip.: Thermoplastic isoprenoidal polymer
Category/Applications: See Chemprene R-10
Form: Viscous liquid to semisolid; Color (Barrett): 2
Gen'l. Prop.: Sol.: Wide range of sol. from min. spirits to ketones and cyclic alcohols; Sp. Gr.: 1.04; Soften. Pt.: 25 C; Stability: Exceptionally stable to oxidation, acids, alkalis, and water; excellent high temp. stability
Std. Pkgs.: Heavy duty, open head drums with clamp lid heads
Ref.: Tech. bulletin

Chemprene R-50. Chemfax Inc.
Chem. Descrip.: Thermoplastic isoprenoidal polymer
Category/Applications: See Chemprene R-10
Form: Viscous liquid to semisolid; Color (Barrett): 2

Gen'l. Prop.: Sol.: Wide range of sol. from min. spirits to ketones and cyclic alcohols; Sp. Gr.: 1.06; Soften. Pt.: 50 C; Stability: Exceptionally stable to oxidation, acids, alkalis, and water; excellent high temp. stability
Std. Pkgs.: Heavy duty, open head drums with clamp lid heads
Ref.: Tech. bulletin

Chemprene R-70. Chemfax Inc.
Chem. Descrip.: Thermoplastic isoprenoidal polymer
Category/Applications: See Chemprene R-10
Form: Solid; Color (Barrett): 2
Gen'l. Prop.: Sol.: Wide range of sol. from min. spirits to ketones and cyclic alcohols; Sp. Gr.: 1.08; Soften. Pt.: 70 C; Stability: Exceptionally stable to oxidation, acids, alkalis, and water; excellent high-temp. stability
Std. Pkgs.: 29 gauge nonreturnable strippable drums
Ref.: Tech. bulletin

Chemprene R-100. Chemfax Inc.
Chem. Descrip.: Thermoplastic isoprenoidal polymer
Category/Applications: See Chemprene R-10
Form: Solid and flakes; flakes will remass on continued storage in a hot climate, but will not lose any original properties; Color (Barrett): 2
Gen'l. Prop.: Sol.: Wide range of sol. from min. spirits to ketones and cyclic alcohols; Sp. Gr.: 1.10; Soften. Pt.: 100 C; Stability: Exceptionally stable to oxidation, acids, alkalis, and water; excellent high temp. stability
Std. Pkgs.: 29 gauge drums (solid); 4-ply natural kraft bags (flakes)
Ref.: Tech. bulletin

Chemprene R-115. Chemfax Inc.
Chem. Descrip.: Thermoplastic isoprenoidal polymer
Category/Applications: See Chemprene R-10
Form: Solid and flakes (flakes will not remass on storage)
Gen'l. Prop.: Sol.: Wide range of sol. from min. spirits to ketones and cyclic alcohols; Sp. Gr.: 1.10; Visc.: Z-1 (G/H tube method, 70% in min. spirits); Soften. Pt.: 115 C; Acid No.: 0; Sapon. No.: 0; Stability: Exceptionally stable to oxidation, acids, alkalis, and water; excellent high temp. stability
Std. Pkgs.: 29 gauge drums (solid); 4-ply natural kraft bags (flakes)
Ref.: Tech. bulletin

CI-2. Bacon Industries Inc.
Chem. Descrip.: Two-part epoxy comp.
Category/Applications: Unfilled, low-visc. coil impregnant for the impregnation of electronic components containing a high percentage of fine wires where complete penetration and freedom from voids is important; also for casting and coating applications where a low-visc. material is required; CI-1 also avail. with same cured properties as CI-2 but with an activated visc. of 15 poise @ R.T.
Gen'l. Prop. (activated impregnant—10.0 phr of Activator BA-1, @ R.T.): Sp. Gr.: 1.18 (cured); Visc.: 5 poise; Surface Tension: 48.8 dyne/cm
Mech. Prop. (cured): Flex. Str.: 13×10^{-3} psi (yield); Flex. Mod. (Young's): 0.5×10^{-6} psi; Hardness (Shore D): 85
Therm. Prop. (cured): Coeff. of Linear Exp.: 49×10^{6} F^{-1} (80–200 F)
Elec. Prop. (cured): Dissip. Factor: 0.002 (1 kHz); Dielec. Const.: 3.5 (1 kHz); Vol. Resist.: 5×10^{16} ohm-cm

Curing Charact.: Pot Life: 8 h @ R.T.; Cure Cycle: 8 h @ 212 F
Toxicity/Handling: May cause injury to skin on prolonged or repeated contact; Activator BA-1 is highly flammable, and may cause skin burns and eye injury; use with adequate ventilation; 2 yr shelf life when stored in unopened cans below 85 F
Std. Pkgs.: Qt and gal containers
Ref.: Data sheets 1071

CI-3. Bacon Industries Inc.
Chem. Descrip.: Two-part epoxy comp.
Category/Applications: See CI-2; CI-3 is used where a shorter pot life or cure cycle is required
Gen'l. Prop. (activated impregnant–5.0 phr of Activator BA-4, @ R.T.): Sp. Gr.: 1.18 (cured); Visc.: 20 poise; Surface Tension: 42.2 dyne/cm
Mech. Prop. (cured): Flex. Str.: 12×10^{-3} psi (yield); Flex. Mod. (Young's): 0.4×10^{-6} psi; Hardness (Shore D): 85
Therm. Prop. (cured): Coeff. of Linear Exp.: 42×10^{6} F^{-1} (80–200 F)
Elec. Prop. (cured): Dissip. Factor: 0.003 (1 kHz); Dielec. Const.: 3.8 (1 kHz); Vol. Resist.: 10^{17} ohm-cm
Curing Charact.: Pot Life: 4 h @ R.T.; Cure Cycle: 2 h @ 200 F
Toxicity/Handling: May cause injury to skin on prolonged or repeated contact; use with adequate ventilation; 2 yr shelf life when stored in unopened cans below 85 F
Std. Pkgs.: Qt containers
Ref.: Data sheet 1071

CI-6. Bacon Industries Inc.
Chem. Descrip.: Two-part epoxy comp.
Category/Applications: See CI-2; CI-6 used in applications requiring a longer pot life
Gen'l. Prop. (activated impregnant–23.0 phr of Activator BA-45, @ R.T.): Sp. Gr.: 1.18 (cured); Visc.: 8.5 poise
Mech. Prop. (cured): Flex. Str.: 12×10^{-3} psi (yield); Flex. Mod. (Young's): 0.4×10^{-6} psi; Hardness (Shore D): 85
Therm. Prop. (cured): Coeff. of Linear Exp.: 40×10^{6} F^{-1} (80–200 F)
Elec. Prop. (cured): Dissip. Factor: 0.002 (1 kHz); Dielec. Const.: 3.7 (1 kHz); Vol. Resist.: 10^{16} ohm-cm
Curing Charact.: Pot Life: 24 h @ R.T.; Cure Cycle: 4 h @ 212 F
Toxicity/Handling: May cause injury to skin on prolonged or repeated contact; Activator BA-45 may cause skin burns and eye injury; use with adequate ventilation; 2 yr shelf life when stored in unopened cans below 85 F
Std. Pkgs.: Qt containers
Ref.: Data sheet 1071

Clear Dip. Isochem Resins Co.
Category/Applications: Clear, transparent thixotropic dip featuring high impact and high gloss; also avail. in sprayable form; mix ratio: 25 phr of Hardener Clear Dip
Form: Color: Clear
Gen'l. Prop.: Sp. Gr.: 1.12; Visc.: 4000 cps (mixed, 27 C)
Mech. Prop.: Tens. Str.: 4800 psi; Compr. Str.: 6800 psi; Hardness (Shore D): 86
Therm. Prop.: Conduct.: 4.5×10^{-4} Cal/s cm^2 C/cm; Distort. Temp.: 140 C; Coeff. of Linear Exp.: 6.2×10^{-5}/C
Elec. Prop.: Dissip. Factor: 0.022 (100 KC, 27 C); Dielec. Str.: 410 V/mil; Dielec. Const.: 4.6 (100 KC, 27 C); Vol. Resist.: 10^{13} ohm-cm
Curing Charact.: Pot Life: 3 h (1 lb mass @ 27 C); Suggested Cure: R.T. or mild heat
Ref.: Bulletin

Clorafin 40. Hercules Inc.
Chem. Descrip.: Chlorinated paraffin
Category/Applications: Used to flameproof and waterproof textiles; as a plasticizer for vinyl and other resins or film-formers; as an additive in cutting oils and drawing compounds
Form: Viscous liquid; Color (Gardner): 6 max. (1933 standards)
Composition: 42% chloride content
Gen'l. Prop.: Sol.: Miscible with all classes of commonly used solvents except lower aliphatic alcohols and water; Sp. Gr.: 1.16; Density: 9.6 lb/gal; Visc.: 26-29 poise; Stability: High degree of stability; good chemical resistance
Ref.: Data sheet 107-2

Clorafin 50. Hercules Inc.
Chem. Descrip.: Chlorinated paraffin
Category/Applications: See Clorafin 40
Form: Viscous liquid; Color (Gardner): 6 max. (1933 standards)
Composition: 50% chloride content
Gen'l. Prop.: Sol.: See Clorafin 40; Sp. Gr.: 1.26; Density: 10.4 lb/gal; Visc.: 430-620 poise; Stability: See Clorafin 40
Ref.: Data sheet 107-2

Clysar ECL. Du Pont Co.
Chem. Descrip.: Irradiated, biaxially oriented polyethylene film
Category/Applications: Shrink film for high profile and irregular shaped packages where a high percentage of shrinkage and/or strong, durable seals are needed; functional on manual, semiautomatic, and most automatic shrink packaging machines; flat film avail. untreated (ECL) and treated on one side (ECL-T); folded film avail. untreated (ECLF) or treated on one side (ECL-FT); FDA and USDA approved for contact with food and meat products respectively
Form: Avail. in 0.60-, 0.75-, 1.00-, and 1.50-mil gauges as flat or folded film
Gen'l. Prop.: Stability: Remains durable at freezer temps.; will not become brittle with age; good burn-through resistance
Mech. Prop.: Tens. Str.: 2700–6800 g/in. (break, based on tens. 10,000 psi, D-882); Tens. Elong.: 115% (D882); Stiffness Mod.: 13.6–34.1 kg/in. (based on mod. 50,000 psi, D882); Tear Str. (Elmendorf): 10–30 g (D1922)
Ref.: Bulletin E-40681

Clysar EH. Du Pont Co.
Chem. Descrip.: Biaxially oriented polyethylene shrink film
Category/Applications: Used for shrink packaging a wide variety of food and nonfood products, esp. overwrapping, multipackaging, bundling, window packages, and tray wraps; FDA and USDA approved; avail. in flat or folded film, untreated or treated on one side
Form: Avail. in 0.60-, 0.75-, 1.00-, 1.50-, and 2.00-mil gauges in flat or folded film
Gen'l. Prop.: Stability: Remains durable at freezer temps.; does not embrittle with age
Mech. Prop.: Tens. Str.: 2000–3000 g/in. (break, based on tens. 10,000 psi, D-882); Tens. Elong.: 130% (D882); Stiffness Mod.: 13.6–45.4 kg/in. (based on mod. 50,000 psi, D882); Tear Str. (Elmendorf): 4–15 g (D1922)
Ref.: Bulletin E-40679

Clysar EHC. Du Pont Co.
Chem. Descrip.: Biaxially oriented copolymer film
Category/Applications: Strong, clear, heat-shrinkable film for applications

where max. optics and a tough, durable wrap are required; produces strong seals and excellent shrinkage on shadow boxes, irregular shapes, and general purpose over-wrapping under a wide range of operating conditions; FDA and USDA approved; avail. in flat or folded film, untreated or treated on one side

Form: Avail. in 0.50-, 0.60-, 0.75-, 1.00-, 1.25-, and 1.50-mil gauges as flat or folded film; Color: Clear

Gen'l. Prop.: Stability: Excellent film durability even at freezer temps.; will not embrittle with age; good burn-through resistance

Mech. Prop.: Tens. Str.: 4310–12,925 g/in. (break, based on tens. 19,000 psi, D-882); Tens. Elong.: 130% (D882); Stiffness Mod.: 25–76 kg/in. (based on mod. 112,000 psi, D882); Tear Str. (Elmendorf): 4.5–8 g (D1922)
Ref.: Bulletin E-40680

CMC-6CTL. Hercules Inc.

Chem. Descrip.: Crude grade of sodium carboxymethylcellulose

Category/Applications: Used as a soil-suspending agent to improve the efficiency and whiteness retention of powder detergents
Form: Granular powder
Composition: 65% purity
Gen'l. Prop.: Sol.: Readily sol. in cold or hot water; solutions have some tolerance for methanol, ethanol, and acetone; Density (Bulk): 0.65 g/ml (tamped); Visc.: 25–150 cps (2% sol'n.); Stability: Films of CMC-6CTL are insol. in organic solvents and resist penetration by oils and greases
Toxicity/Handling: Store in a cool, dry place
Std. Pkgs.: 50 lb net multiwall bag
Ref.: Bulletin No. 484

CMC-T. Hercules Inc.

Chem. Descrip.: Technical grade of sodium carboxymethylcellulose

Category/Applications: Polymer used as a thickener, binder, film-former, stabilizer, and suspending aid in the textile, adhesive, detergent, paper, paint, and other industries; used in applications where its byproduct salts (primarily sodium chloride and sodium glycolate) can be tolerated

Form: Powder, granular, or pellet; Color: Off-white to tan; Odor: Odorless

Composition: 96.0% purity

Gen'l. Prop.: Sol.: Sol. in cold or hot water; Density (Bulk): 0.75 g/ml; Visc. (Brookfield): 20–600 cps (2% sol'n.); Stability: Films of CMC-T are resistant to oils, greases, and organic solvents
Ref.: Bulletin No. 402C

CMC Warp Size. Hercules Inc.

Chem. Descrip.: Sodium carboxymethylcellulose

Category/Applications: Water-soluble "low add-on" warp size for cotton, synthetic fibers, polyester and cotton blends, and other fabric blends; has high film strength, excellent adhesion

Form: Nondusting pellets (CMC Warp Size 8B), granular powder (CMC Warp Size 8), or fine powder (CMC Warp Size Regular); Color: White to tan powder; Odor: Odorless

Gen'l. Prop.: Sol.: Completely sol. in cold or hot water; Density (Bulk): 43 lb/ft^3 (packed pellets), 53 lb/ft^3 (packed powder); Visc. (Brookfield): 1500 cps (60% sol'n.); pH: 7-8.5 (2% solids)
Ref.: Bulletin VC-506

CMHEC-37L. Hercules Inc.
Chem. Descrip.: Sodium salt of carboxymethyl hydroxyethyl cellulose
Category/Applications: Hydrophilic colloid containing ionic and nonionic modifying groups; high water-binding capacity and good flocculating action on suspended solids; forms good films when cast from water solution
Form: Granular powder; Color: Light cream to white
Gen'l. Prop.: Sol.: Sol. in water; broad compatibility with various salts; tolerates ethanol, glycols, and glycerin in certain amounts; Sp. Gr.: 1.58 (film, @ @ 50% r.h.); Density (Bulk): 0.6 g/ml; Visc.: 25–85 cps (2% conc.); Stability: Inert to almost all organic solvents, e.g., ketones, esters, pure alcohols, ethers, hydrocarbons, and chlorinated solvents; broad tolerance for acids, alkalis, and salt; discolors above 230 C; Ref. Index: 1.530 (film, sodium D line)
Mech. Prop.: Tens. Str.: 10,000 psi (50% r.h.); Tens. Elong.: 5% (@ break, 50% r.h.)
Ref.: Bulletin VC-402D

Coagulant CHA. Bayer AG
Chem. Descrip.: Cyclohexyl amine acetate
Category/Applications: Coagulant used in the production of thick-sectioned dipped goods from natural latex by the predip process
Form: Crystalline; Color: White
Gen'l. Prop.: Density: ≈ 1.12 g/cc; pH: ≈ 7.5 (10% aq.)
Std. Pkgs.: 50-kg paper bags
Ref.: Pamphlet KA 32125 e

Coagulant WS. Bayer AG
Chem. Descrip.: Functional polyorganosiloxane
Category/Applications: Heat sensitizer which enables one to obtain a coagulation point of 32–60 C in mixes based on natural or synthetic latex, preferably Perbunan N and Baypren Latices, but also styrene-butadiene and acrylic latices
Form: Liquid; Color: Yellowish
Gen'l. Prop.: Density: ≈ 1.03 g/cc; pH: ≈ 8
Std. Pkgs.: 50-kg plastic cans; 200-kg rolling channel drums
Ref.: Pamphlet KA 32125 e

Conoco 5305 PVC Resin. Conoco Chem. Div.
Chem. Descrip.: Low m.w. PVC homopolymer
Category/Applications: Resin for use in calendering, extrusion, and injection molding applications where low melt viscosity and good physical properties are required; provides increased productivity and shorter molding cycle time due to lower melt viscosity; rigid injection molding applications include pipe fittings, housewares, appliance parts; used in rigid and cellular profile extrusion, and in calendered film applications
Form: Solid
Gen'l. Prop.: Density (bulk): 0.58 g/cc; Stability: Excellent heat stability
Std. Pkgs.: 50-lb bags and bulk
Ref.: Product data sheet

Conoco 5385 PVC Resin. Conoco Chem. Div.
Chem. Descrip.: PVC resin
Category/Applications: PVC resin for use in extrusion, calendering, or injection molding processes for rigid and flexible products; its excellent heat stability, consistency, and processing characteristics are ideal for rigid applications, e.g., PVC pipe, siding, and other profiles; the low gel content, uniform absorption of plasticizer, and high clarity characteristics are ideal for flexible applications,

in the wire and cable, automotive, and shoe industries
Form: Solid
Gen'l. Prop.: Density (bulk): 33.5 lb/ft^3; Stability: Excellent heat stability
Std. Pkgs.: 50-lb bags and bulk
Ref.: Product data sheet

Conoco 5425 PVC Resin. Conoco Chem. Div.
Chem. Descrip.: PVC homopolymer
Category/Applications: General purpose PVC resin for use in extrusion, calendering, or injection molding processes for rigid and flexible products, incl. sheeting, wire, cable, and flexible profiles; uniform absorption of high levels of plasticizer, low gel content, high clarity, excellent heat stability, and good aging characteristics
Form: Solid
Gen'l. Prop.: Density (bulk): 32 lb/ft^3; Stability: Excellent heat stability
Std. Pkgs.: 50-lb bags and bulk
Ref.: Product data sheet

Conoco 5465 PVC Resin. Conoco Chem. Div.
Chem. Descrip.: High m.w. PVC homopolymer
Category/Applications: General purpose PVC resin for calendering, extrusion, and injection molding, with sufficiently high m.w. for "toughness" in finished product, and sufficiently low melt viscosity for ease of manufacture; used in flexible applications (automotive under-the-hood electrical applications), calendering, injection molding (60A durometer), extrusion of sheeting, wire and cable, and flexible profile formulations; allows uniform absorption of high levels of plasticizers, with low gel content and high clarity
Form: Solid
Gen'l. Prop.: Density (bulk): 32 lb/ft^3; Stability: Excellent heat stability
Std. Pkgs.: 50-lb bags and bulk
Ref.: Product data sheet

Conoco 80172. Conoco Chem. Div.
Chem. Descrip.: PVC dryblend
Category/Applications: PVC dryblend for extrusion of pipe and similar products on single-screw extruders; NSF-PPI approved for pressure pipe and DWV applications; also used for irrigation and communications conduit; good processability with less regrind and higher output
Form: Solid
Gen'l. Prop.: Sp. Gr.: 1.41 (D792); Stability: Good heat stability
Mech. Prop.: Tens. Str.: 7800 psi (D-638); Tens. Mod.: 443,000 psi (D638); Impact Str. (Izod): 0.9 ft lb/in. notched (D256); Hardness (Shore D): 78 (D-2240)
Therm. Prop.: Distort. Temp.: 73 C (@ 264 psi, unannealed, D648)
Std. Pkgs.: 1000-lb gaylords, hopper cars, and trucks
Ref.: Product data sheet

Conoco 80273. Conoco Chem. Div.
Chem. Descrip.: PVC dryblend
Category/Applications: PVC dryblend for extrusion of pipe and similar products on twin- and multi-screw extruders; NSF-PPI approved for pressure pipe and DVW applications; also used for irrigation and communications conduit; good processability with less regrind and higher output
Form: Solid
Gen'l. Prop.: Sp. Gr.: 1.41 (D792); Stability: Good heat stability
Mech. Prop.: Tens. Str.: 7900 psi (D-638); Tens. Mod.: 447,000 psi (D638); Impact Str. (Izod): 1.0 ft lb/in. notched

(D256); Hardness (Shore D): 81 (instantaneous, D2240)
Therm. Prop.: Distort. Temp.: 73 C (@ 264 psi, unannealed, D648)
Std. Pkgs.: 1000-lb gaylords, hopper cars and trucks
Ref.: Product data sheet

Conoco 90171. Conoco Chem. Div.
Chem. Descrip.: PVC dryblend
Category/Applications: PVC dryblend providing processing and end-use properties comparable to typical pelletized compounds; for injection molding to produce potable water (PW), and drain, waste, and vent (DVW) fittings (NSF-approved)
Form: Solid; Color: White
Gen'l. Prop.: Sp. Gr.: 1.41 (D792)
Mech. Prop.: Tens. Str.: 7800 psi (D-638); Tens. Mod.: 501,000 psi (D638); Impact Str. (Izod): 0.7 ft lb/in. notched (D256); Hardness (Shore D): 81 (instantaneous, D2240)
Therm. Prop.: Distort. Temp.: 73 C (@ 264 psi, unannealed, D648)
Std. Pkgs.: 1000-lb gaylords, hopper cars, and trucks
Ref.: Product data sheet

Conoco Compound No. 14331. Conoco Chem. Div.
Chem. Descrip.: PVC resin
Category/Applications: PVC resin for general extrusion usage where weather resistance, ultraviolet light stability, and permanence are necessary; exhibits high degree of processing latitude; used for carpet runners, cable coating, and general purpose extrusion
Form: Solid
Gen'l. Prop.: Sp. Gr.: 1.23 (D792-66); Stability: Good ultraviolet light stability, heat stability, and weather resistance
Mech. Prop.: Tens. Str.: 3000 psi (D638-72); Tens. Elong.: 400% (D638-72); 100% Mod.: 1700 psi (D638-72); Hardness (Shore D): 45
Std. Pkgs.: 50-lb bags, hopper cars, hopper trucks, Gaylords
Ref.: Product data sheet

Conoco Compound No. 15041. Conoco Chem. Div.
Chem. Descrip.: PVC resin
Category/Applications: PVC resin for general extrusion usage where weather resistance, ultraviolet light stability, and permanence are necessary; exhibits high degree of processing latitude; used for profiles, carpet runners, cable coating; designed for applications requiring clarity and high gloss
Form: Solid
Gen'l. Prop.: Sp. Gr.: 1.26 (D792-66); Stability: Good ultraviolet light stability, heat stability, and weather resistance
Mech. Prop.: Tens. Str.: 3100 psi (D638-72); Tens. Elong.: 350% (D638-72); 100% Mod.: 2000 psi (D638-72); Hardness (Shore D): 50
Std. Pkgs.: 50-lb bags, hopper cars, hopper trucks, Gaylords
Ref.: Product data sheet

Conoco Compound No. 15071. Conoco Chem. Div.
Chem. Descrip.: PVC resin
Category/Applications: PVC resin for general extrusion usage where weather resistance, ultraviolet light stability, and permanence are necessary; exhibits high degree of processing latitude; used for carpet runners, general purpose extrusion, welting, weather stripping
Form: Solid
Gen'l. Prop.: Sp. Gr.: 1.41 (D792-66); Stability: Good ultraviolet light stability, heat stability, weather resistance
Mech. Prop.: Tens. Str.: 2600 psi (D638-72); Tens. Elong.: 300% (D638-72);

100% Mod.: 1700 psi (D638-72); Hardness (Shore D): 50
Std. Pkgs.: 50-lb bags, hopper cars, hopper trucks, Gaylords
Ref.: Product data sheet

Conoco Compound No. 16081. Conoco Chem. Div.
Chem. Descrip.: PVC resin
Category/Applications: PVC resin for general extrusion usage where weather resistance, ultraviolet light stability, and permanence are necessary; exhibits high degree of processing latitude; used for general purpose extrusion and weather stripping
Form: Solid
Gen'l. Prop.: Sp. Gr.: 1.45 (D792-66); Stability: Good ultraviolet light stability, heat stability, weather resistance
Mech. Prop.: Tens. Str.: 2800 psi (D638-72); Tens. Elong.: 250% (D638-72); 100% Mod.: 2000 psi (D638-72); Hardness (Shore D): 60
Std. Pkgs.: 50-lb bags, hopper cars, hopper trucks, Gaylords
Ref.: Product data sheet

Conoco Compound No. 17461. Conoco Chem. Div.
Chem. Descrip.: PVC resin
Category/Applications: PVC resin for general extrusion usage where weather resistance, ultraviolet light stability, and permanence are necessary; exhibits high degree of processing latitude; used for profiles, general purpose extrusion, and weather stripping
Form: Solid
Gen'l. Prop.: Sp. Gr.: 1.38 (D792-66); Stability: Good ultraviolet light stability, heat stability, weather resistance
Mech. Prop.: Tens. Str.: 1600 psi (D638-72); Tens. Elong.: 350% (D638-72); 100% Mod.: 700 psi (D638-72); Hardness (Shore A): 74
Std. Pkgs.: 50-lb bags, hopper cars, hopper trucks, Gaylords
Ref.: Product data sheet

Conoco Compound No. 18161. Conoco Chem. Div.
Chem. Descrip.: PVC resin
Category/Applications: PVC resin for general extrusion usage where weather resistance, ultraviolet light stability, and permanence are necessary; exhibits high degree of processing latitude; used for profiles, general purpose extrusion, and weather stripping
Form: Solid
Gen'l. Prop.: Sp. Gr.: 1.38 (D792-66); Stability: Good ultraviolet light stability, heat stability, weather resistance
Mech. Prop.: Tens. Str.: 1700 psi (D638-72); Tens. Elong.: 350% (D638-72); 100% Mod.: 700 psi (D638-72); Hardness (Shore A): 81
Std. Pkgs.: 50-lb bags, hopper cars, hopper trucks, Gaylords
Ref.: Product data sheet

Conoco Compound No. 18961. Conoco Chem. Div.
Chem. Descrip.: PVC resin
Category/Applications: PVC resin for general extrusion usage where weather resistance, ultraviolet light stability, and permanence are necessary; exhibits high degree of processing latitude; used for profiles, general purpose extrusion, and weather stripping
Form: Solid
Gen'l. Prop.: Sp. Gr.: 1.38 (D792-66); Stability: Good ultraviolet light stability, heat stability, weather resistance
Mech. Prop.: Tens. Str.: 2300 psi (D638-72); Tens. Elong.: 300% (D638-72); 100% Mod.: 1300 psi (D638-72); Hardness (Shore A): 89
Std. Pkgs.: 50-lb bags, hopper cars, hopper trucks, Gaylords
Ref.: Product data sheet

Conoco Compound No. 26831. Conoco Chem. Div.
Chem. Descrip.: PVC resin
Category/Applications: PVC resin for injection molding applications where high heat stability and permanence are essential; exhibits excellent weather resistance, ultraviolet light stability, and sulfur staining resistance; allows a high degree of processing latitude; can be colored in all shades
Form: Solid
Gen'l. Prop.: Sp. Gr.: 1.23 (D792-66); Stability: Excellent high heat and ultraviolet light stability; good weather and sulfur staining resistance
Mech. Prop.: Tens. Str.: 1500 psi; Tens. Elong.: 450% (D638-72); 100% Mod.: 500 psi (D638-72); Hardness (Shore A): 68
Std. Pkgs.: 50-lb bags, hopper cars and trucks, Gaylords
Ref.: Product data sheet

Conoco Compound No. 27121. Conoco Chem. Div.
Chem. Descrip.: PVC resin
Category/Applications: See Conoco 26831
Form: Solid
Gen'l. Prop.: Sp. Gr.: 1.18 (D792-66); Stability: Excellent high heat and ultraviolet light stability; good weather and sulfur staining resistance
Mech. Prop.: Tens. Str.: 1800 psi; Tens. Elong.: 450% (D638-72); 100% Mod.: 600 psi (D638-72); Hardness (Shore A): 71
Std. Pkgs.: 50-lb bags, hopper cars and trucks, Gaylords
Ref.: Product data sheet

Conoco Compound No. 28131. Conoco Chem. Div.
Chem. Descrip.: PVC resin
Category/Applications: See Conoco 26831
Form: Solid
Gen'l. Prop.: Sp. Gr.: 1.23 (D792-66); Stability: Excellent high heat and ultraviolet light stability; good weather and sulfur staining resistance
Mech. Prop.: Tens. Str.: 2200 psi; Tens. Elong.: 450% (D638-72); 100% Mod.: 900 psi (D638-72); Hardness (Shore A): 81
Std. Pkgs.: 50-lb bags, hopper cars and trucks, Gaylords
Ref.: Product data sheet

Conoco Compound No. 28831. Conoco Chem. Div.
Chem. Descrip.: PVC resin
Category / Applications: See Conoco 26831
Form: Solid
Gen'l. Prop.: Sp. Gr.: 1.23 (D792-66); Stability: Excellent high heat and ultraviolet light stability; good weather and sulfur staining resistance
Mech. Prop.: Tens. Str.: 2400 psi; Tens. Elong.: 350% (D638-72); 100% Mod.: 1300 psi (D638-72); Hardness (Shore A): 88
Std. Pkgs.: 50-lb bags, hopper cars and trucks, Gaylords
Ref.: Product data sheet

Conoco Compound No. 34431. Conoco Chem. Div.
Chem. Descrip.: PVC compound
Category/Applications: PVC wire and cable compound with excellent heat stability, tens. strength, and vol. resistivity; good processability; U.L. applications are: general purpose jacket, UF insulation, T, TW, NMC (UF) 80 C appliance wire, SPT–1,2,3
Form: Solid
Gen'l. Prop.: Sp. Gr.: 1.37 ± .02 (D792); Stability: Excellent heat stability
Mech. Prop.: Tens. Str.: 2500 psi (D-638); Tens. Elong.: 300% (D638); 100%

Mod.: 1500 psi (D638); Hardness (Shore D): 44 ± 3
Elec. Prop.: Vol. Resist.: 7 X 10^{13} ohm-cm (stabilized @ 50 C, D257)
Std. Pkgs.: 50-lb bags, hopper cars and trucks, Gaylords
Ref.: Product data sheet

Conoco Compound No. 34551. Conoco Chem. Div.
Chem. Descrip.: PVC compound
Category/Applications: PVC wire and cable compound with excellent heat stability, tens. strength, and vol. resistivity; good processability; U.L. applications are: insulation, SJT, ST, 105 C appliance wire
Form: Solid
Gen'l. Prop.: Sp. Gr.: 1.36 ± .02 (D792); Stability: Excellent heat stability
Mech. Prop.: Tens. Str.: 2700 psi (D-638); Tens. Elong.: 300% (D638); 100% Mod.: 1700 psi (D638); Hardness (Shore D): 45 ± 3 (D2240)
Elec. Prop.: Vol. Resist.: 5 X 10^{12} ohm-cm (stabilized @ 50 C, D257)
Std. Pkgs.: 50-lb bags, hopper cars and trucks, Gaylords
Ref.: Product data sheet

Conoco Compound No. 34641. Conoco Chem. Div.
Chem. Descrip.: PVC compound
Category/Applications: PVC wire and cable compound with excellent heat stability, tens. strength, and vol. resistivity; good processability; U.L. applications are: machine tool, radiant heating wire 1000 V; TW, THW 90 C appliance wire, 60 C oil; also used for IPCEA
Form: Solid
Gen'l. Prop.: Sp. Gr.: 1.37 ± .02 (D792); Stability: Excellent heat stability
Mech. Prop.: Tens. Str.: 2400 spi (D-638); Tens. Elong.: 300% (D638); 100% Mod.: 1600 psi (D638); Hardness (Shore D): 46 ± 3 (D2240)
Elec. Prop.: Vol. Resist.: 3 X 10^{12} ohm-cm (stabilized @ 50 C, D257)
Std. Pkgs.: 50-lb bags, hopper cars and trucks, Gaylords
Ref.: Product data sheet

Conoco Compound No. 35441. Conoco Chem. Div.
Chem. Descrip.: PVC compound
Category/Applications: PVC wire and cable compound with excellent heat stability, tens. strength, and vol. resistivity; good processability; U.L. applications are: insulation, appliance wire 1000 V; TW, THW, 90 C appliance wire, 75 C oil
Form: Solid
Gen'l. Prop.: Sp. Gr.: 1.40 ± .02 (D792); Stability: Excellent heat stability
Mech. Prop.: Tens. Str.: 2700 psi (D-638); Tens. Elong.: 300% (D638); 100% Mod.: 1900 psi (D638); Hardness (Shore D): 54 ± 3 (D2240)
Elec. Prop.: Vol. Resist.: 3 X 10^{13} ohm-cm (stabilized @ 50 C, D257)
Std. Pkgs.: 50-lb bags, hopper cars and trucks, Gaylords
Ref.: Product data sheet

Conoco Compound No. 35851. Conoco Chem. Div.
Chem. Descrip.: PVC compound
Category/Applications: PVC wire and cable compound with excellent heat stability, tens. strength, and vol. resistivity; good processability; U.L. applications are: insulation, appliance, building wire 1000 V; TW, THW, THWN, THHN, 105 C appliance wire; also used for IPCEA
Form: Solid
Gen'l. Prop.: Sp. Gr.: 1.38 ± .02 (D-792); Stability: Excellent heat stability
Mech. Prop.: Tens. Str.: 2900 psi (D-638); Tens. Elong.: 300% (D638); 100% Mod.: 2300 psi (D638); Hardness (Shore D): 58 ± 3 (D2240)

Elec. Prop.: Vol. Resist.: 3 X 10^{13} ohm-cm (stabilized @ 50 C, D257)
Std. Pkgs.: 50-lb bags, hopper cars and trucks, Gaylords
Ref.: Product data sheet

Conoco Compound No. 38551. Conoco Chem. Div.
Chem. Descrip.: PVC compound
Category/Applications: PVC wire and cable compound with excellent heat stability, tens. strength, and vol. resistivity; good processability; U.L. applications are: primary insulation, jacket SPT–1,2,3; 105 C appliance wire; ST, SVT, SJT, CSA blanket wire
Form: Solid
Gen'l. Prop.: Sp. Gr.: 1.33 ± .02 (D-792); Stability: Excellent heat stability
Mech. Prop.: Tens. Str.: 2300 psi (D-638); Tens. Elong.: 400% (D638); 100% Mod.: 1000 psi (D638); Hardness (Shore A): 85 ± 3 (D2240)
Elec. Prop.: Vol. Resist.: 2 X 10^{11} ohm-cm (stabilized @ 50 C, D257)
Std. Pkgs.: 50-lb bags, hopper cars and trucks, Gaylords
Ref.: Product data sheet

Conoco Compound No. 38601. Conoco Chem. Div.
Chem. Descrip.: PVC compound
Category/Applications: PVC wire and cable compound with excellent heat stability, tens. strength, and vol. resistivity; good processability; U.L. applications are: general purpose jacket, 1/32 in. wall service entry, 60 C, 80 C appliance wire, tubing, gen. purpose extrusion
Form: Solid
Gen'l. Prop.: Sp. Gr.: 1.37 ± .02 (D-792); Stability: Excellent heat stability
Mech. Prop.: Tens. Str.: 2200 psi (D-638); Tens. Elong.: 350% (D638); 100% Mod.: 1000 psi (D638); Hardness (Shore A): 86 ± 3 (D2240)
Elec. Prop.: Vol. Resist.: 6 X 10^{10} ohm-cm (stabilized @ 50 C, D257)
Std. Pkgs.: 50-lb bags, hopper cars and trucks, Gaylords
Ref.: Product data sheet

Conoco Compound No. 38631. Conoco Chem. Div.
Chem. Descrip.: PVC compound
Category/Applications: PVC wire and cable compound with excellent heat stability, tens. strength, and vol. resistivity; good processability; U.L. applications are: insulation, general purpose jacket, flexible cords; 60 C, 80 C appliance wire; oil, abrasion resistant, nonelectrical cable jacket, tube
Form: Solid
Gen'l. Prop.: Sp. Gr.: 1.40 ± .02 (D-792); Stability: Excellent heat stability
Mech. Prop.: Tens. Str.: 2300 psi (D-638); Tens. Elong.: 350% (D638); 100% Mod.: 1100 psi (D638); Hardness (Shore A): 86 ± 3 (D2240)
Elec. Prop.: Vol. Resist.: 3 X 10^{10} ohm-cm (stabilized @ 50 C, D257)
Std. Pkgs.: 50-lb bags, hopper cars and trucks, Gaylords
Ref.: Product data sheet

Conoco Compound No. 38671. Conoco Chem. Div.
Chem. Descrip.: PVC compound
Category/Applications: PVC wire and cable compound with excellent heat stability, tens. strength, and vol. resistivity; good processability; U.L. applications are: low tension applications, SPT–1,2,3 parallel cord; also used for molded parts, plugs
Form: Solid
Gen'l. Prop.: Sp. Gr.: 1.45 ± .02 (D-792); Stability: Excellent heat stability
Mech. Prop.: Tens. Str.: 2100 psi (D-638); Tens. Elong.: 350% (D638); 100% Mod.: 1100 psi (D638); Hardness (Shore A): 86 ± 3 (D2240)

Elec. Prop.: Vol. Resist.: 3×10^{11} ohm-cm (stabilized @ 50 C, D257)
Std. Pkgs.: 50-lb bags, hopper cars and trucks, Gaylords
Ref.: Product data sheet

Conoco Compound No. 38931. Conoco Chem. Div.
Chem. Descrip.: PVC compound
Category/Applications: PVC wire and cable compound with excellent heat stability, tens. strength, and vol. resistivity; good processability; U.L. applications are: blanket wire insulation, 80 C appliance wire; also used for low odor EF conduit
Form: Solid
Gen'l. Prop.: Sp. Gr.: 1.37 ± .02 (D-792); Stability: Excellent heat stability
Mech. Prop.: Tens. Str.: 2200 psi (D-638); Tens. Elong.: 350% (D638); 100% Mod.: 1200 psi (D638); Hardness (Shore A): 89 ± 3 (D2240)
Elec. Prop.: Vol. Resist.: 1×10^{13} ohm-cm (stabilized @ 50 C, D257)
Std. Pkgs.: 50-lb bags, hopper cars and trucks, Gaylords
Ref.: Product data sheet

Conoco Compound No. 44171. Conoco Chem. Div.
Chem. Descrip.: PVC resin
Category/Applications: See Conoco 26831
Form: Solid
Gen'l. Prop.: Sp. Gr.: 1.44 (D792-66); Stability: Excellent high heat and ultraviolet light stability; good weather and sulfur staining resistance
Mech. Prop.: Hardness (Shore D): 41
Std. Pkgs.: 50-lb bags, hopper cars and trucks, Gaylords
Ref.: Product data sheet

Conoco Compound No. 44331. Conoco Chem. Div.
Chem. Descrip.: PVC resin
Category/Applications: PVC resin for general extrusion usage where weather resistance, ultraviolet light stability, and permanence are necessary; exhibits high degree of processing latitude; designed for applications requiring clarity and high gloss; used for clear uv stabilized jacket, fencing, lawn furniture
Form: Solid
Gen'l. Prop.: Sp. Gr.: 1.24 (D792-66); Stability: Good ultraviolet light stability, heat stability, weather resistance
Mech. Prop.: Tens. Str.: 2900 psi (D638-72); Tens. Elong.: 400% (D638-72); 100% Mod.: 1700 psi (D638-72); Hardness (Shore D): 44
Std. Pkgs.: 50-lb bags, hopper cars, hopper trucks, Gaylords
Ref.: Product data sheet

Conoco Compound No. 45541. Conoco Chem. Div.
Chem. Descrip.: PVC resin
Category/Applications: PVC resin for general extrusion usage where weather resistance, ultraviolet light stability, and permanence are necessary; exhibits high degree of processing latitude; used for clear uv stabilized jacket, fencing
Form: Solid
Gen'l. Prop.: Sp. Gr.: 1.25 (D792-66); Stability: Good ultraviolet light stability, heat stability, weather resistance
Mech. Prop.: Tens. Str.: 3200 psi (D638-72); Tens. Elong.: 350% (D638-72); 100% Mod.: 2200 psi (D638-72); Hardness (Shore D): 55
Std. Pkgs.: 50-lb bags, hopper cars, hopper trucks, Gaylords
Ref.: Product data sheet

Conoco Compound No. 48561. Conoco Chem. Div.
Chem. Descrip.: PVC resin
Category/Applications: See Conoco 26831
Form: Solid
Gen'l. Prop.: Sp. Gr.: 1.36 (D792-66); Stability: Excellent high heat and ultraviolet light stability; good weather and sulfur staining resistance
Mech. Prop.: Tens. Str.: 1900 psi; Tens. Elong.: 300% (D638-72); 100% Mod.: 1000 psi (D638-72); Hardness (Shore A): 85
Std. Pkgs.: 50-lb bags, hopper cars and trucks, Gaylords
Ref.: Product data sheet

Conoco Compound No. 48562. Conoco Chem. Div.
Chem. Descrip.: PVC resin
Category/Applications: See Conoco 26831
Form: Solid
Gen'l. Prop.: Sp. Gr.: 1.36 (D792-66); Stability: Excellent high heat and ultraviolet light stability; good weather and sulfur staining resistance
Mech. Prop.: Tens. Str.: 1900 psi; Tens. Elong.: 300% (D638-72); 100% Mod.: 1000 psi (D638-72); Hardness (Shore A): 85
Std. Pkgs.: 50-lb bags, hopper cars and trucks, Gaylords
Ref.: Product data sheet

Copolymer 845. GAF Corp.
Chem. Descrip.: Vinylpyrrolidone/dimethylaminoethylmethacrylate copolymer
Category/Applications: Mildly substantive to hair without buildup problems; esp. for use in blow-dry conditioners, setting lotions, and hair-stying gels; also used in skin creams, hand lotions, soaps, shaving products, colognes, deodorants, antiperspirants, hair colors, creme rinses, and shampoos
Form: Liquid
Composition: 20% aq. sol'n.
Ref.: Catalogue 2303-036

Corvic C4988. AECI Ltd./Plastics Div.
Chem. Descrip.: Copolymer of vinyl chloride with 15% vinyl acetate (suspension polymer)
Category/Applications: Resin used for gramophone records and flooring
Composition: 1.8% volatile content
Gen'l. Prop.: Density (Apparent): 780 kg/m^3 (ISO-R60); Visc. (Inherent): 0.52 (D1243)
Ref.: Booklet

Corvic C-6297. AECI Ltd./Plastics Div.
Chem. Descrip.: Copolymer with 10% vinyl acetate (suspension polymer)
Category/Applications: Resin for use in calendering rigid film and foil, esp. for foil intended for deep thermoforming, and for rigid extrusion of film and foil where these are to be deep-drawn
Composition: 0.2% volatile content
Gen'l. Prop.: Density (Apparent): 510 kg/m^3 (ISO-R60); Visc. (Inherent): 0.75 (D1243)
Ref.: Booklet

Corvic E7236. AECI Ltd./Plastics Div.
Chem. Descrip.: Emulsion polymer of vinyl chloride
Category/Applications: Used in the manufacture of battery separators by the sintering process; contains a surfactant to ensure wetting of the separator by the battery acid
Gen'l. Prop.: Density (Apparent): 560 kg/m^3 (ISO-R60); Visc. (Inherent): 1.04 (D1243)
Ref.: Booklet

Corvic P7250. AECI Ltd./Plastics Div.
Chem. Descrip.: Vinyl paste polymer
Category/Applications: Suitable for the production of high visc. plastisols used for coating and dipping for flexible cellular and noncellular artificial leathers and textile-backed gloves
Gen'l. Prop.: Density (Apparent): 370 kg/m^3 (ISO-R60); Visc. (Inherent): 1.04 (D1243)
Ref.: Booklet

Corvic S5716. AECI Ltd./Plastics Div.
Chem. Descrip.: Vinyl suspension polymer
Category/Applications: Used for blow molding of rigid articles (bottles, stadium seating, etc.), rigid injection molding (e.g., large pipe fittings, record cartridge holder, transistor radio case or component, electrical applications), and high-speed calendering of rigid film and foil (e.g., packaging, etc.)
Composition: 0.2% volatile content
Gen'l. Prop.: Density (Apparent): 580 kg/m^3 (ISO-R60); Visc. (Inherent): 0.70 (D1243)
Ref.: Bulletin

Corvic S6611. AECI Ltd./Plastics Div.
Chem. Descrip.: Vinyl suspension polymer
Category/Applications: Used in general purpose flexible extrusions (also suitable for sheathing cable at high speed); flexible injection molding (e.g., shoes, shoe soles); flexible film and foil (e.g., food packaging, seat coverings, shower curtains, floor coverings)
Composition: 0.2% volatile content
Gen'l. Prop.: Density (Apparent): 530 kg/m^3 (ISO-R60); Visc. (Inherent): 0.91 (D1243)
Ref.: Booklet

Corvic S6617. AECI Ltd./Plastics Div.
Chem. Descrip.: Med. molecular mass homopolymer of vinyl chloride, made by a suspension polymerization process
Category/Applications: General purpose resin for rigid and flexible extrusions; to be mixed with suitable stabilizers, lubricants, etc. before processing; can produce unplasticized granulated compounds and dry blends which can be extruded directly into UPVC pressure pipe, sewer and drain-pipe, soil waste and vent pipe, rainwater guttering and downpipe, and profiles
Composition: 0.2% volatile content
Gen'l. Prop.: Density (Apparent): 560 kg/m^3 (ISO-R60); Visc. (Inherent): 0.91 (D1243)
Ref.: Booklet

Corvic S6818. AECI Ltd./Plastics Div.
Chem. Descrip.: Vinyl suspension polymer
Category/Applications: Used for rigid extrusion on twin-screw extruders; esp. useful for pressure pipe
Composition: 0.2% volatile content
Gen'l. Prop.: Density (Apparent): 580 kg/m^3 (ISO-R60); Visc. (Inherent): 0.94 (D1243)
Ref.: Booklet

Corvic S7106. AECI Ltd./Plastics Div.
Chem. Descrip.: High molecular mass homopolymer of vinyl chloride, made by a suspension polymerization process
Category/Applications: General purpose polymer esp. recommended for the manufacture of plasticized compositions for extrusion, film blowing, calendering, and injection molding, e.g., electric cable, footwear, hose pipe, and flexible sections; a special grade is avail. for the manufacture of film and sheeting containing a very low level of "fish eyes"
Composition: 0.2% volatile content

Gen'l. Prop.: Density (Apparent): 500 kg/m^3 (ISO-R60); Visc. (Inherent): 1.06 (D1243)
Ref.: Booklet

Cosden Polystyrene 500. Cosden Oil & Chem. Co.
Chem. Descrip.: Polystyrene
Category/Applications: Easy-flowing polystyrene with fast set-up and excellent clarity; can be used in thin and med. wall sections requiring long distances of flow; broad molding range
Gen'l. Prop.: Sp. Gr.: 1.05 (D792-66); Soften. Pt. (Vicat): 196 F (D1525-70)
Mech. Prop.: Tens. Str.: 5600 psi (D-638-72); Tens. Mod.: 330,000 psi (D-638-72); Flex. Str.: 10,000 psi (D790-71); Flex. Mod.: 390,000 psi (D790-71); Hardness (Rockwell): M 75 (D785-65)
Therm. Prop.: Distort. Temp.: 155 F (unannealed, D648-72)
Ref.: Product information

Cosden Polystyrene 525. Cosden Oil & Chem. Co.
Chem. Descrip.: Polystyrene
Category/Applications: Standard flow polystyrene with a higher heat distortion than Cosden 500, providing faster set-up required in heavier section molding; good clarity; broader molding range at a somewhat higher temp. than Cosden 500
Gen'l. Prop.: Sp. Gr.: 1.05 (D792-66); Soften. Pt. (Vicat): 200 F (D1525-70)
Mech. Prop.: Tens. Str.: 5800 psi (D-638-72); Tens. Mod.: 350,000 psi (D-638-72); Flex. Str.: 9800 psi (D790-71); Flex. Mod.: 380,000 psi (D790-71); Hardness (Rockwell): M 65 (D785-65)
Therm. Prop.: Distort. Temp.: 160 F (unannealed, D648-72)
Ref.: Product information

Cosden Polystyrene 550. Cosden Oil & Chem. Co.
Chem. Descrip.: Polystyrene
Category/Applications: High heat polymer for injection molding applications requiring max. heat distortion and increased strength, e.g., high quality cutlery; also used in foam sheet; good flow and excellent clarity
Gen'l. Prop.: Sp. Gr.: 1.05 (D792-66); Soften. Pt. (Vicat): 224 F (D1525-70)
Mech. Prop.: Tens. Str.: 7500 psi (D-638-72); Tens. Mod.: 360,000 psi (D-638-72); Flex. Str.: 14,600 psi (D790-71); Flex. Mod.: 430,000 psi (D790-71); Hardness (Rockwell): M 60 (D785-65)
Therm. Prop.: Distort. Temp.: 174 F (unannealed, D648-72)
Ref.: Product information

Cosden Polystyrene 625. Cosden Oil & Chem. Co.
Chem. Descrip.: Polystyrene
Category/Applications: Medium impact resin with very easy flow; used for thin wall containers, toys, and industrial parts; an excellent low cost, impact material
Gen'l. Prop.: Sp. Gr.: 1.04 (D792-66); Soften. Pt. (Vicat): 192 F (D1525-70)
Mech. Prop.: Tens. Str.: 3100 psi (D-638-72); Tens. Elong.: 30% (D638-72); Tens. Mod.: 290,000 psi (D638-72); Flex. Str.: 5600 psi (D790-71); Flex. Mod.: 360,000 psi (D790-71); Impact Str. (Izod): 0.9 ft lb/in. notched (D256-73); Hardness (Rockwell): M 35 (D785-65)
Therm. Prop.: Distort. Temp.: 150 F (unannealed, D648-72)
Ref.: Product information

Cosden Polystyrene 710. Cosden Oil & Chem. Co.
Chem. Descrip.: Polystyrene
Category/Applications: Medium impact

extrusion grade polystyrene, esp. for packaging applications which will be formed in ridges or waffle patterns
Gen'l. Prop.: Sp. Gr.: 1.04 (D792-66); Soften. Pt. (Vicat): 210 F (D1525-70)
Mech. Prop.: Tens. Str.: 4200 psi (D-638-72); Tens. Elong.: 30% (D638-72); Tens. Mod.: 300,000 psi (D638-72); Flex. Str.: 7700 psi (D790-71); Flex. Mod.: 280,000 psi (D790-71); Impact Str. (Izod): 1.2 ft lb/in. notched (D-256-73); Hardness (Rockwell): M 40 (D785-65)
Therm. Prop.: Distort. Temp.: 165 F (unannealed, D648-72)
Ref.: Product information

Cosden Polystyrene 825. Cosden Oil & Chem. Co.
Chem. Descrip.: Polystyrene
Category/Applications: High impact polystyrene with outstanding shock resistance, flex properties (resists cracking and splitting); used in appliance parts, hobby horses, tackle boxes, furniture applications, containers, and blow molded parts
Gen'l. Prop.: Sp. Gr.: 1.04 (D792-66); Soften. Pt. (Vicat): 200 F (D1525-70)
Mech. Prop.: Tens. Str.: 2700 psi (D-638-72); Tens. Elong.: 50% (D638-72); Tens. Mod.: 240,000 psi (D638-72); Flex. Str.: 5300 psi (D790-71); Flex. Mod.: 300,000 psi (D790-71); Impact Str. (Izod): 2.0 ft lb/in. notched (D-256-73); Hardness (Rockwell): M 10 (D785-65)
Therm. Prop.: Distort. Temp.: 155 F (unannealed, D648-72)
Ref.: Product information

Cosden Polystyrene 825 E. Cosden Oil & Chem. Co.
Chem. Descrip.: Polystyrene
Category/Applications: Highly specialized polystyrene formulation that extrudes into uniform gauge sheeting that forms easily into thin-wall coffee cups or thick-wall refrigerator door liners; high heat distortion and impact strength and outstanding stress crack resistance; good processability
Gen'l. Prop.: Sp. Gr.: 1.04 (D792-66); Soften. Pt. (Vicat): 210 F (D1525-70)
Mech. Prop.: Tens. Str.: 3800 psi (D-638-72); Tens. Elong.: 55% (D638-72); Tens. Mod.: 250,000 psi (D638-72); Flex. Str.: 6500 psi (D790-71); Flex. Mod.: 300,000 psi (D790-71); Impact Str. (Izod): 2.0 ft lb/in. notched (D256-73); Hardness (Rockwell): M 20 (D785-65)
Therm. Prop.: Distort. Temp.: 162 F (unannealed, D648-72)
Ref.: Product information

CP-51. Continental Polymers Inc.
Chem. Descrip.: Acrylic plastic pellets
Category/Applications: Injection molding and extrusion grade pellets; med. heat resistance; easiest flow of CP series (melt flow rate = 21.5 g/10 min, D1238 I)
Form: Pellets
Gen'l. Prop.: Sp. Gr.: 1.18 (D792); Ref. Index: 1.49 (D542)
Mech. Prop.: Tens. Str.: 10,000 psi (D-638); Tens. Mod.: 430,000 psi (D638); Flex. Str.: 13,000 psi (D790); Compr. Str.: 14,000 psi (D695); Impact Str. (Izod): 0.5 ft lb/in. (D256); Hardness (Rockwell): M 89 (D785)
Therm. Prop.: Distort. Temp.: 190 F (D648); Flow Temp.: 280 F (D569); Flamm.: 0.8 in./min (D635)
Elec. Prop.: Power Factor: 0.05 (60 Hz, D150); Dielec. Str.: 500 V/mil (D149); Dielec. Const.: 3.3 (60 Hz, D150); Arc Resist.: No track (D495); Loss Factor: 0.17 (60 Hz, D150)
Toxicity/Handling: Combustible thermoplastics; observe fire precautions

Std. Pkgs.: 300 lb net drums, 1200 lb net cartons, rail cars, bulk trucks; also avail. in 55 lb net specialized bags for overseas export
Ref.: Product information

CP-61. Continental Polymers Inc.
Chem. Descrip.: Acrylic plastic pellets
Category/Applications: Injection molding and extrusion grade pellets; med. heat resistance, med. flow [melt flow rate = 10.0 g/10 min (D1238-I)]
Form: Pellets
Gen'l. Prop.: Sp. Gr.: 1.18 (D792); Ref. Index: 1.49 (D542)
Mech. Prop.: Tens. Str.: 10,000 psi (D-638); Tens. Mod.: 430,000 psi (D638); Flex. Str.: 14,000 psi (D790); Compr. Str.: 15,000 psi (D695); Impact Str. (Izod): 0.5 ft lb/in. (D256); Hardness (Rockwell): M 92 (D785)
Therm. Prop.: Distort. Temp.: 196 F (D648); Flow Temp.: 305 F (D569); Flamm.: 0.8 in./min (D635)
Elec. Prop.: Power Factor: 0.06 (60 Hz, D150); Dielec. Str.: 500 V/mil (D149), Dielec. Const.: 3.3 (60 Hz, D150); Arc Resist.: No track (D495); Loss Factor: 0.18 (60 Hz, D150)
Toxicity/Handling: Combustible thermoplastic; observe fire precautions
Std. Pkgs.: See CP-51
Ref.: Product information

CP-80. Continental Polymers Inc.
Chem. Descrip.: Acrylic plastic pellets
Category/Applications: Extrusion grade pellets; unlubricated version of CP-81; highest m.w.; max. heat resistance; melt flow rate = 2.8 g/10 min (D1238-I)
Form: Pellets
Gen'l. Prop.: Sp. Gr.: 1.19 (D792); Ref. Index: 1.49 (D542)
Mech. Prop.: Tens. Str.: 10,500 psi (D-638); Tens. Mod.: 430,000 psi (D638); Flex. Str.: 16,000 psi (D790); Compr. Str.: 16,000 psi (D695); Impact Str. (Izod): 0.5 ft lb/in. (D256); Hardness (Rockwell): M 95 (D785)
Therm. Prop.: Distort. Temp.: 198 F (D648); Flow Temp.: 319 F (D569); Flamm.: 0.9 in./min (D635)
Elec. Prop.: Power Factor: 0.06 (60 Hz, D150); Dielec. Str.: 500 V/mil (D149); Dielec. Const.: 3.3 (60 Hz, D150); Arc Resist.: No track (D495); Loss Factor: 0.19 (60 Hz, D150)
Toxicity/Handling: Combustible thermoplastic; observe fire precautions
Std. Pkgs.: See CP-51
Ref.: Product information

CP-81. Continental Polymers Inc.
Chem. Descrip.: Acrylic plastic pellets
Category/Applications: Extrusion grade pellets; highest m.w., max. heat resistance; melt flow rate = 2.9 g/10 min (D-1238-I)
Form: Pellets
Gen'l. Prop.: Sp. Gr.: 1.19 (D792); Ref. Index: 1.49 (D542)
Mech. Prop.: Tens. Str.: 10,500 psi (D-638); Tens. Mod.: 430,000 psi (D638); Flex. Str.: 16,000 psi (D790); Compr. Str.: 16,000 psi (D695); Impact Str. (Izod): 0.5 ft lb/in. (D256); Hardness (Rockwell): M 95 (D785)
Therm. Prop.: Distort. Temp.: 198 F (D648); Flow Temp.: 319 F (D569); Flamm.: 0.9 in./min (D635)
Elec. Prop.: Power Factor: 0.06 (60 Hz, D150); Dielec. Str.: 500 V/mil (D149); Dielec. Const.: 3.3 (60 Hz, D150); Arc Resist.: No track (D495); Loss Factor: 0.19 (60 Hz, D150)
Toxicity/Handling: Combustible thermoplastic; observe fire precautions
Std. Pkgs.: See CP-51
Ref.: Product information

CP-82. Continental Polymers Inc.
Chem. Descrip.: Acrylic plastic pellets
Category/Applications: Injection molding grade pellets esp. designed for superior mold release properties; highest m.w., max. heat resistance; melt flow rate = 3.2 g/10 min (D1238-I)
Form: Pellets
Gen'l. Prop.: Sp. Gr.: 1.19 (D792); Ref. Index: 1.49 (D542)
Mech. Prop.: Tens. Str.: 10,500 psi (D-638); Tens. Mod.: 430,000 psi (D638); Flex. Str.: 16,000 psi (D790); Compr. Str.: 16,000 psi (D695); Impact Str. (Izod): 0.5 ft lb/in. (D256); Hardness (Rockwell): M 95 (D785)
Therm. Prop.: Distort. Temp.: 198 F (D648); Flow Temp.: 319 F (D569); Flamm.: 1.0 in./min (D635)
Elec. Prop.: Power Factor: 0.06 (60 Hz, D150); Dielec. Str.: 500 V/mil (D149); Dielec. Const.: 3.3 (60 Hz, D150); Arc Resist.: No track (D495); Loss Factor: 0.19 (60 Hz, D150)
Toxicity/Handling: Combustible thermoplastic; observe fire precautions
Std. Pkgs.: See CP-51
Ref.: Product information

CP-343-1 (25% in xylene). Eastman Chem. Products, Inc.
Chem. Descrip.: Chlorinated polyolefins
Category/Applications: Used in the coatings industry to formulate primers or topcoats for polypropylene, polyallomer, aluminum foil, and paper substrates; excellent adhesion characteristics
Form: Viscous liquid; Color: Amber; Odor: Xylene
Composition: 25% solids in xylene
Gen'l. Prop.: Sol.: Sol. in toluene, xylene, high-flash aromatic naphtha, methyl chloride, trichloroethylene, tetrahydrofuran; Sp. Gr.: 0.898 (sol'n.), 0.99–1.03 (as molded); Visc. (Brookfield): 100 cps; B.P.: 145 C; Flash Pt. (TCC): 27 C; Fire Pt.: 30 C
Mech. Prop. (as molded): Hardness (Tukon): < 1 knoops (D1474)
Toxicity/Handling: Avoid eye and skin contact and inhalation of vapor; use with adequate ventilation; flammable liquid; avoid contact with oxidizing materials
Ref.: Bulletin GN-315A

CP-343-1 (50% in xylene). Eastman Chem. Products, Inc.
Chem. Descrip.: Chlorinated polyolefins
Category/Applications: See CP-343-1 (25% in xylene)
Form: Gel; Color: Light amber; Odor: Xylene
Composition: 50% solids in xylene
Gen'l. Prop.: Sol.: See CP-343-1 (25%); Sp. Gr.: 0.935 (sol'n.), 0.99–1.03 (as molded); Visc. (Brookfield): > 100,000 cps; B.P.: 145 C; Flash Pt. (TCC): 28 C; Fire Pt.: 52 C
Mech. Prop. (as molded): Hardness (Tukon): < 1 knoops (D1474)
Toxicity/Handling: See CP-343-1 (25%)
Ref.: Bulletin GN-315A

CP-515-2. Eastman Chem. Products, Inc.
Chem. Descrip.: Chlorinated polyolefins
Category/Applications: See CP-343-1 (25%)
Form: Viscous liquid; Color: Translucent; Odor: Xylene
Composition: 40% solids in xylene
Gen'l. Prop.: Sol.: See CP-343-1 (25%); Sp. Gr.: 0.955 (sol'n.), 1.04–1.09 (as molded); Visc. (Brookfield): 80 cps; B.P.: 145 C; Flash Pt. (TCC): 28 C; Fire Pt.: 43 C
Mech. Prop. (as molded): Hardness (Tukon): 2–4 knoops (D1474)
Toxicity/Handling: See CP-343-1 (25%)
Ref.: Bulletin GN-315A

Crodym MX. Crosby Chem., Inc.
Chem. Descrip.: C_{36}–C_{54} di and tri carboxylic aliphatic acids
Category/Applications: Dimer acid used as a chemical intermediate for the preparation of polyamide resins, architectural enamels, urethane foams, alkyd resins, epoxy resins, oil field chemicals, and other compounds where the complexity of high m.w. polycarboxylic acids is advantageous
Form: Liquid; Color (Gardner): 12 max.
Composition: 80% dimer acid, 14% trimer acid, 6% monomer acid
Gen'l. Prop.: Sp. Gr.: 0.950 (25/20 C); Density: 7.91 lb/gal; Visc. (Gardner-Holdt): Z_3; Acid Value: 175–185; Sapon. Value: 188–195; Ref. Index: 1.4832 (26 C)
Therm. Prop.: Flash Pt.: 535 F
Std. Pkgs.: 55 gal (410 lb net) lined drums; 4000, 8000, 10,000, and 20,000 gal aluminum tankcars
Ref.: Product information

Crodym MX-Special. Crosby Chem., Inc.
Chem. Descrip.: C_{36}–C_{54} di and tri carboxylic aliphatic acids
Category/Applications: See Crodym MX
Form: Liquid; Color (Gardner): 14 max.
Composition: 82% dimer acid, 15% trimer acid, 3% monomer acid
Gen'l. Prop.: Sp. Gr.: 0.950 (25/20 C); Density: 7.91 lb/gal; Visc. (Gardner-Holdt): Z_4; Acid Value: 175–185; Sapon. Value: 178–183; Ref. Index: 1.4832 (26 C)
Therm. Prop.: Flash Pt.: 535 F
Std. Pkgs.: See Crodym MX
Ref.: Product information

Crodym T-2. Crosby Chem., Inc.
Chem. Descrip.: C_{36}–C_{54} di and tri carboxylic aliphatic acids
Category/Applications: See Crodym MX
Form: Liquid; Color (Gardner): 5 max.
Composition: 95% dimer acid, 3% monomer acid, 2% trimer acid
Gen'l. Prop.: Sp. Gr.: 0.945 (25/20 C); Density: 7.85 lb/gal; Visc.: 4520 cps; Acid Value: 188–195; Sapon. Value: 193–197; Ref. Index: 1.4860 (26 C)
Therm. Prop.: Flash Pt.: 540 F
Std. Pkgs.: See Crodym MX
Ref.: Product information

Crodym T-18. Crosby Chem., Inc.
Chem. Descrip.: C_{36}–C_{54} di and tri carboxylic aliphatic acids
Category/Applications: See Crodym MX
Form: Liquid; Color (Gardner): 8 max.
Composition: 79% dimer acid, 18% trimer acid, 3% monomer acid
Gen'l. Prop.: Sp. Gr.: 0.953 (25/20 C); Density: 7.92 lb/gal; Visc.: 8300 cps; Acid Value: 185–195; Sapon. Value: 190–197; Ref. Index: 1.4852 (26 C)
Therm. Prop.: Flash Pt.: 530 F
Std. Pkgs.: See Crodym MX
Ref.: Product information

Crodym T-22. Crosby Chem., Inc.
Chem. Descrip.: C_{36}–C_{54} di and tri carboxylic aliphatic acids
Category/Applications: See Crodym MX
Form: Liquid; Color (Gardner): 9 max.
Composition: 77% dimer acid, 20% trimer acid, 3% monomer acid
Gen'l. Prop.: Sp. Gr.: 0.951 (25/20 C); Density: 7.90 lb/gal; Visc.: 8400 cps; Acid Value: 188–193; Sapon. Value: 190–195; Ref. Index: 1.4847 (26 C)
Therm. Prop.: Flash Pt.: 542 F
Std. Pkgs.: See Crodym MX
Ref.: Product information

Cropolamid 18W26. Crosby Chem., Inc.
Chem. Descrip.: Polyamide resin derived from reaction of linear diamines with Crodyms and acidic resins

Category/Applications: Resin for use in aq. flexographic ink systems where water reducibility, instant drying speed, and water resistance are important
Form: Solid, flakes; Color (Gardner): 12 max. (molten)
Gen'l. Prop.: Sol.: Sol. in alcohol-water-ammonia solvent systems; Visc.: 32 poise (kinematic, 160 C); Soften. Pt.: 105 C; Acid Value: 90; Amine Value: 5; Stability: Excellent water, grease, and oil resistance
Std. Pkgs.: 29 gauge galvanized rosin drums (467 lb net); flakes in 50 lb net multiwall kraft bags
Ref.: Product information

Cropolamid 18W58. Crosby Chem., Inc.
Chem. Descrip.: Polyamide resin derived from reaction of linear diamines with Crodyms
Category/Applications: See Cropolamid 18W26
Form: Solid, flakes; Color (Gardner): 12 max.
Gen'l. Prop.: Sol.: Sol. in ammonia-water systems at pH 7.5–8.5; Soften. Pt. (Ring & Ball): 110 C; Acid Value: 130; Amine Value: 8; Stability: Excellent water, grease, and oil resistance
Std. Pkgs.: 29 gauge galvanized rosin drums (467 lb net); flakes in 50 lb net multiwall kraft bags
Ref.: Product information

Cropolamid 18W59. Crosby Chem., Inc.
Chem. Descrip.: Polyamide resin derived from reaction of linear diamines with Crodyms
Category/Applications: See Cropolamid 18W26
Form: Flakes; Color (Gardner): 12 max.
Gen'l. Prop.: Sol.: Sol. in an alcohol-water-ammonia system at pH 7.5–8.5; Soften. Pt. (Ring & Ball): 125 C; Acid Value: 100; Amine Value: 8; Stability: Excellent water, grease, and oil resistance
Std. Pkgs.: 50 lb net multiwall bags
Ref.: Product information

Cropolamid 480-B-75. Crosby Chem., Inc.
Chem. Descrip.: Reactive polyamide/epoxy adduct solution
Category/Applications: Epoxy hardener for both solid epoxy resin solutions and liquid epoxy resins
Form: Liquid; Color (Gardner): 10
Composition: 75% solids in n-butanol solvent
Gen'l. Prop.: Sp. Gr.: 0.97; Density: 8.1 lb/gal; Visc. (Brookfield): 15–30 poise (40 C); Amine Value: 235–260
Therm. Prop.: Flash Pt.: 140 F
Std. Pkgs.: 55 gal (425 lb net) black steel drums
Ref.: Product information

Cropolamid 1803. Crosby Chem., Inc.
Chem. Descrip.: Thermoplastic polyamide resin derived from reaction of Crodyms with linear diamines
Category/Applications: Resin for use in lacquers, heat seal coatings, flexographic and rotogravure inks; alcohol/aliphatic cosolvent solubility permits preparation of flexographic and gravure inks with uniform viscosity and low-temp. storage properties
Form: Solid, flakes; Color (Gardner): 12 max.
Gen'l. Prop.: Sol.: Forms gel in n-propanol, isopropanol, ethanol; Visc.: 24 poises (kinematic, 160 C); Soften. Pt. (Ring & Ball): 108 C; Acid Value: 4; Amine Value: 4
Std. Pkgs.: 29 gauge galvanized rosin

drums (467 lb net); flakes in 50 lb net multiwall kraft bags
Ref.: Product information

Cropolamid 1804. Crosby Chem., Inc.
Chem. Descrip.: Thermoplastic polyamine resin derived from reaction of Crodyms with linear diamines
Category/Applications: See Cropolamid 1803; 1804 exhibits cosolvent properties similar to 1803 at lower viscosities due to slightly lower m.w.
Form: Solid, flakes; Color (Gardner): 12 max.
Gen'l. Prop.: Sol.:Sol. in toluene-ethanol (1:1); forms gel in n-propanol, isopropanol, and ethanol; Visc.: 15 poises (kinematic, 160 C); Soften. Pt. (Ring & Ball): 108 C; Acid Value: 4; Amine Value: 4; Stability: Excellent water, grease, and oil resistance
Std. Pkgs.: See Cropolamid 1803
Ref.: Product information

Cropolamid 1805. Crosby Chem., Inc.
Chem. Descrip.: Thermoplastic polyamide resin derived from reaction of Crodyms with linear diamines
Category/Applications: Resin for use in heat seal coatings, flexographic and rotogravure inks; its alcohol/aliphatic cosolvent solubility and good freeze thaw properties make it ideal for flexographic and rotogravure inks for low-temp. storage
Form: Solid, flakes; Color (Gardner): 12 max.
Gen'l. Prop.: Sol.: Sol. in n-propanol, isopropanol, toluene-ethanol (1:1); Visc.: 8 poises (kinematic, 160 C); Soften. Pt. (Ring & Ball): 101 C; Acid Value: 4; Amine Value: 4; Stability: Excellent water, grease, and oil resistance
Std. Pkgs.: See Cropolamid 1803
Ref.: Product information

Cropolamid 1807. Crosby Chem., Inc.
Chem. Descrip.: Thermoplastic polyamide resin derived from reaction of linear diamines with Crodyms in conjunction with various monomeric acids
Category/Applications: Resin for use in flexographic and rotogravure inks where alcohol solubility is required (eliminates the need for synthetic rubber plates); has excellent nitrocellulose compatibility with properties of high gloss, good pigment wetting, blocking, and adhesion resistance
Form: Solid, flakes; Color (Gardner): 10 max.
Gen'l. Prop.: Sol.: Sol. in n-propanol, isopropanol, ethanol, toluene-ethanol (1:1); Visc.: 4 poise (kinematic, 160 C); Soften. Pt. (Ring & Ball): 112 C; Acid Value: 4; Amine Value: 4; Stability: Excellent water, grease, and oil resistance
Std. Pkgs.: See Cropolamid 1803
Ref.: Product information

Cropolamid 1808. Crosby Chem., Inc.
Chem. Descrip.: Thermoplastic polyamide resin produced by reaction of Crodyms with various monomeric acids and linear diamines
Category/Applications: Resin for use in flexographic inks with good nitrocellulose compatibility, gloss, pigment wetting, abrasion resistance, and ethanol solubility
Form: Solid, flakes; Color (Gardner): 10 max.
Gen'l. Prop.: Sol.: Sol. in n-propanol, isopropanol, ethanol, toluene-ethanol (1:1); Visc.: 6 poise (kinematic, 160 C); Soften. Pt. (Ring & Ball): 103 C; Acid Value: 4; Amine Value: 5; Stability: Excellent water, grease, and oil resistance
Std. Pkgs.: See Cropolamid 1803
Ref.: Product information

Cropolamid 1833. Crosby Chem., Inc.
Chem. Descrip.: Thermoplastic polyamide resin derived from reaction of Crodyms with linear diamines
Category/Applications: Resin for use in lacquers, heat seal coatings, flexographic and rotogravure inks; alcohol/aliphatic cosolvent solubility permits preparation of flexographic and gravure inks with uniform viscosity and low-temp. storage properties
Form: Solid, flakes; Color (Gardner): 12 max.
Gen'l. Prop.: Sol.: Sol. in n-propanol, isopropanol, toluene-ethanol (1:1); Visc.: 30 poise (kinematic, 160 C); Soften. Pt. (Ring & Ball): 102 C; Acid Value: 4; Amine Value: 4; Stability: Excellent water, grease, and oil resistance
Std. Pkgs.: See Cropolamid 1803
Ref.: Product information

Cropolamid 1860. Crosby Chem., Inc.
Chem. Descrip.: Thermoplastic polyamide resin derived by reaction of Crodyms with linear diamines
Category/Applications: Alcohol-sol., high melting resin for use as a heat seal coating (with excellent adhesion to paper, metals, foils, and plastics), and in lacquers, flexographic and rotogravure inks
Form: Solid; Color (Gardner): 12 max.
Composition: 30% resin, 60% isopropanol, 10% N-propylacetate
Gen'l. Prop.: Sol.: Alcohol-sol.; Visc.: 2–4 poise (170 C); Soften. Pt. (Ring & Ball): 148–160 C; Acid Value: 5 max.; Amine Value: 5 max.; Stability: Excellent water, grease, and oil resistance
Std. Pkgs.: 50 lb net multiwall kraft bags
Ref.: Product information

Cropolamid 1862. Crosby Chem., Inc.
Chem. Descrip.: Thermoplastic polyamide resin derived from reaction of Crodyms with diamines
Category/Applications: Resin for use in lacquers, heat seal coatings, and flexographic and gravure inks where pigment wetting, cold pack resistance, and good gloss are important; the alcohol soluble/reducible properties permit manufacture of coating vehicles of uniform viscosity
Form: Solid, flakes; Color (Gardner): 12 max.
Gen'l. Prop.: Sol.: Sol. in n-propanol, isopropanol, ethanol, toluene-ethanol (1:1); Visc.: 8 poise (kinematic, 160 C); Soften. Pt. (Ring & Ball): 110 C; Acid Value: 4; Amine Value: 4; Stability: Excellent water, grease, and oil resistance
Std. Pkgs.: 29 gauge galvanized rosin drums (467 lb net); flakes in 50 lb net multiwall kraft bags
Ref.: Product information

Cropolamid 1878-96. Crosby Chem., Inc.
Chem. Descrip.: Thermoplastic polyamide resin derived from reaction of Crodyms with linear diamines
Category/Applications: Resin for use in lacquers, heat seal coatings, and flexographic and gravure inks where pigment wetting, cold pack resistance, and good gloss are important; the alcohol soluble/reducible properties permit manufacture of coating vehicles of uniform viscosity
Form: Solid, flakes; Color (Gardner): 9 max.
Gen'l. Prop.: Sol.: Alcohol soluble/reducible; Visc.: 4.5 poise (kinematic, 160 C); Soften. Pt. (Ring & Ball): 93 C; Acid Value: 4; Amine Value: 4
Std. Pkgs.: See Cropolamid 1862
Ref.: Product information

Cropolamid 1889. Crosby Chem., Inc.
Chem. Descrip.: Thermoplastic polyamide resin derived from reaction of Crodyms with diamines
Category/Applications: Resin for use in lacquers, heat seal coatings, and flexographic and gravure inks where gel recovery, pigment wetting, cold pack resistance, and good gloss are important; the alcohol soluble/reducible properties permit manufacture of coating vehicles with uniform viscosity
Form: Solid, flakes; Color (Gardner): 10 max.
Gen'l. Prop.: Sol.: Alcohol soluble/reducible; Visc.: 5 poise (kinematic, 160 C); Soften. Pt. (Ring & Ball): 103 C; Acid Value: 4; Amine Value: 4
Std. Pkgs.: See Cropolamid 1862
Ref.: Product information

Cropolamid L-100. Crosby Chem., Inc.
Chem. Descrip.: Polyamide resin derived from the reaction of Crodym with polyethylene amines
Category/Applications: Reactive resin used in combination with epoxies to produce thermoplastic resinous compositions useful as adhesives, laminating coating, potting, and protective coating compounds; these thermoset adhesives are resistant to most chemicals, solvents, and environmental conditions
Form: Semisolid; Color (Gardner): 12 max.
Composition: 100% solids
Gen'l. Prop.: Visc. (Brookfield): 10 poise (150 C); Acid Value: 5; Amine Value: 90
Std. Pkgs.: Lever-lock open top drums
Ref.: Product information

Cropolamid L-100 CX. Crosby Chem., Inc.
Chem. Descrip.: Solution of Cropolamid L-100 in Cellusolve/xylene (1:9)
Category/Applications: See Cropolamid L-100
Form: Liquid; Color (Gardner): 12 max.
Composition: 60% solids
Gen'l. Prop.: Visc.: 10–15 poise (40 C); Acid Value: 4; Amine Value: 55
Std. Pkgs.: 55 gal (400 lb net) lined drums; 4000 and 8000 gal aluminum tankcars
Ref.: Product information

Cropolamid L-100 IT. Crosby Chem., Inc.
Chem. Descrip.: Solution of Cropolamid L-100 in isopropanol/toluene (1:1)
Category/Applications: See Cropolamid L-100
Form: Liquid; Color (Gardner): 12 max.
Composition: 60% solids
Gen'l. Prop.: Visc.: 5–9 poise (40 C); Acid Value: 4; Amine Value: 55
Std. Pkgs.: See Cropolamid L-200 CX
Ref.: Product information

Cropolamid L-250X. Crosby Chem., Inc.
Chem. Descrip.: Solution of Cropolamid L-250 in xylene
Category/Applications: See Cropolamid L-250
Form: Liquid; Color (Gardner): 12 max.
Composition: 70% solids
Gen'l. Prop.: Visc.: 6 poise (40 C); Acid Value: 4; Amine Value: 170
Std. Pkgs.: 55 gal (400 lb net) lined drums; 4000 and 8000 gal aluminum tankcars
Ref.: Product information

Cropolamid L-350, LM-350. Crosby Chem., Inc.
Chem. Descrip.: Reaction product of Crodym with polyethylene amines which yields amido-amine resins with free primary and secondary amino groups
Category/Applications: L-350 is used in

combination with epoxies to produce thermoplastic resinous compositions useful as adhesives, laminating coating, potting and protective coating compounds; these thermoset adhesives are resistant to most chemicals, solvents, and environmental conditions; LM-350 requires ≈ twice as long to cure epoxy resins as L-350; the two resins can be blended to attain any desired reactivity
Form: Liquid; Color (Gardner): 12 max.
Composition: 100% solids
Gen'l. Prop.: Visc. (Brookfield): 8 poise (75 C); Acid Value: 4; Amine Value: 340
Std. Pkgs.: 55 gal (445 lb net) black steel drums
Ref.: Product information

Cropolamid L-400 and LM-400. Crosby Chem., Inc.
Chem. Descrip.: Reaction product of Crodym with polyethylene amines which yields amido-amine resins with free primary and secondary amino groups
Category/Applications: L-400 used in combination with epoxies to produce thermoplastic resinous compositions useful as adhesives, laminating coating, potting and protective coating compounds; these thermoset adhesives are resistant to most chemicals, solvents, and environmental conditions; LM-400 requires ≈ twice as long to cure epoxy resins as L-400; the two resins can be blended to attain any desired reactivity
Form: Liquid; Color (Gardner): 12 max.
Composition: 100% solids
Gen'l. Prop.: Visc. (Brookfield): 4 poise (75 C); Acid Value: 3; Amine Value: 380
Std. Pkgs.: 55 gal (445 lb net) black steel drums
Ref.: Product information

Cropolamid L-520. Crosby Chem., Inc.
Chem. Descrip.: Amido-amine resin produced by reaction of Crodym with polyethylene amines yielding free primary and secondary amino groups
Category/Applications: Resin used in combination with epoxies to produce thermoplastic resinous compositions useful as adhesives, laminating coating, potting and protective coating compounds; these thermoset adhesives are resistant to most chemicals, solvents, and environmental conditions
Form: Liquid; Color (Gardner): 12 max.
Gen'l. Prop.: Visc. (Brookfield): 7 poise; Acid Value: 2; Amine Value: 540
Std. Pkgs.: 55 gal (445 lb net) black steel drums
Ref.: Product information

Cropolamid L-600. Crosby Chem., Inc.
Chem. Descrip.: Fatty amidoamine resin
Category/Applications: Low viscosity resin for use with epoxy resins to formulate coatings, adhesives, and laminates with good chemical and water resistance
Form: Liquid; Color (Gardner): 10
Gen'l. Prop.: Sp. Gr.: 0.98; Density: 8.15 lb/gal; Visc. (Brookfield): 12–25 poise; Amine Value: 580–620
Therm. Prop.: Flash Pt.: 430 F (COC)
Std. Pkgs.: 55 gal (425 lb net) black steel drums
Ref.: Product information

Cropolamid LM-250. Crosby Chem., Inc.
Chem. Descrip.: Reaction product of Crodym with polyethylene amines which yields amido-amine resins with free primary and secondary amino groups
Category/Applications: Reactive resin used in combination with epoxies to produce thermoplastic resinous compositions useful as adhesives, laminating coatings, potting and protective coating

compounds; these thermoset adhesives are resistant to most chemicals, solvents, and environmental conditions
Form: Liquid; Color (Gardner): 12 max.
Composition: 100% solids
Gen'l. Prop.: Visc. (Brookfield): 35 poise (75 C); Acid Value: 3; Amine Value: 240
Std. Pkgs.: Lever-lock open top drums
Ref.: Product information

Cropolamid LM-250-X-70. Crosby Chem., Inc.
Chem. Descrip.: Solution of Cropolamid L-250 in xylene
Category/Applications: See Cropoalmid L-250
Form: Liquid; Color (Gardner): 12 max.
Composition: 70% solids
Gen'l. Prop.: Density: 7.8 lb/gal; Visc.: 3–8 poise (40 C); Amine Value: 160–175
Therm. Prop.: Flash Pt.: 45 C
Std. Pkgs.: 55 gal (425 lb net) black steel drums
Ref.: Product information

Cropolamid LV-450, LVM-450. Crosby Chem., Inc.
Chem. Descrip.: Polyamide resin
Category/Applications: Resin used in combination with epoxies to produce thermoplastic resinous compositions useful as adhesives, laminating coating, potting and protective coating compounds; these thermoset adhesives are resistant to most chemicals, solvents, and environmental conditions; LVM-450 requires ≈ twice as long to cure epoxy resins as LV-450; the two resins can be blended to attain any desired reactivity
Form: Liquid; Color (Gardner): 12 max.
Composition: 100% solids
Gen'l. Prop.: Visc. (Brookfield): 10 poise; Acid Value: 3; Amine Value: 445
Std. Pkgs.: 55 gal (445 lb net) black steel drums
Ref.: Product information

Crosby 68-F. Crosby Chem., Inc.
Chem. Descrip.: Blend of tall oil rosins
Category/Applications: Rosin for use in paper sizing, linoleum and roofing materials, greases, gloss oils, core binders, printing inks, and various adhesives
Form: Solid; Color: F
Gen'l. Prop.: Sol.: Sol. in isopropanol, mineral spirits, hexane, toluene, ethyl acetate, acetone; hazy sol. in ethanol; M.P. (Drop): 80 C; Acid Value: 150
Std. Pkgs.: 29 gauge galvanized steel drums (510 lb net); 8000 gal insulated tankcars
Ref.: Product information

Crosby 700M. Crosby Chem., Inc.
Chem. Descrip.: Heat modified tall oil rosin
Category/Applications: Rosin for the manufacture of paper sizes, limed rosins, gloss oils, spirit varnishes, ester gums, kettle esters, resinates, emulsifiers, greases, adhesives, core oils, disinfectants, synthetic resins, solders, batik, etc., esp. in products in which regular rosins tend to crystallize
Form: Solid; Color: X
Gen'l. Prop.: M.P. (Drop): 82 C; Acid Value: 157
Std. Pkgs.: 515–525 lb net galvanized rosin drums
Ref.: Product information

Crosby 730-M. Crosby Chem., Inc.
Chem. Descrip.: Heat modified Mirares rosin
Category/Applications: Rosin for the manufacture of paper size, limed rosin, gloss oils, spirit varnishes, ester gum, kettle esters, resinates, emulsifiers,

greases, adhesives, core oils, disinfectants, synthetic resins, solder, batik, etc., esp. in products in which regular rosins tend to crystallize; also avail. in terpene and petroleum solvent solutions at 60–80% concentrations
Form: Solid; Color: X
Gen'l. Prop.: M.P. (Drop): 83 C; Acid Value: 155
Std. Pkgs.: 515–525 lb net galvanized steel drums; 8000 gal insulated tankcars
Ref.: Product information

Crosby 730-ML-65. Crosby Chem., Inc.
Chem. Descrip.: Limed heat modified rosin
Category/Applications: Rosin for use in printing inks, adhesives, and similar products that do not require color grades of FF and paler limed rosins
Form: Solid, flakes; Color (Grade): WG
Gen'l. Prop.: Sol.: Sol. in aliphatics, aromatics; partly sol. in esters and glycols; Visc. (Gardner): 2-R (60% in min. spirits); M.P. (Ring & Ball): 124 C; Acid Value: 70
Std. Pkgs.: 515–525 lb net galvanized steel drums; flakes in 50 lb net multiwall kraft bags
Ref.: Product information

Crosby 800M. Crosby Chem., Inc.
Chem. Descrip.: Tall oil rosin
Category/Applications: Rosin which has been heat modified to reduce crystallization tendency and to induce adhesive properties; used in paper sizings, core oils, printing inks, synthetic rosin oils, spirit varnishes, gloss oils, brewer's pitches, disinfectants, roofing cements, animal dips, solders, adhesives, fly papers, belt dressings, linoleums, matches, greases, pyrotechnics, cutting oils, insulating compounds, sealing compounds, glues, etc.
Form: Solid; Color (Grade): FF
Gen'l. Prop.: Sol.: Sol. in isopropanol, min. spirits, hexane, toluene, ethyl acetate, acetone; hazy sol. in ethanol; M.P. (Drop): 72 C; Acid Value: 140
Std. Pkgs.: 515–525 lb net galvanized rosin drums
Ref.: Product information

Crosby Calbrite. Crosby Chem., Inc.
Chem. Descrip.: Limed rosin
Category/Applications: Rosin with high melting pt., low acid number, and freedom from unreacted lime; used for gloss oils, cold cut varnish and printing inks
Form: Solid; Color (Grade): WG
Gen'l. Prop.: Visc. (Gardner): L-O (60% in min. spirits); M.P. (Drop): 144 C; Acid Value: 72
Std. Pkgs.: 515–535 lb net galvanized steel drums
Ref.: Product information

Crosby C.R.O. Crosby Chem., Inc.
Chem. Descrip.: Blend of low polymeric terpene hydrocarbons removed by distillation from terpene hydrocarbon resins
Category/Applications: Used in waterproofing compounds, sealants, putties, paints, and rubber adhesives; good tack, adhesive, and water repellency properties
Form: Viscous liquid; Color: Orange
Gen'l. Prop.: Sp. Gr.: 0.925 min.; Visc. (Gardner): A; B.P.: 165 C min. (initial); Acid Value: 1.0 max.
Std. Pkgs.: 55 gal open head drums; 4000, 8000, and 10,000 gal tankcars
Ref.: Product information

Crosby Cromon D-1. Crosby Chem., Inc.
Chem. Descrip.: Monobasic volatile fatty component removed by vacuum distillation from a dimerized tall oil fatty acid composite
Category/Applications: High quality oleic acid type fatty acid used as a raw material for the manufacture of dry

cleaning soaps, emulsifiers, plastics, carbon paper, lubricants, polishes, and rubber compounding chemicals; sulfated derivatives used as wetting agents and textile treating compounds; sulfurized derivatives used as high quality cutting oils; metallic soaps of lead and aluminum may be incorporated in greases and gear oils to increase film strength and impart rust preventive properties; the copper soap us used as a fungicide for the preservation of cellulosic materials
Form: Color (Gardner): 6 max.
Gen'l. Prop.: Acid Value: 175; Sapon. Value: 185; Iodine Value: 85
Std. Pkgs.: 55 gal (410 lb net) lined drums; 4000, 8000, 10,000, and 20,000 gal aluminum tankcars
Ref.: Product information

Crosby Crosdim. Crosby Chem., Inc.
Chem. Descrip.: Thermoplastic resin concentrate of light-colored polymerized tall oil rosin acids
Category/Applications: Resin with high soften. pt. and excellent oxidative resistance used in rubber tackifiers, adhesives, hot melt coatings, lacquers, varnishes, printing inks, and in the preparation of various rosin esters
Form: Solid, flakes; Color: M
Gen'l. Prop.: Sol.: Sol. in ethanol, isopropanol, min. spirits, hexane, toluene, ethyl acetate, acetone; M.P. (Ring & Ball): 140 C; Acid Value: 145
Std. Pkgs.: 515–525 lb net galvanized steel drums; flakes in 50 lb multiwall bags
Ref.: Product information

Crosby Croturez B-85. Crosby Chem., Inc.
Chem. Descrip.: Terpene resin produced by the polymerization of highly refined terpene hydrocarbons consisting mainly of Beta Pinene
Category/Applications: Resin characterized by its uniformity, low acid no., pale color, solubility, and low ash content; used in adhesives, binders, coatings, tapes, sizing, waterproofing, inks, paint and varnish, rubber goods, paper, waxes, concrete, plasticizer, etc.
Form: Solid, flakes; Color (Grade): WW
Gen'l. Prop.: Sp. Gr.: 0.981; M.P. (Ring & Ball): 85 C; Acid Value: 1.0
Std. Pkgs.: 515 lb net galvanized 29 gauge steel drums; flakes in 50 lb net kraft bags
Ref.: Product information

Crosby Croturez B-100. Crosby Chem., Inc.
Chem. Descrip.: Terpene resin produced by the polymerization of highly refined terpene hydrocarbons consisting mainly of Beta Pinene
Category/Applications: See Crosby Croturez B-85
Form: Solid, flakes; Color (Grade): WW
Gen'l. Prop.: Sp. Gr.: 0.981; M.P. (Ring & Ball): 100 C; Acid Value: 1.0
Std. Pkgs.: 515 lb net galvanized 29 gauge steel drums; flakes in 50 lb net kraft bags
Ref.: Product information

Crosby Croturez B-115. Crosby Chem., Inc.
Chem. Descrip.: Terpene resin produced by the polymerization of highly refined terpene hydrocarbons consisting mainly of Beta Pinene
Category/Applications: See Crosby Croturez B-85
Form: Solid, flakes; Color (Grade): WW
Gen'l. Prop.: Sp. Gr.: 0.969; M.P. (Ring & Ball): 115 C; Acid Value: 1.0
Std. Pkgs.: 515 lb net galvanized 29 gauge steel drums; flakes in 50 lb net kraft bags
Ref.: Product information

Crosby Ester Gum. Crosby Chem., Inc.
Chem. Descrip.: Pure glycerol ester of pale rosin
Category/Applications: General purpose ester gum characterized by its uniformity, low acid no., freedom from insol. ingred., pale color, low ash content, and absence of free glycerin; used for general purpose varnishes, nitrocellulose lacquers, interior varnishes, paint vehicles, lacquer sealers, paper coatings, emulsion paints, aluminum liquids, adhesives, industrial finishes, interior enamels, rubber cements, cold cuts, wax compositions, printing inks, linoleum, floor tile, etc.
Form: Solid, flakes; Color (Gardner): 8 (60% in min. spirits)
Gen'l. Prop.: Sol.: Sol. in aliphatics, aromatics, esters, ketones; partly sol. in glycols; Visc. (Gardner): A-C (60% in min. spirits); M.P. (Drop): 88 C; Acid Value: 7
Std. Pkgs.: 515–525 lb net galvanized steel drums; flakes in 50 lb net multiwall kraft bags
Ref.: Product information

Crosby Ester Gum 721. Crosby Chem., Inc.
Chem. Descrip.: Glycerol ester of a modified tall oil rosin acid fraction
Category/Applications: Ester gum fcr use in solvent, emulsion and hot melt coatings, printing inks, and adhesives
Form: Solid, flakes; Color (Gardner): 11 max. (60% in min. spirits)
Gen'l. Prop.: Sol.: Sol. in aliphatics, aromatics, esters, ketones; partly sol. in glycols; Visc. (Gardner): A (60% in min. spirits); M.P. (Drop): 82 C; Acid Value: 8
Std. Pkgs.: 515–525 lb net galvanized steel drums; flakes in 50 lb net multiwall kraft bags
Ref.: Product information

Crosby Ester Gum CP. Crosby Chem., Inc.
Chem. Descrip.: Hard resin produced by esterification of pentaerythritol with Crosdim
Category/Applications: Resin with high melting pt., high gloss, excellent sol. in low KB solvents; used in heat-set offset inks, in rotogravure and silk screen inks, and in specialty varnishes where fast drying and hard films are required
Form: Solid, flakes; Color: Amber
Gen'l. Prop.: Sol.: Sol. in aliphatics, aromatics, esters; partly sol. in glycols; Visc.: U (50% in min. spirits); M.P. (Ring & Ball): 165 C; Acid No.: 16
Std. Pkgs.: 515–525 lb net galvanized steel drums; flakes in 50 lb net multiwall kraft bags
Ref.: Product information

Crosby Ester Gum H. Crosby Chem., Inc.
Chem. Descrip.: Glycerol ester of polymerized rosin
Category/Applications: Ester gum used in adhesives, lacquers, emulsion paints, varnishes, wax compositions, floor tile, and printing inks
Form: Solid; Color (Gardner): 9 (60% in min. spirits)
Gen'l. Prop.: M.P. (Ring & Ball): 107 C; Acid No.: 7
Std. Pkgs.: 515–525 lb net galvanized steel drums
Ref.: Product information

Crosby Ester Gum HE. Crosby Chem., Inc.
Chem. Descrip.: Ethylene glycol ester of polymerized rosin
Category/Applications: Ester gum for use in adhesives, lacquers, varnishes, wax compositions, and printing inks

Form: Solid; Color (Gardner): 9 (60% in min. spirits)
Gen'l. Prop.: M.P. (Drop): 85 C; Acid No.: 8
Std. Pkgs.: 515–525 lb net galvanized steel drums
Ref.: Product information

Crosby Ester Gum HV. Crosby Chem., Inc.
Chem. Descrip.: Pure glycerol ester of pale rosin
Category/Applications: See Crosby Ester Gum; Ester Gum HV is recommended where a higher melting pt. or viscosity than Crosby Ester Gum is required
Form: Solid, flakes; Color (Gardner): 7 (60% in min. spirits)
Gen'l. Prop.: Sol.: Sol. in aliphatics, aromatics, esters, ketones; partly sol. in glycols; Visc. (Gardner): D-E (60% in min. spirits); M.P. (Ring & Ball): 87 C; Acid Value: 7
Std. Pkgs.: 515–525 lb net galvanized steel drums; flakes in 50 lb net kraft bags
Ref.: Product information

Crosby Ester Gum S. Crosby Chem., Inc.
Chem. Descrip.: Pure glycerol ester of Crosby Mirares
Category/Applications: Ester gum meeting FDA requirements for chewing gum base, characterized by low acid value, pale color, low ash content, and freedom from volatile glyceride and terpene oils; also used as general purpose ester gum (see Crosby Ester Gum)
Form: Solid, flakes; Color (Gardner): 6
Gen'l. Prop.: Visc.: B (60% in min. spirits); M.P. (Drop): 90 C; Acid Value: 6
Std. Pkgs.: 515–525 lb net galvanized steel drums; flakes in 50 lb net multiwall kraft bags
Ref.: Product information

Crosby L755. Crosby Chem., Inc.
Chem. Descrip.: Calcium resinate prepared by liming a blend of tall oil rosin and rosin pitch
Category/Applications: Resinate with low acidity, uniform melting pt., and freedom from unreacted lime; used as dry core binders, and ingredients of floor tiles, waterproof adhesives, roofing felts, gasket cements, and ink vehicles
Form: Solid; Color: Dark
Gen'l. Prop.: M.P. (Ring & Ball): 120 C; Acid Value: 40
Std. Pkgs.: 520–540 lb net black steel drums
Ref.: Product information

Crosby Mirares. Crosby Chem., Inc.
Chem. Descrip.: Tall oil rosin
Form: Solid; Color (Grade): X max.
Gen'l. Prop.: Sol.: Sol. in ethanol, isopropanol, min. spirits, hexane, toluene, ethyl acetate, acetone; M.P. (Drop): 85 C; Acid No.: 163
Std. Pkgs.: Bulk, drum
Ref.: Product information

Crosby P E Ester Gum. Crosby Chem., Inc.
Chem. Descrip.: Pure pentaerythritol ester of pale rosin
Category/Applications: Ester gum characterized by its uniformity, pale color, low ash content, freedom from insol. ingred., absence of free pentaerythritol, and high melting pt.; used in paints and varnishes, emulsion paints, wax compositions, adhesives, printing inks, floor tile, linoleum, etc.
Form: Solid; Color (Gardner): 8 (60% in min. spirits)
Gen'l. Prop.: Sol.: Sol. in aliphatics, aromatics, esters, ketones; partly sol. in glycols; Visc. (Gardner): G-H (60% in min. spirits); M.P. (Drop): 101 C; Acid Val-

ue: 14
Std. Pkgs.: 515–525 lb net galvanized steel drums
Ref.: Product information

Crosby P E Ester Gum 721. Crosby Chem., Inc.
Chem. Descrip.: Pentaerythritol ester derivative of a chemically and heat modified tall oil rosin fraction
Category/Applications: Ester gum used in solvent, emulsion and hot melt coatings, printing inks, and adhesives
Form: Solid, flakes; Color (Gardner): 9 (60% in min. spirits)
Gen'l. Prop.: M.P. (Drop): 105 C; Acid Value: 14
Std. Pkgs.: 515–525 lb net galvanized steel drums; flakes in 50 lb net multiwall kraft bags
Ref.: Product information

Crosby P E Ester Gum H. Crosby Chem., Inc.
Chem. Descrip.: Hard resin produced by the esterification of pentaerythritol with polymerized tall oil rosin
Category/Applications: Resin for use in antifouling paints, emulsion paints, varnishes, and adhesives; aids viscosity stability on storage in emulsion adhesives; important in rubber cements, and Neoprene adhesives as a tackifier resin; used in combination with Elvax and paraffin waxes to produce hot melt adhesives of good pouring characteristics, cohesive strength, and flexibility
Form: Solid, flakes; Color (Gardner): 8 (60% in min. spirits)
Gen'l. Prop.: Soften. Pt. (Ring & Ball): 128 C; Acid No.: 15
Std. Pkgs.: 515–525 lb net galvanized steel drums; flakes in 50 lb net multiwall bags
Ref.: Product information

Crosby Polros A. Crosby Chem., Inc.
Chem. Descrip.: Polymerized pale rosin produced from tall oil rosin
Category/Applications: Rosin for use in the manufacture of synthetic resins, ester gums, rosin modified resins, limed rosin, gloss oils, adhesives, varnishes, papercoating varnishes, printing inks, core oils, greases, solder flux, and similar products requiring high melting, noncrystallizing rosin
Form: Solid; Color (Grade): WG
Gen'l. Prop.: Sol.: Sol. in ethanol, isopropanol, min. spirits, hexane, toluene, ethyl acetate, acetone; Sp. Gr.: 1.07; M.P. (Drop): 107 C; Acid Value: 153; Sapon. Value: 155
Std. Pkgs.: 515–525 lb net galvanized steel drums
Ref.: Product information

Crosby Polros L-6. Crosby Chem., Inc.
Chem. Descrip.: Metallic resinate produced by reaction of hydrated lime with Crosby Polymerized Rosin
Category/Applications: Hard thermoplastic resin with rapid solvent release and good pigment wetting properties; used in rotogravure and heat-set inks, overprint varnishes, core oils, and waterproof adhesives
Form: Solid, flakes; Color (Grade): WG
Gen'l. Prop.: M.P. (Ring & Ball): 173 C
Std. Pkgs.: 515–525 lb net galvanized steel drums; flakes in 50 lb net multiwall kraft bags
Ref.: Product information

Crosby Polros L-7. Crosby Chem., Inc.
Chem. Descrip.: Metallic resinate produced by reaction of hydrated lime with Crosby Polymerized Rosin
Category/Applications: See Crosby Polros L-6
Form: Solid; Color (Grade): N

Gen'l. Prop.: Sol.: Sol. in aliphatics, aromatics; partly sol. in esters and glycols; Visc. (Gardner): Z_6 (60% in min. spirits); M.P. (Ring & Ball): 184 C
Std. Pkgs.: 515–525 lb net galvanized steel drums; flakes in 50 lb net multiwall kraft bags
Ref.: Product information

Crosby Polros LF-6. Crosby Chem., Inc.
Chem. Descrip.: Clear metallic resinate produced by reaction of hydrated lime with Crosby Polymerized Rosin
Category/Applications: See Crosby Polros L-6
Form: Solid; Color (Grade): K
Gen'l. Prop.: Sol.: Sol. in aliphatics, aromatics, esters; partly sol. in glycols; Visc. (Gardner): Z_3–Z_4 (60% in min. spirits); M.P. (Ring & Ball): 170 C; Acid Value: 60
Std. Pkgs.: 515–525 lb net galvanized steel drums
Ref.: Product information

Crosby Polros LF-8. Crosby Chem., Inc.
Chem. Descrip.: Metallic resinate
Category/Applications: Resin used in printing inks incl. gravure, flexographic, heat set, steam set, metallic no scratch, soap and alkali resistant inks, and overprint varnishes
Form: Solid; Color (Grade): K
Gen'l. Prop.: Visc. (Gardner): J-T (50% in min. spirits); M.P. (Ring & Ball): 185 C; Acid Value: 30
Std. Pkgs.: 525 lb net galvanized and black rosin drums
Ref.: Product information

Crosby P.R.O. Crosby Chem., Inc.
Chem. Descrip.: Blend of low polymeric tall oil rosin oils produced as the volatile constituents from rosin polymers
Category/Applications: Used in sealants, putties, rubber adhesives, solder flux, varnishes, and core and gloss oils; embodies the tackifier, adhesive, water repellency, and carboxylic reactive properties of rosin polymers
Form: Viscous liquid
Gen'l. Prop.: Sp. Gr.: 1.006; Visc.: Z_4–Z_6; Acid Value: 80–95
Std. Pkgs.: 55 gal open head drums; 4000, 8000, and 10,000 gal tank cars
Ref.: Product information

Crosby Resin 721. Crosby Chem., Inc.
Chem. Descrip.: Resin produced by fractional distillation of stabilized tall oil rosin
Category/Applications: Pale colored resin with improved heat, light, and oxidative stability compared to regular rosin
Form: Solid; Color (USDA Rosin Scale): WW
Gen'l. Prop.: Sol.: Sol. in ethanol, isopropanol, min. spirits, hexane, toluene, ethyl acetate, and acetone; Sp. Gr.: 1.04; M.P. (Ring & Ball): 67 C; Acid Value: 155
Std. Pkgs.: 515–525 lb net galvanized steel drums
Ref.: Product information

Crosby Resin No. 2035. Crosby Chem., Inc.
Chem. Descrip.: Calcium-zinc resinate
Category/Applications: Resinate with high melting pt., low acid number, clarity, and freedom from unreacted metals; used in paint, varnish, printing ink, adhesives, etc.
Form: Solid; Color (Grade): M
Gen'l. Prop.: Sol.: Sol. in aliphatics, aromatics; partly sol. in esters and glycols; Visc. (Gardner): T-U (60% in min. spirits); M.P. (Ring & Ball): 152 C; Acid Value: 40 (calc.)
Std. Pkgs.: 515–535 lb net galvanized steel drums
Ref.: Product information

Crosby Valros. Crosby Chem., Inc.
Chem. Descrip.: Polymerized rosin
Category/Applications: High quality, general purpose rosin with the uniformity, clarity, brightness, and heat bleaching properties of pale wood rosin, and the reactivity, high melting pt. and viscosities of the end products of gum rosin; used as alkyd modifier, metallic resinates, ester gums, adhesives, rubber, soap, paper size, and synthetic resins
Form: Solid; Color (Grade): M max.
Gen'l. Prop.: Sol.: Sol. in ethanol, isopropanol, min. spirits, hexane, toluene, ethyl acetate, acetone; Sp. Gr.: 1.06; M.P. (Drop): 94 C; Acid Value: 157; Sapon. Value: 160
Std. Pkgs.: 515–525 lb net galvanized steel drums
Ref.: Product information

Crosby Valros LF-6. Crosby Chem., Inc.
Chem. Descrip.: Calcium resinate
Category/Applications: General purpose, clear pale calcium resinate for use in ink resins and coatings, gloss oils, and varnishes
Form: Solid, flakes; Color (Grade): K
Gen'l. Prop.: Sol.: Sol. in aliphatics, aromatics; partly sol. in esters and glycols; Visc. (Gardner): VW (60% in min. spirits); M.P. (Ring & Ball): 155 C; Acid Value: 65
Std. Pkgs.: 475–510 lb net 29 gauge galvanized rosin drums; flakes in 50 lb net kraft bags
Ref.: Product information

Crosby Valros LF-65. Crosby Chem., Inc.
Chem. Descrip.: Metallic resinate
Category/Applications: All purpose metallic resinate for use in ink manufacture, due to high melting pt., compatibility with ethyl cellulose, and clarity; also used in gloss oils, floor tiles, matches, cements, varnishes, etc.
Form: Solid, flakes; Color (Grade): K
Gen'l. Prop.: Sol.: Sol. in aliphatics, aromatics; partly sol. in esters and glycols; Visc. (Gardner): Z_2–Z_3 (60% in min. spirits); M.P. (Ring & Ball): 170 C; Acid Value: 40
Std. Pkgs.: 525 lb net 29 gauge galvanized rosin drums; flakes in 50 lb net multiwall kraft bags
Ref.: Product information

Crystic 57. Scott Badder Co. Ltd.
Chem. Descrip.: Polyester resin
Category/Applications: Color dispersing agent of unsaturated polyester
Form: Liquid; Color (Gardner): 3 max.
Gen'l. Prop.: Sp. Gr.: 1.20; Visc.: 6–10 poise; Acid No.: 14–22 mg KOH/g
Toxicity/Handling: Keep away from flames; shelf life of no less than 12 mos.
Std. Pkgs.: 200-kg containers
Ref.: Tech. leaflet 138F

Crystic 113PA. Scott Bader Co. Ltd.
Chem. Descrip.: Polyester resin
Category/Applications: Preaccelerated resin designed to minimize loss of styrene from laminate surface during molding; suitable for use on thick multilayer laminates with either chopped strand mat or chopped strand mat/woven rovings
Ref.: Tech. leaflet 961

Crystic 189LV. Scott Bader Co. Ltd.
Chem. Descrip.: Polyester resin
Category/Applications: Resin for the construction of boathulls, meeting requirements of BS 3532:1962 Type A; suitable for all contact and low pressure molding applications where outstanding toughness and water resistance is required; LV grade is ideal for most cold curing applications where woven glass cloth is used
Form: Liquid; Color: Straw

Composition: 38% volatile content
Gen'l. Prop.: Sp. Gr.: 1.15; Visc.: 3.6 poise; Acid Value: 24 mg KOH/g; Stability: 6 mo. in the dark @ 20 C; Ref. Index: 1.569 (cured, unfilled casting); Flash Pt.: < 32 C
Mech. Prop. (cured, unfilled casting): Tens. Str.: 70 MPa (BS 2782:1976); Tens. Elong.: 4% (@ break, 20 C, filtered resin, void-free casting); Tens. Mod.: 3.2 GPa (BS 2782:1976); Hardness (Barcol): 45 (Model GYZJ-934-1)
Therm. Prop. (cured, unfilled casting): Distort. Temp.: 60 C (@ 1.80 MPa, BS 2782:1976)
Curing Charact.: Gel Time: 9 min (@ 25 C with Crystic 189 100 pbw, Catalyst Paste H 4 pbw, Accelerator E 4 pbw
Toxicity/Handling: Highly flammable; keep away from naked flames; recommended storage @ ⩽ 20 C
Std. Pkgs.: 25-kg and 225-kg steel containers and bulk
Ref.: Tech. leaflet 119

Crystic 189MV. Scott Bader Co. Ltd.
Chem. Descrip.: Polyester resin
Category/Applications: See Crystic 189-LV; MV grade is suitable for contact molding with chopped strand mat
Form: Liquid; Color: Straw
Composition: 33% volatile content
Gen'l. Prop.: Sp. Gr.: 1.17; Visc.: 9 poise; Acid Value: 26 mg KOG/g; Stability: 6 mo. in the dark @ 20 C; Ref. Index: 1.566 (cured, unfilled casting); Flash Pt.: < 32 C
Mech. Prop. (cured, unfilled casting): Tens. Str.: 70 MPa (BS 2782-1976); Tens. Elong.: 4% (@ break, 20 C, filtered resin, void-free casting); Tens. Mod.: 3.2 GPa (BS 2782:1976); Hardness (Barcol): 45 (Model GYZJ-934-1)
Therm. Prop. (cured, unfilled casting): Distort. Temp.: 56 C (1.80 MPa, BS 2782:1976)
Curing Charact.: See Crystic 189LV
Toxicity/Handling: See Crystic 189LV
Ref.: Tech. leaflet 119

Crystic 191LV. Scott Bader Co. Ltd.
Chem. Descrip.: Styrene-modified polyester resin
Category/Applications: Light-stabilized resin meeting requirements of BS 3532: 1962 Type A; used for the manufacture of translucent sheeting by machine or hand methods, for the manufacture of flat decorative laminates using printed paper, fabric, or other carriers, and for the lamination of wall charts and posters; higher viscosity version (Crystic 191 MV) also avail.
Form: Liquid; Color: Water clear
Composition: 36% volatile content
Gen'l. Prop.: Sp. Gr.: 1.11; Visc.: 3.5 poise; Acid Value: 32 mg KOH/g; Stability: 6 mo. in the dark @ 20 C; fully cured sheeting has excellent weather resistance; Ref. Index: 1.560 (cured cast sheet); Flash Pt.: < 32 C
Mech. Prop. (cured cast sheet): Tens. Str.: 60 MPa (BS 2782:1976); Tens. Elong.: 2% (@ break, 20 C, filtered resin, void-free casting); Tens. Mod.: 3.6 GPa (BS 2782:1976); Hardness (Barcol): 48 (Model GYZJ-934-1)
Therm. Prop. (cured cast sheet): Distort. Temp.: 66 C (1.80 MPa, BS 2782:1976)
Curing Charact.: Gel Time: 17 min (@ 25 C using Crystic 191LV 100 pbw, Catalyst Paste H 4pbw, Accelerator E 4 pbw)
Toxicity/Handling: See Crystic 189LV
Std. Pkgs.: 25-kg and 225-kg steel containers and bulk
Ref.: Tech. leaflet 111

Crystic 193 HR. Scott Bader Co. Ltd.
Chem. Descrip.: Methyl methacrylate-modified polyester resin
Category/Applications: Light-stabilized

resin for the manufacture of high-clarity machine-made GRP sheeting; used with powder bonded 'E' glass mat for maximum clarity sheeting with good weathering properties
Form: Liquid; Color: Light straw
Composition: 40% volatile content
Gen'l. Prop.: Sp. Gr.: 1.11; Visc.: 1.8 poise; Acid Value: 30 mg KOH/g; Stability: 6 mo. in the dark @ 20 C; Ref. Index: 1.552 (cured, unfilled casting); Flash Pt.: < 32 C
Mech. Prop. (cured, unfilled casting): Tens. Str.: 65 MPa (BS 2782:1976); Tens. Elong.: 2.3% (@ break, filtered resin, void-free casting); Tens. Mod.: 3.8 GPa (BS 2782:1976); Hardness (Barcol): 48 (Model GYZJ-934-1)
Therm. Prop. (cured, unfilled casting): Distort. Temp.: 64 C (1.80 MPa, BS 2782:1976)
Curing Charact.: Gel Time: 11 min (@ 25 C using Crystic 193 HR 100 pbw, Catalyst Paste H 4 pbw, Accelerator E 4 pbw)
Toxicity/Handling: See Crystic 189LV
Std. Pkgs.: 25-kg and 225-kg steel containers and bulk
Ref.: Tech. leaflet 143

Crystic 195. Scott Bader Co. Ltd.
Chem. Descrip.: Polyester resin containing methyl methacrylate
Category/Applications: Light-stabilized resin for the manufacture of hand-made GRP sheeting of maximum clarity
Form: Liquid; Color: Water white to pale blue
Composition: 38% volatile content
Gen'l. Prop.: Sp. Gr.: 1.12; Visc.: 1.7 poise; Acid Value: 31 mg KOH/g; Stability: 6 mo. in the dark @ 20 C; fully cured laminates made with powder bonded 'E' glass chopped strand mat have excellent weather resistance; Ref. Index: 1.548 (cured, unfilled casting); Flash Pt.: < 32 C
Mech. Prop. (cured, unfilled casting): Tens. Str.: 55 MPa (BS 2782:1976); Tens. Elong.: 1.2% (@ break, filtered resin, void-free casting); Tens. Mod.: 4.5 GPa (BS 2782:1976); Hardness (Barcol): 47 (Model GYZJ-934-1)
Therm. Prop. (cured, unfilled casting): Distort. Temp.: 60 C (1.80 MPa, BS 2782:1976)
Curing Charact.: Gel Time: 25 min (@ 25 C using Crystic 195 100 pbw, Catalyst M 2 pbw, Accelerator E 4pbw)
Toxicity/Handling: See Crystic 189LV
Std. Pkgs.: 25-kg and 225-kg steel containers and bulk
Ref.: Tech. leaflet 128

Crystic 196. Scott Bader Co. Ltd.
Chem. Descrip.: Polyester resin
Category/Applications: General purpose, rapid-curing resilient resin with good mechanical and electrical properties for use in hand lay-up or machine molding applications; meets BS 3532:1962 Type B; Crystic 196 laminates suited for the manufacture of vehicle bodies, ISO container bodies, railway coaches, food and potable liquid containers
Form: Liquid; Color: Light straw
Composition: 33% volatile content
Gen'l. Prop.: Sp. Gr.: 1.12; Visc.: 9 poise; Acid Value: 21 mg KOH/g; Stability: 6 mo. in the dark @ 20 C; Ref. Index: 1.560 (cured, unfilled casting); Flash Pt.: < 32 C
Mech. Prop. (cured, unfilled casting): Tens. Str.: 69 MPa (BS 2782:1976); Tens. Elong.: 2.3% (@ break, filtered resin, void-free casting), Tens. Mod.: 3.8 GPa (BS 2782:1976); Hardness (Barcol): 45 (Model GYZJ-934-1)
Therm. Prop. (cured, unfilled casting): Distort. Temp.: 72 C (1.80 MPa, BS 2782:1976)
Elec. Prop. (cured, unfilled casting):

Dielec. Const.: 3.0 (1000 Hz, BS 2782: 1976); Dielec. Loss: 0.006 (1000 Hz, BS 2782:1976)
Curing Charact.: Gel Time: 8 min (@ 25 C using Crystic 196 100 pbw, Catalyst Paste H 4 pbw, Accelerator E 4 pbw)
Toxicity/Handling: See Crystic 189LV
Std. Pkgs.: 25-kg and 225-kg steel containers and bulk
Ref.: Tech. leaflet 123

Crystic 196E. Scott Bader Co. Ltd.
Chem. Descrip.: Polyester resin
Category/Applications: General purpose, rapid-curing resin for use in tropical climates; used in hand lay-up, spray deposition, and machine molding processes; meets BS 3532: 1962 Type B; cured resin has good mechanical and electrical properties; glass fiber-reinforced laminates are resilient with good mechanical strength, rigidity, and impact resistance; used for the manufacture of boats, vehicle bodies, ISO container bodies, railway coaches, caravans, water tanks, building panels, machine fairings, etc.
Form: Liquid; Color: Light straw
Composition: 33% volatile content
Gen'l. Prop.: Sp. Gr.: 1.12; Visc.: 9 poise; Acid Value: 21 mg KOH/g; Stability: 6 mo. in the dark @ 25 C; Ref. Index: 1.560 (fully-cured, unfilled casting); Flash Pt.: < 32 C
Mech. Prop. (unfilled casting, fully-cured 24 h @ 20 C, followed by 3 h @ 80 C): Tens. Str.: 69 MPa (BS 2782: 1976); Tens. Elong.: 2.3% (@ break, 20 C, filtered resin, void-free casting); Tens. Mod.: 3.8 GPa (BS 2782:1976); Hardness (Barcol): 45 (Model GYZJ-934-1)
Therm. Prop. (unfilled casting, fully-cured 24 h @ 20 C, 5 h @ 80 C, 3 h @ 120 C): Distort. Temp.: 72 C (1.80 MPa, BS 2782:1976)
Elec. Prop. (unfilled casting, fully-cured, see Mech. Prop.): Dielec. Const.: 3.0 (1000 Hz, BS 2782:1976); Dielec. Loss: 0.006 (1000 Hz, BS 2782:1976)
Curing Charact.: Gel Time: 15 min (@ 25 C using Crystic 196E 100 pbw, Catalyst M 2 pbw, Accelerator E 4 pbw)
Toxicity/Handling: Flammable; store in the dark in closed containers at temps. ⩽ 25 C
Std. Pkgs.: 225-kg steel containers
Ref.: Tech. leaflet 154

Crystic 196 PALV. Scott Bader Co. Ltd.
Chem. Descrip.: Polyester resin
Category/Applications: Preaccelerated, low viscosity version of Crystic 196 for use in resin-injection, cold press or vacuum impregnation processes when used with a suitable curing system
Form: Liquid; Color: Pinkish
Composition: 38% volatile content
Gen'l. Prop.: Sp. Gr.: 1.11; Visc.: 3.5 poise; Acid Value: 19 mg KOH/g; Stability: 6 mo. in the dark @ 20 C
Curing Charact.: Gel Time: 14 min (@ 25 C using Crystic 196 PALV 100 pbw, Catalyst M 2 pbw)
Ref.: Tech. leaflet 123

Crystic 198. Scott Bader Co. Ltd.
Chem. Descrip.: Orthophthalic-based polyester resin
Category/Applications: Med.-visc. resin meeting BS 3532:1962 Type C; used for the manufacture of laminates requiring heat and chemical resistance, construction of tanks, pipes, and fumestacks to be operated in environments at elevated temps., to back up other specialist Crystic resins, to laminate on to Celmar (glass fiber backed polypropylene), to reinforce uPVC, for the fabrication of molds for heat-assisted contact molding and hot press molding processes, etc.
Form: Liquid; Color: Light straw
Composition: 36% volatile content
Gen'l. Prop.: Sp. Gr.: 1.11; Visc.: 6.5

poise (Ferranti shear rate 37.35 s^{-1}); Acid Value: 24 mg KOH/g; Stability: 6 mos. in the dark @ 20 C; Ref. Index: 1.556 (cured, unfilled casting); Flash Pt.: < 32 C
Mech. Prop. (cured, unfilled casting): Tens. Str.: 60 MPa (BS 2782:1976); Tens. Elong.: 2.1% (@ break, filtered resin, void-free casting); Tens. Mod.: 3.8 GPa (BS 2782:1976); Hardness (Barcol): 50 (Model GYZJ-934-1)
Therm. Prop. (cured, unfilled casting): Distort. Temp.: 110 C (1.80 MPa)
Curing Charact.: Gel Time: 12 min (@ 25 C using Crystic 198 100 pbw, Catalyst Paste H 4 pbw, Acceleratoor E 4 pbw)
Toxicity/Handling: See Crystic 189LV
Std. Pkgs.: 25-kg and 225-kg containers and bulk
Ref.: Tech. leaflet 126

Crystic 272. Scott Bader Co. Ltd.
Chem. Descrip.: Isophthalic polyester resin
Category/Applications: Resin for high performance applications using filament winding, extrusion, centrifugal molding, or contact molding methods; esp. suited for use with continuous roving due to good wetting characteristics with glass fiber; fully cured moldings have high mechanical strength and good strength retention in wet environments at temps. up to 60 C
Form: Liquid; Color: Light straw
Composition: 41% volatile content
Gen'l. Prop.: Sp. Gr.: 1.10; Visc.: 3.5 poise; Acid Value: 18 mg KOH/g; Stability: 6 mos. in the dark @ 20 C; Ref. Index: 1.570 (cured, unfilled casting); Flash Pt.: < 32 C
Mech. Prop. (cured, unfilled casting): Tens. Str.: 75 MPa (BS 2782:1976); Tens. Elong.: 3.8% (@ break, filtered resin, void-free casting); Tens. Mod.: 3.5 GPa (BS 2782:1976); Hardness (Barcol): 45 (Model GYZJ-934-1)
Therm. Prop. (cured, unfilled casting): Distort. Temp.: 75 C (1.80 MPa, BS 2782:1976)
Curing Charact.: Gel Time: 8 min @ 25 C using Crystic 272 100 pbw, Catalyst Paste H 4 pbw, Accelerator E 4 pbw
Toxicity/Handling: See Crystic 189LV
Std. Pkgs.: 25-kg and 225-kg steel containers and bulk
Ref.: Tech. leaflet 107

Crystic 272E. Scott Bader Co. Ltd.
Chem. Descrip.: Isophthalic polyester resin
Category/Applications: Resin for use in tropical climates for high-performance applications such as pipes, tanks, and ducting using filament winding, extrusion, centrifugal molding, or contact molding methods; good wetting characteristics with a range of reinforcements make it suitable for use with continuous rovings, carbon fibers, and Kevlar; fully-cured moldings have high mechanical strength and strength retention in wet environments up to 60 C
Form: Liquid; Color: Light straw
Composition: 40% volatile content
Gen'l. Prop.: Sp. Gr.: 1.10; Visc.: 3.5 poise (Ferranti shear rate 37.35 s^{-1}); Acid Value: 18 mg KOH/g; Stability: 6 mos. in the dark @ 25 C; Ref. Index: 1.570 (cured, unfilled casting); Flash Pt.: < 32 C
Mech. Prop. (unfilled casting, cured 24 h @ 20 C, 3 h @ 80 C): Tens. Str.: 75 MPa (BS 2782:1976); Tens. Elong.: 3.5% (@ break, 20 C, filtered resin, void-free casting); Tens. Mod.: 3.5 GPa (BS 2782: 1976); Hardness (Barcol): 45 (Model GYZJ-934-1)
Therm. Prop. (unfilled casting, cured 24 h @ 20 C, 5 h @ 80 C, 3 h @ 120 C): Distort. Temp.: 75 C (1.80 MPa, BS

2782:1976)
Curing Charact.: Gel Time: 18 min @ 25 C using Crystic 272 E 100 pbw, Catalyst M 2 pbw, Accelerator E 4 pbw
Toxicity/Handling: See Crystic 196E
Std. Pkgs.: 225-kg steel containers
Ref.: Tech. leaflet 151

Crystic 302. Scott Bader Co. Ltd.
Chem. Descrip.: Het acid polyester resin
Category/Applications: High performance resin for the production of reduced fire hazard laminates, e.g., for the manufacture of marine structures, boats, building components, cladding panels, radar scanner housings, radomes, and industrial moldings
Form: Liquid; Color: Straw
Composition: 27% volatile content
Gen'l. Prop.: Sp. Gr.: 1.33; Visc.: 10.5 poise; Acid Value: 24 mg KOH/g; Stability: 6 mos. in the dark @ 20 C; Ref. Index: 1.568 (cured, unfilled casting); Flash Pt.: < 32 C
Mech. Prop. (cured, unfilled casting): Tens. Str.: 55 MPa (BS 2782:1976); Tens. Elong.: 2% (@ break, filtered resin, void-free casting); Tens. Mod.: 3.5 GPa (BS 2782:1976); Hardness (Barcol): 43 (Model GYZJ-934-1)
Therm. Prop. (cured, unfilled casting): Distort. Temp.: 65 C (1.80 MPa, BS 2782:1976)
Curing Charact.: Gel Time: 17 min @ 25 C using Crystic 302 100 pbw, Catalyst Paste H 4 pbw, Accelerator E 4 pbw
Toxicity/Handling: See Crystic 189LV
Std. Pkgs.: 25-kg and 225-kg steel containers and bulk
Ref.: Tech. leaflet 105

Crystic 306. Scott Bader Co. Ltd.
Chem. Descrip.: Het acid polyester resin
Category/Applications: Light-stabilized resin for the production of machine-made translucent GRP roof sheeting, laminates with high clarity, good weathering, and low fire hazard properties
Form: Liquid; Color: Light straw
Composition: 31% volatile content
Gen'l. Prop.: Sp. Gr.: 1.18; Visc.: 3.5 poise; Acid Value: 24 mg KOH/g; Stability: 6 mo. in the dark @ 20 C; Ref. Index: 1.550 (cured, unfilled casting); Flash Pt.: < 32 C
Mech. Prop. (cured, unfilled casting): Tens. Str.: 48 MPa (BS 2782:1976); Tens. Elong.: 4.0% (@ break, filtered resin, void-free casting); Tens. Mod.: 3.0 GPa (BS 2782:1976); Hardness (Barcol): 38 (Model GYZJ-934-1)
Therm. Prop. (cured, unfilled casting): Distort. Temp.: 52 C (1.80 MPa, BS 2782:1976)
Curing Charact.: Gel Time: 23 min (@ 25 C using Crystic 306 100 pbw, Catalyst Paste H 4 pbw, Accelerator E 4 pbw)
Toxicity/Handling: See Crystic 189LV
Std. Pkgs.: 25-kg and 225-kg steel containers and bulk
Ref.: Tech. leaflet 110

Crystic 323 PA. Scott Bader Co. Ltd.
Chem. Descrip.: Polyester resin
Category/Applications: Versatile, low-cost, preaccelerated, reduced fire hazard resin for general contact molding applications incl. automotive and industrial moldings; laminates have excellent mechanical properties and good water resistance and durability; also avail. in non-accelerated version, Crystic 323
Form: Liquid; Color: Opaque white
Composition: 33% volatile content
Gen'l. Prop.: Sp. Gr.: 1.20; Visc.: 8 poise (Ferranti shear rate 37.35 s^{-1}); Acid Value: 18 mg KOH/g; Stability: 6 mo. in the dark @ 20 C; Flash Pt.: < 32 C
Mech. Prop. (cured casting): Tens. Str.: 43 MPa (BS 2782:1976); Tens. Elong.: 1.4% (@ break, filled resin, unfiltered);

Tens. Mod.: 3.8 GPa (BS 2782:1976); Hardness (Barcol): 48 (Model GYZJ-934-1)
Therm. Prop. (cured casting): Distort. Temp.: 70 C (1.80 MPa, BS 2782:1976)
Curing Charact.: Gel Time: 8.5 min (@ 25 C using Crystic 323 PA 100 pbw, Catalyst M 2 pbw)
Toxicity/Handling: Highly flammable; keep away from naked flames; store ⩽ 20 C; dust from machining laminates may be harmful if inhaled
Std. Pkgs.: 25-kg and 225-kg steel containers
Ref.: Tech. leaflet 101

Crystic 328PA. Scott Bader Co. Ltd.
Chem. Descrip.: Polyester resin
Category/Applications: Filled, low-cost, preaccelerated, low fire hazard resin for opaque general purpose moldings
Ref.: Tech. leaflet 961

Crystic 328 PA LV. Scott Bader Co. Ltd.
Chem. Descrip.: Polyester resin
Category/Applications: Filled, preaccelerated, low-cost, reduced fire hazard resin for the fabrication of GRP components for the building industry; fully cured laminates can obtain a Class 2 surface spread of flame rating, and is antimony-free producing less smoke in fire test conditions
Form: Liquid; Color: Pinkish opaque
Composition: 24% volatile content
Gen'l. Prop.: Sp. Gr.: 1.38; Visc.: 13 poise (Ferranti shear rate 8.73 s^{-1}); Acid Value: 12 mg KOH/g; Stability: 6 mo. in the dark @ 20 C; Flash Pt.: < 32 C
Mech. Prop. (cured casting): Tens. Str.: 35 MPa (BS 2782:1976); Tens. Elong.: 1.3% (@ break, filled resin, unfiltered); Tens. Mod.: 3.7 GPa (BS 2782:1976); Hardness (Barcol): 42 (Model GYZJ-934-1)
Therm. Prop. (cured casting): Distort. Temp.: 72 C (1.80 MPa, BS 2782:1976)
Curing Charact.: Gel Time: 18 min (@ 25 C using Crystic 328 PA LV 100 pbw, Catalyst M 2 pbw)
Toxicity/Handling: See Crystic 323 PA
Std. Pkgs.: 25-kg and 225-kg steel containers
Ref.: Tech. leaflet 165

Crystic 345PA. Scott Bader Co. Ltd.
Chem. Descrip.: Polyester resin
Category/Applications: Filled, low-cost, preaccelerated, low fire hazard laminating resin; suitable for GRP builidng applications
Ref.: Tech. leaflet 961

Crystic 345 PA LV. Scott Bader Co. Ltd.
Chem. Descrip.: Polyester resin
Category/Applications: Preaccelerated, low-cost, Class 1, reduced fire hazard resin for use in contact molding applications
Form: Liquid; Color: Opaque pinkish
Composition: 25% volatile content
Gen'l. Prop.: Sp. Gr.: 1.45; Visc.: 18 poise; Acid Value: 13 mg KOH/g; Stability: 3 mo. in the dark @ 20 C; Flash Pt.: < 32 C
Mech. Prop. (cured casting): Tens. Str.: 30 MPa (BS 2782:1976); Tens. Elong.: 0.7% (@ break, filled resin, unfiltered); Tens. Mod.: 5.2 GPa (BS 2782:1976); Hardness (Barcol): 43 (Model GYZJ-934-1)
Therm. Prop. (cured casting): Distort. Temp.: 55 C (1.80 MPa, BS 2782:1976)
Curing Charact.: Gel Time: 18 min (@ 25 C using Crystic 345 PA LV 100 pbw, Catalyst M 2 pbw)
Toxicity/Handling: Flammable—keep away from naked flames; store ⩽ 20 C;

dust from machining laminates may be harmful if inhaled; this filled resin is prone to separation and settling on storage—mix thoroughly before use
Std. Pkgs.: 25-kg and 225-kg steel containers
Ref.: Tech. leaflet 103

Crystic 346 PA. Scott Bader Co. Ltd.
Chem. Descrip.: Polyester resin
Category/Applications: Preaccelerated, low fire hazard resin for use in GRP structural applications, e.g., fabrication of building components, cladding, and certain land transport components
Form: Liquid; Color: Opaque pinkish
Composition: 21% volatile content
Gen'l. Prop.: Sp. Gr.: 1.48; Visc.: 20 poise (Ferranti shear rate 8.73 s^{-1}); Acid Value: 12 mg KOH/g; Stability: 3 mo. in the dark @ 20 C; Flash Pt.: < 32 C
Mech. Prop. (cured casting): Tens. Str.: 35 MPa (BS 2782:1976); Tens. Elong.: 0.5% (@ break, 20 C, filled resin, unfiltered); Tens. Mod.: 7.3 GPa (BS 2782: 1976); Flex. Str.: 58 MPa (BS 2782: 1976); Hardness (Barcol): 51 (Model GYZJ-934-1)
Therm. Prop. (cured casting): Distort. Temp.: 64 C (1.80 MPa, BS 2782:1976)
Curing Charact.: Gel Time: 18 min (@ 25 C using Crystic 346 PA 100 pbw, Catalyst M 2 pbw)
Toxicity/Handling: See Crystic 345 PA LV
Std. Pkgs.: 25-kg and 225-kg steel containers
Ref.: Tech. leaflet 149

Crystic 385E. Scott Bader Co. Ltd.
Chem. Descrip.: Isophthalic/Het acid polyester resin
Category/Applications: High-performance, light-stabilized, reduced fire hazard, thixotropic resin of high translucency for contact molding applications in tropical climates; used in the manufacture of complete roof structures, roof domes, canopies, and awnings
Form: Liquid; Color: Cloudy
Composition: 34% volatile content
Gen'l. Prop.: Sp. Gr.: 1.15; Visc.: 4 poise (Ferranti shear rate 37.35 s^{-1}); Acid Value: 27 mg KOH/g; Stability: 6 mo. in the dark @ 25 C; Ref. Index: 1.553 (cured, unfilled casting); Flash Pt.: < 32 C
Mech. Prop. (cured, unfilled casting): Tens. Str.: 60 MPa (BS 2782:1976); Tens. Elong.: 3.0% (@ break, filtered resin, void-free casting); Tens. Mod.: 3.0 GPa (BS 2782:1976); Hardness (Barcol): 45 (Model GYZJ-934-1)
Therm. Prop. (cured, unfilled casting): Distort. Temp.: 70 C (1.80 MPa, BS 2782:1976)
Curing Charact.: Gel Time: 30 min (@ 35 C using Crystic 385E 100 pbw, Catalyst M 1 pbw, Accelerator E 3 pbw)
Toxicity/Handling: See Crystic 196E
Std. Pkgs.: 225-kg steel containers
Ref.: Tech. leaflet 141

Crystic 385 PA. Scott Bader Co. Ltd.
Chem. Descrip.: Isophthalic/Het acid polyester resin
Category/Applications: High performance, preaccelerated, light-stabilized, reduced fire hazard, thixotropic resin of high translucency, with outstanding wetting properties impregnating glass fiber reinforcement rapidly and thoroughly; used in the manufacture of complete roof structures, roof domes, canopies, and awnings; also avail. in nonaccelerated version, Crystic 385
Form: Liquid; Color: Cloudy pink
Composition: 34% volatile content
Gen'l. Prop.: Sp. Gr.: 1.15; Visc.: 5 poise (Ferranti shear rate 37.35 s^{-1}); Acid Value: 27 mg KOH/g; Stability:

3 mo. in the dark @ 20 C; Ref. Index: 1.553 (cured, unfilled casting); Flash Pt.: < 32 C
Mech. Prop. (cured, unfilled casting): Tens. Str.: 55 MPa (BS 2782:1976); Tens. Elong.: 2.9% (@ break, filtered resin, void-free casting); Tens. Mod.: 3.2 GPa (BS 2782:1976); Hardness (Barcol): 45 (Model GYZJ-934-1)
Therm. Prop. (cured, unfilled casting): Distort. Temp.: 70 C (1.80 MPa, BS 2782:1976)
Curing Charact.: Gel Time: 10 min (@ 25 C using Crystic 385 PA 100 pbw, Catalyst M 2 pbw)
Toxicity/Handling: See Crystic 189LV
Std. Pkgs.: 25-kg and 225-kg steel containers and bulk
Ref.: Tech. leaflet 163

Crystic 387. Scott Bader Co. Ltd.
Chem. Descrip.: Polyester resin
Category/Applications: Light-stabilized resin for the manufacture of reduced fire hazard high clarity machine-made GRP sheeting suitable for internal use and can be used for external applications if suitably protected
Form: Liquid; Color: Hazy light straw
Composition: 32% volatile content
Gen'l. Prop.: Sp. Gr.: 1.32; Visc.: 2.0 poise; Acid Value: 20 mg KOH/g; Stability: 3 mo. in the dark @ 20 C; Ref. Index: 1.570 (cured, unfilled casting); Flash Pt.: < 32 C
Mech. Prop. (cured, unfilled casting): Tens. Str.: 50 MPa (BS 2782:1976); Tens. Elong.: 1.8% (@ break, filtered resin, void-free casting); Tens. Mod.: 3.1 GPa (BS 2782:1976); Hardness (Barcol): 35 (Model GYZJ-934-1)
Therm. Prop. (cured, unfilled casting): Distort. Temp.: 55 C (1.80 MPa, BS 2782:1976)
Curing Charact.: Gel Time: 11 min (@ 25 C using Crystic 387 100 pbw, Catalyst Paste H 4 pbw, Accelerator E 4 pbw)
Toxicity/Handling: See Crystic 323 PA
Std. Pkgs.: 25-kg and 225-kg steel containers and bulk
Ref.: Tech. leaflet 169

Crystic 392. Scott Bader Co. Ltd.
Chem. Descrip.: Isophthalic-neopentyl glycol polyester resin
Category/Applications: Resin for GRP and unplasticized PVC laminates, for use in the chemical, food, and process industries; fully-cured laminates have good heat and chemical resistance, esp. resistant to alkaline environments;
Form: Liquid; Color: Light straw
Composition: 41% volatile content
Gen'l. Prop.: Sp. Gr.: 1.07; Visc.: 8 poise; Acid Value: 14 mg KOH/g; Stability: 12 mo. in the dark @ 20 C; Ref. Index: 1.557 (cured, unfilled casting); Flash Pt.: < 32 C
Mech. Prop. (cured, unfilled casting): Tens. Str.: 70 MPa (BS 2782:1970, BS 3532:1962); Tens. Elong.: 3.0% (@ break, 20 C, filtered resin, void-free casting); Tens. Mod.: 3.4 GPa (BS 2782:1970, BS 3532:1962); Hardness (Barcol): 45 (Model GYZJ-934-1)
Therm. Prop. (cured, unfilled casting): Distort. Temp.: 90 C (1.80 MPa, BS 2782:1970, BS 3532:1962)
Curing Charact.: Gel Time: 9 min (@ 25 C using Crystic 392 100 pbw, Catalyst M 2 pbw, Accelerator E 4 pbw)
Toxicity/Handling: See Crystic 189LV
Std. Pkgs.: 25-kg and 225-kg steel containers
Ref.: Tech. leaflet 142

Crystic 392E. Scott Bader Co. Ltd.
Chem. Descrip.: Isophthalic-neopentyl glycol polyester resin
Category/Applications: See Crystic 392; 392E is designed for use in tropical climates

Form: Liquid; Color: Light straw
Composition: 41% volatile content
Gen'l. Prop.: Sp. Gr.: 1.07; Visc.: 8 poise (Ferranti shear rate 37.35 s^{-1}); Acid Value: 13 mg KOH/g; Stability: 6 mo. in the dark @ 25 C; Ref. Index: 1.557 (cured, unfilled casting); Flash Pt.: < 32 C
Mech. Prop. (cured, unfilled casting): Tens. Str.: 70 MPa (BS 2782:1976); Tens. Elong.: 3.0% (@ break, 20 C, filtered resin, void-free casting); Tens. Mod.: 3.4 GPa (BS 2782:1976); Hardness (Barcol): 45 (Model GYZJ-934-1)
Therm. Prop. (cured, unfilled casting): Distort. Temp.: 90 C (1.80 MPa, BS 2782:1976)
Curing Charact.: Gel Time: 30 min (@ 35 C using Crystic 392E 100 pbw, Catalyst M 1 pbw, Accelerator E 3 pbw)
Toxicity/Handling: See Crystic 196 E
Std. Pkgs.: 225-kg steel containers
Ref.: Tech. leaflet 136

Crystic 405E. Scott Bader Co. Ltd.
Chem. Descrip.: Polyester resin
Form: Liquid
Gen'l. Prop.: Flash Pt.: 31 C
Curing Charact.: Gel Time: 45 min (@ 25 C using Crystic 405E 100 pbw, Cata lyst M 2 pbw, Accelerator E 1 pbw)
Toxicity/Handling: See Crystic 196E
Std. Pkgs.: 225-kg steel containers
Ref.: Tech. leaflet

Crystic 405 PA. Scott Bader Co. Ltd.
Chem. Descrip.: Polyester resin
Category/Applications: Thixotropic, preaccelerated resin for the boatbuilding industry; used for the construction of small and med.-sized boats, and other applications requiring good wetting and impregnation of the reinforcement; fully-cured laminates have good water resistance and weathering properties; also avail. in a nonaccelerated version, Crystic 405
Form: Liquid; Color: Cloudy mauvish
Composition: 37% volatile content
Gen'l. Prop.: Sp. Gr.: 1.12; Visc.: 5 poise (Ferranti shear rate 37.35 s^{-1}); Acid Value: 24 mg KOH/g; Stability: 6 mo. in the dark @ 20 C; Flash Pt.: < 32 C
Mech. Prop. (cured, unfilled casting): Tens. Str.: 65 MPa (BS 2782:1976); Tens. Elong.: 2.0% (@ break, filtered resin, void-free casting); Tens. Mod.: 3.8 GPa (BS 2782:1976); Hardness (Barcol): 50 (Model GYZJ-934-1)
Therm. Prop. (cured, unfilled casting): Distort. Temp.: 69 C (1.80 MPa, BS 2782:1976)
Curing Charact.: Gel Time: 18 min (@ 25 C using Crystic 405 PA 100 pbw, Catalyst M 2 pbw)
Toxicity/Handling: See Crystic 189LV
Std. Pkgs.: 25-kg and 225-kg steel containers and bulk
Ref.: Tech. leaflet 115

Crystic 406 PA. Scott Bader Co. Ltd.
Chem. Descrip.: Polyester resin
Category/Applications: Preaccelerated, thixotropic resin for contact molding applications, with rapid wetting and impregnation of the reinforcement; fully-cured laminates have high water resistance and excellent mechanical properties; used in the construction of boat hulls, vehicle bodies, and industrial and general purpose moldings
Form: Liquid; Color: Cloudy mauvish
Composition: 40% volatile content
Gen'l. Prop.: Sp. Gr.: 1.10; Visc.: 11 poise (Ferranti shear rate 1.1 s^{-1}); Acid Value: 17 mg KOH/g; Stability: 3 mo. in the dark @ 20 C; Flash Pt.: < 32 C
Mech. Prop. (cured, unfilled casting): Tens. Str.: 65 MPa (BS 2782:1976); Tens. Elong.: 2.0% (@ break, filtered

resin, void-free casting); Tens. Mod.: 4.0 GPa (BS 2782:1976); Hardness (Barcol): 45 (Model GYZJ-934-1)
Therm. Prop. (cured, unfilled casting): Distort. Temp.: 62 C (1.80 MPa, BS 2782:1976)
Curing Charact.: Gel Time: 16 min (@ 25 C using Crystic 406 PA 100 pbw, Catalyst M 2 pbw)
Toxicity/Handling: See Crystic 189LV
Std. Pkgs.: 25-kg and 225-kg steel containers and bulk
Ref.: Tech. leaflet 118

Crystic 471 PA LV. Scott Bader Co. Ltd.
Chem. Descrip.: Polyester resin
Category/Applications: Preaccelerated, low-viscosity, rapid-hardening resin with fast wetting and impregnation of the reinforcement and short mold-release time, for contact molding applications and for use in catalyst injection spray equipment; recommended for automotive, marine, industrial, and general purpose applications
Form: Liquid; Color: Cloudy mauvish
Composition: 41% volatile content
Gen'l. Prop.: Sp. Gr.: 1.11; Visc.: 3 poise (Ferranti shear rate 37.35 s^{-1}); Acid Value: 19 mg KOH/g; Stability: 6 mo. in the dark @ 20 C; Flash Pt.: < 32 C
Mech. Prop. (cured, unfilled casting): Tens. Str.: 65 MPa (BS 2782:1976); Tens. Elong.: 2.2% (@ break, filtered resin, void-free casting); Tens. Mod.: 3.6 GPa (BS 2782:1976); Hardness (Barcol): 45 (Model GYZJ-934-1)
Therm. Prop. (cured, unfilled casting): Distort. Temp.: 75 C (1.80 MPa, BS 2782:1976)
Curing Charact.: Gel Time: 10 min (@ 25 C using Crystic 471 PA LV 100 pbw, Catalyst M 2 pbw)
Toxicity/Handling: See Crystic 189LV
Std. Pkgs.: 25-kg and 225-kg steel containers and bulk
Ref.: Tech. leaflet 127

Crystic 474 PA. Scott Bader Co. Ltd.
Chem. Descrip.: Orthophthalic-based polyester resin
Category/Applications: Preaccelerated thixotropic version of Crystic 198
Form: Liquid; Color: Cloudy mauvish
Composition: 38% volatile content
Gen'l. Prop.: Sp. Gr.: 1.10; Visc.: 5 poise (Ferranti shear rate 37.35 s^{-1}); Acid Value: 23 mg KOH/g; Stability: 6 mos. in the dark @ 20 C; Flash Pt.: < 32 C
Mech. Prop. (cured, unfilled casting): Tens. Str.: 60 MPa (BS 2782:1976); Tens. Elong.: 2.1% (@ break, filtered resin, void-free casting); Tens. Mod.: 3.8 GPa (BS 2782:1976); Hardness (Barcol): 50 (Model GYZJ-934-1)
Therm. Prop. (cured, unfilled casting): Distort. Temp.: 110 C (1.80 MPa, BS 2782:1976)
Curing Charact.: Gel Time: 12 min (@ 25 C using Crystic 474 PA 100 pbw, Catalyst M 2 pbw)
Toxicity/Handling: See Crystic 189LV
Std. Pkgs.: 25-kg and 225-kg containers and bulk
Ref.: Tech. leaflet 126

Crystic 489 PA. Scott Bader Co. Ltd.
Chem. Descrip.: Isophthalic polyester resin
Category/Applications: Preaccelerated, thixotropic resin developed for use with Crystic Gelcoat 65 PA to provide a matched isophthalic gelcoat/laminating resin system for molding boat hulls with outstanding durability and performance; fully-cured boat hulls and moldings have high mechanical performance and impact resistance, and good strength retention

in wet environments up to 40 C
Form: Liquid; Color: Pinkish
Composition: 44% volatile content
Gen'l. Prop.: Sp. Gr.: 1.11; Visc.: 4.1 poise (Ferranti shear rate 37.35 s^{-1}); Acid Value: 19 mg KOH/g; Stability: 3 mo. in the dark @ 20 C; Flash Pt.: $<$ 32 C
Mech. Prop. (unfilled casting, cured 24 h @ 20 C, 3 h @ 80 C): Tens. Str.: 75 MPa (BS 2782:1976); Tens. Elong.: 3.5% (@ break, filtered resin, void-free casting); Tens. Mod.: 3.5 GPa (BS 2782:1976); Hardness (Barcol): 42 (Model GYZJ-934-1)
Therm. Prop. (unfilled casting, cured 24 h @ 20 C, 5 h @ 80 C, 3 h @ 120 C): Distort. Temp.: 75 C (1.80 MPa, BS 2782:1976)
Curing Charact.: Gel Time: 23 min (@ 25 C using Crystic 489 PA 100 pbw, Catalyst M 1 pbw)
Toxicity/Handling: See Crystic 189LV
Std. Pkgs.: 200-kg steel containers and bulk
Ref.: Tech. leaflet 172

Crystic 491. Scott Bader Co. Ltd.
Chem. Descrip.: Isophthalic polyester resin
Category/Applications: Nonaccelerated version of Crystic 491 PA
Form: Liquid; Color: Cloudy
Composition: 40% volatile content
Gen'l. Prop.: Sp. Gr.: 1.11; Visc.: 8 poise (Ferranti shear rate 37.35 s^{-1}); Acid Value: 18 mg KOH/g; Stability: 6 mo. in the dark @ 20 C; Flash Pt.: $<$ 32 C
Curing Charact.: Gel Time: 9 min (@ 25 C using Crystic 491 100 pbw, Catalyst Paste H 4 pbw, Accelerator E 4 pbw)
Toxicity/Handling: See Crystic 189LV
Ref.: Tech. leaflet 164

Crystic 491E. Scott Bader Co. Ltd.
Chem. Descrip.: Isophthalic polyester resin
Category/Applications: Thixotropic resin for the manufacture of laminates in applications requiring high performance, toughness, and water- and chemical-resistance; for use in tropical climates; used for chemical plant, tanks, pipes, and containers for fuels, hydrocarbons, glycols, and all types of foodstuffs; for taint-free, chemical-resistant liners for vessels; in the fabrication of high performance boat hulls esp. when used with Crystic Gelcoat 65E
Form: Liquid; Color: Cloudy
Composition: 40% volatile content
Gen'l. Prop.: Sp. Gr.: 1.11; Visc.: 8 poise (Ferranti shear rate 8.73 s^{-1}); Acid Value: 18 mg KOH/g; Stability: 6 mo. in the dark @ 25 C; Flash Pt.: $<$ 32 C
Mech. Prop. (cured, unfilled casting): Tens. Str.: 75 MPa (BS 2782:1976); Tens. Elong.: 2.9% (@ break, filtered resin, void-free casting); Tens. Mod.: 3.5 GPa (BS 2782:1976); Hardness (Barcol): 45 (Model GYZJ-934-1)
Therm. Prop. (cured, unfilled casting): Distort. Temp.: 75 C (1.80 MPa, BS 2782:1976)
Curing Charact.: Gel Time: 30 min (@ 35 C using Crystic 491E 100 pbw, Catalyst M 1 pbw, Accelerator E 3 pbw)
Toxicity/Handling: See Crystic 196E
Std. Pkgs.: 225-kg steel containers
Ref.: Tech. leaflet 166

Crystic 491 PA. Scott Bader Co. Ltd.
Chem. Descrip.: Isophthalic polyester resin
Category/Applications: Preaccelerated, thixotropic resin for the manufacture of laminates in applications requiring high performance, toughness, and water- and chemical-resistance; used for chemical

plant, tanks, pipes, and containers for fuels, hydrocarbons, glycols, and all types of foodstuffs; for taint-free, chemical-resistant liners for vessels; in the fabrication of high performance boat hulls esp. when used with Crystic Gelcoat 65 PA;
Form: Liquid; Color: Cloudy mauvish
Composition: 42% volatile content
Gen'l. Prop.: Sp. Gr.: 1.11; Visc.: 5 poise (Ferranti shear rate 37.35 s^{-1}); Acid Value: 18 mg KOH/g; Stability: 6 mo. in the dark @ 20 C; Flash Pt.: < 32 C
Mech. Prop. (cured, unfilled casting): Tens. Str.: 75 MPa (BS 2782:1976); Tens. Elong.: 2.9% (@ break, filtered resin, void-free casting); Tens. Mod.: 3.5 GPa (BS 2782:1976); Hardness (Barcol): 45 (Model GYZJ-934-1)
Therm. Prop. (cured, unfilled casting): Distort. Temp.: 75 C (1.80 MPa, BS 2782:1976)
Curing Charact.: Gel Time: 12 min (@ 25 C using Crystic 491 PA 100 pbw, Catalyst M 2 pbw)
Toxicity/Handling: See Crystic 189LV
Std. Pkgs.: 25-kg and 225-kg steel containers and bulk
Ref.: Tech. leaflet 164

Crystic 625E. Scott Bader Co. Ltd.
Chem. Descrip.: Isophthalic polyester resin
Category/Applications: Resin for use in tropical climates, for the manufacture of large GRP structures; laminates have excellent long term water and weather resistance, and retain high rigidity and strength up to 120 C; suitable for marine applications
Form: Liquid; Color: Light straw
Composition: 34% volatile content
Gen'l. Prop.: Sp. Gr.: 1.13; Visc.: 8 poise (Ferranti shear rate 37.35 s^{-1}); Acid Value: 17 mg KOH/g; Stability: 6 mo. in the dark @ 25 C; Ref. Index: 1.556 (cured, unfilled casting); Flash Pt.: < 32 C
Mech. Prop. (cured, unfilled casting): Tens. Str.: 65 MPa (BS 2782:1976); Tens. Elong.: 2.2% (@ break, filtered resin, void-free casting); Tens. Mod.: 3.8 GPa (BS 2782:1976); Hardness (Barcol): 50 (Model GYZJ-934-1)
Therm. Prop. (cured, unfilled casting): Distort. Temp.: 94 C (1.80 MPa, BS 2782:1976)
Curing Charact.: Gel Time: 30 min (@ 35 C using Crystic 625E 100 pbw, Catalyst M 1 pbw, Accelerator E 3 pbw)
Toxicity/Handling: See Crystic 196E
Std. Pkgs.: 225-kg steel containers
Ref.: Tech. leaflet 162

Crystic 625 MV. Scott Bader Co. Ltd.
Chem. Descrip.: Isophthalic polyester resin
Category/Applications: Resin for the manufacture of large GRP structures, e.g., marine applications requiring good heat resistance and long-term water resistance; laminates retain high rigidity and strength up to 120 C
Form: Liquid; Color: Light straw
Composition: 34% volatile content
Gen'l. Prop.: Sp. Gr.: 1.13; Visc.: 8 poise; Acid Value: 17 mg KOH/g; Stability: 6 mo. in the dark @ 20 C; Ref. Index: 1.556 (cured, unfilled casting); Flash Pt.: < 32 C
Mech. Prop. (cured, unfilled casting): Tens. Str.: 65 MPa (BS 2782:1976); Tens. Elong.: 2.2% (@ break, filtered resin, void-free casting); Tens. Mod.: 4.0 GPa (BS 2782:1976); Hardness (Barcol): 50 (Model GYZJ-934-1)
Therm. Prop. (cured, unfilled casting): Distort. Temp.: 94 C (1.80 MPa, BS 2782:1976)
Curing Charact.: Gel Time: 10 min (@ 25 C using Crystic 625 MV 100 pbw,

Catalyst M 2 pbw, Accelerator E 4 pbw)
Toxicity/Handling: See Crystic 189LV
Std. Pkgs.: 25-kg and 225-kg steel containers and bulk
Ref.: Tech. leaflet 147

Crystic 17449. Scott Bader Co. Ltd.
Chem. Descrip.: Styrene-modified polyester resin
Category/Applications: High performance, heat-resistance, low viscosity resin with good wetting properties for use with high strength reinforcements esp. at elevated temps.; used in all conventional molding processes, incl. hand layup esp. when glass cloth or woven rovings are employed; used in the aircraft industry for radomes, ductings and fairings for high-speed aircraft; in the electrical industry for bus-bar supports, ducting, switchgear, and generator insulation; in the chemical industry for chimneys and ducting for hot corrosive gases
Form: Liquid; Color: Straw
Composition: 33% volatile content
Gen'l. Prop.: Sp. Gr.: 1.11; Visc.: 3.8 poise; Acid Value: 20 mg KOH/g; Stability: 6 mo. in the dark @ 20 C; Flash Pt.: $<$ 32 C
Mech. Prop. (cured, unfilled casting): Tens. Str.: 60 MPa (BS 2782:1976); Tens. Elong.: 2% (@ break, filtered resin, void-free casting); Tens. Mod.: 3.8 GPa (BS 2782:1976); Hardness (Barcol): 48 (Model GYZJ-934-1)
Therm. Prop. (cured, unfilled casting): Distort. Temp.: 115 C (1.80 MPa, BS 2782:1976)
Elec. Prop. (cured, unfilled casting): Dielec. Const.: 3.12 (@ 1 MHz, BS 2782: 1976); Dielec. Loss.: 0.018 (@ 1 MHz, BS 2782:1976)
Curing Charact.: Gel Time: 38 min (@ 25 C using Crystic 17449 100 pbw, Catalyst O 4 pbw, Accelerator G 3 pbw)
Toxicity/Handling: See Crystic 189LV
Std. Pkgs.: 25-kg and 225-kg steel containers and bulk
Ref.: Tech. leaflet 170

Crystic 2-406 PA. Scott Bader Co. Ltd.
Chem. Descrip.: Polyester resin
Category/Applications: Preaccelerated thixotropic contact molding resin with low styrene emission for the construction of boat hulls, vehicle bodies, and general purpose moldings; fully-cured laminates have good water resistance and mechanical properties
Form: Liquid; Color: Pinkish-mauve
Composition: 41% volatile content
Gen'l. Prop.: Sp. Gr.: 1.10; Visc.: 4.0 poise (Ferranti shear rate 37.35 s^{-1}); Acid Value: 16 mg KOH/g; Stability: 3 mo. in the dark @ 20 C; Flash Pt.: $<$ 32 C
Mech. Prop. (unfilled casting cured 24 h @ 20 C, 3 h @ 80 C): Tens. Str.: 64 MPa (BS 2782:1976); Tens. Elong.: 2.4% (@ break, filtered resin, void-free casting); Tens. Mod.: 3.4 GPa (BS 2782:1976); Hardness (Barcol): 46 (Model GYZJ-934-1)
Therm. Prop. (unfilled casting cured 24 h @ 20 C, 5 h @ 80 C, 3 h @ 120 C): Distort. Temp.: 60 C (1.80 MPa, BS 2782:1976)
Curing Charact.: Gel Time: 16 min (@ 25 C using Crystic 2-406 PA 100 pbw, Catalyst M 1 pbw)
Toxicity/Handling: See Crystic 189LV
Std. Pkgs.: 200-kg steel containers and bulk
Ref.: Tech. leaflet 174

Crystic 2-489 PA. Scott Bader Co. Ltd.
Chem. Descrip.: Isophthalic polyester resin
Category/Applications: Similar to Crystic 489 PA but with low styrene emission
Form: Liquid; Color: Pinkish
Composition: 44% volatile content

Gen'l. Prop.: Sp. Gr.: 1.11; Visc.: 4.1 poise (Ferranti shear rate 37.35 s^{-1}); Acid Value: 19 mg KOH/g; Stability: 3 mo. in the dark @ 20 C; Flash Pt.: < 32 C
Mech. Prop. (unfilled casting cured 24 h @ 20 C, 3 h @ 80 C): Tens. Str.: 75 MPa (BS 2782:1976); Tens. Elong.: 3.5% (@ break, filtered resin, void-free casting); Tens. Mod.: 3.5 GPa (BS 2782:1976); Hardness (Barcol): 40 (Model GYZJ-934-1)
Therm. Prop. (unfilled casting cured 24 h @ 20 C, 5 h @ 80 C, 3 h @ 120 C): Distort. Temp.: 75 C (1.80 MPa, BS 2782:1976)
Curing Charact.: Gel Time: 23 min (@ 25 C using Crystic 2-489 PA 100 pbw, Catalyst M 1 pbw)
Toxicity/Handling: See Crystic 189LV
Std. Pkgs.: 200-kg steel containers and bulk
Ref.: Tech. leaflet 173

Crystic 2-491 PA. Scott Bader Co. Ltd.
Chem. Descrip.: Isophthalic polyester resin
Category/Applications: Similar to Crystic 491 PA but with low styrene emission
Form: Liquid; Color: Cloudy mauvish
Composition: 42% volatile content
Gen'l. Prop.: Sp. Gr.: 1.11; Visc.: 5 poise (Ferranti shear rate 37.35 s^{-1}); Acid Value: 18 mg KOH/g; Stability: 6 mo. in the dark @ 20 C; Flash Pt.: < 32 C
Mech. Prop. (unfilled casting, cured 24 h @ 20 C, 3 h @ 80 C): Tens. Str.: 75 MPa (BS 2782:1976); Tens. Elong.: 2.9% (@ break, filtered resin, void-free casting); Tens. Mod.: 3.5 GPa (BS 2782:1976); Hardness (Barcol): 43 (Model GYZJ-934-1)
Therm. Prop. (unfilled casting, cured 24 h @ 20 C, 5 h @ 80 C, 3 h @ 120 C): Distort. Temp.: 75 C (1.80 MPa, BS 2782:1976)
Curing Charact.: Gel Time: 12 min (@ 25 C using Crystic 2-491 PA 100 pbw, Catalyst M 2 pbw)
Toxicity/Handling: See Crystic 189LV
Std. Pkgs.: 200-kg steel containers and bulk
Ref.: Tech. leaflet 175

Crystic Fireguard 75 PA. Scott Bader Co. Ltd.
Chem. Descrip.: Polyester resin
Category/Applications: Thixotropic, intumescent surface coating resin giving fire protection to GRP laminates accidentally exposed to direct flaming (produces inert gases causing a carbonaceous foam to form when a flame is applied); also used as a fire-retardant coating on timber and other slightly porous surfaces; ready to apply by brush after addition of catalyst
Form: Liquid; Color: Opaque white, black, or grey
Composition: 20% volatile content
Gen'l. Prop.: Sp. Gr.: 1.35; Acid Value: 18 mg KOH/g; Stability: 2 mo. in the dark @ 20 C; Flash Pt.: < 32 C
Mech. Prop. (cured resin): Hardness (Barcol): 40 (Model GYZJ-934-1)
Curing Charact.: Pot Life: 10 min (@ 25 C using Crystic Fireguard 75 PA 100 pbw, Catalyst M 2 pbw)
Toxicity/Handling: Flammable—keep away from naked flames; store ≤ 20 C; resin must be stirred well before adding catalyst
Std. Pkgs.: 25-kg steel containers
Ref.: Tech. leaflet 132

Crystic Fireguard Gelcoat 70 PA. Scott Bader Co. Ltd.
Chem. Descrip.: Polyester resin
Category/Applications: Specially formulated gelcoat version of Crystic Fireguard 75 PA; used for applications in the build-

ing industry where an intumescent finish is required on the molded surface
Ref.: Tech. leaflet 132

Crystic Gelcoat 33E. Scott Bader Co. Ltd.
Chem. Descrip.: Isophthalic polyester resin
Category/Applications: Thixotropic gelcoat resin with excellent impact resistance and good water resistance, esp. for use in warm climates; used for vehicle body moldings, and small and med.-sized boat hulls
Form: Liquid; Color: Cloudy
Composition: 33% volatile content
Gen'l. Prop.: Sp. Gr.: 1.15; Acid Value: 19 mg KOH/g; Stability: 4mo. in the dark @ 25 C; Flash Pt.: 31 C
Mech. Prop. (cured resin): Tens. Str.: 60 MPa (BS 2782:1970, BS 3532:1962); Tens. Elong.: 4.0% (@ break, 20 C, BS 2782:1970, BS 3532:1962); Tens. Mod.: 3.0 GPa (BS 2782:1970, BS 3532:1962); Hardness (Barcol): 38 (Model GYZJ-934-1)
Therm. Prop. (cured resin): Distort. Temp.: 52 C (1.81 MPa, BS 2782:1970, BS 3532:1962)
Curing Charact.: Gel Time: 14 min (@ 25 C using Crystic Gelcoat 33E 100 pbw, Catalyst M 2 pbw, Accelerator E 4 pbw)
Toxicity/Handling: See Crystic 196E
Std. Pkgs.: 25-kg and 225-kg steel containers
Ref.: Tech. leaflet 952

Crystic Gelcoat 39PA. Scott Bader Co. Ltd.
Chem. Descrip.: Polyester resin
Category/Applications: Light-stabilized, resilient, preaccelerated, low fire hazard gelcoat with high impact strength; can be used with Crystic 385PA for translucent roof domes and canopies
Ref.: Tech. leaflet 961

Crystic Gelcoat 46PA. Scott Bader Co. Ltd.
Chem. Descrip.: Polyester resin
Category/Applications: Light-stabilized, preaccelerated, low fire hazard, filled gelcoat normally used with Crystic 323PA for building, where fire resistance is important
Ref.: Tech. leaflet 961

Crystic Gelcoat 47PA. Scott Bader Co. Ltd.
Chem. Descrip.: Polyester resin
Category/Applications: Light-stabilized, preaccelerated, low fire hazard, filled gelcoat esp. suitable for use internally in building applications; protect with Crystic LS7 for external applications
Ref.: Tech. leaflet 961

Crystic Gelcoat 65E. Scott Bader Co. Ltd.
Chem. Descrip.: Isophthalic polyester resin
Category/Applications: Thixotropic, gelcoat resin with excellent water resistance, for use in warm climates; used for boat hulls which remain in water for long periods, and other moldings requiring good chemical and water resistance
Form: Liquid; Color: Cloudy
Composition: 36% volatile content
Gen'l. Prop.: Sp. Gr.: 1.11; Acid Value: 19 mg KOH/g; Stability: 4 mo. in the dark @ 25 C; Flash Pt.: < 32 C
Mech. Prop. (unfilled casting cured 24 h @ 20 C, 3 h @ 80 C): Tens. Str.: 75 MPa (BS 2782:1976); Tens. Elong.: 3.0% (@ break, 20 C, filled resin, unfiltered); Tens. Mod.: 3.5 GPa (BS 2782:1976); Hardness (Barcol): 42 (Model GYZJ-934-1)
Therm. Prop. (unfilled casting cured 24 h @ 20 C, 5 h – 80 C, 3 h @ 120 C): Distort. Temp.: 75 C (1.80 MPa, BS 2782:1976)

Curing Chract.: Gel Time: 14 min (@ 25 C using Crystic Gelcoat 65E 100 pbw, Catalyst M 2 pbw, Accelerator E 4 pbw)
Toxicity/Handling: See Crystic 196E
Std. Pkgs.: 25-kg and 225-kg steel containers
Ref.: Tech. leaflet 153

Crystic Gelcoat 65PA. Scott Bader Co. Ltd.
Chem. Descrip.: Polyester resin
Category/Applications: Resilient, preaccelerated gelcoat with excellent weather resistance, suitable for GRP builidng applications; also suitable for flow coat on reverse side of laminate
Ref.: Tech. leaflet 961

Crystic Gelcoat 68E. Scott Bader Co. Ltd.
Chem. Descrip.: Isophthalic polyester resin
Category/Applications: Similar to Crystic Gelcoat 68 PA but esp. designed for use in tropical climates
Form: Cloudy liquid; Color: Avail. in a range of colors
Composition: 42% volatile content
Gen'l. Prop.: Sp. Gr.: 1.10; Visc.: 2.7 poise (Ferranti shear rate 4500 s^{-1}); Acid Value: 18 mg KOH/g; Stability: 3 mo. in the dark @ 25 C; Flash Pt.: < 32 C
Mech. Prop. (cast resin cured 24 h @ 20 C, 3 h @ 80 C): Tens. Str.: 70 MPa (BS 2782:1976); Tens. Elong.: 4.0% (@ break, 20 C, filled resin, unfiltered); Tens. Mod.: 3.5 GPa (BS 2782:1976); Hardness (Barcol): 40 (Model GYZJ-934-1)
Therm. Prop. (cast resin cured 24 h @ 20 C, 5 h @ 80 C, 3 h @ 120 C): Distort. Temp.: 70 C (1.80 MPa, BS 2782:1976)
Curing Charact.: Gel Time: 10 min (@ 25 C using Crystic Gelcoat 68E 100 pbw, Catalyst M 2 pbw, Accelerator E 4 pbw)
Toxicity/Handling: See Crystic 196E
Std. Pkgs.: 25-kg and 220-kg steel containers
Ref.: Tech. leaflet 156

Crystic Gelcoat 68 PA. Scott Bader Co. Ltd.
Chem. Descrip.: Isophthalic polyester resin
Category/Applications: Thixotropic, preaccelerated gelcoat resin for application by pressure-fed spray equipment; suitable for the fabrication of vehicle bodies and boat hulls; also avail. in a nonaccelerated version, Crystic Gelcoat 68, for use in two-pot spray equipment
Form: Cloudy liquid; Color: Avail. in a range of colors
Composition: 42% volatile content
Gen'l. Prop.: Sp. Gr.: 1.10; Visc.: 2.7 poise (Ferranti shear rate 4500 s^{-1}); Acid Value: 18 mg KOH/g; Stability: 3 mo. in the dark @ 20 C; Flash Pt.: < 32 C
Mech. Prop. (cured, cast resin): Tens. Elong.: 4.0% (@ break, filled resin, unfiltered); Tens. Mod.: 3.5 GPa (BS 2782:1976); Hardness (Barcol): 40 (Model GYZJ-934-1)
Therm. Prop. (cured, cast resin): Distort. Temp.: 70 C (1.80 MPa, BS 2782:1976)
Curing Charact.: Gel Time: 10 min (@ 25 C using Crystic Gelcoat 68 PA 100 pbw, Catalyst M 2 pbw)
Toxicity/Handling: See Crystic 189LV
Std. Pkgs.: 25-kg and 225-kg steel containers
Ref.: Tech. leaflet 131

Crystic LS7. Scott Bader Co. Ltd.
Chem. Descrip.: Solution of a tough acrylic resin in xylene/MEK solvents
Category/Application: Transparent lacquer for GRP sheeting; protects against

discoloration and surface erosion, and contains an ultra-violet screening agent of good stability and long life which absorbs light of wavelengths < 340 mμ; has excellent adhesion to all cured polyester resin surfaces
Form: Liquid; Color: Pale yellow
Composition: 18% solids content
Gen'l. Prop.: Sp. Gr.: 0.89 (20 C); Visc.: 18 cps; Stability: Several years storage life @ 20 C; Flash Pt. (Abel CC): −1 C
Toxicity/Handling: Flammable—keep away from naked flames; store ≤ 20 C in tightly closed containers
Std. Pkgs.: 20-kg containers
Ref.: Tech. leaflet 116

Crystic Prefil F. Scott Bader Co. Ltd.
Chem. Descrip.: Dispersion of antimony trioxide and a chlorinated organic comp. in a special polyester resin
Category/Applications: Resin additive to facilitate the manufacture of reduced fire hazard moldings costing little more than conventional moldings
Form: Paste
Gen'l. Prop.: Flash Pt.: < 32 C
Toxicity/Handling: See Crystic 323 PA
Std. Pkgs.: 25-kg and 225-kg containers
Ref.: Tech. leaflet 122

Crystic Pregel 17. Scott Bader Co. Ltd.
Chem. Descrip.: Dispersion of silica in a general purpose polyester resin
Category/Applications: Highly thixotropic resin additive to confer thixotropic properties to general purpose polyester resins for laminating or gelcoat applications, enabling the resin to be laid up on vertical or inclined surfaces without drainage
Form: Paste; Color: Pale yellow
Gen'l. Prop.: Stability: 6 mo. in the dark @ 20 C; Flash Pt.: < 32 C
Toxicity/Handling: See Crystic LS7
Std. Pkgs.: 25-kg and 225-kg containers
Ref.: Tech. leaflet 135

Crystic Pregel 27. Scott Bader Co. Ltd.
Chem. Descrip.: Dispersion of silica in an isophthalic polyester resin
Category/Applications: Highly thixotropic resin additive to confer thixotropic properties to Crystic 199 and Crystic 625 MV, and to general purpose Crystic resins wherever a higher heat resistance or higher heat deflection temp. is required
Form: Paste; Color: Pale yellow
Gen'l. Prop.: See Crystic Pregel 17
Toxicity/Handling: See Crystic LS7
Std. Pkgs.: 25-kg and 225-kg containers
Ref.: Tech. leaflet 135

Crystic Release Agent No. 1. Scott Bader Co. Ltd.
Chem. Descrip.: Cellulose acetate sol'n.
Category/Applications: Quick-drying solution for use as a primary sealer on wood, plaster, or other porous surfaces; a secondary release agent should be applied over it; before molding, ensure that all solvent has evaporated or difficulty may be experience when attempting separation
Form: Liquid; Color: Avail. colored or colorless
Gen'l. Prop.: Flash Pt.: < 32 C
Toxicity/Handling: See Crystic 189LV
Std. Pkgs.: 5-kg, 25-kg, and 200-kg containers
Ref.: Tech. leaflet 121

Crystic Release Agent No. 2. Scott Bader Co. Ltd.
Chem. Descrip.: Polyvinyl alcohol sol'n.
Category/Applications: Quick-drying sol'n. for use on metal or reinforced polyester molds, or as a secondary release

agent on molds already treated with Crystic Release Agent No. 1; may be applied by sponge, soft brush, or spray; should be used when the mold has a matt finish
Form: Liquid; Color: Avail. colored or colorless
Gen'l. Prop.: Flash Pt.: < 32 C
Toxicity/Handling: See Crystic 189LV
Std. Pkgs.: 5-kg, 25-kg, and 200-kg containers
Ref.: Tech. leaflet 121

Crystic Stopper. Scott Bader Co. Ltd.
Chem. Descrip.: Two-part filler: Crystic Stopper Base (liquid polyester resin) and Crystic Stopper Powder
Category/Applications: Two-part, fast-setting filler for use in the repair of glass fiber reinforced plastic moldings and motor car bodies, caravans, and similar sheet metal structures; has excellent adhesion to metal; mix ratio: 1 part Base to 2 parts Powder to form paste
Form: Liquid, powder
Gen'l. Prop.: Stability: Resistant to oil, grease, petrol, and similar solvents
Toxicity/Handling: Liquid base is flammable–keep away from naked flames; store ⩽ 20 C
Std. Pkgs.: 25-kg containers
Ref.: Tech. leaflet 133

Cumar LX-509. Neville Chem. Co.
Chem. Descrip.: Coumarone-indene resin
Category/Applications: Resin for use in adhesives (hot melt, hot melt pressure sensitive, mastic, pressure sensitive), coatings (aluminum, can & drum, citrus fruit, emulsion, epoxy, industrial, marine, paper, traffic, wire & cable), rubber (cements, mechanical goods, molded goods, tires), and caulking compounds
Form: Solid, flake; Color (Gardner): 11 (50% in toluene, D1544)
Gen'l. Prop.: M.W.: 1090 (D2503); Sol.: Sol. in esters, ethers, ketones (except acetone), chlorinated, aromatic, naphthenic, and terpene hydrocarbons; partly sol. in paraffinic hydrocarbons; Sp. Gr.: 1.140 (D71); Soften. Pt. (Ring & Ball): 155 C (E28); Iodine No. (Wijs): 35 (D1959); Stability: Excellent resistance to alkalis, dilute acids, and moisture
Therm. Prop.: Flash Pt. (COC): 560 F (D92)
Ref.: Product pamphlet

Cumar P-10. Neville Chem. Co.
Chem. Descrip.: Coumarone-indene resin
Category/Applications: See Cumar LX-509
Form: Liquid; Color (Gardner): 16 (50% in toluene, D1544)
Gen'l. Prop.: M.W.: 330 (D2503); Sol.: Sol. in esters, ethers, ketones (except acetone), chlorinated, aromatic, naphthenic, and terpene hydrocarbons; partly sol. in paraffinic hydrocarbons; Sp. Gr.: 1.040 (D71); Soften. Pt. (Ring & Ball): 10 C (E28); Iodine No. (Wijs): 44 (D1959); Stability: Excellent resistance to alkalis, dilute acids, and moisture
Therm. Prop.: Flash Pt. (COC): 350 F (D92)
Ref.: Product pamphlet

Cumar P-25. Neville Chem. Co.
Chem. Descrip.: Coumarone-indene resin
Category/Applications: See Cumar LX-509
Form: Liquid; Color (Gardner): 16 (50% in toluene, D1544)
Gen'l. Prop.: M.W.: 350 (D2503); Sol.: Sol. in esters, ethers, ketones (except acetone), chlorinated, aromatic, naphthenic, and terpene hydrocarbons; partly sol. in paraffinic hydrocarbons; Sp.

Gr.: 1.060 (D71); Soften. Pt. (Ring & Ball): 25 C (E28); Iodine No. (Wijs): 36 (D1959); Stability: Excellent resistance to alkalis, dilute acids, and moisture
Therm. Prop.: Flash Pt. (COC): 330 F (D92)
Ref.: Product pamphlet

Cumar R-1. Neville Chem. Co.
Chem. Descrip.: Coumarone-indene resin
Category/Applications: Resin for use in adhesives (hot melt, hot melt pressure sensitive, mastic, pressure sensitive), coatings (aluminum, can & drum, citrus fruit, emulsion, epoxy, industrial, marine, paper, traffic, wire & cable), rubber (cements, mechanical goods, molded goods, tires), and caulking compounds
Form: Solid, flakes; Color (Gardner): 9 (50% in toluene, D1544)
Gen'l. Prop.: M.W.: 730 (D2503); Sol.: Sol. in esters, ethers, ketones (except acetone), chlorinated, aromatic, naphthenic, and terpene hydrocarbons; partly sol. in paraffinic hydrocarbons; Sp. Gr.: 1.140 (D71); Soften. Pt. (Ring & Ball): 126 C; Iodine No. (Wijs): 58 (D-1959); Stability: Excellent resistance to alkalis, dilute acids, and moisture
Therm. Prop.: Flash Pt. (COC): 540 F (D92)
Ref.: Product pamphlet

Cumar R-3 (W-1). Neville Chem. Co.
Chem. Descrip.: Coumarone-indene resin
Category/Applications: See Cumar R-1
Form: Solid, flakes; Color (Gardner): 11 (50% in toluene, D1544)
Gen'l. Prop.: M.W.: 800 (D2503); Sol.: Sol. in esters, ethers, ketones (except acetone), chlorinated, aromatic, naphthenic, and terpene hydrocarbons; partly sol. in paraffinic hydrocarbons; Sp. Gr.: 1.140 (D71); Soften. Pt. (Ring & Ball): 126 C (E28); Iodine No. (Wijs): 43 (D1959); Stability: Excellent resistance to alkalis, dilute acids, and moisture
Therm. Prop.: Flash Pt. (COC): 540 F (D92)
Ref.: Product pamphlet

Cumar R-5 (W-1½). Neville Chem. Co.
Chem. Descrip.: Coumarone-indene resin
Category/Applications: See Cumar R-1
Form: Solid, flakes; Color (Gardner): 12 (50% in toluene, D1544)
Gen'l. Prop.: M.W.: 750 (D2503); Sol.: Sol. in esters, ethers, ketones (except acetone), chlorinated, aromatic, naphthenic, and terpene hydrocarbons; partly sol. in paraffinic hydrocarbons; Sp. Gr.: 1.130 (D71); Soften. Pt. (Ring & Ball): 126 C (E28); Iodine No. (Wijs): 50 (D1959); Stability: Excellent resistance to alkalis, dilute acids, and moisture
Therm. Prop.: Flash Pt. (COC): 540 F (D92)
Ref.: Product pamphlet

Cumar R-6 (W-2, 2½). Neville Chem. Co.
Chem. Descrip.: Coumarone-indene resin
Category/Applications: See Cumar R-1
Form: Solid, flakes; Color (Gardner): 14 (50% in toluene, D1544)
Gen'l. Prop.: M.W.: 755 (D2503); Sol.: Sol. in esters, ethers, ketones (except acetone), chlorinated, aromatic, naphthenic, and terpene hydrocarbons; partly sol. in paraffinic hydrocarbons; Sp. Gr.: 1.140 (D71); Soften. Pt. (Ring & Ball): 126 C (E28); Iodine No. (Wijs): 45 (D1959); Stability: Excellent resistance to alkalis, dilute acids, and moisture
Therm. Prop.: Flash Pt. (COC): 550 F (D92)
Ref.: Product pamphet

Cumar R-7. Neville Chem. Co.
Chem. Descrip.: Coumarone-indene resin
Category/Applications: See Cumar R-1
Form: Solid, flakes; Color (Gardner): 11 (50% in toluene, D1544)
Gen'l. Prop.: M.W.: 580 (D2503); Sol.: Sol. in esters, ethers, ketones (except acetone), chlorinated, aromatic, naphthenic, and terpene hydrocarbons; partly sol. in paraffinic hydrocarbons; Sp. Gr.: 1.140 (D71); Soften. Pt. (Ring & Ball): 106 C (E28); Iodine No. (Wijs): 44 (D1959); Stability: Excellent resistance to alkalis, dilute acids, and moisture
Therm. Prop.: Flash Pt. (COC): 580 F (D92)
Ref.: Product pamphlet

Cumar R-9. Neville Chem. Co.
Chem. Descrip.: Coumarone-indene resin
Category/Applications: See Cumar R-1
Form: Solid, flakes; Color (Gardner): 9 (50% in toluene, D1544)
Gen'l. Prop.: M.W.: 610 (D2503); Sol.: Sol. in esters, ethers, ketones (except acetone), chlorinated, aromatic, naphthenic, and terpene hydrocarbons; partly sol. in paraffinic hydrocarbons; Sp. Gr.: 1.140 (D71); Soften. Pt. (Ring & Ball): 112 C (E28); Iodine No. (Wijs): 55 (D1959); Stability: Excellent resistance to alkalis, dilute acids, and moisture
Therm. Prop.: Flash Pt. (COC): 515 F (D92)
Ref.: Product pamphlet

Cumar R-10 (V-1). Neville Chem. Co.
Chem. Descrip.: Coumarone-indene resin
Category/Applications: See Cumar R-1
Form: Solid, flakes; Color (Gardner): 11 (50% in toluene, D1544)
Gen'l. Prop.: M.W.: 640 (D2503); Sol.: Sol. in esters, ethers, ketones (except acetone), chlorinated, aromatic, naphthenic, and terpene hydrocarbons; partly sol. in paraffinic hydrocarbons; Sp. Gr.: 1.140 (D71); Soften. Pt. (Ring & Ball): 112 C (E28); Iodine No. (Wijs): 44 (D1959); Stability: Excellent resistance to alkalis, dilute acids, and moisture
Therm. Prop.: Flash Pt. (COC): 520 F (D92)
Ref.: Product pamphlet

Cumar R-11 (V-1½). Neville Chem. Co.
Chem. Descrip.: Coumarone-indene resin
Category/Applications: See Cumar R-1
Form: Solid, flakes; Color (Gardner): 12 (50% in toluene, D1544)
Gen'l. Prop.: M.W.: 700 (D2503); Sol.: Sol. in esters, ethers, ketones (except acetone), chlorinated, aromatic, naphthenic, and terpene hydrocarbons; Sp. Gr.: 1.140 (D71); Soften. Pt. (Ring & Ball): 112 C (E28); Iodine No. (Wijs): 50 (D1959); Stability: Excellent resistance to alkalis, dilute acids, and moisture
Therm. Prop.: Flash Pt (COC): 515 F (D92)
Ref.: Product pamphlet

Cumar R-12 (V-2, V-2½). Neville Chem. Co.
Chem. Descrip.: Coumrone-indene resin
Category/Applications: See Cumar R-1
Form: Solid, flakes; Color (Gardner): 14 (50% in toluene, D1544)
Gen'l. Prop.: M.W.: 580 (D2503); Sol.: Sol. in esters, ethers, ketones (except acetone), chlorinated, aromatic, naphthenic, and terpene hydrocarbons; partly sol. in paraffinic hydrocarbons; Sp. Gr.: 1.140 (D71); Soften. Pt. (Ring & Ball): 112 C (E28); Iodine No. (Wijs): 40 (D1959); Stability: Excellent resistance to alkalis, dilute acids, and moisture

Therm. Prop.: Flash Pt. (COC): 515 F (D92)
Ref.: Product pamphlet

Cumar R-12A (V-3). Neville Chem. Co.
Chem. Descrip.: Coumarone-indene resin
Category/Applications: See Cumar R-1
Form: Solid, flakes; Color (Gardner): 16 (50% in toluene, D1544)
Gen'l. Prop.: M.W.: 600 (D2503); Sol.: Sol. in esters, ethers, ketones (except acetone), chlorinated, aromatic, naphthenic, and terpene hydrocarbons; Sp. Gr.: 1.060 (D71); Soften. Pt. (Ring & Ball): 112 C (E28); Iodine No. (Wijs): 60 (D1959); Stability: Excellent resistance to alkalis, dilute acids, and moisture
Therm. Prop.: Flash Pt. (COC): 510 F (D92)
Ref.: Product pamphlet

Cumar R-13. Neville Chem. Co.
Chem. Descrip.: Coumarone-indene resin
Category/Applications: See Cumar R-1
Form: Solid, flakes; Color (Gardner): 9 (50% in toluene, D1544)
Gen'l. Prop.: M.W.: 830 (D2503); Sol.: Sol. in esters, ethers, ketones (except acetone), chlorinated, aromatic, naphthenic, and terpene hydrocarbons; partly sol. in paraffinic hydrocarbons; Sp. Gr.: 1.150 (D71); Soften. Pt. (Ring & Ball): 100 C (E28); Iodine No. (Wijs): 57 (D1959); Stability: Excellent resistance to alkalis, dilute acids, and moisture
Therm. Prop.: Flash Pt. (COC): 515 F (D92)
Ref.: Product pamphlet

Cumar R-14 (MH-1). Neville Chem. Co.
Chem. Descrip.: Coumarone-indene resin
Category/Applications: See Cumar R-1
Form: Solid, flakes; Color (Gardner): 11 (50% in toluene, D1544)
Gen'l. Prop.: M.W.: 675 (D2503); Sol.: Sol. in esters, ethers, ketones (except acetone), chlorinated, aromatic, naphthenic, and terpene hydrocarbons; partly sol. in paraffinic hydrocarbons; Sp. Gr.: 1.100 (D71); Soften. Pt. (Ring & Ball): 100 C (E28); Iodine No. (Wijs): 50 (D1959); Stability: Excellent resistance to alkalis, dilute acids, and moisture
Therm. Prop.: Flash Pt. (COC): 520 F (D92)
Ref.: Product pamphlet

Cumar R-15 (MH-1½). Neville Chem. Co.
Chem. Descrip.: Coumarone-indene resin
Category/Applications: See Cumar R-1
Form: Solid, flakes; Color (Gardner): 12 (50% in toluene, D1544)
Gen'l. Prop.: M.W.: 680 (D2503); Sol.: Sol. in esters, ethers, ketones (except acetone), chlorinated, aromatic, naphthenic, and terpene hydrocarbons; partly sol. in paraffinic hydrocarbons; Sp. Gr.: 1.120 (D71); Soften. Pt. (Ring & Ball): 100 C (E28); Iodine No. (Wijs): 55 (D1959); Stability: Excellent resistance to alkalis, dilute acids, and moisture
Therm. Prop.: Flash Pt. (COC): 515 F (D92)
Ref.: Product pamphlet

Cumar R-16 (MH-2, 2½). Neville Chem. Co.
Chem. Descrip.: Coumarone-indene resin
Category/Applications: See Cumar R-1
Form: Solid, flakes; Color (Gardner): 14 (50% in toluene, D1544)
Gen'l. Prop.: M.W.: 770 (D2503); Sol.: Sol. in esters, ethers, ketones (except acetone), chlorinated, aromatic, naphthenic, and terpene hydrocarbons; partly sol. in paraffinic hydrocarbons; Sp. Gr.: 1.075 (D71); Soften. Pt. (Ring & Ball): 103 C (E28); Iodine No. (Wijs):

31 (D1959); Stability: Excellent resistance to alkalis, dilute acids, and moisture
Therm. Prop.: Flash Pt. (COC): 415 F (D92)
Ref.: Product pamphlet

Cumar R-16A (MH-3). Neville Chem. Co.
Chem. Descrip.: Coumarone-indene resin
Category/Applications: See Cumar R-1
Form: Solid, flakes; Color (Gardner): 16 (50% in toluene, D1544)
Gen'l. Prop.: M.W.: 760 (D2503); Sol.: Sol. in esters, ethers, ketones (except acetone), chlorinated, aromatic, naphthenic, and terpene hydrocarbons; partly sol. in paraffinic hydrocarbons; Sp. Gr.: 1.060 (D71); Soften. Pt. (Ring & Ball): 103 C (E28); Iodine No. (Wijs): 55 (D1959); Stability: Excellent resistance to alkalis, dilute acids, and moisture
Therm. Prop.: Flash Pt. (COC): 420 F (D92)
Ref.: Product pamphlet

Cumar R-17 (RH-17). Neville Chem. Co.
Chem. Descrip.: Coumarone-indene resin
Category/Applications: See Cumar R-1
Form: Solid; Color (Gardner): 16 (50% in toluene, D1544)
Gen'l. Prop.: M.W.: 400 (D2503); Sol.: Sol. in esters, ethers, ketones (except acetone), chlorinated, aromatic, naphthenic, and terpene hydrocarbons; partly sol. in paraffinic hydrocarbons; Sp. Gr.: 1.120 (D71); Soften. Pt. (Ring & Ball): 76 C (E28); Iodine No. (Wijs): 65 (D1959); Stability: Excellent resistance to alkalis, dilute acids, and moisture
Therm. Prop.: Flash Pt. (COC): 225 F (D92)
Ref.: Product pamphlet

Cumar R-19. Neville Chem. Co.
Chem. Descrip. Coumarone-indene resin
Category/Applications: See Cumar R-1
Form: Solid; Color (Gardner): 16 (50% in toluene, D1544)
Gen'l. Prop.: M.W.: 590 (D2503); Sol.: Sol. in esters, ethers, ketones (except acetone), chlorinated, aromatic, naphthenic, and terpene hydrocarbons; partly sol. in paraffinic hydrocarbons; Sp. Gr.: 1.020 (D71); Soften. Pt. (Ring & Ball): 58 C (E28); Iodine No. (Wijs): 65 (D1959); Stability: Excellent resistance to alkalis, dilute acids, and moisture
Therm. Prop.: Flash Pt. (COC): 420 F (D92)
Ref.: Product pamphlet

Cumar R-21. Neville Chem. Co.
Chem. Descrip. Coumarone-indene resin
Category/Applications: See Cumar R-1
Form: Solid; Color (Gardner): 16 (50% in toluene, D1544)
Gen'l. Prop.: M.W.: 530 (D2503); Sol.: Sol. in esters, ethers, ketones (except acetone), chlorinated, aromatic, naphthenic, and terpene hydrocarbons; partly sol. in paraffinic hydrocarbons; Sp. Gr.: 1.050 (D71); Soften. Pt. (Ring & Ball): 45 C (E28); Iodine No. (Wijs): 50 (D1959); Stability: Excellent resistance to alkalis, dilute acids, and moisture
Therm. Prop.: Flash Pt. (COC): 525 F (D92)
Ref.: Product pamphlet

Cumar R-27. Neville Chem. Co.
Chem. Descrip.: Coumarone-indene resin
Category/Applications: See Cumar R-1
Form: Liquid; Color (Gardner): 11 (50% in toluene, D1544)
Gen'l. Prop.: M.W.: 290 (D2503); Sol.: Sol. in esters, ethers, ketones (except acetone), chlorinated, aromatic, naph-

thenic, and terpene hydrocarbons; partly sol. in paraffinic hydrocarbons; Sp. Gr.: 1.080 (D71); Soften. Pt. (Ring & Ball): 10 C (E28); Iodine No. (Wijs): 58 (D1959); Stability: Excellent resistance to alkalis, dilute acids, and moisture
Therm. Prop.: Flash Pt. (COC): 350 F (D92)
Ref.: Product pamphlet

Cumar R-28. Neville Chem. Co.
Chem. Descrip.: Coumarone-indene resin
Category/Applications: See Cumar R-1
Form: Liquid; Color (Gardner): 14 (50% in toluene, D1544)
Gen'l. Prop.: M.W.: 345 (D2503); Sol.: Sol. in esters, ethers, ketones (except acetone), chlorinated, aromatic, naphthenic, and terpene hydrocarbons; partly sol. in paraffinic hydrocarbons; Sp. Gr.: 1.080 (D71); Soften. Pt. (Ring & Ball): 33 C (E28); Iodine No. (Wijs): 55 (D1959); Stability: Excellent resistance to alkalis, dilute acids, and moisture
Therm. Prop.: Flash Pt. (COC): 380 F (D92)
Ref.: Product pamphlet

Cumar R-29 (25°). Neville Chem. Co.
Chem. Descrip.: Coumarone-indene resin
Category/Applications: See Cumar R-1
Form: Liquid; Color (Gardner): 16 (50% in toluene, D1544)
Gen'l. Prop.: M.W.: 455 (D2503); Sol.: Sol. in esters, ethers, ketones (except acetone), chlorinated, aromatic, naphthenic, and terpene hydrocarbons; partly sol. in paraffinic hydrocarbons; Sp. Gr.: 1.030 (D71); Soften. Pt. (Ring & Ball): 28 C (E28); Iodine No. (Wijs): 45 (D1959); Stability: Excellent resistance to alkalis, dilute acids, and moisture
Therm. Prop.: Flash Pt. (COC): 380 F (D92)
Ref.: Product pamphlet

Cure-rite Accelerator. BF Goodrich Co.
Chem. Descrip.: Thiocarbamyl sulfenamide
Category/Applications: Accelerators for SBR, EPDM, and other general purpose rubbers, and for special elastomers incl. nitriles and butyls; provides fast cure time and compound processing safety to minimize scorching
Gen'l. Prop.: Stability: Excellent storage stability
Ref.: Product information

Cutless 50. GAF Corp.
Chem. Descrip.: Ampholytic polymer
Category/Applications: Stable, nonthickening, bulk storable, highly durable, reactive softener-finish for natural and synthetic fibers; applied by padding or exhaustion to produce a soft, luxurious hand and optimum lubricity for napping, cutting, and sewing; improves abrasion resistance and tear strength of permanent press fabrics
Form: Liquid
Ref.: Catalogue 2303-036

CXA 1025. Du Pont Co.
Chem. Descrip.: Ethylene vinyl acetate terpolymer
Category/Applications: Coextrudable adhesive resin for flexible packaging; provides good adhesion in coextrusions with acrylonitrile, PET polyester, high-impact styrene, styrene-acrylonitrile, styrene/maleic anhydride, ABS, polycarbonate, acrylic, hydrolyzed vinyl acetate resins, etc.; used in conventional extrusion and coextrusion equipment designed to process polyethylene resins
Form: Pellets

Gen'l. Prop.: Density: 1.00 g/cc (D792); Soften. Pt. (Vicat): 35 C (D1525); Melt Index: 35 dg/min (D1238)
Toxicity/Handling: Low concs. of fumes may evolve at temps. above 230 C–use with adequate ventilation; protect the hands and other exposed parts of the body when handling molten polymer
Ref.: Bulletin E-28703

CXA 1064. Du Pont Co.
Chem. Descrip.: Ethylene vinyl acetate terpolymer
Category/Applications: Coextrudable adhesive resin for flexible packaging; provides good adhesion in coextrusions with polypropylene, PET polyester, high-impact styrene, styrene/maleic anhydride, styrene/acrylonitrile resins, etc.; for use in conventional extrusion and coextrusion equipment designed to process polyethylene resins
Form: Pellets
Gen'l. Prop.: Density: 0.944 g/cc (D-792); Soften. Pt. (Vicat): 42 C (D-1525); Melt Index: 20 dg/min (D1238)
Toxicity/Handling: See CXA 1025
Ref.: Bulletin E-28704

CXA 1104. Du Pont Co.
Chem. Descrip.: Ethylene vinyl acetate terpolymer
Category/Applications: Coextrudable adhesive resin for flexible packaging; provides good adhesion in coextrusions with polypropylene, polyethylene, EVA, Surlyn® ionomer, high-impact styrene, PET polyester, PVdC, polycarbonate, styrene/maleic anhydride, styrene/acrylonitrile resins, etc.; for use in conventional extrusion and coextrusion equipment designed to process polyethylene resins
Form: Pellets
Gen'l. Prop.: Density: 0.955 g/cc (D-792); Soften. Pt. (Vicat): 43 C (D1525); Melt Index: 6 dg/min (D1238)
Toxicity/Handling: See CXA 1025
Ref.: Bulletin E-28705

CXA 1202. Du Pont Co.
Chem. Descrip.: Ethylene vinyl acetate terpolymer
Category/Applications: Coextrudable adhesive resin for flexible packaging; provides good adhesion in coextrusions with polypropylene, polyethylene, EVA, Surlyn® ionomer resins, etc.; for use in conventional extrusion and coextrusion equipment designed to process polyethylene resins
Form: Pellets
Gen'l. Prop.: Density: 0.935 g/cc (D-792); Soften. Pt. (Vicat): 61 C (D1525); Melt Index: 3.5 dg/min (D1238)
Toxicity/Handling: See CXA 1025
Ref.: Bulletin E-28706

CXA 2002. Du Pont Co.
Chem. Descrip.: Ethylene terpolymer with acidic functionality
Category/Applications: Coextrudable adhesive resin for flexible packaging; provides good adhesion in coextrusions with nylon, polyethylene, EVA, Surlyn® ionomer, polypropylene resins, etc.; for use in conventional extrusion and coextrusion equipment designed to process polyethylene resins
Gen'l. Prop.: Density: 0.941 g/cc (D-792); Soften. Pt. (Vicat): 48 C (D1525); Melt Index: 10 dg/min (D1238)
Toxicity/Handling: Low concs. of fumes may evolve at temps. above 300 C–use with adequate ventilation; protect the hands and exposed parts when handling molten polymer
Ref.: Bulletin E-28707

CXA 2022. Du Pont Co.
Chem. Descrip.: Ethylene terpolymer with acidic functionality
Category/Applications: See CXA 2002
Form: Pellets
Gen'l. Prop.: Density: 0.939 g/cc (D-792); Soften. Pt. (Vicat): 55 C (D1525); Melt Index: 35 dg/min (D1238)
Toxicity/Handling: See CXA 2002
Ref.: Bulletin E-28708

CXA 3095. Du Pont Co.
Chem. Descrip.: Modified polyolefin resin containing vinyl acetate and other functional groups
Category/Applications: Coextrudable adhesive resin for flexible packaging; provides good adhesion in coextrusions with polyethylene, EVA, nylon, and vinyl alcohol copolymers; for use in conventional extrusion and coextrusion equipment designed to process polyethylene resins
Form: Pellets
Gen'l. Prop.: Density: 0.937 g/cc (D-792); Soften. Pt. (Vicat): 82 C (D1525); Melt Index: 2.3 dg/min (D1238)
Toxicity/Handling: See CXA 1025
Ref.: Bulletin E-28709

CXA 3101. Du Pont Co.
Chem. Descrip.: Ethylene-based polymer with vinyl acetate and other functional groups
Category/Applications: Coextrudable adhesive resin for flexible packaging; provides good adhesion in coextrusions with polyethylene, EVA, Surlyn® ionomers, polypropylene, nylon, saran, polystyrenes, polyester, and polycarbonate resins; for use in conventional extrusion and coextrusion equipment designed to process polyethylene resins
Form: Pellets
Gen'l. Prop.: Density: 0.948 g/cc (D-792); Soften. Pt. (Vicat): 55 C (D1525); Melt Index: 3.5 dg/min (D1238)
Toxicity/Handling: See CXA 1025
Ref.: Bulletin E-28710

Cyanacryl C. Amer. Cyanamid Co./Elastomers Dept.
Chem. Descrip.: Acrylic elastomer
Category/Applications: Low temp. elastomer with good processing characteristics for mill and banbury mixing, good extrusion characteristics, mold flow, and mold release; capable of being compression, transfer, and injection molded, and blended with Cyanacryl R or L; can be used with Cyanamid soap-sulfur cure systems and ammonium salt cure systems; used for O-rings, valve stem oil deflectors, transmission lip seals, shaft seals, hose, gaskets, spark plug boots, mechanical goods, and extruded goods
Form: Slab; Color: Off-white
Gen'l. Prop.: Sp. Gr.: 1.13; Visc. (Mooney): 30–40 (212 F, ML-4); Stability: Good high and low temp. resistance; excellent oil and compression set resistance; acceptable corrosion resistance by the GMC test procedure
Ref.: Product pamphlet

Cyanacryl L. Amer. Cyanamid Co./Elastomers Dept.
Chem. Descrip.: Acrylic elastomer
Category/Applications: Elastomer with good flex retention at temps. in the range of −15 F and maintains good performance in hot air and oils at temps. > 300 F; excellent processability; easily cured with amine and soap-sulfur systems; good electrical properties, and extrusion characteristics and mold definition; used in the manufacture of automotive parts, e.g., shaft seals, lip seals, O-rings, gaskets
Composition: 0.8–1.1% chlorine content

Gen'l. Prop. (post-cured): Visc. (Mooney): 40–52
Mech. Prop. (post-cured): Tens. Str.: 1250 psi min.; Tens. Elong.: 150% min.; 100% Mod.: 550 psi min.; Hardness (Shore A): 73 ± 5
Ref.: Product information

Cyanacryl R. Amer. Cyanamid Co./Elastomers Dept.
Chem. Descrip.: Acrylic elastomer
Category/Applications: Cyanacryl R is inferior to Cyanacryl L in low temp. flex but has better performance in hot air and oil @ 350 F; recommended for use in seals, gaskets, O-rings, valve stem seals, spark plug covers, ignition wire and general electrical gaskets, hose, adhesives, emission control gasket, sponge gaskets, rolls, construction sealants, conveyor belts, and tank lining
Composition: 0.8–1.1% chlorine content
Gen'l. Prop.: Sp. Gr.: 1.10; Visc. (Mooney): 40–52; Stability: Good resistance to oxygen, ozone, and weathering; good oil resistance @ 350 F
Mech. Prop. (post-cured): Tens. Str.: 1550 psi min.; Tens. Elong.: 150% min.; 100% Mod.: 500 psi min.; Hardness (Shore A): 68 ± 5
Ref.: Product information

Cycolac CG. Borg-Warner Chem.
Chem. Descrip.: ABS resin
Category/Applications: Low gloss extrusion grade polymer with matte finish and excellent scuff resistance; good "hot guts" and toughness for forming large parts
Gen'l. Prop.: Sp. Gr.: 1.02 (23 C, D792/A)
Mech. Prop. (23 C): Tens. Str.: 30 MPa (D638); Tens. Mod.: 1.7 GPa (D638); Flex. Str.: 54 MPa (yield, D790); Flex. Mod.: 1.7 GPa (D790); Impact Str. (Izod): 401 J/m notch (D256/A); Hardness (Rockwell): R 82 (D785)
Therm. Prop.: Distort. Temp.: 91 C (@ 1820 kPa, 0.25 mm defl., unannealed, D648); Coeff. of Linear Exp.: 10.3 × 10^{-5} cm/cm/C (D696); Flamm.: 3.1 cm/min (3.2 mm, D635)
Ref.: Pamphlet SR-604-B

Cycolac CGA. Borg-Warner Chem.
Chem. Descrip.: ABS resin
Category/Applications: Extrusion grade ABS for producing low gloss sheet offering good impact strength and ductility, and good color retention as virgin material or blended regrind; also for thermoforming general purpose parts
Gen'l. Prop.: Sp. Gr.: 1.03 (23 C, D792/A)
Mech. Prop. (23 C): Tens. Str.: 29 MPa (D638); Tens. Mod.: 1.7 GPa (D638); Flex. Str.: 54 MPa (yield, D790); Flex. Mod.: 1.7 GPa (D790); Impact Str. (Izod): 374 J/m notch (D256/A); Hardness (Rockwell): R 81 (D785)
Therm. Prop.: Distort. Temp.: 88 C (@ 1820 kPa, 0.25 mm defl., unannealed, D648); Coeff. of Linear Exp.: 11.0 × 10^{-5} cm/cm/C (D696); Flamm.: 3.6 cm/min (23 C, D635)
Ref.: Data sheet P-503A

Cycolac CTB. Borg-Warner Chem.
Chem. Descrip.: ABS resin
Category/Applications: Med.-impact, transparent ABS material for injection molding applications, esp. where light transmission and abuse resistance are required, e.g., refrigerator crisper trays and interior trim, toys, dust covers, video and audio cassette housings, display stands, and industrial packaging; not recommended for parts requiring continuous outdoor exposure
Gen'l. Prop.: Sp. Gr.: 1.07 (D792/A); Ref. Index: 1.536 (73 F, D542)

Mech. Prop. (73 F): Tens. Str.: 6300 psi (D638); Tens. Mod.: 330,000 psi (D-638); Flex. Str.: 10,500 psi (yield, D-790); Flex. Mod.: 350,000 psi (D790); Impact Str. (Izod): 2.5 ft lb/in. notch (D256/A); Hardness (Rockwell): R 105 (D785)
Therm. Prop.: Distort. Temp.: 170 F (@ 264 psi, 10 mil defl., unannealed, D648); Coeff. of Linear Exp.: 8.9×10^{-5} in./in./C (D696)
Ref.: Data sheet P-511

Cycolac DFA-R. Borg-Warner Chem.
Chem. Descrip.: ABS resin
Category/Applications: Economical, molding grade ABS with high modulus, med. impact resistance, and high flow; used for housewares, appliances, decorative office and home furnishings, hardware, and toys
Gen'l. Prop.: Sp. Gr.: 1.04 (D792/A)
Mech. Prop. (23 C): Tens. Str.: 43 MPa (D638); Tens. Mod.: 2.3 GPa (D638); Flex. Str.: 79 MPa (yield, D790); Flex. Mod.: 2.5 GPa (D790); Impact Str. (Izod): 214 J/m notch (D256/A); Hardness (Rockwell): R 108 (D785)
Therm. Prop.: Distort. Temp.: 84 C (@ 1820 kPa, 0.25 mm defl., unannealed, D648); Coeff. of Linear Exp.: 8.8×10^{-5} cm/cm/C (D696); Flamm.: HB (@ 1.47 mm, UL94)
Elec. Prop. (23 C): Dielec. Str.: 32 kV/mm (D149); Vol. Resist.: 2.7×10^{9} megohm-cm (D257); Arc Resist.: 83 s (D495)
Ref.: Pamphlet SR-605-C

Cycolac DH. Borg-Warner Chem.
Chem. Descrip.: ABS resin
Category/Applications: Economical, high heat, high modulus and tensile, med. impact grade polymer for injection molding applications; offers better flow than Cycolac X-11 or X-17 high heat grades
Gen'l. Prop.: Sp. Gr.: 1.05 (D792/A)
Mech. Prop. (23 C): Tens. Str.: 52 MPa (D638); Tens. Mod.: 2.6 GPa (D638); Flex. Str.: 90 MPa (yield, D790); Flex. Mod.: 2.8 GPa (D790); Impact Str. (Izod): 235 J/m notch (D256/A); Hardness (Rockwell): R 111 (D785)
Therm. Prop.: Distort. Temp.: 94 C (@ 1820 kPa, 0.25 mm defl., unannealed, D-648); Flamm.: HB (1.47 mm, UL94)
Elec. Prop. (23 C): Dielec. Str.: 15 kV/mm (D149); Vol. Resist.: 9.3×10^{7} megohm-cm (D257); Arc Resist.: 86 s (D495)
Ref.: Pamphlet SR-605-C

Cycolac EP-3510. Borg-Warner Chem.
Chem. Descrip.: ABS resin
Category/Applications: Electroplating product with high melt flow designed to mold complex, stress-free parts; used in small appliance housings, interior and exterior automotive components, cosmetic cases, appliance knobs, etc.
All properties—gray pigmented samples:
Gen'l. Prop.: Sp. Gr.: 1.06 (D792/A)
Mech. Prop. (23 C): Tens. Str.: 38 MPa (D638); Tens. Mod.: 2.3 GPa (D638); Flex. Str.: 69 MPa (yield, D790); Flex. Mod.: 2.3 GPa (D790); Impact Str. (Izod): 374 J/m notch (D256/A); Hardness (Rockwell): R 103 (D785)
Therm. Prop.: Distort. Temp.: 89 C (@ 1820 kPa, 0.25 mm defl., unannealed, D-648); Coeff. of Linear Exp.: 7.0×10^{-5} cm/cm/C (D696); Flamm.: HB (1.47 mm, UL94)
Elec. Prop. (23 C): Dielec. Str.: 17 kV/mm (D149); Vol. Resist.: 2.15×10^{9} megohm-cm (D257); Arc Resist.: 103 s (D495)
Ref.: Pamphlet SR-605-C

Cycolac EPB-3570. Borg-Warner Chem.
Chem. Descrip.: ABS resin
Category/Applications: Plating grade

ABS designed to provide max. thermocycling performance in plating applications; used for automotive exterior trim, large and small appliances, plumbing fixtures, and marine equipment, and for applications requiring high performance over a wide temp. range in varied environments
All properties–gray pigmented samples:
Gen'l. Prop.: Sp. Gr.: 1.06 (23 C, D792)
Mech. Prop. (23 C): Tens. Str.: 44 MPa (D638); Tens. Mod.: 2.6 GPa (D638); Flex. Str.: 80 MPa (yield, D790); Flex. Mod.: 2.7 GPa (D790); Impact Str. (Izod): 267 J/m notch (D256); Hardness (Rockwell): R 109 (D785)
Therm. Prop.: Distort. Temp.: 92 C (@ 1820 kPa, 0.25 mm defl., unannealed, D648); Coeff. of Linear Exp.: 6.5×10^{-5} cm/cm/C (–30 to 30 C, D696); Flamm.: HB (1.47 mm, UL94)
Elec. Prop. (23 C): Dielec. Str.: 37 kV/mm (D149); Vol. Resist.: 3.2×10^9 megohm-cm (D257); Arc Resist.: 53 s (D495)
Ref.: Pamphlet SR-605-C, DS P-502D

Cycolac FBK. Borg-Warner Chem.
Chem. Descrip.: Flame-retardant ABS structural foam polymer
Category/Applications: Flame-retardant structural foam material for injection molding parts that exhibit high strength with good stiffness-to-weight ratio; offers high flow and fast set-up; used for television components, business machine housings, small appliances, communication devices, automotive applications, industrial tools, industrial floor care units, etc.; good replacement for die casting or sheet metal
Form: Pellets
All properties–natural, injection molded specimens:
Mech. Prop. (23 C): Tens. Str.: 13–27 MPa (D638); Tens. Mod.: 1.10–1.72 GPa (D638); Flex. Str.: 29–54 MPa (yield, D790); Flex. Mod.: 1.45–1.93 GPa (D-790); Compr. Str.: 30 MPa (@ 20% density reduct'n., D695); Impact Str. (Dart): 14 J; Hardness (Rockwell): R 60–70 (D-785)
Therm. Prop.: Distort. Temp.: 67–78 C (@ 1820 kPa, 0.25 mm defl., unannealed, D648); Flamm.: V-O/5V (6.10 mm, UL-94)
Ref.: Data sheet P-515B

Cycolac GSE. Borg-Warner Chem.
Chem. Descrip.: ABS resin
Category/Applications: All purpose, high-impact extrusion grade polymer with excellent toughness, good formability, high gloss, and good stiffness
Gen'l. Prop.: Sp. Gr.: 1.04 (23 C, D792/A)
Mech. Prop. (23 C): Tens. Str.: 45 MPa (D638); Tens. Mod.: 2.2 GPa (D638); Flex. Str.: 76 MPa (yield, D790); Flex. Mod.: 2.3 GPa (D790); Impact Str. (Izod): 374 J/m notch (D256/A); Hardness (Rockwell): R 103 (D785)
Therm. Prop.: Distort. Temp.: 92 C (@ 1820 kPa, 0.25 mm defl., unannealed, D648); Coeff. of Linear Exp.: 9.5×10^{-5} cm/cm/C (D696); Flamm.: HB (@ 1.47 mm, UL94)
Ref.: Pamphlet SR-604-B

Cycolac GSM. Borg-Warner Chem.
Chem. Descrip.: ABS resin
Category/Applications: General purpose injection molding grade polymer offering high impact and tensile properties, and a Relative Temp. Index of 70 C; used in applications where additional abuse resistance is required, e.g., golf club heads, vacuum cleaner components, power tools housings, lawn and garden equipment
Gen'l. Prop.: Sp. Gr.: 1.04 (D792/A)
Mech. Prop. (23 C): Tens. Str.: 43 MPa

(D638); Tens. Mod.: 2.2 GPa (D638); Flex. Str.: 74 MPa (yield, D790); Flex. Mod.: 2.3 GPa (D790); Impact Str. (Izod): 374 J/m notch (D256/A); Hardness (Rockwell): R 102 (D785)
Therm. Prop.: Distort. Temp.: 89 C (@ 1820 kPa, 0.25 mm defl., unannealed, D648); Coeff. of Linear Exp.: 9.5 × 10^{-5} cm/cm/C (D696); Flamm.: HB (@ 1.47 mm, UL94)
Elec. Prop. (23 C): Dielec. Str.: 17 kV/mm (D149); Vol. Resist.: 3.7 × 10^9 megohm-cm (D257); Arc Resist.: 130 s (D495)
Ref.: Pamphlet SR-605-C

Cycolac KJB. Borg-Warner Chem.
Chem. Descrip.: Flame-retardant ABS resin
Category/Applications: Flame-retardant, med. impact ABS thermoplastic for injection molding applications offering a wide processing latitude and good engineering properties; used in electrical applications requiring regulatory agency approval, e.g., business machine and computer components, electrical appliances, and power tools, and in applications where uv resistance to color change is important in indoor use
All properties—natural specimens:
Gen'l. Prop.: Sp. Gr.: 1.22 (23 C, D792/A)
Mech. Prop. (23 C): Tens. Str.: 40 MPa (D638); Tens. Mod.: 2.2 GPa (D638); Flex. Str.: 69 MPa (yield, D790); Flex. Mod.: 2.3 GPa (D790); Compr. Str.: 53 MPa (D695); Compr. Mod.: 1.4 GPa (D-695); Impact Str. (Izod): 214 J/m notch (D256/A); Hardness (Rockwell): R 97 (D785)
Therm. Prop.: Distort. Temp.: 88 C (@ 1820 kPa, 0.25 mm defl., unannealed, D648); Flamm.: V-O @ 1.47 mm, V-O/5V @ 3.05 mm (UL94)
Elec. Prop. (23 C): Dielec. Str.: 16 kV/mm (D149); Vol. Resist.: 1.42 × 10^8 megohm-cm (D257); Arc Resist.: 19 s (D495)
Ref.: Pamphlet SR-605-C, DS P-509-C

Cycolac KJT. Borg-Warner Chem.
Chem. Descrip.: Flame-retardant ABS resin
Category/Applications: Flame-retardant injection molding grade polymer for use in business machine housings and components, electrical appliances, power tools, and television parts
Gen'l. Prop.: Sp. Gr.: 1.20 (D792/A)
Mech. Prop. (23 C): Tens. Str.: 43 MPa (D638); Tens. Mod.: 2.3 GPa (D638); Flex. Str.: 69 MPa (yield, D790); Flex. Mod.: 2.3 GPa (D790); Impact Str. (Izod): 187 J/m notch (D256/A); Hardness (Rockwell): R 102 (D785)
Therm. Prop.: Distort. Temp.: 88 C (@ 1820 kPa, 0.25 mm defl., unannealed, D648); Coeff. of Linear Exp.: 8.3 × 10^{-5} cm/cm/C (D696); Flamm.: V-O (@ 3.05 mm, UL94)
Elec. Prop. (23 C): Dielec. Str.: 25 kV/mm (D149); Vol. Resist.: 1.1 × 10^9 megohm-cm (D257); Arc Resist.: 28 s (D495)
Ref.: Pamphlet SR-605-C

Cycolac KJW. Borg-Warner Chem.
Chem. Descrip.: Flame-ratardant ABS resin
Category/Applications: Flame-retardant injection molding grade polymer with a Relative Temp. Index of 60 C
Gen'l. Prop.: Sp. Gr.: 1.22 (D792/A)
Mech. Prop. (23 C): Tens. Str.: 41 MPa (D638); Tens. Mod.: 2.2 GPa (D638); Flex. Str.: 69 MPa (yield, D790); Flex. Mod.: 2.3 GPa (D790); Impact Str. (Izod): 215 J/m notch (D256/A); Hardness (Rockwell): R 99 (D785)
Therm. Prop.: Distort. Temp.: 88 C (@ 1820 kPa, 0.25 mm defl., unannealed,

D648); Flamm.: V-O @ 1.47 mm and 5V @ 3.05 mm (UL94)
Ref.: Pamphlet SR-605-C

Cycolac L. Borg-Warner Chem.
Chem. Descrip.: ABS resin
Category/Applications: High impact injection molding grade ABS offering max. low temp. toughness and impact resistance; used in applications requiring durability and toughness, e.g., electrical connectors, safety helmets, small appliances, video tape housings, and snowmobile components; ideal for thick-wall applications
Gen'l. Prop.: Sp. Gr.: 1.02 (D792/A)
Mech. Prop. (23 C): Tens. Str.: 34 MPa (D638); Tens. Mod.: 1.8 GPa (D638); Flex. Str.: 59 MPa (yield, D790); Flex. Mod.: 1.9 GPa (D790); Compr. Str.: 40 MPa (D695); Compr. Mod.: 1.2 GPa (D695); Impact Str. (Izod): 400 J/m notch (D256/A); Hardness (Rockwell): R 90 (D785)
Therm. Prop.: Distort. Temp.: 87 C (@ 1820 kPa, 0.25 mm defl., unannealed, D648); Coeff. of Linear Exp.: 11.0 X 10^{-5} cm/cm/C (D696); Flamm.: HB (1.47 mm, UL94)
Elec. Prop. (23 C): Dielec. Str.: 32 kV/mm (D149); Vol. Resist.: 1.8 X 10^9 megohm-cm (D257); Arc Resist.: 85 s (D495)
Ref.: Pamphlet SR-605-C, DS P-517B

Cycolac LS. Borg-Warner Chem.
Chem. Descrip.: ABS resin
Category/Applications: High impact extrusion grade polymer with superior toughness for parts that are subject to abuse; offers high gloss and excellent "hot guts" for forming large, complex shapes
Gen'l. Prop.: Sp. Gr.: 1.02 (23 C, D792/A)
Mech. Prop. (23 C): Tens. Str.: 36 MPa (D638); Tens. Mod.: 1.7 GPa (D638); Flex. Str.: 59 MPa (yield, D790); Flex. Mod.: 1.8 GPa (D790); Impact Str. (Izod): 427 J/m notch (D256/A); Hardness (Rockwell): R 91 (D785)
Therm. Prop.: Distort. Temp.: 89 C (@ 1820 kPa, 0.25 mm defl., unannealed, D648); Coeff. of Linear Exp.: 10.8 X 10^{-5} cm/cm/C (D696); Flamm.: HB (@ 1.47 mm, UL94)
Ref.: Pamphlet SR-604-B

Cycolac SF. Borg-Warner Chem.
Chem. Descrip.: ABS resin
Category/Applications: Semiflexible, nonstress whitening ABS comp. for profile extrusion with excellent dimensional stability
Gen'l. Prop.: Sp. Gr.: 0.97 (23 C, D792/A)
Mech. Prop. (23 C): Tens. Str.: 16 MPa (D638); Tens. Mod.: 0.83 GPa (D638); Flex. Str.: 26 MPa (yield, D790); Flex. Mod.: 0.9 GPa (D790); Impact Str. (Izod): 192 J/m notch (D256/A); Hardness (Shore D): 65
Therm. Prop.: Distort. Temp.: 66 C (@ 1820 kPa, 0.25 mm defl., unannealed, D648); Coeff. of Linear Exp.: 13.9 X 10^{-5} cm/cm/C (D696); Flamm.: 4.3 cm/min (3.2 mm, D635)
Ref.: Pamphlet SR-604-B

Cycolac T. Borg-Warner Chem.
Chem. Descrip.: ABS resin
Category/Applications: General purpose injection molding ABS thermoplastic material offering an optimum balance of toughness, rigidity, high flow and gloss at an economical price; used for applications requiring good aesthetic qualities, scratch resistance, and good dimensional stability, e.g., telephone housings, headsets and desk equipment, kitchen appliance housings, toys, and furniture components
Gen'l. Prop.: Sp. Gr.: 1.04 (D792/A)

Mech. Prop. (23 C): Tens. Str.: 41 MPa (D638); Tens. Mod.: 2.3 GPa (D638); Flex. Str.: 72 MPa (yield, D790); Flex. Mod.: 2.3 GPa (D790); Compr. Str.: 5.5 MPa (D695); Compr. Mod.: 1.3 GPa (D695); Impact Str. (Izod): 347 J/m notch (D256/A); Hardness (Rockwell): R 103 (D785)

Therm. Prop.: Distort. Temp.: 87 C (@ 1820 kPa, 0.25 mm defl., unannealed, D648); Coeff. of Linear Exp.: 9.5 × 10^{-5} cm/cm/C (D696); Flamm.: HB (1.47 mm, UL94)

Elec. Prop.: Dielec. Str.: 16 kV/mm (D-149); Vol. Resist.: 1.1 × 10^{8} megohm-cm (D257); Arc Resist.: 89 s (D495)

Ref.: Pamphlet SR-605-C, DS P-525A

Cycolac X-11. Borg-Warner Chem.

Chem. Descrip.: ABS resin

Category/Applications: High heat, injection molding grade polymer exhibiting better heat resistance than Cycolac T with equivalent abuse resistance, and offering better processability than Cycolac X-17; used in building products and interior automotive trim

Gen'l. Prop.: Sp. Gr.: 1.04 (D792/A)

Mech. Prop. (23 C): Tens. Str.: 41 MPa (D638); Tens. Mod.: 2.1 GPa (D638); Flex. Str.: 69 MPa (yield, D790); Flex. Mod.: 2.1 GPa (D790); Impact Str. (Izod): 320 J/m notch (D256/A); Hardness (Rockwell): R 100 (D785)

Therm. Prop.: Distort. Temp.: 96 C (@ 1820 kPa, 0.25 mm defl., unannealed, D648); Coeff. of Linear Exp.: 9.2 × 10^{-5} cm/cm/C (D696); Flamm.: HB (1.47 mm, UL94)

Elec. Prop. (23 C): Dielec. Str.: 33 kV/mm (D149); Vol. Resist.: 8.8 × 10^{9} megohm-cm (D257); Arc Resist.: 66 s (D495)

Ref.: Pamphlet SR-605-C

Cycolac X-17. Borg-Warner Chem.

Chem. Descrip.: ABS resin

Category/Applications: High heat, injection molding grade polymer offering higher heat resistance than Cycolac X-11 but less impact strength; recommended for applications where impact resistance is secondary to heat resistance; used in automotive parts, small appliances, power tools, etc.

Gen'l. Prop.: Sp. Gr.: 1.05 (D792/A)

Mech. Prop. (23 C): Tens. Str.: 50 MPa (D638); Tens. Mod.: 2.4 GPa (D638); Flex. Str.: 86 MPa (yield, D790); Flex. Mod.: 2.6 GPa (D790); Impact Str. (Izod): 123 J/m notch (D256/A); Hardness (Rockwell): R 111 (D785)

Therm. Prop.: Distort. Temp.: 104 C (@ 1820 kPa, 0.25 mm defl., unannealed, D648); Coeff. of Linear Exp.: 6.7 × 10^{-5} cm/cm/C (D696); Flamm.: HB (1.47 mm, UL94)

Elec. Prop. (23 C): Dielec. Str.: 36 kV/mm (D149); Vol. Resist.: 3.7 × 10^{9} megohm-cm (D257); Arc Resist.: 85 s (D495)

Ref.: Pamphlet SR-605-C

Cycolac X-37. Borg-Warner Chem.

Chem. Descrip.: ABS resin

Category/Applications: High heat, med. impact injection molding ABS material formulated to withstand temps. to 115 C, and surpassing Cycolac X-17 in both heat resistance and impact strength; offers excellent surface color and gloss, abuse resistance, and improved structural/engineering properties; used in appliances (vacuum cleaner components, personal care housings, power tool casings, air conditioners, dishwashers, kitchen appliances, etc.), communication equipment, business machine componentry, photographic equipment, etc. All properties—natural specimens:

Gen'l. Prop.: Sp. Gr.: 1.06 (D792/A)

Mech. Prop. (23 C): Tens. Str.: 48 MPa (D638); Tens. Mod.: 2.4 GPa (D638); Flex. Str.: 83 MPa (yield, D790); Flex. Mod.: 2.4 GPa (D790); Compr. Str.: 59 MPa (D695); Compr. Mod.: 2.3 GPa (D695); Impact Str. (Izod): 160 J/m notch (D256/A); Hardness (Rockwell): R 109 (D785)
Therm. Prop.: Distort. Temp.: 110 C (@ 1820 kPa, 0.25 mm defl., unannealed, D648); Coeff. of Linear Exp.: 7.9 X 10^{-5} cm/cm/C (D696); Flamm.: HB (1.47 mm, UL94)
Elec. Prop. (23 C): Dielec. Str.: 37 kV/mm (D149); Vol. Resist.: 7.0 X 10^9 megohm-cm (D257); Arc Resist.: 93 s (D495)
Ref.: Pamphlet SR-605-C, DS P-540

Cycolac Z-36. Borg-Warner Chem.
Chem. Descrip.: ABS resin
Category/Applications: High heat extrusion polymer; offers good impact resistance, rigidity, and outstanding resistance to heat distortion
Gen'l. Prop.: Sp. Gr.: 1.06 (23 C, D792/A)
Mech. Prop. (23 C): Tens. Str.: 45 MPa (D638); Tens. Mod.: 2.3 GPa (D638); Flex. Str.: 76 MPa (yield, D790); Flex. Mod.: 2.4 GPa (D790); Impact Str. (Izod): 214 J/m notch (D256/A); Hardness (Rockwell): R 106 (D785)
Therm. Prop.: Distort. Temp.: 102 C (@ 1820 kPa, 0.25 mm defl., unannealed, D648); Coeff. of Linear Exp.: 8.6 X 10^{-5} cm/cm/C (D696); Flamm.: 4.6 cm/min (3.2 mm, D635)
Ref.: Pamphlet SR-604-B

Cycoloy EHA. Borg-Warner Chem.
Chem. Descrip.: ABS/polycarbonate alloy
Category/Applications: High heat, high impact electroplating alloy for injection molding applications; offers an excellent balance of physical characteristics, with good impact strength, superior thermal cycle performance, and resistance to warpage; used for automotive plated exterior trim applications, e.g., head lamp bezels, mirrors, and wheel covers, and in institutional lighting, plumbing, major appliance trim, and plated structural applications requiring load-bearing capabilities
All properties—gray specimens:
Gen'l. Prop.: Sp. Gr.: 1.09 (23 C, D792/A)
Mech. Prop. (23 C): Tens. Str.: 45 MPa (D638); Tens. Mod.: 2.4 GPa (D638); Flex. Str.: 79 MPa (yield, D790); Flex. Mod.: 2.5 GPa (D790); Impact Str. (Izod): 368 J/m notch (D256/A); Hardness (Rockwell): R 111 (D785)
Therm. Prop.: Distort. Temp.: 111 C (@ 1820 kPa, 0.25 mm defl., unannealed, D648); Coeff. of Linear Exp.: 6.2 X 10^{-5} cm/cm/C (D696)
Ref.: Pamphlet SR-605-C, DS P-548A

Cycovin KAB. Borg-Warner Chem.
Chem. Descrip.: Flame-retardant PVC/ABS alloy
Category/Applications: Flame-retardant extrusion polymer with excellent toughness, good gloss, and low price, for general purpose flame retardant applications
Gen'l. Prop.: Sp. Gr.: 1.20 (23 C, D792/A)
Mech. Prop. (23 C): Tens. Str.: 39 MPa (D638); Tens. Mod.: 2.1 GPa (D638); Flex. Str.: 64 MPa (yield, D790); Flex. Mod.: 2.1 GPa (D790); Impact Str. (Izod): 774 J/m notch (D256/A); Hardness (Rockwell): R 96 (D785)
Therm. Prop.: Distort. Temp.: 66 C (@ 1820 kPa, 0.25 mm defl., unannealed, D648); Coeff. of Linear Exp.: 8.3 X 10^{-5} cm/cm/C (D696); Flamm.: V-1 (1.47 mm, UL94)
Ref.: Pamphlet SR-604-B

Cymel 303. American Cyanamid Co./ Resins Dept.
Chem. Descrip.: Hexamethoxymethylmelamine
Category/Applications: Cross-linking agent for use in melamine resin coating systems providing films with good hardness, initial gloss, flexibility, impact resistance, chemical resistance; used in general industrial finishes, coil-coating enamels, appliance finishes, paper coatings, and electrodeposition finishes; also useful with alkyd, polyester, thermosetting acrylic, epoxy, and cellulose resins
Form: Clear, viscous liquid; Color (Gardner 1963): 2 max.
Composition: 98.0% min. nonvolatiles
Gen'l. Prop.: Sol.: Limited water sol.; Density: 10.0 lb/gal; Visc. (Gardner-Holdt): X-Z_2; Flash Pt.: > 180 F
Ref.: Pamphlet CRT 403 A

Cymel 370. American Cyanamid Co./ Resins Dept.
Chem. Descrip.: Methylated melamine resin
Category/Applications: Cross-linking agent for both solvent- and water-based coatings, e.g., general purpose spray coatings, coil coatings, metal decorating enamels, and water-based conversion varnishes, imparting excellent gloss, hardness, flexibility, impact and chemical resistance; requires only weak acid catalysis for cure; compatible with alkyds, polyester, thermosetting acrylic, vinyl, epoxy, and cellulose resins
Form: Clear liquid; Color (Gardner 1963): 1 max.
Composition: 88 ± 2% solids in isopropanol/butanol solvent
Gen'l. Prop.: Sol.: Sol. in methanol, ethanol, isobutanol, 2-ethoxyethanol, 2-butoxyethanol, ethylene and propylene glycols, MEK, MIBK, isophorone, toluene, and xylene; partially sol. in water; Sp. Gr.: 1.18; Density: 9.8 lb/gal; Visc. (Gardner-Holdt): Z_2-Z_4; Flash Pt. (CC): 72 F
Ref.: Pamphlets CRT 403 A, CRT-45

Cymel 373. American Cyanamid Co./ Resins Dept.
Chem. Descrip.: Partially methylated melamine formaldehyde resin
Category/Applications: Cross-linking agent for water-based coatings, e.g., emulsions or solutions, imparting high gloss, good stability, fast cure response, and chemical resistance; requires only weak acid catalysis for cure
Form: Clear liquid; Color (Gardner 1963): 1 max.
Composition: 85 ± 2% solids in water
Gen'l. Prop.: Sol.: Sol. in methanol, ethanol, isopropanol, 2-ethoxyethanol, 2-butoxyethanol, ethylene and propylene glycols, MEK; Sp. Gr.: 1.26; Density: 10.5 lb/gal; Visc. (Gardner-Holdt): Z-Z_4
Ref.: Pamphlet CRT-45

Cymel 380. American Cyanamid Co./ Resins Dept.
Chem. Descrip.: Partially methylated melamine formaldehyde resin
Category/Applications: See Cymel 370
Form: Clear liquid; Color (Gardner 1963): 1 max.
Composition: 80 ± 2% solids in isopropanol/butanol solvent
Gen'l. Prop.: Sol.: See Cymel 370; Sp. Gr.: 1.15; Density: 9.6 lb/gal; Visc. (Gardner-Holdt): V-Z; Flash Pt. (CC): 62 F
Ref.: Pamphlet CRT-45

Cymel 1141. American Cyanamid Co./ Resins Dept.
Chem. Descrip.: Alkylated carboxyl-modified melamine formaldehyde resin

Category/Applications: Cross-linking agent for cationic resins with a pK_B value > 5.5; it is reactive with carboxyl, hydroxyl, and amide groups, and with cationic polymers, with improved adhesion and salt-spray resistance; used for emulsions, water-soluble coatings, and cationic electrocoating applications
Form: Clear, viscous liquid; Color (Gardner 1963): 3 max.
Composition: 85 ± 2% solids in isobutanol
Gen'l. Prop.: Sol.: Water-disp.; sol. in all common organic solvents; Sp. Gr.: 1.07; Density: 8.95 lb/gal; Visc. (Gardner-Holdt): W–Y; Acid No.: 22 ± 3 (resin solids); Stability: Stable under acidic conditions; Flash Pt. (Seta CC): 93 F
Ref.: Pamphlet CRT-108

Cytor 7040. Amer. Cyanamid Co./Elastomers Dept.
Chem. Descrip.: Thermoplastic polyether polyurethane
Category/Applications: Polymer with good hydrolytic stability, grease and oil resistance, low temp. properties, good abrasion resistance, and toughness for applications requiring unusually high performance and long life in the automotive industry, etc.
Form: Pellets
Gen'l. Prop.: Sp. Gr.: 1.24 (D792); Stability: Good hydrolytic stability, grease, oil, fungal, and abrasion resistance; excellent resistance to high temp. exposure in both air and oils
Mech. Prop.: Tens. Str.: 4000 psi (D-412); Tens. Elong.: 600% (D412); Tens. Mod.: 900 psi (@ 100%, D412); Compr. Mod.: 200 psi (@ 5% deflect'n., D575); Hardness (Shore): 82A (2240)
Therm. Prop.: Coeff. of Linear Exp.: 1.25, 10^{-4}/F (D696)
Ref.: Product information

Cytor 7045. Amer. Cyanamid Co./Elastomers Dept.
Chem. Descrip.: Thermoplastic polyether polyurethane
Category/Applications: See Cytor 7040
Form: Pellets
Gen'l. Prop.: Sp. Gr.: 1.24 (D792); Stability: See Cytor 7040
Mech. Prop.: Tens. Str.: 4540 psi (D-412); Tens. Elong.: 550% (D412); Tens. Mod.: 1330 psi (@ 100%, D412); Compr. Mod.: 325 psi (@ 5% deflect'n., D575); Hardness (Shore): 90A (2240)
Therm. Prop.: Coeff. of Linear Exp.: 1.10, 10^{-5}/F (D696)
Ref.: Product information

Cytor 7050. Amer. Cyanamid Co./Elastomers Dept.
Chem. Descrip.: Thermoplastic polyether polyurethane
Category/Applications: See Cytor 7040
Form: Pellets
Gen'l. Prop.: Sp. Gr.: 1.24 (D792); Stability: See Cytor 7040
Mech. Prop.: Tens. Str.: 4800 psi (D-412); Tens. Elong.: 500% (D412): Tens. Mod.: 2000 psi (@ 100%, D412); Compr. Mod.: 475 psi (@ 5% deflectn., D575); Hardness (Shore): 52D
Therm. Prop.: Coeff. of linear Exp.: 1.08, 10^{-4}/F (D696)
Ref.: Product information

D

Daran 220. W.R. Grace & Co./Organic Chem. Div.
Chem. Descrip.: Polyvinylidene chloride emulsion
Category/Applications: All purpose coating material exhibiting high barrier protection against water, moisture, grease, oils, and gases; can be applied as supplied using conventional multipurpose coating equipment; avail. in two grades with heat sealability ranges of 285 F and 350+ F respectively
Form: Emulsion; particle size 0.10 μ and 0.14 μ respectively; Color: Cream, white respectively
Composition: 60% and 62% solids
Gen'l. Prop.: Sp. Gr.: 1.326 and 1.340 respectively; Density: 11.0 and 11.2 lb/gal; Visc. (Brookfield): 75 cps max. (both, LVF, #1 spindle, 60 rpm); F.P.: 36 F (both); Stability: No freeze-thaw stability; emulsions exhibit excellent mechanical and storage stability, but recommended use is within 6 mos.; pH: 2.5 and 4.5 respectively
Ref.: Tech. information

Daran 229. W.R. Grace & Co./Organic Chem. Div.
Chem. Descrip.: Polyvinylidene chloride emulsion
Category/Applications: Coating providing high adhesion to paper and board, low heat seal temps., high barriers, nonblocking and excellent slip characteristics; produces odor-free coatings; avail. in two grades offering heat sealability ranges of 250 F and 350+ F respectively
Form: Emulsion; particle size 0.15 μ and 0.19 μ respectively; Color: Cream and white respectively
Composition: 60% and 62% solids respectively
Gen'l. Prop.: Sp. Gr.: 1.326 and 1.340 respectively; Density: 11.0 and 11.2 lb/gal respectively; Visc. (Brookfield): 50 cps max. (both, LVF, #1 spindle, 60 rpm); F.P.: 36 F (both); Stability: See Daran 220; pH: 1.5 and 3.5 respectively
Ref.: Tech. information

Daran 820. W.R. Grace & Co./Organic Chem. Div.
Chem. Descrip.: Polyvinylidene chloride emulsion
Category/Applications: PVDC emulsion providing the barrier advantages of other Daran coatings with primerless adhesion to many plastic films; also offers low foaming, improved whiteness and gloss, excellent flow and leveling; not recommended for mixing with other Daran emulsions; should be tested on specific substrates; avail. in two grades offering heat sealability ranges as low as 215 F
Form: Emulsion; particle size 0.08 μ

and 0.12 μ respectively; Color: Cream, white respectively
Composition: 49% and 51% solids respectively
Gen'l. Prop.: Sp. Gr.: 1.252 and 1.265 respectively; Density: 10.4 and 10.5 lb/gal respectively; Visc. (Brookfield): 30 cps max. (both, LVF, #1 spindle, 60 rpm); F.P.: 36 F (both); Stability: See Daran 220; pH: 1.5 and 2.5 respectively
Ref.: Tech. information

Daratak 17-200, 17-230, 17-300. W.R. Grace & Co./Organic Chem. Div.
Chem. Descrip.: PVAc homopolymer emulsions
Category/Applications: Med. m.w., general purpose adhesive emulsion noted for its superior compounding and machining stability; produces water white, slightly hazy, med. gloss films with good water resistance; avail. in low-med., med-high, and high visc. versions respectively
Form: Emulsion; particle size 1.5 μ; Color: White; Odor: Slight, characteristic
Composition: 54–56% solids
Gen'l. Prop.: Density: 9.93 lb/gal (all); Visc.: 1500–2000 cps, 2400–2800 cps, and 3000–4000 cps respectively; Stability: See Daratak 52L; pH: 4.5–6.0 (all)
Ref.: Tech. data

Daratak A, B. W.R. Grace & Co./Organic Chem. Div.
Chem. Descrip.: Copolymer emulsion
Category/Applications: Adhesive emulsion combining excellent compounding properties with extremely good macining; esp. useful in bonding flexible, nonporous substrates; produces water white, clear, high gloss films with excellent water resistance; Daratak A offers greater hardness and less flexibility than Daratak B
Form: Emulsion; particle size 1.5 μ; Color: White; Odor: Slight, characteristic
Composition: 54–56% solids
Gen'l. Prop.: Density: 9.76 and 9.57 lb/gal respectively; Visc.: 1000–1400 cps (both); Stability: See Daratak 52L; pH: 4.0–5.0 (both)
Ref.: Tech. data

Daratak SP1065. W.R. Grace & Co./Organic Chem. Div.
Chem. Descrip.: Externally plasticized PVAc emulsion
Category/Applications: Adhesive emulsion with exceptional compounding stability; used in highly pigmented mastics, adhesives, coatings, dry wall cement, and patching compounds; produces water white, clear, high gloss films with excellent water resistance
Form: Emulsion; particle size 1.0 μ; Color: White; Odor: Slight, characteristic
Composition: 55–57% solids
Gen'l. Prop.: Density: 9.93 lb/gal; Visc.: 1500–3000 cps; Stability: See Daratak 52L; pH: 4.5–5.5
Ref.: Tech. data

Daratak SP1500. W.R. Grace & Co./Organic Div.
Category/Applications: Adhesive emulsion that meets critical requirements for many film and foil adhesive applications; produces water white, slightly hazy, med. gloss films with excellent water resistance
Form: Emulsion; particle size 0.5 μ; Color: White; Odor: Slight, characteristic
Composition: 51–53% solids
Gen'l. Prop.: Density: 8.7 lb/gal; Visc.: 1000–3000 cps; Stability: See Daratak 52L; pH: 6.0–7.5
Ref.: Tech. data

Daratak 52L. W.R. Grace & Co./Organic Chem. Div.
Chem. Descrip.: PVAc emulsion
Category/Applications: Low m.w., low visc., fast breaking adhesive emulsion used in adhesives, textiles, surface coatings, and heat sealing applications; produces water white, clear, high gloss films with good water resistance and no odor
Form: Emulsion; particle size 1.5 μ; Color: White; Odor: Slight, characteristic
Composition: 54–56% solids
Gen'l. Prop.: Density: 9.93 lb/gal; Visc.: 300–600 cps; Stability: Emulsions exhibit 20+ min mechanical stability; films exhibit excellent light stability and aging characteristics; pH: 5.5–7.0
Ref.: Tech. data

Daratak 56L. W.R. Grace & Co./Organic Chem. Div.
Chem. Descrip.: PVAc emulsion
Category/Applications: Very high m.w., low visc., fast drying adhesive emulsion used in adhesives, primer sealers, textile finishes, and paper saturation; produces water white, slightly hazy, low gloss films with good water resistance and no odor
Form: Emulsion; particle size 1.0 μ; Color: White; Odor: Slight, characteristic
Composition: 54–56% solids
Gen'l. Prop.: Density: 9.93 lb/gal; Visc.: 300–600 cps; Stability: See Daratak 52L; pH: 5.5–7.0
Ref.: Tech. data

Daratak 61L. W.R. Grace & Co./Organic Chem. Div.
Chem. Descrip.: PVAc homopolymer emulsion
Category/Applications: High visc., borax tolerant adhesive emulsion used in heat sealing, paper and foil adhesives, and in fire-retardant mastics; produces water white, clear, med. gloss films with excellent water resistance
Form: Emulsion; particle size 1.0 μ; Color: White; Odor: Slight, characteristic
Composition: 54–56% solids
Gen'l. Prop.: Density: 9.93 lb/gal; Visc.: 2000–4000 cps; Stability: See Daratak 52L; pH: 4.5–5.5
Ref.: Tech. data

Daratak 62L. W.R. Grace & Co./Organic Chem. Div.
Chem. Descrip.: PVAc homopolymer emulsion
Category/Applications: High m.w., high visc. adhesive emulsion characterized by excellent adhesion, fast setting, and good penetration into coated surfaces; produces water white, slightly hazy, med. gloss films with good water resistance
Form: Emulsion; particle size 1.0 μ; Color: White; Odor: Slight, characteristic
Composition: 54–56% solids
Gen'l. Prop.: Density: 9.93 lb/gal; Visc.: 2300–2700 cps; Stability: See Daratak 52L; pH: 5.0–7.0
Ref.: Tech. data

Daratak 65L. W.R. Grace & Co./Organic Chem. Div.
Chem. Descrip.: PVAc homopolymer emulsion
Category/Applications: Med. m.w. and visc. adhesive emulsion with strong wet tack; compatible with fully hydrolyzed polyvinyl alcohol; ideal for paper packaging applications; produces water white, slightly hazy, med. gloss films with good water resistance
Form: Emulsion; particle size 1.5 μ; Color: White; Odor: Slight, characteristic
Composition: 54–56% solids
Gen'l. Prop.: Density: 9.93 lb/gal; Visc.: 1400–2000 cps; Stability: See Daratak 52L; pH: 4.5–5.5
Ref.: Tech. data

Daratak 71L. W.R. Grace & Co./Organic Chem. Div.
Chem. Descrip.: PVAc homopolymer emulsion
Category/Applications: Med. m.w. adhesive emulsion that imparts fast grab to adhesive formulations; tolerant of large amounts of solvents and plasticizers; used in packaging applications; produces water white, slightly hazy, med. gloss films with good water resistance
Form: Emulsion; particle size 1.5 μ; Color: White; Odor: Slight, characteristic
Composition: 54–56% solids
Gen'l. Prop.: Density: 9.93 lb/gal; Visc.: 1500–2200 cps; Stability: See Daratak 52L; pH: 5.0–6.0
Ref.: Tech. data

Daratak 74L. W.R. Grace & Co./Organic Chem. Div.
Chem. Descrip.: Acrylic terpolymer emulsion
Category/Applications: Adhesive emulsion which dries to a permanently tacky film; designed for adhering to a wide range of nonporous surfaces, e.g., foil, polyolefins, cellulose acetate, cellophane, etc.; produces water white, clear, high gloss films with excellent water resistance
Form: Emulsion; particle size 1.0 μ; Color: White; Odor: Slight, characteristic
Composition: 54–56% solids
Gen'l. Prop.: Density: 9.35 lb/gal; Visc.: 2000–4000 cps; Stability: See Daratak 52L; pH: 4.5–5.5
Ref.: Tech. data

Daratak 78L. W.R. Grace & Co./Organic Chem. Div.
Chem. Descrip.: Carboxylated PVAc emulsion
Category/Applications: Alkali soluble, high reactive emulsion offering excellent adhesion to aluminum foil and many coated surfaces; produces water white, clear, med. gloss films with good water resistance; may be crosslinked with formaldehyde resins to provide glossy coatings
Form: Emulsion; particle size 1.0 μ; Color: White; Odor: Slight, characteristic
Composition: 54–56% solids
Gen'l. Prop.: Density: 9.93 lb/gal; Visc.: 2000–6000 cps; Stability: See Daratak 52L; pH: 4.0–6.0
Ref.: Tech. data

Daratak 79L. W.R. Grace & Co./Organic Chem. Div.
Chem. Descrip.: Emulsion polymer
Category/Applications: Emulsion polymer possessing excellent compatibility with various dextrins, increasing speed and reducing stringing in remoistenable glues for envelopes and labels; produces water white, slightly hazy, high gloss films with poor water resistance
Form: Emulsion; particle size 1.0 μ; Color: White; Odor: Slight, characteristic
Composition: 55–57% solids
Gen'l. Prop.: Density: 9.9 lb/gal; Visc.: 700–1600 cps; Stability: See Daratak 52L; pH: 4.0–5.0
Ref.: Tech. data

Daratak 84L. W.R. Grace & Co./Organic Chem. Div.
Chem. Descrip.: Emulsion polymer
Category/Applications: High m.w., pressure sensitive laminating adhesive base supplied as an aq. emulsion; dries to form an aggressively tacky film with good resistance to cold flow, good clarity, high gloss, excellent water resistance, and water white color
Form: Emulsion; particle size 1.0 μ; Color: White; Odor: Slight, characteristic
Composition: 48–50% solids

Gen'l. Prop.: Density: 8,8 lb/gal; Visc.: 1500–2500 cps; Stability: See Daratak 52L; pH: 4.5–5.5
Ref.: Tech. data

Darex 41. W.R. Grace & Co./Organic Chem. Div.
Chem. Descrip.: Synthetic polymer
Category/Applications: Polymer for the treatment of water for scale inhibition and conditioning of suspended matter; disperses particulate matter producing a nonadherent precipitate which is easily carried through to blowdown; inhibits deposition of new scale and sludge; noncorrosive to metals
Form: Free-flowing powder; Color: Pure white
Gen'l. Prop.: Sol.: Sol. in water systems; Density (Bulk): 29 lb/ft^3 (untamped); Stability: Stable over wide pH range and at high temps. and pressures; pH: 7.5 (1% sol'n.); Surface Tension: 70 dynes/cm (1% sol'n.)
Ref.: Tech. data

Darex 3333. W.R. Grace & Co./Organic Chem. Div.
Chem. Descrip.: Modified styrene butadiene latex
Category/Applications: Latex for use as a coal crusting agent; exhibits excellent application, penetration, and wetting characteristics; can be used with high dilution by water without hindering its crusting capabilities
Form: Particle size 0.2 μ
Composition: 50% total solids
Gen'l. Prop.: Density: 8.5 lb/gal; Visc. (Brookfield): 750 cps; Stability: Excellent mechanical stability; pH: 7.5
Ref.: Tech. data

Darex SP-1566. W.R. Grace & Co./Organic Chem. Div.
Chem. Descrip.: Modified vinyl latex
Category/Applications: Biodegradable latex designed for use in the reclammation of surface-mined lands and for the replanting of steep slopes along new highway right of ways; excellent binder for the mulches of shredded leaves, straw, and wood chips used as humus, thus stabilizing the soil against wind and rain erosion; compatible with most liquid or dry fertilizers; does not inhibit germination of grain or grass seeding; its films are water white with good clarity, med. gloss, and no odor, with very good water resistance, and excellent light stability and aging characteristics
Form: Emulsion; particle size 1 μ; Color: White; Odor: Slight
Composition: 54–56% total solids
Gen'l. Prop.: Density: 9.93 lb/gal; Visc. (Brookfield): 900–1500 cps (#3 spindle, 60 rpm); Stability: Good mechanical stability; pH: 4.5–5.5
Ref.: Tech. data

Darex X442. W.R. Grace & Co./Organic Chem. Div.
Chem. Descrip.: Styrene acrylate latex
Category/Applications: Nonfilm forming latex at R.T., but does form tough, stiff films at elevated temps.; at temps. as low as 90 C, cures to form films that are so thoroughly crosslinked that they exhibit the excellent solvent resistance of thermosetting resins; webs saturated with Darex X442 can be heat molded to form permanent shapes that are resistant to water, solvents, and high temps.
Form: Particle size 0.15 μ
Composition: 50% total solids
Gen'l. Prop.: Density: 8.8 lb/gal; Visc. (Brookfield): 130 cps (LVF, #2 spindle, 60 rpm); Glass Trans. Temp.: 33 C; Stability: Good mechanical stability; pH: 7; Surface Tension: 41 dynes/cm
Ref.: Tech. data

Darex 43S. W.R. Grace & Co./Organic Chem. Div.
Chem. Descrip.: High styrene copolymer resin
Category/Applications: Thermoplastic resin which gives stiffness, hardness, and high abrasion resistance to natural and synthetic rubbers; improves mold flow, calendering, and extruding properties of rubber stocks; used for nonstaining and nonmarking shoe soles, floor tiles, and rubber household goods; used in wire and cable manufacture, e.g., submarine cable and specialty wire insulation with low power factor and low dielec. constant
Form: Free-flowing, low-dusting granular form
Gen'l. Prop.: Sp. Gr.: 1.04; Soften. Pt.: 96–108 F; Stability: Good resistance to acids, alkalies, and oils; poor resistance to aromatic solvents; molded sheets remain transparent and light colored at temps. as high as 350 F
Mech. Prop. (cast film): Tens. Str.: 7700 psi; Tens. Elong.: 2% (ultimate); Hardness (Shore D): > 80
Ref.: Tech. data

Darex 110L. W.R. Grace & Co./Organic Chem. Div.
Chem. Descrip.: Nitrile latex
Category/Applications: Copolymer latex for paper and felt saturation offering high delamination resistance and edge tear making it esp. effective in tape and sandpaper bases; curable with heat; reacts with formaldehyde to give higher delamination resistance and excellent wet tens. strength
Form: Particle size 0.2 μ
Composition: 47% solids
Gen'l. Prop.: Density: 8.4 lb/gal; Visc. (Brookfield): 30 cps (LVF, #1 spindle, 60 rpm); Stability: Excellent mechanical stability; offers better oil, solvent, and plasticizer resistance than styrene butadiene latices; pH: 7.0; Surface Tension: 40 dynes/cm
Ref.: Bulletin L-173-2

Darex 165L. W.R. Grace & Co./Organic Chem. Div.
Chem. Descrip.: Styrene-acrylonitrile latex
Category/Applications: Water-based latex designed for bonding and impregnating fibers, webs, and films; forms stiff films by itself; can produce flexible films by plasticizing with softer latices or with compatible plasticizers such as the phthalyl glycollates
Form: Particle size 0.3 μ
Composition: 50% total solids; ≈ 71% styrene, 29% acrylonitrile
Gen'l. Prop.: Density: 8.7 lb/gal; Visc. (Brookfield): 50 cps (LVF, #2 spindle, 60 rpm); Stability: Excellent mechanical stability; pH: 9; Surface Tension: 47 dynes/cm (42% solids)
Ref.: Tech. data

Darex 508L. W.R. Grace & Co./Organic Chem. Div.
Chem. Descrip.: Carboxylated styrene-butadiene latex
Category/Applications: Latex which forms clear, flexible but stiff films of high toughness; esp. suitable for use in modified cement mortars because of its unusual compatibility with highly alkaline Portland cement; improves bond strength, compressive strength, tensile strength, and flexural strength; can be used to improve tile adhesives, tile grouts, precast panel adhesives, adhesives to bond new concrete to old, concrete pipe repair compounds, concrete and Portland cement plasters, and compounds to repair "Tuck Pointing" of unit masonry
Form: Particle size 0.2 μ

Composition: 49.5% total solids; ≈ 67% styrene, 33% butadiene
Gen'l. Prop.: Density: 8.5 lb/gal; Visc. (Brookfield): 150 cps (LVF, #2 spindle, 60 rpm); Stability: Excellent mechanical stability; pH: 10.5; Surface Tension: 40 dynes/cm
Ref.: Bulletin L-178-1

Darex 510L. W.R. Grace & Co./Organic Chem. Div.
Chem. Descrip.: Carboxylated, styrene-butadiene latex
Category/Applications: Latex which forms clear, flexible but stiff films of high toughness; has excellent dry filler tolerance at high solids and outstanding freedom from discoloration and deterioration on exposure to heat, light, or gas; esp. useful in carpet backing compounds for both foam and jute lamination
Form: Particle size 0.2 μ
Composition: 55% total solids; ≈ 67% styrene, 33% butadiene
Gen'l. Prop.: Density: 8.5 lb/gal; Visc. (Brookfield): 200 cps (LVF, #2 spindle, 60 rpm); Stability: Excellent mechanical stability; pH: 8.5; Surface Tension: 45 dynes/cm
Ref.: Bulletin L-167-2

Darex 515L. W.R. Grace & Co./Organic Chem. Div.
Chem. Descrip.: Copolymer latex
Category/Applications: Pigment binder developed for use in surface coatings, e.g., interior flat wall paints and primer sealers; offers excellent pigment binding efficiency without sacrifice of pigment compatibility, allowing wide compounding latitude
Form: Particle size 0.2 μ
Composition: 47–49% total solids
Gen'l. Prop.: Density: 8.5 lb/gal; Visc. (Brookfield): 20–60 cps (LVF, #2 spindle, 60 rpm); Stability: Excellent mechanical stability; pH: 7.0–8.5; Surface Tension: 33–40 dynes/cm
Ref.: Bulletin L172-1

Darex 526L. W.R. Grace & Co./Organic Chem. Div.
Chem. Descrip.: Carboxylated, styrene-butadiene latex
Category/Applications: Latex which forms soft, flexible films; useful in making compounds for textile backing applications where a relatively soft hand is desired; has excellent filler tolerance and compounding stability, permitting dry loading to high solids; can be blended with Darex 510L to gives films of intermediate flexibility
Form: Particle size 0.2 μ
Composition: 53% total solids; ≈ 50% styrene, 50% butadiene
Gen'l. Prop.: Density: 8.3 lb/gal; Visc. (Brookfield): 100 cps (LVF, #2 spindle, 60 rpm); Stability: Excellent mechanical stability; pH: 7.5; Surface Tension: 45 dynes/cm
Ref.: Bulletin L-168-2

Darex 529L. W.R. Grace & Co./Organic Chem. Div.
Chem. Descrip.: Carboxylated terpolymer of butadiene, styrene, and vinylidene chloride;
Category/Applications: Used as a binder-saturant to give products flexible hand, freedom from tackiness, and good abrasion resistance; offers improved flame resistance, reducing or eliminating the need for fire retardant additives; has good stability to the addition of fillers and other compounding ingredients; compatible with most latex binder systems
Form: Particle size 0.2 μ
Composition: 53% total solids
Gen'l. Prop.: Density: 9.1 lb/gal; Visc. (Brookfield): 50 cps (LVF, #1 spindle,

60 rpm); Stability: Excellent mechanical stability; pH: 9.5; Surface Tension: 40 dynes/cm
Ref.: Tech. data

Darex 535L. W.R. Grace & Co./Organic Chem. Div.
Chem. Descrip.: Carboxylated terpolymer of butadiene, styrene, and vinylidene chloride
Category/Applications: Similar to Darex 529L but containing more vinylidene chloride and less styrene making it softer and providing more chlorine in the polymer structure for improved flame resistance
Form: Particle size 0.2 μ
Composition: 52% total solids
Gen'l. Prop.: Density: 9.3 lb/gal; Visc. (Brookfield): 40 cps (LVF, #1 spindle, 60 rpm); Stability: Excellent mechanical stability; pH: 8.5; Surface Tension: 40 dynes/cm
Ref.: Tech. data

Darex 537L. W.R. Grace & Co./Organic Chem. Div.
Chem. Descrip.: Carboxylated styrene-butadiene latex
Category/Applications: Used as a paper saturant to impart high internal bond strength, excellent edge tear, and superior elongation; offers ease of compounding; contains a nonstaining antioxidant for good aging at ambient and elevated temps.
Composition: 48% total solids; ≈ 37% styrene, 63% butadiene
Gen'l. Prop.: Density: 8.4 lb/gal; Visc. (Brookfield): 35 cps (LVF, #1 spindle, 60 rpm); Stability: Excellent mechanical stability; pH: 8.5; Surface Tension: 35 dynes/cm
Ref.: Tech. data

Darex 620L. W.R. Grace & Co./Organic Chem. Div.
Chem. Descrip.: SBR 200 type latex
Category/Applications: Used for beater addition, paper saturation, adhesives, fabric backings, and binders for foam scarp, leather, and curled hair; used in compounds both with and without sulfur curing systems
Form: Particle size 0.15 μ
Composition: 42% total solids; ≈ 47% styrene, 53% butadiene
Gen'l. Prop.: Density: 8.3 lb/gal; Visc. (Brookfield): 40 cps (LVF, #1 spindle, 60 rpm); Stability: Fair mechanical stability; pH: 10.5; Surface Tension: 50 dynes/cm
Ref.: Bulletin L-150-3

Darex 632L. W.R. Grace & Co./Organic Chem. Div.
Chem. Descrip.: Styrene-butadiene latex
Category/Applications: High styrene latex designed for blending with natural and synthetic latices to give increased stiffness, hardness, and tear resistance; used in compounds for backing rugs, carpets, and other textile goods; to reinforce dipped goods, foam, and other rubber products; to saturate felt for box toes to give stiffness and hardness without brittleness
Form: Particle size 0.15 μ
Composition: 52% total solids; ≈ 90% sytrene, 10% butadiene
Gen'l. Prop.: Density: 8.6 lb/gal; Visc. (Brookfield): 60 cps (LVF, #1 spindle, 60 rpm); Stability: Good mechanical stability; pH: 11.0; Surface Tension: 50 dynes/cm
Ref.: Bulletin L-24-3

Darex 636L. W.R. Grace & Co./Organic Chem. Div.
Chem. Descrip.: Styrene-butadiene latex
Category/Applications: Nonfilm forming

latex for use as a box toe fabric saturant; film coalescence is accomplished by heat or solvent fusion; molding takes place concurrently with fusion and firmness develops rapidly so that min. dwell time is required; antioxidants are not required

Form: Particle size 0.15 μ

Composition: 50–52% total solids; ≈ 97% styrene, 3% butadiene

Gen'l. Prop.: Density: 8.6 lb/gal; Visc. (Brookfield): 30–110 cps (LVF, #1 spindle, 60 rpm); Stability: Good mechanical stability; pH: 10.5–12.0; Surface Tension: 35–45 dynes/cm

Ref.: Bulletin L-170-1

Darex 637L. W.R. Grace & Co./Organic Chem. Div.

Chem. Descrip.: Styrene-butadiene latex

Category/Applications: Latex developed for coating and saturation applications requiring a relatively firm, yet flexible, polymer film; useful in dry saturation applications where penetration characteristics are important

Form: Partice size 0.2 μ

Composition: 48% total solids; ≈ 67% styrene, 33% butadiene

Gen'l. Prop.: Density: 8.4 lb/gal; Visc. (Brookfield): 30 cps (LVF, #1 spindle, 60 rpm); Stability: Excellent mechanical stability; outstanding resistance to discoloration or deterioration by heat, gas fumes, or aging; pH: 8.0; Surface Tension: 33 dynes/cm

Ref.: Bulletin L-177-2

Darex 643L. W.R. Grace & Co./Organic Chem. Div.

Chem. Descrip.: Styrene-butadiene latex

Category/Applications: High styrene latex designed for blending with natural or synthetic rubber latices to give increased stiffness, hardness, and tear resistance; classed as a nonfilm forming latex, but has minimal effect on film formation when blended with film forming latices; useful in saturation and coating compounds

Composition: 39% total solids; ≈ 85% styrene, 15% butadiene

Gen'l. Prop.: Density: 8.6 lb/gal; Visc. (Brookfield): 10 cps (LVF, #1 spindle, 60 rpm); Stability: Good mechanical stability; pH: 10.5; Surface Tension: 63 dynes/cm

Ref.: Bulletin 671L-2

Darex 670L. W.R. Grace & Co./Organic Chem. Div.

Chem. Descrip.: Polystyrene latex

Category/Applications: Nonfilm forming latex for use as a saturant; film coalescence is accomplished by either heat or solvent fusion; antioxidants are not required

Form: Particle size 0.15 μ

Composition: 50.5% total solids

Gen'l. Prop.: Density: 8.6 lb/gal; Visc. (Brookfield): 80 cps (LVF, #1 spindle, 60 rpm); Stability: Good mechanical stability; pH: 11.5; Surface Tension: 40 dynes/cm

Ref.: Tech. data

Darex 5281L. W.R. Grace & Co./Organic Chem. Div.

Chem. Descrip.: Styrene butadiene latex

Category/Applications: Low emulsifier latex providing high gloss, water-resistant coatings with good resistance to blocking; the latex dries rapidly, with little heat, to a clear, glossy, smooth finish; paper and paperboard substrates exhibit excellent gross water holdout after application; FDA approved for use in contact with aqueous and/or fatty foods

Form: Particle size 0.2 μ *(latex)*; clear, glossy appearance with low odor (film)

Composition: 46–48% total solids

Gen'l. Prop.: Density: 8.4 lb/gal; Visc. (Brookfield): 20–60 cps (LVF, #3

spindle, 60 rpm); Stability: Good mechanical stability; films exhibit good aging characteristics; pH: 7.5–8.5; Surface Tension: 60–70 dynes/cm
Ref.: Tech. data

DC 2900 HW. Hardman Inc.
Chem. Descrip.: Two-part epoxy comp.
Category/Applications: Elevated temp. curing comp. for electrical/electronic applications; mix ratio: 100A/45B by wt.
Gen'l. Prop.: Visc.: 16,300 cps (mixed)
Mech. Prop.: Hardness (Shore D): 90
Therm. Prop.: Conduct.: 8.9×10^{-4} Cal/cm/s/cm^2/C; Coeff. of Linear Exp.: 147×10^{-6} in./in./C
Elec. Prop.: Dissip. Factor: 0.010 (100 kHz); Dielec. Const.: 3.4 (100 kHz); Vol. Resist.: 10^{15} ohm-cm
Curing Charact.: Pot Life: 73 min @ 121 C (200 g); Cure Schedule: 16 h @ 121 C
Ref.: Data sheet

Delrin 100. Du Pont Co.
Chem. Descrip.: Acetal resin
Category/Applications: Engineering thermoplastic homopolymer offering a high balance of strength and stiffness, high fatigue endurance, natural lubricity, corrosion resistance, and resilience in applications requiring spring-like recovery and toughness under a wide range of temp. and humidity conditions; used in machinery, agricultural equipment, interior automotive door handles, clock mechanisms, ballcock valves, disposable cigarette lighters, videocassettes, and other molding applications; Delrin 100 is the toughest acetal in series
Gen'l. Prop.: Sp. Gr.: 1.42 (D792); M.P.: 175 C (crystalline, D2133); Stability: Resists weak acids (pH 4) and weak bases (pH 9); not recommended in strong acids or alkalies; excellent resistance to a wide variety of solvents, ethers, oil, greases, gasoline, and other petroleum hydrocarbons
Mech. Prop.: Tens. Str.: 68.9 MPa (23 C, D638); Tens. Elong.: 75% (23 C, D638); Flex. Mod.: 2620 MPa (23 C, D790); Compr. Str.: 35.9 MPa (23 C @ 1% def., D695); Shear Str.: 65.5 MPa (23 C, D-732); Impact Str. (Izod): 123 J/m notch (23 C, D256); Hardness (Rockwell): M 94 (D785)
Therm. Prop.: Conduct.: 0.37 W/m•K; Distort. Temp.: 124 C (@ 1.8 MPa, D-648); Coeff. of Linear Exp.: 12.2×10^{-5} m/m/C (29–60 C, D696); Flamm.: HB (UL94)
Elec. Prop.: Dissip. Factor: 0.005 (23 C, 50% r.h., 10^6 Hz, D150); Dielec. Str.: 19.7 kV/mm (S/T, 2.3 mm, D149); Dielec. Const.: 3.7 (23 C, 50% r.h., 10^2 to 10^6 Hz, D150); Vol. Resist.: 10^{15} ohm-cm (23 C, D257); Arc Resist.: 220 s, no tracking (3.1 mm, D495)
Ref.: Booklet E-32986

Delrin 500. Du Pont Co.
Chem. Descrip.: Acetal resin
Category/Applications: General purpose grade; see Delrin 100
Gen'l. Prop.: Sp. Gr.: 1.42 (D792); M.P.: 175 C (crystalline, D2133); Stability: See Delrin 100
Mech. Prop.: Tens. Str.: 68.9 MPa (23 C, D638); Tens. Elong.: 40% (23 C, D638); Flex. Mod.: 2830 MPa (23 C, D790); Compr. Str.: 35.9 MPa (23 C @ 1% def., D695); Shear Str.: 65.5 MPa (23 C, D-732); Impact Str. (Izod): 74.7 J/m notch (23 C, D256); Hardness (Rockwell): M 94 (D785)
Therm. Prop.: Conduct.: 0.37 W/m•K; Distort. Temp.: 124 C (@ 1.8 MPa, D-648); Coeff. of Linear Exp.: 12.2×10^{-5} m/m/C (29–60 C, D696); Flamm.: HB (UL94)
Elec. Prop.: Dissip. Factor: 0.005 (23 C, 50% r.h., 10^6 Hz, D150); Dielec. Str.: 19.7 kV/mm (S/T, 2.3 mm, D149);

Dielec. Const.: 3.7 (23 C, 50% r.h., 10^2 to 10^6 Hz, D150); Vol. Resist.: 10^{15} ohm-cm (23 C, D257); Arc Resist.: 220 s, no tracking (3.1 mm, D495)
Ref.: Booklet E-32986

Delrin 570. Du Pont Co.
Chem. Descrip.: Acetal resin
Category/Applications: High stiffness grade; see Delrin 100
Gen'l. Prop.: Sp. Gr.: 1.56 (D792); M.P.: 175 C (crystalline, D2133); Stability: See Delrin 100
Mech. Prop.: Tens. Str.: 58.6 MPa (23 C, D638); Tens. Elong.: 12% (23 C, D638); Flex. Mod.: 5030 MPa (23 C, D790); Compr. Str.: 35.9 MPa (23 C @ 1% def., D695); Shear Str.: 66.5 MPa (23 C, D-732); Impact Str. (Izod): 42.7 J/m notch (23 C, D256); Hardness (Rockwell): M 90 (D785)
Therm. Prop.: Distort. Temp.: 157 C (@ 1.8 MPa, D648); Coeff. of Linear Exp.: 3.6–8.1 $\times$ 10^{-5} m/m/C (–40 to 29 C, D696); Flamm.: HB (UL94)
Elec. Prop.: Dissip. Factor: 0.005 (23 C, 50% r.h., 10^6 Hz, D150); Dielec. Str.: 19.3 kV/mm (S/T, 3.2 mm, D149); Dielec. Const.: 3.9 (23 C, 50% r.h., 10^2 to 10^6 Hz, D150); Vol. Resist.: 5 $\times$ 10^{14} ohm-cm (23 C, D257); Arc Resist.: 168 s, no tracking (3.1 mm, D495)
Ref.: Booklet E-32986

Delrin 900. Du Pont Co.
Chem. Descrip.: Acetal resin
Category/Applications: High flow grade; see Delrin 100
Gen'l. Prop.: Sp. Gr.: 1.42 (D792); M.P.: 175 C (crystalline, D2133); Stability: See Delrin 100
Mech. Prop.: Tens. Str.: 68.9 MPa (23 C, D638); Tens. Elong.: 25% (23 C, D638); Flex. Mod.: 2960 MPa (23 C, D790); Compr. Str.: 34.5 MPa (23 C @ 1% def., D695); Shear Str.: 68.9 MPa (23 C, D-732); Impact Str. (Izod): 69.4 J/m notch (23 C, D256); Hardness (Rockwell): M 94 (D785)
Therm. Prop.: Conduct.: 0.37 W/m•K; Distort. Temp.: 124 C (@ 1.8 MPa, D648); Coeff. of Linear Exp.: 12.2 $\times$ 10^{-5} m/m/C (29–60 C, D696); Flamm.: HB (UL94)
Elec. Prop.: Dissip. Factor: 0.005 (23 C, 50% r.h., 10^6 Hz, D150); Dielec. Str.: 19.7 kV/mm (S/T, 2.3 mm, D149); Dielec. Const.: 3.7 (23 C, 50% r.h., 10^2 to 10^6 Hz, D150); Vol. Resist.: 10^{15} ohm-cm (23 C, D257); Arc Resist.: 220 s, no tracking (3.1 mm, D495)
Ref.: Booklet E-32986

Delrin 100 AF. Du Pont Co.
Chem. Descrip.: Acetal resin with Teflon fibers
Category/Applications: Low friction/low wear grade; see Delrin 100
Gen'l. Prop.: Sp. Gr.: 1.54 (D792); M.P.: 175 C (crystalline, D2133); Stability: See Delrin 100
Mech. Prop.: Tens. Str.: 52.4 MPa (23 C, D638); Tens. Elong.: 22% (23 C, D638); Flex. Mod.: 2340 MPa (23 C, D790); Compr. Str.: 31.0 MPa (23 C @ 1% def., D695); Shear Str.: 55.2 MPa (23 C, D-732); Impact Str. (Izod): 64.1 J/m notch (23 C, D256); Hardness (Rockwell): M 78 (D785)
Therm. Prop.: Distort. Temp.: 100 C (@ 1.8 MPa, D648); Coeff. of Linear Exp.: 12.2 $\times$ 10^{-5} m/m/C (29–60 C, D696); Flamm.: HB (UL94)
Elec. Prop.: Dissip. Factor: 0.005 (23 C, 50% r.h., 10^6 Hz, D150); Dielec. Str.: 15.7 kV/mm (S/T, 3.2 mm, D149); Dielec. Const.: 3.1 (23 C, 50% r.h., 10^2 to 10^6 Hz, D150); Vol. Resist.: 3 $\times$ 10^{16} ohm-cm (23 C, D257); Arc Resist.: 183 s, no tracking (3.1 mm, D495)
Ref.: Booklet E-32986

Delrin 500 AF. Du Pont Co.
Chem. Descrip.: Acetal resin with Teflon fibers
Category/Applications: Low friction/low wear grade; see Delrin 100
Gen'l. Prop.: Sp. Gr.: 1.54 (D792); M.P.: 175 C (crystalline, D2133); Stability: See Delrin 100
Mech. Prop.: Tens. Str.: 47.6 MPa (23 C, D638); Tens. Elong.: 15% (23 C, D638); Flex. Mod.: 2410 MPa (23 C, D790); Compr. Str.: 31.0 MPa (23 C @ 1% def., D695); Shear Str.: 55.2 MPa (23 C, D-732); Impact Str. (Izod): 37.4 J/m notch (23 C, D256); Hardness (Rockwell): M 78 (D785)
Therm. Prop.: Distort. Temp.: 100 C (@ 1.8 MPa, D648); Coeff. of Linear Exp.: 12.2×10^{-5} m/m/C (29–60 C, D696); Flamm.: HB (UL94)
Elec. Prop.: Dissip. Factor: 0.005 (23 C, 50% r.h., 10^6 Hz, D150); Dielec. Str.: 15.7 kV/mm (S/T, 3.2 mm, D149); Dielec. Const.: 3.1 (23 C, 50% r.h., 10^2 to 10^6 Hz, D150); Vol. Resist.: 3×10^{16} ohm-cm (23 C, D257); Arc Resist.: 183 s, no tracking (3.1 mm, D495)
Ref.: Booklet E-32986

Delrin 500 CL. Du Pont Co.
Chem. Descrip.: Chemically lubricated acetal resin
Category/Applications: Low friction/low wear grade; see Delrin 100
Gen'l. Prop.: Sp. Gr.: 1.42 (D792); M.P.: 175 C (crystalline, D2133); Stability: See Delrin 100
Mech. Prop.: Tens. Str.: 65.5 MPa (23 C, D638); Tens. Elong.: 40% (23 C, D638); Flex. Mod.: 2760 MPa (23 C, D790); Compr. Str.: 31.0 MPa (23 C @ 1% def., D695); Shear Str.: 66.5 MPa (23 C, D-732); Impact Str. (Izod): 74.7 J/m notch (23 C, D256); Hardness (Rockwell): M 90 (D2785)
Therm. Prop.: Distort. Temp.: 121 C (@ 1.8 MPa, D648); Coeff. of Linear Exp.: 12.2×10^{-5} m/m/C (29–60 C, D696); Flamm.: HB (UL94)
Elec. Prop.: Dissip. Factor: 0.005 (23 C, 50% r.h., 10^6 Hz, D150); Dielec. Str.: 15.7 kV/mm (S/T, 3.2 mm, D149); Dielec. Const.: 3.5 (23 C, 50% r.h., 10^2 to 10^6 Hz, D150); Vol. Resist.: 5×10^{14} ohm-cm (23 C, D257); Arc Resist.: no tracking (3.1 mm, D495)
Ref.: Booklet E-32986

Delrin 500 F. Du Pont Co.
Chem. Descrip.: Acetal resin
Category/Applications: High productivity grade; see Delrin 100
Gen'l. Prop.: Sp. Gr.: 1.42 (D792); M.P.: 175 C (crystalline, D2133); Stability: See Delrin 100
Mech. Prop.: Tens. Str.: 66.9 MPa (23 C, D638); Tens. Elong.: 25% (23 C, D638); Flex. Mod.: 2760 MPa (23 C, D790); Compr. Str.: 31.0 MPa (23 C @ 1% def., D695); Shear Str.: 68.9 MPa (23 C, D-732); Impact Str. (Izod): 74.7 J/m notch (23 C, D256); Hardness (Rockwell): M 92 (D785)
Therm. Prop.: Distort. Temp.: 121 C (@ 1.8 MPa, D648); Coeff. of Linear Exp.: 12.2×10^{-5} m/m/C (29–60 C, D696); Flamm.: HB (UL94)
Elec. Prop.: Dissip. Factor: 0.005 (23 C, 50% r.h., 10^6 Hz, D150); Dielec. Str.: 19.7 kV/mm (S/T, 2.3 mm, D149); Dielec. Const.: 3.7 (23 C, 50% r.h., 10^2 to 10^6 Hz, D150); Vol. Resist.: 10^{15} ohm-cm (23 C, D257); Arc Resist.: 200 s, no tracking (3.1 mm, D495)
Ref.: Booklet E-32986

Delrin 900 F, 907 F. Du Pont Co.
Chem. Descrip.: Acetal resin
Category/Applications: High productivity grades; see Delrin 100

Gen'l. Prop.: Sp. Gr.: 1.42 (D792); M.P.: 175 C (crystalline, D2133); Stability: See Delrin 100
Mech. Prop.: Tens. Str.: 66.9 MPa (23 C, D638); Tens. Elong.: 30% (23 C, D638); Flex. Mod.: 2760 MPa (23 C, D790); Compr. Str.: 31.0 MPa (23 C @ 1% def., D695); Shear Str.: 66.5 MPa (23 C, D-732); Impact Str. (Izod): 69.4 J/m notch (23 C, D256); Hardness (Rockwell): M 92 (D785)
Therm. Prop.: Distort. Temp.: 121 C (@ 1.8 MPa, D648); Coeff. of Linear Exp.: 12.2 $\times 10^{-5}$ m/m/C (29–60 C, D696); Flamm.: HB (UL94)
Elec. Prop.: Dissip. Factor: 0.005 (23 C, 50% r.h., 10^6 Hz, D150); Dielec. Str.: 19.7 kV/mm (S/T, 2.3 mm, D149); Dielec. Const.: 3.7 (23 C, 50% r.h., 10^2 to 10^6 Hz, D150); Vol. Resist.: 10^{15} ohm-cm (23 C, D257); Arc Resist.: 200 s, no tracking (3.1 mm, D495)
Ref.: Booklet E-32986

Delta "Deck-Color." Delta Plastics Co.
Category/Applications: Cement, sand, and polymer coating for masonry type materials which contains built-in colors; usually used as a heavy-build decorative coating surface for swimming pool decks

D.E.R. 331. Dow Chem. Co.
Chem. Descrip.: Bisphenol A/epichlorohydrin epoxy resin
Form: Liquid
Mech. Prop. (after cure: curing agent D.E.H. 24): Tens. Str.: 11,600 psi (D-638-60); Tens. Elong. (ultimate): 4.70% (D638-60); Tens. Mod.: 461,000; Flex. Str.: 15,680 psi (D790-58T; Flex. Mod.: 425,000 (D790-58T); Compr. Str.: 16,460 psi (D695-54); Compr. Mod.: 270,000 (D695-54); Impact Str. (Izod): 0.50 ft lb/in. notched (D256-56); Hardness (Rockwell): M110 (D785-60)
Therm. Prop. (after cure, see above): Distort. Temp.: 111 C (D648-56)
Elec. Prop. (after cure, see above): Dielec. Const.: 4.03 (@ 60 cps, D150-59T); Vol. Resist.: > 1.2 $\times 10^{16}$ (D257-58); Surf. Resist.: > 7.85 $\times 10^{15}$ (D-257-58); Dissip. Factor: 0.14 (@ 60 cps, D150-59T)
Curing Charact. (curing agent D.E.H. 24): Initial Gel: 16 h @ 25 C; Post Cure: 3 h @ 100 C
Ref.: Form No. 296-002-79

D.E.R. 383. Dow Chem. Co.
Chem. Descrip.: Epoxy resin
Category/Applications: Liquid epoxy resin designed for reduced viscosity and extended pot life of catalyzed formulations; used for filament winding, potting and encapsulation, etc.; provides improved handling, higher filler loadings, better process control, and reduced resin/hardener waste
Form: Liquid; Color (Gardner): 2 max. (D1544)
Gen'l. Prop.: Density: 9.7 lb/gal (D-1963); Visc.: 9000–11,500 cps (D445)
Mech. Prop. (cured with 27 parts MDA/100 parts resin): Tens. Str.: 10,300 psi; Tens. Mod.: 360,000 psi; Flex. Str.: 18,300 psi; Flex. Mod.: 390,000 psi
Therm. Prop. (cured with 27 parts MDA/100 parts resin): Distort. Temp.: 163 C
Toxicity/Handling: Observe precautions as for other liquid bis-A epoxy resins; may cause severe burns, sensitization; harmful if swallowed; burning produces carbon monoxide
Ref.: Form No. 190-288-78

D.E.R. 642U. Dow Chem. Co.
Chem. Descrip.: Epoxy resin produced by reaction of epichlorohydrin and bisphenol-A

Category/Applications: Resin used for protective coatings and fnishes incl. maintenance coatings, industrial floor coatings, air-dry spar varnishes, wire enamels, collapsible tube coatings, automotive primers, appliance primers, baked drum and tank linings for storing corrosive chemicals, electrical insulating varnishes, electrical and structural prepregs, etc.; excellent toughness and adhesion; designed esp. for powder formulations

Form: Solid; Color (Gardner): 2 max.

Gen'l. Prop.: Sp. Gr.: 1.19; Density: 9.9 lb/gal (cast); Soften. Pt. (Durran mercury method): 87–94 C; Stability: Excellent resistance to corrosion under most extreme conditions, to solvents and chemicals (even hot caustic and hot acetone), and to shock

Toxicity/Handling: Very slightly irritating to eyes; low in acute oral toxicity; may cause skin rashes due to sensitization; organic material which may burn in appropriate conditions; dusts can be explosive

Ref.: Form No. 296-290-79

D.E.R. 652-A75. Dow Chem. Co.

Chem. Descrip.: Epoxy resin in acetone

Category/Applications: Solution epoxy resin for industrial and maintenance coatings

Form: Liquid; Color (Gardner): 3 max.

Composition: 75 ± 1% solids

Gen'l. Prop.: Sp. Gr.: 1.07; Density: 8.9 lb/gal; Visc.: 1500–4500 cps; Stability: See D.E.R. 642U

Toxicity/Handling: May cause irritation to skin and eyes, and sensitization responses; may present a hazard from inhalation; organic resin burns under appropriate conditions; flammable solvent

Ref.: Form No. 296-290-76

D.E.R. 661. Dow Chem. Co.

Chem. Descrip.: See D.E.R. 642U

Category/Applications: Resin used for protective coatings and finishes incl. maintenance coatings, industrial floor coatings, air-dry spar varnishes, wire enamels, collapsible tube coatings, automotive primers, appliance primers, baked drum and tank linings for storing corrosive chemicals, electrical insulating varnishes, electrical and structural prepregs, etc.; excellent toughness and adhesion

Form: Solid; Color (Gardner): 1 max.

Gen'l. Prop.: Sp. Gr.: 1.19; Density: 9.9 lb/gal (cast); Soften. Pt. (Durran mercury method): 70–80 C; Stability: See D.E.R. 642U

Toxicity/Handling: See D.E.R. 642U

Ref.: Form No. 296-290-79

D.E.R. 661-A80. Dow Chem. Co.

Chem. Descrip.: Epoxy resin in acetone

Category/Applications: See D.E.R. 652-A75

Form: Liquid; Color (Gardner): 2 max.

Composition: 80 ± 1% solids

Gen'l. Prop.: Sp. Gr.: 1.09; Density: 9.1 lb/gal; Visc.: 3500–8500 cps; Stability: See D.E.R. 642U

Toxicity/Handling: See D.E.R. 652-A75

Ref.: Form No. 296-290-76

D.E.R. 662. Dow Chem. Co.

Chem. Descrip.: See D.E.R. 642U

Category/Applications: See D.E.R. 642U

Form: Solid; Color (Gardner): 1 max.

Gen'l. Prop.: Sp. Gr.: 1.19; Density: 9.9 lb/gal (cast); Soften. Pt. (Durran mercury method): 80–90 C; Stability: See D.E.R. 642U

Toxicity/Handling: See D.E.R. 642U

Ref.: Form No. 296-290-79

D.E.R. 663U. Dow Chem. Co.
Chem. Descrip.: See D.E.R. 642U
Category/Applications: See D.E.R. 642U
Form: Solid; Color (Gardner): 2 max.
Gen'l. Prop.: Sp. Gr.: 1.19; Density: 9.9 lb/gal (cast); Soften. Pt. (Durran mercury method): 88–98 C; Stability: See D.E.R. 642U
Toxicity/Handling: See D.E.R. 642U
Ref.: Form No. 296-290-79

D.E.R. 664. Dow Chem. Co.
Chem. Descrip.: See D.E.R. 642U
Category/Applications: See D.E.R. 661
Form: Solid; Color (Gardner): 3 max.
Gen'l. Prop.: Sp. Gr.: 1.19; Density: 9.9 lb/gal (cast); Soften. Pt. (Durran mercury method): 95–105 C; Stability: See D.E.R. 642U
Toxicity/Handling: See D.E.R. 642U
Ref.: Form No. 296-290-76

D.E.R. 664U. Dow Chem. Co.
Chem. Descrip.: See D.E.R. 642U
Category/Applications: See D.E.R. 642U
Form: Solid; Color (Gardner): 2 max.
Gen'l. Prop.: Sp. Gr.: 1.19; Density: 9.9 lb/gal (cast); Soften. Pt. (Durran mercury method): 95–105 C; Stability: See D.E.R. 642U
Toxicity/Handling: See D.E.R. 642U
Ref.: Form No. 296-290-76

D.E.R. 667. Dow Chem. Co.
Chem. Descrip.: See D.E.R. 642U
Category/Applications: See D.E.R. 642U
Form: Solid; Color (Gardner): 3 max.
Gen'l. Prop.: Sp. Gr.: 1.19; Density: 9.9 lb/gal (cast); Soften. Pt. (Durran mercury method): 115–130 C; Stability: See D.E.R. 642U
Toxicity/Handling: See D.E.R. 642U
Ref.: Form No. 296-290-76

D.E.R. 671-EE75. Dow Chem. Co.
Chem. Descrip.: Epoxy resin in Dowanol EE
Category/Applications: Solution epoxy resin for industrial and maintenance coatings, esp. for applications where pigments are used; provides improved pigment dispersion, pigment suspension, and tinting strength; develops greater thixotropy which enhances brushing and nonsagging properties
Form: Liquid; Color (Gardner): 2 max.
Composition: 75 ± 1% solids
Gen'l. Prop.: Sp. Gr.: 1.14; Density: 9.5 lb/gal; Visc.: 4000–20,000 cps; Stability: See D.E.R. 642U
Toxicity/Handling: See D.E.R. 652-A75
Ref.: Form No. 296-290-76

D.E.R. 671-EEA75. Dow Chem. Co.
Chem. Descrip.: Epoxy resin in Dowanol EEA
Category/Applications: See D.E.R. 671-EE75
Form: Liquid; Color (Gardner): 2 max.
Composition: 75 ± 1% solids
Gen'l. Prop.: Sp. Gr.: 1.12; Density: 9.4 lb/gal; Visc.: 6000–30,000 cps; Stability: See D.E.R. 642U
Toxicity/Handling: See D.E.R. 652-A75
Ref.: Form No. 296-290-76

D.E.R. 671-EEK75. Dow Chem. Co.
Chem. Descrip.: Epoxy resin in Dowanol EE-70 MEK-30
Category/Applications: See D.E.R. 671-EE75
Form: Liquid; Color (Gardner): 2 max.
Composition: 75 ± 1% solids
Gen'l. Prop.: Sp. Gr.: 1.10; Density: 9.2 lb/gal; Visc.: 2000–10,000 cps; Stability: See D.E.R. 642U
Toxicity/Handling: See D.E.R. 652-A75
Ref.: Form No. 296-290-76

D.E.R. 671-MK75. Dow Chem. Co.
Chem. Descrip.: Epoxy resin in MIBK
Category/Applications: See D.E.R. 671-EE75
Form: Liquid; Color (Gardner): 2 max.
Composition: 75 ± 1% solids
Gen'l. Prop.: Sp. Gr.: 1.07; Density: 8.9 lb/gal; Visc.: 2200–10,000 cps; Stability: See D.E.R. 642U
Toxicity/Handling: See D.E.R. 652-A75
Ref.: Form No. 296-290-76

D.E.R. 671-T75. Dow Chem. Co.
Chem. Descrip.: Epoxy resin in toluene
Category/Applications: See D.E.R. 671-EE75
Form: Liquid; Color (Gardner): 2 max.
Composition: 75 ± 1% solids
Gen'l. Prop.: Sp. Gr.: 1.07; Density: 8.9 lb/gal; Visc.: 2200–10,000 cps; Stability: See D.E.R. 642U
Toxicity/Handling: See D.E.R. 652-A75
Ref.: Form No. 296-290-76

D.E.R. 671-X75. Dow Chem. Co.
Chem. Descrip.: Epoxy resin in xylene
Category/Applications: See D.E.R. 671-EE75
Form: Liquid; Color (Gardner): 2 max.
Composition: 75 ± 1% solids
Gen'l. Prop.: Sp. Gr.: 1.07; Density: 9.1 lb/gal; Visc.: 4500–10,000 cps; Stability: See D.E.R. 642U
Toxicity/Handling: See D.E.R. 652-A75
Ref.: Form No. 296-290-76

D.E.R. 671-XM75. Dow Chem. Co.
Chem. Descrip.: Epoxy resin in xylene-35, MIBK-65
Category/Applications: See D.E.R. 671-EE75
Form: Liquid; Color (Gardner): 2 max.
Composition: 75 ± 1% solids
Gen'l. Prop.: Sp. Gr.: 1.07; Density: 8.9 lb/gal; Visc.: 2500–9000 cps; Stability: See D.E.R. 642U
Toxicity/Handling: See D.E.R. 652-A75
Ref.: Form No. 296-290-76

D.E.R. 673MF. Dow Chem. Co.
Chem. Descrip.: See D.E.R. 642U
Category/Applications: See D.E.R. 642U
Form: Solid
Gen'l. Prop.: Sp. Gr.: 1.18; Soften. Pt. (Durran mercury method): 85–95 C; Stability: See D.E.R. 642U
Toxicity/Handling: See D.E.R. 642U
Ref.: Form No. 296-290-76

D.E.R. 684-EK40. Dow Chem. Co.
Chem. Descrip.: Epoxy resin in MEK
Category/Applications: See D.E.R. 652-A75
Form: Liquid; Color (Gardner): 3 max.
Composition: 40 ± 1% solids
Gen'l. Prop.: Sp. Gr.: 0.96; Density: 8.0 lb/gal; Visc. (Gardner-Holdt): U-Y; Stability: See D.E.R. 642U
Toxicity/Handling: See D.E.R. 652-A75
Ref.: Form No. 296-290-76

D.E.R. 732. Dow Chem. Co.
Chem. Descrip.: Epichlorohydrin-polyglycol reaction product
Category/Applications: Flexible epoxy resin for use as additives to base epoxy systems; provides increased flexibility in adhesives, construction and civil engineering applications (esp. aggregates, terrazzo, seamless floors, putties, and machine grouts), industrial maintenance coatings, and structural prepregs
Form: Liquid; Color (Gardner): 1 max. (D1544)
Gen'l. Prop.: Sol.: Insol. in water; Sp.

Gr.: 1.06; Density: 8.9 lb/gal (D1963); Visc.: 55–100 cps (D445); Flash Pt. (COC): 205 C (D92); Stability: Avoid amine compounds; incompatible with acid, base
Mech. Prop. (after cure: D.E.R. 732/ D.E.R. 331 (3:7)–curing agent D.E.H. 24): Tens. Str.: 5985 psi (D638-60); Tens. Elong. (ultimate): 5.53% (D638-60); Tens. Mod.: 354,000 psi; Flex. Str.: 10,825 psi (D790-58T); Flex. Mod.: 325,000 (D790-58T); Compr. Str.: 12,200 psi (D695-54); Compr. Mod.: 299,000 (D695-54); Impact Str. (Izod): 0.54 ft lb/in. notched (D256-56); Hardness (Rockwell): M87 (D785-60)
Therm. Prop. (after cure, see above): Distort. Temp.: 58 C (D648-56)
Elec. Prop. (after cure, see above): Dielec. Const.: 3.87 (@ 60 cps, D150-59T); Vol. Resist.: 1.97×10^{15} (D257-58); Surf. Resist.: 3.14×10^{15} (D257-58); Dissip. Factor: 0.014 (@ 60 cps, D-150-59T)
Curing Charact. (curing agent D.E.H. 24): Initial Gel: 16 h @ 25 C; Post Cure: 3 h @ 100 C
Toxicity/Handling: Low acute oral toxicity; may cause moderate eye irritation, minor to moderate skin irritation; vapors may be irritating
Ref.: Form No. 296-002-79

D.E.R. 736. Dow Chem. Co.
Chem. Descrip.: Polyglycol diepoxide epoxy resin
Category/Applications: See D.E.R. 732
Form: Liquid; Color (Gardner): 1 max. (D1544)
Gen'l. Prop.: Sol.: Very slight sol. in water; Sp. Gr.: 1.14; Density: 9.5 lb/gal (D1963); Visc.: 30–60 cps (D445); Flash Pt. (COC): 170 C (D92); Stability: Incompatible with bases
Mech. Prop. (after cure: D.E.R. 736/ D.E.R. 331 (3:6)–curing agent D.E.H. 24): Tens. Str.: 9225 psi (D638-60); Tens. Elong. (ultimate): 6% (D638-60); Tens. Mod.: 475,000; Flex. Str.: 14,400 psi (D790-58T); Flex. Mod.: 421,000 (D790-58T); Compr. Str.: 12,040 psi (D-695-54); Compr. Mod.: 351,000 (D695-54); Impact Str. (Izod): 0.61 ft lb/in. notched (D256-56); Hardness (Rockwell): M91 (D785-60)
Therm. Prop. (after cure, see above): Distort. Temp.: 76 C (D648-56)
Elec. Prop. (after cure, see above): Dielec. Const.: 3.95 (@ 60 cps, D150-59T); Vol. Resist.: 1.24×10^{15} (D257-58); Surf. Resist.: 1.02×10^{15} (D257-58); Dissip. Factor: 0.013 (@ 60 cps, D-150-59T)
Curing Charact. (curing agent D.E.H. 24): Initial Gel: 16 h @ 25 C; Post Cure: 3 h @ 100 C
Toxicity/Handling: Low single dose oral toxicity; may cause moderate eye irritation; prolonged or repeated exposure may cause moderate skin irritation; may be a sensitizer; vapors from heated material may be irritating
Ref.: Form No. 296-002-79

Derakane 411 Resin (45% styrene). Dow Chem. Co.
Chem. Descrip.: Vinyl ester resin
Category/Applications: Thermosetting resin with outstanding resistance to corrosion by chemicals (incl. acids and alkalis), resistance to impact, and good electrical and thermal insulation properties; used in fabricating a range of fiber-reinforced plastic structures and equipment by all conventional methods, incl. absorption towers, process vessels, storage tanks, piping, hood scrubbers, ducts, and exhaust stacks; Derakane 411 is used esp. for contact molding and filament winding
Form: Med.-visc. liquid
Composition: 45% styrene content

Gen'l. Prop.: Sp. Gr.: 1.04 (uncured), 1.12 (R.T. clear casting); Visc. (Brookfield): 500 cps; Flash Pt. (TOC): 93 F
Mech. Prop. (R.T. clear casting): Tens. Str.: 11,800 psi; Tens. Elong.: 5%; Tens. Mod.: 4.9 X 10^{-5} psi; Flex. Str.: 18,000 psi; Flex. Mod.: 4.5 X 10^{-5} psi; Compr. Str.: 16,600 psi; Compr. Mod.: 3.5 X 10^{-5} psi; Hardness (Barcol): 35
Therm. Prop. (R.T. clear casting): Distort. Temp.: 215 F
Toxicity/Handling: May cause significant eye and skin effects, and hypersensitivity; avoid inhalation; cured resin considered toxicologically inert; resin is flammable–take precautions against fire and explosion hazards
Ref.: Form No. 190-197-77

Derakane 411 Resin (50% styrene). Dow Chem. Co.
Chem. Descrip.: Vinyl ester resin
Category/Applications: See Derakane 411 Resin (45% styrene)
Form: Med.-visc. liquid
Compostion: 50% styrene content
Gen'l. Prop.: Sp. Gr.: 1.03; Visc. (Brookfield): 300 cps; Flash Pt. (TOC): 90 F
Toxicity/Handling: See Derakane 411 Resin (45% styrene)
Ref.: Form No. 190-197-77

Derakane 411-C Resin. Dow Chem. Co.
Chem. Descrip.: Vinyl ester resin
Category/Applications: See Derakane 411 Resin (45% styrene); Derakane 411-C is a lower visc. version of 411 and is used primarily for centrifugal casting and other applications requiring extremely fast wet-out
Form: Liquid
Gen'l. Prop.: Sp. Gr.: 1.023 (uncured), 1.12 (R.T. clear casting); Visc. (Brookfield): 125 cps; Flash Pt. (TOC): 90 F
Mech. Prop. (R.T. clear casting): Tens. Str.: 11,900 psi; Tens. Elong.: 6%; Tens. Mod.: 5.1 X 10^{-5} psi; Flex. Str.: 19,000 psi; Flex. Mod.: 4.7 X 10^{-5} psi; Compr. Str.: 16,700 psi; Compr. Mod.: 3.5 X 10^{-5} psi; Hardness (Barcol): 35
Therm. Prop. (R.T. clear casting): Distort. Temp.: 215 F
Toxicity/Handling: See Derakane 411 Resin (45% styrene)
Ref.: Form No. 190-197-77

Derakane 470 Resin (36% styrene). Dow Chem. Co.
Chem. Descrip.: Vinyl ester resin
Category/Applications: See Derakane 411 Resin (45% styrene); 470 Resin has exceptional oxidation and solvent resistance, and thermal stability, and is used for fabricating handling equipment that is used where manufacturers must concentrate and combine corrosive materials to meet EPA pollution-control requirements
Form: Liquid
Gen'l. Prop.: Sp. Gr.: 1.07 (uncured), 1.17 (R.T. clear casting); Visc. (Brookfield): 200 cps; Flash Pt. (TOC): 90 F
Mech. Prop. (R.T. clear casting): Tens. Str.: 11,000 psi; Tens. Elong: 3–4%; Tens. Mod.: 5.1 X 10^{-5} psi; Flex. Str.: 20,000 psi; Flex. Mod.: 5.5 X 10^{-5} psi; Compr. Str.: 18,500 psi; Compr. Mod.: 3.2 X 10^{-5} psi; Hardness (Barcol): 40
Therm. Prop. (R.T. clear casting): Distort. Temp.: 290–300 F
Toxicity/Handling: See Derakane 411 Resin (45% styrene)
Ref.: Form No. 190-197-77

Derakane 470 Resin (45% styrene). Dow Chem. Co.
Chem. Descrip.: Vinyl ester resin
Category/Applications: See Derakane 470 Resin (36% styrene)
Form: Liquid
Gen'l. Prop.: Sp. Gr.: 1.03 (uncured), 1.15 (R.T. clear casting); Visc. (Brook-

field): 75 cps; Flash Pt. (TOC): 90 F
Mech. Prop. (R.T. clear casting): Tens. Str.: 11,000 psi; Tens. Elong.: 3–4%; Tens. Mod.: 5.1 × 10^{-5} psi; Flex. Str.: 20,000 psi; Flex. Mod.: 5.5 × 10^{-5} psi; Compr. Str.: 18,500 psi; Compr. Mod.: 3.2 × 10^{-5} psi; Hardness (Barcol): 40
Therm. Prop. (R.T. clear casting): Distort. Temp.: 280 F
Toxicity/Handling: See Derakane 411 Resin (45% styrene)
Ref.: Form No. 190-197-77

Derakane 510-A Resin. Dow Chem. Co.
Chem. Descrip.: Vinyl ester resin
Category/Applications: See Derakane 411 Resin (45% styrene); 510-A Resin has additionally a degree of fire resistance
Form: Liquid
Composition: 40% styrene content
Gen'l. Prop.: Sp. Gr.: 1.22 (uncured), 1.32 (R.T. clear casting); Visc. (Brookfield): 250 cps; Flash Pt. (TOC): 93 F
Mech. Prop. (R.T. clear casting): Tens. Str.: 10,600 psi; Tens. Elong.: 4–5%; Flex. Str.: 18,000 psi; Flex. Mod.: 5.2 × 10^{-5} psi; Compr. Str.: 17,000 psi; Compr. Mod.: 3.3 × 10^{-5} psi; Hardness (Barcol): 40
Therm. Prop. (R.T. clear casting): Distort. Temp.: 230 F
Toxicity/Handling: See Derakane 411 Resin (45% styrene)
Ref.: Form No. 190-197-77

Desmocoll 110. Bayer AG
Chem. Descrip.: High m.w. hydroxyl polyurethane
Category/Applications: Elastomer for use in the manufacture of contact and heat-activated adhesives for the footwear, fancy goods, and packaging industries; Desmocoll adhesives adhere to materials which are normally difficult to bond; may be used with or without Desmodur polyisocyanates
Gen'l. Prop. (Sol'n. in MEK 15% by wt. Desmodur): Visc. (Brookfield): 0.2 ± 0.1 Pa·s (23 C, spindle 3, 60 rpm); Crystallization Rate: Low; Thermoplasticity: High
Ref.: Product pamphlet

Desmocoll 130. Bayer AG
Chem. Descrip.: High m.w. hydroxyl polyurethane
Category/Applications: See Desmocol 110
Gen'l. Prop. (Sol'n. in MEK 15% by wt. Desmodur): Visc. (Brookfield): 0.2 ± 0.1 Pa·s (23 C, spindle 3, 60 rpm); Crystallization Rate: Med.; Thermoplasticity: High
Ref.: Product pamphlet

Desmocoll 176. Bayer AG
Chem. Descrip.: High m.w. hydroxyl polyurethane
Category/Applications: See Desmocoll 110
Gen'l. Prop. (Sol'n. in MEK 15% by wt. Desmodur): Visc. (Brookfield): 0.6 ± 0.2 Pa·s (23 C, spindle 3, 60 rpm); Crystallization Rate: Med.; Thermoplasticity: High
Ref.: Product pamphlet

Desmocoll 400. Bayer AG
Chem. Descrip.: High m.w. hydroxyl polyurethane
Category/Applications: See Desmocoll 110; adhesives produced with this grade adhere to heavily plasticized PVC with bonds of outstanding stability
Form: Liquid
Gen'l. Prop. (Sol'n. in MEK 15% by wt. Desmodur): Visc. (Brookfield): 0.6 ± 0.2 Pa·s (23 C, spindle 3, 60 rpm); Crystallization Rate: High; Thermoplasticity: High
Ref.: Pamphlet KA 30536 e

Desmocoll 406. Bayer AG
Chem. Descrip.: High m.w. hydroxyl polyurethane
Category/Applications: See Desmocoll 400
Gen'l. Prop. (Sol'n. in MEK 15% by wt. Desmodur): Visc. (Brookfield): 0.6 ± 0.2 Pa·s (23 C, spindle 3, 60 rpm); Crystallization Rate: Low; Thermoplasticity: High
Ref.: Pamphlet KA 30536 e

Desmocoll 420. Bayer AG
Chem. Descrip.: High m.w. hydroxyl polyurethane
Category/Applications: See Desmocoll 400
Gen'l. Prop. (Sol'n. in MEK 15% by wt. Desmodur): Visc. (Brookfield): 1.2 ± 0.2 Pa·s (23 C, spindle 3, 60 rpm); Crystallization Rate: High; Thermoplasticity: Low
Ref.: Pamphlet KA 30536 e

Desmocoll 500. Bayer AG
Chem. Descrip.: High m.w. hydroxyl polyurethane
Category/Applications: See Desmocoll 400
Gen'l. Prop. (Sol'n. in MEK 15% by wt. Desmodur): Visc. (Brookfield): 0.6 ± 0.2 Pa·s (23 C, spindle 3, 60 rpm); Crystallization Rate: Very high; Thermoplasticity: High
Ref.: Pamphlet KA 30536 e

Desmocoll 510. Bayer AG
Chem. Descrip.: High m.w. hydroxyl polyurethane
Category/Applications: See Desmocoll 400
Gen'l. Prop. (Sol'n. in MEK 15% by wt. Desmodur): Visc. (Brookfield): 1.2 ± 0.2 Pa·s (23 C, spindle 3, 60 rpm); Crystallization Rate: Very high; Thermoplasticity: Med.
Ref.: Pamphlet KA 30536 e

Desmocoll 530. Bayer AG
Chem. Descrip.: High m.w. hydroxyl polyurethane
Category/Applications: See Desmocoll 400
Gen'l. Prop. (Sol'n. in MEK 15% by wt. Desmodur): Visc. (Brookfield): 1.2 ± 0.2 Pa·s (23 C, spindle 3, 60 rpm); Crystallization Rate: Very high; Thermoplasticity: Low
Ref.: Pamphlet KA 30536 e

Desmodur KN. Bayer AG
Chem. Descrip.: Isocyanato-urethane
Category/Applications: Moisture-setting adhesive for plastic film
Form: Liquid; Color: Yellowish
Composition: 70% sol'n. in ethyl acetate
Ref.: Pamphlet KA 30536 e

Desmodur L 75. Bayer AG
Chem. Descrip.: Toluene diisocyanate-polyol reaction product
Category/Applications: Synthetic material used in the production of low-solvent or solventless Baycoll adhesives
Form: Liquid; Color: Yellowish
Composition: 75% sol'n. in ethyl acetate
Ref.: Pamphlet KA 30536 e

Desmodur N. Bayer AG
Chem. Descrip.: Polyfunctional aliphatic isocycnate
Category/Applications: Two-part bonding agent used in the coating of synthetic fiber fabrics with PVC pastes; esp. used for coatings that must withstand light
Form: Liquid
Ref.: Pamphlet KL 42911 e

Desmodur R. Bayer AG
Chem. Descrip.: Triphenylmethane triisocyanate
Category/Applications: Crosslinking agent to improve the adhesion of rubber to other materials, e.g., fabrics or metals; adhesives crosslinked with Desmodur have outstanding strength, high thermal stability, and resistance to oils, greases, and solvents; also crosslinking agent for self-curing cements
Form: Liquid; Color: Green to violet
Composition: 20% sol'n. in methylene chloride
Gen'l. Prop.: Density: 1.32 g/cc
Toxicity/Handling: Avoid contact with atmospheric humidity while in use
Std. Pkgs.: 1-kg aluminum bottles in boxes, 60-kg metal cans
Ref.: Pamphlet KA 30536 e, KA 32125 e

Desmodur RC. Bayer AG
Chem. Descrip.: Aromatic polyisocyanate
Category/Applications: Crosslinking agent to improve the adhesion of rubber to other materials; adhesives crosslinked with Desmodur have outstanding strength, high thermal stability, and resistance to oils, greases, and solvents
Form: Liquid; Color: Pale yellowish
Composition: 30% sol'n. in ethyl acetate
Ref.: Pamphlet KA 30536 e

Desmodur RF. Bayer AG
Chem. Descrip.: Thiophosphoric acid-tris-(p-isocyanatophenylester)
Category/Applications: See Desmodur R
Form: Liquid; Color: Brownish yellow
Composition: 20% sol'n. in methylene chloride
Gen'l. Prop.: Density: 1.32 g/cc
Toxicity/Handling: Avoid contact with atmospheric humidity while in use
Std. Pkgs.: 1-kg aluminum bottles in boxes, 15-kg and 60-kg metal cans
Ref.: Pamphlet KA 30536 e, KA 32125 e

Desmodur RU. Bayer AG
Chem. Descrip.: Aromatic polyisocyanate
Category/Applications: See Desmodur RC
Form: Liquid; Color: Brownish
Composition: 20% sol'n. in methylene chloride
Ref.: Pamphlet KA 30536 e

Desmodur TT. Bayer AG
Chem. Descrip.: Aromatic diisocyanate
Category/Applications: Two-part bonding agent used in the coating of synthetic fiber fabrics with PVC pastes; exp. for coatings needing extremely good adhesion; highly reactive at R.T.
Form: Powder
Ref.: Pamphlet KL 42911 e

Desmodur VK. Bayer AG
Chem. Descrip.: Mixture of aromatic polyisocyanates
Category/Applications: See Desmodur L 75
Composition: 100% active ingred.
Ref.: Pamphlet KA 30536 e

Dion Cor-Res 6693FR. Koppers Co., Inc.
Chem. Descrip.: Polyester resin
Category/Applications: Flame retardant (class I filled with 3% Sb_2O_3, E84), corrosion resistant (premium and fume handling), promoted resin for use in transportation, construction, and corrosion control applications by sprayup/layup and filament winding processes; used for chemical process equipment (tanks, hoods, pipe, ducts, etc.), and pollution control equipment
Form: Low-visc. liquid

Gen'l. Prop.: Sp. Gr.: 1.16; Visc. (Brookfield): 500 cps
Curing Charact.: R.T. Gel Time: 20 min
Ref.: Bulletin OM-535-28

Dion Cor-Res 6693HV. Koppers Co., Inc.
Chem. Descrip.: Polyester resin
Category/Applications: Flame retardant (class I filled with 3% Sb_2O_3, E84), corrosion resistant resin for transportation, construction, electrical, and corrosion control applications by compression molding/SMC/BMC processes; high heat resistance; used for molded parts requiring fire retardancy and corrosion resistance; chemically thickenable
Form: High-visc.
Gen'l. Prop.: Sp. Gr.: 1.3; Visc. (Brookfield): 3500 cps
Curing Charact.: 180 F Gel Time: 12 min
Ref.: Bulletin OM-535-28

Dion Cor-Res 6694. Koppers Co., Inc.
Chem. Descrip.: Polyester resin
Category/Applications: Corrosion resistant, promoted resin for use in sprayup/layup and filament winding processes; used for chemical process equipment (tanks, hoods, pipe, etc.), chlorine cell head covers; high heat resistance
Form: Low-visc. liquid
Gen'l. Prop.: Sp. Gr.: 1.02; Visc. (Brookfield): 400 cps
Curing Charact.: R.T. Gel Time: 15 min
Ref.: Bulletin OM-535-28

Dion Cor-Res 6694HV. Koppers Co., Inc.
Chem. Descrip.: Polyester resin
Category/Applications: Corrosion resistant resin for use in construction and corrosion control applications by compression molding/SMC/BMC processes; used for molded parts requiring high temp. and corrosion resistant properties
Form: High-visc.
Gen'l. Prop.: Sp. Gr.: 1.06; Visc. (Brookfield): 8000 cps
Curing Charact.: 180 F Gel Time: 16 min
Ref.: Bulletin OM-535-28

Dion Cor-Res 7000. Koppers Co., Inc.
Chem. Descrip.: Polyester resin
Category/Applications: Corrosion resistant, promoted resin for sprayup/layup and filament winding processes, with high heat resistance; used for chemical process equipment (tanks, hoods, pipe, etc.)
Form: Low-visc. liquid
Gen'l. Prop.: Sp. Gr.: 1.02; Visc. (Brookfield): 400 cps
Curing Charact.: R.T. Gel Time: 25 min
Ref.: Bulletin OM-535-28

Dion FR 6114. Koppers Co., Inc.
Chem. Descrip.: Polyester resin
Category/Applications: Light stabilized, flame retardant (class I unfilled), promoted resin for use in construction applications, esp. corrugated and flat building panels
Form: Low visc. liquid
Gen'l. Prop.: Sp. Gr.: 1.30; Visc. (Brookfield): 600 cps; Stability: Excellent weathering
Curing Charact.: 180 F Gel Time: 2 min
Ref.: Bulletin OM-535-28

Dion FR 6124. Koppers Co., Inc.
Chem. Descrip.: Polyester resin
Category/Applications: Flame retardant (class I with antimony trioxide, class II unfilled), promoted resin for use in transportation and construction applications, esp. for reefer panels, corrugated, and flat building panels
Form: Low visc. liquid
Gen'l. Prop.: Sp. Gr.: 1.30; Visc. (Brook-

field): 600 cps
Curing Charact.: 180 F Gel Time: 3 min
Ref.: Bulletin OM-535-28

Dion FR 6125. Koppers Co., Inc.
Chem. Descrip.: Polyester resin
Category/Applications: Flame retardant (class I unfilled), light stabilized, promoted resin for use in construction applications, esp. for corrugated and flat building panels; excellent weatherability
Form: Low visc. liquid
Gen'l. Prop.: Sp. Gr.: 1.30; Visc. (Brookfield): 600 cps
Curing Charact.: 180 F Gel Time: 4–5 min
Ref.: Bulletin OM-535-28

Dion FR 6127. Koppers Co., Inc.
Chem. Descrip.: Polyester resin
Category/Applications: Flame retardant (class I with fillers, E84), light stabilized, promoted resin for use in construction applications, esp. corrugated and flat building panels
Form: Low visc. liquid
Gen'l. Prop.: Sp. Gr.: 1.17; Visc. (Brookfield): 100 cps
Curing Charact.: 180 F Gel Time: 4–5 min
Ref.: Bulletin OM-535-28

Dion FR 6308. Koppers Co., Inc.
Chem. Descrip.: Polyester resin
Category/Applications: Flame retardant (class II unfilled, class I with 3% Sb_2O_3), for use in corrosion applications in filament winding and compression molding/SMC/BMC processes; used for molded parts incl. industrial grating, computer housings, etc.; high physical strengths (meets MIL-R-7575), good fume handling corrosion resistance, and high heat resistance; ambient or elev. temp. gel and cure
Form: Med.-visc.
Gen'l. Prop.: Sp. Gr.: 1.24; Visc. (Brookfield): 2000 cps
Curing Charact.: 180 F Gel Time: 8 min
Ref.: Bulletin OM-535-28

Dion FR 6399. Koppers Co., Inc.
Chem. Descrip.: Polyester resin
Category/Applications: Flame retardant (class I with fillers, class II unfilled, meets MIL-R-21607) resin for use in transportation and electrical applications by compression molding/SMC/BMC processes, esp. for molded parts requiring good fire retardant properties, and electrical parts; also used as blending resin; meets MIL-R-7575
Form: High visc.
Gen'l. Prop.: Sp. Gr.: 1.37; Visc. (Brookfield): 3500 cps
Curing Charact.: 180 F Gel Time: 9 min
Ref.: Bulletin OM-535-28

Dion FR 6604T. Koppers Co., Inc.
Chem. Descrip.: Polyester resin
Category/Applications: Flame retardant (class I with filler (3% Sb_2O_3), class II unfilled, E84), thixotropic, promoted resin for use in transportation, marine, construction, and corrosion applications by sprayup/layup and filament winding processes; has fume handling corrosion resistance, high heat resistance, and meets MIL-R-7575; used for boats, computer housings, factory built homes, fume handling equipment, hood and duct applications
Form: Low visc. liquid
Gen'l. Prop.: Sp. Gr.: 1.2; Visc. (Brookfield): 500 cps
Curing Charact.: R.T. Gel Time: 20 min
Ref.: Bulletin OM-535-28

Dion FR 6655T. Koppers Co., Inc.
Chem. Descrip.: Polyester resin
Category/Applications: Flame retardant

(class I unfilled, E84), thixotropic resin for use in transportation, marine, and construction applications by sprayup/layup process
Form: Low-visc. liquid
Gen'l. Prop.: Sp. Gr.: 1.28; Visc. (Brookfield): 500 cps
Curing Charact.: R.T. Gel Time: 20 min
Ref.: Bulletin OM-535-28

Dion FR 6657. Koppers Co., Inc.
Chem. Descrip.: Polyester resin
Category/Applications: Flame retardant (class I with fillers, E84), promoted resin for transportation and construction applications, esp. sprayup and layup parts requiring low smoke; designed only for use with alumina trihydrate
Form: Low visc. liquid
Gen'l. Prop.: Sp. Gr.: 1.10; Visc. (Brookfield): 50 cps
Curing Charact.: R.T. Gel Time: 16 min (filled system)
Ref.: Bulletin OM-535-28

Dion FR 6692T. Koppers Co., Inc.
Chem. Descrip.: Polyester resin
Category/Applications: Flame retardant (class I with fillers, class II unfilled, meets MIL-R-21607), thixotropic, promoted resin for transportation, marine, and construction applications in sprayup/layup and filament winding processes; also meets MIL-R-7575; used for military boats, survival craft, aircraft, facia panels, pleasure boats
Form: Low-visc. liquid
Gen'l. Prop.: Sp. Gr.: 1.25; Visc. (Brookfield): 500 cps
Curing Charact.: R.T. Gel Time: 20 min
Ref.: Bulletin OM-535-28

Dion FR 8300. Koppers Co., Inc.
Chem. Descrip.: Polyester resin
Category/Applications: Flame retardant (class I filled with 5% $Sb_2)_3$, E84) resin for pultrusion use in transportation, marine, construction, and electrical applications; used for pultruded parts, grating, plates, strips, electrical sheet, rods, channels, etc.; smooth fiber-free surfaces at fast line speeds; low exotherm
Form: Low-visc. liquid
Gen'l. Prop.: Sp. Gr.: 1.26; Visc. (Brookfield): 500 cps
Curing Charact.: SPI Gel Time: 4 min
Ref.: Bulletin OM-535-28

Dion-Iso 6246. Koppers Co., Inc.
Chem. Descrip.: Isophthalic polyester resin
Category/Applications: Resin for use in transportation, marine, and corrosion control applications, esp. filament wound pipe, moderate corrosion service, parts requiring impact resistance; used in sprayup/layup and filament winding processes; meets MIL-R-7575
Form: Low-visc. liquid
Gen'l. Prop.: Sp. Gr.: 1.10; Visc. (Brookfield): 500 cps
Mech. Prop.: Tens. Elong.: 4.2%
Curing Charact.: R.T. Gel Time: 10 min
Ref.: Bulletin OM-535-28

Dion-Iso 6314. Koppers Co., Inc.
Chem. Descrip.: Isophthalic polyester resin
Category/Applications: Resin for use in transportation and corrosion applications in compression molding/SMC/BMC and pultrusion processes; used for molded parts requiring good corrosion resistance and physicals; chemically thickenable for SMC and BMC; designed for fast molding cycles; good color properties
Form: High-visc.
Gen'l. Prop.: Sp. Gr.: 1.13; Visc. (Brookfield): 7000 cps

Curing Charact.: 180 F Gel Time: 12 min; SPI Gel Time: 4 min
Ref.: Bulletin OM-535-28

Dion-Iso 6315HV. Koppers Co., Inc.
Chem. Descrip.: Isophthalic polyester resin
Category/Applications: Resin for use in transportation, construction, electrical, and corrosion applications, esp. for molded parts, premix, gel coats; meets MIL-R-7575C and CS-222-59 (sanitary gel coat requirement); high heat distortion; excellent water resistance and abrasion resistance; outstanding hot strength
Form: High visc.
Gen'l. Prop.: Sp. Gr.: 1.13; Visc. (Brookfield): 4500 cps
Curing Charact.: R.T. Gel Time: 10 min
Ref.: Bulletin OM-535-28

Dion-Iso 6631T. Koppers Co., Inc.
Chem. Descrip.: Isophthalic polyester resin
Category/Applications: Thixotropic, promoted resin for use in transportation, marine, construction, and corrosion applications in sprayup/layup and filament winding processes; used for parts requiring high physicals, good corrosion resistance, high heat resistance, excellent wet-out and handling; good tooling resin; meets MIL-R-7575
Form: Low-visc. liquid
Gen'l. Prop.: Sp. Gr.: 1.08; Visc. (Brookfield): 500 cps
Curing Charact.: R.T. Gel Time: 15 min
Ref.: Bulletin OM-535-28

Dion-Iso 8101. Koppers Co., Inc.
Chem. Descrip.: Isophthalic polyester resin
Category/Applications: Resin for use in transportation, marine, construction, and corrosion applications, for pultruded parts up to 3/8 in. diameter (e.g., rods, channels, sheet, etc.); fast line speeds, low reject rates; smooth, glossy surface; pigmentable; suitable for thermal only or radio frequency augmented cure
Form: Low-visc. liquid
Gen'l. Prop.: Sp. Gr.: 1.1; Visc. (Brookfield): 500 cps
Curing Charact.: SPI Gel Time: 4 min
Ref.: Bulletin OM-535-28

Dion-Iso 8200. Koppers Co., Inc.
Chem. Descrip.: Isophthalic polyester resin
Category/Applications: Resin for use in transportation, marine, construction, and corrosion applications, for pultruded parts with thick cross sections (⩾ ½ in.), e.g., rods, channels, sheet, bars, etc.; fast line speeds; lower reject rates; less internal cracking due to controlled cure and low exotherm; smooth, glossy surfaces; thermal only or radio frequency augmented cure
Form: Low-visc. liquid
Gen'l. Prop.: Sp. Gr.: 1.06; Visc. (Brookfield): 500 cps
Curing Charact.: SPI Gel Time: 3 min
Ref.: Bulletin OM-535-28

Dow ABS 213 Resin. Dow Chem. Co.
Chem. Descrip.: Acrylonitrile-butadiene-styrene resin
Category/Applications: Resin available in molding, extrusion, and pipe extrusion grades for use in toys, sports equipment, tools, small appliances, automotive parts, refrigeration, DWV, conduit, and other pipe, housings, luggage, etc.; 213 is noted for easy flow, matte finish, and FDA compliance
Form: Solid; Color: Light
Gen'l. Prop.: Sp. Gr.: 1.05 (D792);

Soften. Pt. (Vicat): 102 C (D1525); Stability: Excellent thermal stability; excellent chem. resistance to organic acids, aliphatic amines, bases, polyglycols, pharmaceuticals; good resistance to weak and strong inorganic acids, alcohols
Mech. Prop. (compression molded and injection molded, respectively): Tens. Str. (ultimate): 240 kgf/cm^2, 350 kgf/cm^2 (D638); Tens. Elong. (ultimate): 40%, 50% (D638); Tens. Mod.: 22,500 kgf/cm^2, 25,300 kgf/cm^2 (D638); Flex. Str.: 490 kgf/cm^2, 665 kgf/cm^2 (D790); Flex. Mod.: 23,200 kgf/cm^2, 26,700 kgf/cm^2 (D790); Impact Str. (Izod): 8.1 cm kgf/cm notched, 9.8 cm kgf/cm notched (@ 23 C, D256); Hardness (Rockwell): R 105 (D785)
Therm. Prop.: Distort. Temp.: 93 F (annealed, @ 18.6 kgf/cm^2, D648)
Elec. Prop.: Dielec. Const.: 2.83 (1000 Hz, D150); Dissip. Factor: 0.0050 (1000 Hz, D150)
Toxicity/Handling: Inhalation of dusts can be hazardous; resin will burn
Ref.: Form No. 301-431-79

Dow ABS 350 Resin. Dow Chem. Co.
Chem. Descrip.: Acrylonitrile-butadiene-styrene resin
Category/Applications: See Dow ABS 213; 350 Resin is noted for high gloss, refrigeration grade, FDA compliance
Form: Solid; Color: Light
Gen'l. Prop.: Sp. Gr.: 1.06 (D792); Soften. Pt. (Vicat): 102 C; Stability: See Dow ABS 213 Resin
Mech. Prop. (compression molded and injection molded, respectively): Tens. Str. (ultimate): 355 kgf/cm^2, 410 kgf/cm^2 (D638); Tens. Elong. (ultimate): 20%, 25% (D638); Tens. Mod.: 23,900 kgf/cm^2, 28,100 kgf/cm^2 (D638); Flex. Str.: 740 kgf/cm^2, 905 kgf/cm^2 (D-790); Flex. Mod.: 24,600 kgf/cm^2, 30,900 kgf/cm^2 (D790); Impact Str. (Izod): 10.4 cm kgf/cm notched, 17.5 cm kgf/cm notched (@ 23 C, D256); Hardness (Rockwell): R 112 (D785)
Therm. Prop.: Distort. Temp.: 96 F (annealed, @ 18.6 kgf/cm^2, D648)
Elec. Prop.: Dielec. Const.: 2.92 (1000 Hz, D150); Dissip. Factor: 0.0071 (1000 Hz, D150)
Toxicity/Handling: See Dow ABS 213 Resin
Ref.: Form No. 301-432-79

Dow ABS 500 Resin. Dow Chem. Co.
Chem. Descrip.: Acrylonitrile-butadiene-styrene resin
Category/Applications: See Dow ABS 213 Resin; 500 Resin is noted for impact resistance, medium gloss, and FDA compliance
Form: Solid; Color: Light
Gen'l. Prop.: Sp. Gr.: 1.05 (D792); Soften. Pt. (Vicat): 102 C (D1525); Stability: See Dow ABS 213 Resin
Mech. Prop. (compression molded and injection molded, respectively): Tens. Str. (ultimate): 330 kgf/cm^2, 365 kgf/cm^2 (D638); Tens. Elong. (ultimate): 5%, 10% (D638); Tens. Mod.: 21,800 kgf/cm^2, 23,900 kgf/cm^2; Flex. Str.: 575 kgf/cm^2, 760 kgf/cm^2 (D790); Flex. Mod.: 22,500 kgf/cm^2, 26,000 kgf/cm^2 (D790); Impact Str. (Izod): 14.7 cm kgf/cm notched, 21.8 cm kgf/cm notched (@ 23 C, D256); Hardness (Rockwell): R 104 (D785)
Therm. Prop.: Distort. Temp.: 92 F (annealed, @ 18.6 kgf/cm^2, D648)
Elec. Prop.: Dielec. Const.: 2.85 (1000 Hz, D150); Dissip. Factor: 0.0062 (1000 Hz, D150)
Toxicity/Handling: See Dow ABS 213 Resin
Ref.: Form No. 301-433-79

Dow ABS PG 912 Resin. Dow Chem. Co.
Chem. Descrip.: Acrylonitrile-butadiene-styrene resin
Category/Applications: Resin for pipe extrusion, esp. used for blending and sewer pipe
Form: Solid
Gen'l. Prop.: Soften. Pt. (Vicat): 220 F (D1525); Stability: See Dow ABS 213 Resin
Mech. Prop. (compression molded and injection molded, respectively): Tens. Str. (ultimate): 3550 psi, 4600 psi (73 F, D638); Tens. Elong.: 40%, 50% (73 F, D638); Tens. Mod.: 280,000 psi, 290,000 psi (73 F, D638); Impact Str. (Izod): 1.8 ft lb/in. notched, 2.5 ft lb/in. notched (73 F, D256)
Therm. Prop.: Distort. Temp.: 180 F compression molded, unannealed, 264 psi, 1/8 in. × 1/2 in. × 5 in. bar, D648)
Toxicity/Handling: Vapors can be hazardous; resin will burn; products of combustion can be toxic
Ref.: Form No. 301-435-79

Dow ABS PG 940 Resin. Dow Chem. Co.
Chem. Descrip.: Acrylonitrile-butadiene-styrene resin
Category/Applications: Resin for pipe extrusion, esp. for DWV pipe fittings and well casings
Form: Solid
Gen'l. Prop.: Soften. Pt (Vicat): 222 F (compression molded, D1525); Stability: See Dow ABS 213 Resin
Mech. Prop. (compression molded and injection molded, respectively): Tens. Str. (ultimate): 4600 psi, 5200 psi (73 F, D638); Tens. Elong.: 5%, 11% (73 F, D638); Tens. Mod.: 310,000 psi, 320,000 psi (73 F, D638); Impact Str. (Izod): 5.3 ft lb/in. notched, 5.9 ft lb/in. notched (73 F, D256)
Therm. Prop.: Distort. Temp.: 190 F (compression molded, unannealed, 264 psi, 1/8 in. × 1/2 in. × 5 in. bar, D648)
Toxicity/Handling: Vapors can be hazardous; resin will burn; products of combustion can be toxic
Ref.: Form No. 301-435-79

Drapex 3.2. Argus
Chem. Descrip.: Octyl epoxy stearate
Category/Applications: Epoxy plasticizer for vinyl compounds; compatible with primary plasticizers; used with Argus Ba/Cd stabilizers to synergistically improve heat and light stability; imparts excellent low temp. flexibility to vinyl compounds; due to their low visc. and solvating action, suitable for use in plastisols and organosols

Drapex 4.4. Argus
Chem. Descrip.: Octyl epoxy tallate
Category/Applications: See Drapex 3.2

Drapex 6.8, 10.4. Argus
Chem. Descrip.: Epoxidized soybean and linseed oils, respectively
Category/Applications: Plasticizers for use with vinyl compounds; compatible with primary plasticizers; used with Argus Ba/Cd stabilizers to synergistically improve heat and light stability; resistant to extraction and migration; FDA-approved for food packaging use

Dresinate 81. Hercules Inc.
Chem. Descrip.: Modified aq. sol'n. of sodium soap of pale processed or modified rosins
Category/Applications: Liquid soap for use as a component of the emulsifier and/or stabilizer systems for soluble oils, solvent cleaners, emulsions, disinfectants, oil-well drilling muds, and drawing and grinding compounds; as plasticizing agents; in specialty coatings
Form: Liquid; Color: Pale
Composition: 87% solids

Gen'l. Prop.: Sol.: Dilutible with water and aq. sol'ns. of alcohols and glycols; compatible in sol.-oil formulations and solvent cleaners with mineral oil, petroleum sulfonates, glycols, butanol, fatty acid soaps, and nonionic surfactants; Density: 1.04 kg/l (21 C); Visc.: 5.7 poise (60 C); Acid No.: 15
Std. Pkgs.: 211-kg net bung-type and open-head metal drums
Ref.: Data sheet 7246-2

Dresinate 90. Hercules Inc.
Chem. Descrip.: Modified aq. sol'n. of potassium soap of pale processed or modified rosins
Category/Applications: See Dresinate 81; also FDA-cleared for use in food-packaging and processing applications
Form: Liquid; Color: Pale
Composition: 90% solids
Gen'l. Prop.: Sol.: See Dresinate 81; Density: 1.08 kg/l (21 C); Visc.: 14.7 poise (60 C); Acid No.: 12
Std. Pkgs.: See Dresinate 81
Ref.: Data sheet 7246-2

Dresinate 91. Hercules Inc.
Chem. Descrip.: Modiifed aq. sol'n. of potassium soap of pale processed or modified rosins
Category/Applications: See Dresinate 81
Form: Liquid; Color: Pale
Composition: 88% solids
Gen'l. Prop.: Sol.: See Dresinate 81; Density: 1.06 kg/l (21 C); Visc.: 5.3 poise (60 C); Acid No.: 15
Std. Pkgs.: See Dresinate 81
Ref.: Data sheet 7246-2

Dresinate 92. Hercules Inc.
Chem. Descrip.: Modified aq. sol'n. of potassium soap of dark rosin
Category/Applications: See Dresinate 81
Form: Liquid; Color: Dark
Composition: 88% solids
Gen'l. Prop.: Sol.: See Dresinate 81; Density: 1.08 kg/l (21 C); Visc.: 9.2 poise (60 C); Acid No.: 15
Std. Pkgs.: See Dresinate 81
Ref.: Data sheet 7246-2

Dresinate 93. Hercules Inc.
Chem. Descrip.: Modified aq. sol'n. of potassium soap of dark rosin
Category/Applications: See Dresinate 90
Form: Liquid; Color: Dark
Composition: 87% solids
Gen'l. Prop.: Sol.: See Dresinate 81; Density: 1.14 kg/l (21 C); Visc.: 7.3 poise (60 C); Acid No.: 3
Std. Pkgs.: See Dresinate 81
Ref.: Data sheet 7246-2

Dresinate 94. Hercules Inc.
Chem. Descrip.: Modified aq. sol'n. of potassium soap of pale processed or modified rosins
Category/Applications: See Dresinate 90
Form: Liquid; Color: Pale
Composition: 87% solids
Gen'l. Prop.: Sol.: See Dresinate 81; Density: 1.12 kg/l (21 C); Visc.: 12.4 poise (60 C); Acid No.: 5
Std. Pkgs.: See Dresinate 81
Ref.: Data sheet 7246-2

Dresinate 95. Hercules Inc.
Chem. Descrip.: Modified aq. sol'n. of potassium soap of dark rosin
Category/Applications: See Dresinate 90
Form: Liquid; Color: Dark
Composition: 88% solids
Gen'l. Prop.: Sol.: See Dresinate 81; Density: 1.09 kg/l (21 C); Visc.: 4.5 poise (60 C); Acid No.: 15
Std. Pkgs.: See Dresinate 81
Ref.: Data sheet 7246-2

Dresinate 214, 515, 731. Hercules Inc.
Chem. Descrip.: Soaps of rosins
Category/Applications: Used mainly

with rubber latices and other polymer emulsions as sizing agents, and for pigment-wetting and foaming operations
Form: Paste; Color: Pale to dark
Composition: 70–80% solids
Ref.: Products catalogue

Dresinate DS-60W, TX-60W. Hercules Inc.
Chem. Descrip.: Sodium soaps of tall oil derivatives
Category/Applications: Used for flotation beneficiation of glass sand and phosphates and as emulsifier for asphalt
Form: Liquid
Composition: 60% solids aq. sol'ns.
Ref.: Products catalogue

Dresinate TX, X, XX. Hercules Inc.
Chem. Descrip.: Sodium soaps of rosins and tall oil
Category/Applications: Used as modifiers of heavy-duty metal and other industrial cleaners, dispersants for pigments and resinated colors, and as emulsifier for vegetable oils, pine oil, and mineral oils
Form: Free-flowing powder
Composition: 95% min. solids
Ref.: Products catalogue

Dresinol 40. Hercules Inc.
Chem. Descrip.: Aq.-based dispersion of a pale rosin
Category/Applications: Solvent-free modifier for polymeric film-formers in aq. systems, e.g., adhesives, coatings, sizings, etc.; improves stiffness or reinforcement of coating films, wetting of pigments and fillers, and penetration and adhesion to substrates; improves surface tack and bond strength between substrates in adhesives systems when blended with elastomer latices
Form: Liquid; Color: Cream
Composition: 40% solids
Gen'l. Prop.: Sol.: Compatible with water systems of animal glues, starches, dextrins, corn and soybean proteins, casein, phenol- and urea-formaldehyde resins, CMC, and with latices of PVAc, natural rubber, neoprene, SBR, and other synthetic elastomeric and thermoplastic high polymers; Visc.: 550 cps; Soften. Pt.: 87 C (resin solids); pH: 9.1
Toxicity/Handling: Can be damaged irreparably if frozen or subjected to prolonged heat; store > 7 C
Std. Pkgs.: 208-l (205-kg net) drums
Ref.: Data sheet 7252-3

Dresinol 42. Hercules Inc.
Chem. Descrip.: Aq.-based dispersion of a partially decarboxylated rosin
Category/Applications: See Dresinol 40
Form: Liquid; Color: Cream
Composition: 40% solids
Gen'l. Prop.: Sol.: See Dresinol 40; Visc.: 600 cps; Soften. Pt.: 46 C (rosin solids); pH: 9.4
Toxicity/Handling: See Dresinol 40
Std. Pkgs.: 208-l (205-kg net) drums
Ref.: Data sheet 7252-3

Dry Pexol 200. Hercules Inc.
Chem. Descrip.: Fortified pale rosin
Category/Applications: Fortified dry size for use with alum (aluminum sulfate) in paper and paperboard to produce a high level of resistance to water and aq. sol'ns.; suitable for use in high brightness grades of paper and paperboard; can be used at high levels without causing brightness, foam, or other operating problems on the paper machine
Form: Powder; Color: Light tan
Composition: 96% solids by weight
Gen'l. Prop.: Sol.: Readily sol. in water at R.T.; Density: 28 lb/ft^3; Stability: 1 yr shelf life
Toxicity/Handling: Hygroscopic product—bags must be kept intact and stored

in a cool, dry place
Std. Pkgs.: 50-lb net multiwall paper bags containing a vapor barrier
Ref.: Data sheet 7316-1

Dry Pexol 243. Hercules Inc.
Chem. Descrip.: Fortified dark rosin
Category/Applications: Fortified dark dry size for lower brightness paper grades (GE brightness < 60)
Form: Powder
Ref.: Data sheet 7308

Duolite A-7. Diamond Shamrock/Functional Polymers Div.
Chem. Descrip.: Polyfunctional amine cross-linked phenol-formaldehyde resin
Category/Applications: Low-cost, macroporous, weak-base anion-exchange resin with good operating capacity and high porosity; used as an acid adsorbent, an organic scavenger, in sugar deionization, decolorization, removal of high m.w. colorants (e.g., in wine processing); effectively operates at pH 0–6 and max. temp. of 40 C
Form: Granules (16–50 mesh); Color: Cream
Gen'l. Prop.: Sp. Gr.: 1.12 (free-base); Density: 320–400 g/l (BS&D); Stability: Highly resistant to physical attrition
Std. Pkgs.: Shipped in salt form
Ref.: Tech. sheet 130, data leaflet no. 2

Duolite A-30B. Diamond Shamrock/Functional Polymers Div.
Chem. Descrip.: Polyfunctional amine epoxy-amine resin
Category/Applications: Microporous, weak-base anion-exchange resin with high capacity for removal of strong acids and relatively high sp. gr.; used in multiple-bed deionization in water treatment and hydrometallurgical separations of transition metals; operates at pH 0–9 and max. temp. of 50 C
Form: Granules (20–50 mesh)
Gen'l. Prop.: Sp. Gr.: 1.15 (free-base)
Std. Pkgs.: Shipped in free-base form
Ref.: Data leaflet no. 2

Duolite A-101D. Diamond Shamrock/Functional Polymers Div.
Chem. Descrip.: Quat. ammonium (type 1) crosslinked polystyrene resin
Category/Applications: Microporous, strong-base anion exchanger for water deionization, incl. silica removal; used in multiple- and mixed-bed systems; high bead strength; good regeneration efficiency; effectively operates at pH 0–14 and max. temps. of 90 C (chloride form) and 60 C (hydroxide form)
Form: Moist beads (mesh range: 20–50); Color: Light, translucent
Gen'l. Prop.: Sp. Gr.: 1.08 (Cl); Density: 710–740 g/l net; Stability: Unaffected by acids, alkalis, and solvents
Std. Pkgs.: Chloride form shipped in 2 ft^3 polyethylene-lined bags, 5 ft^3 polyethylene-lined fiber drums
Ref.: Tech. sheet, data leaflet no. 2

Duolite A-102D. Diamond Shamrock/Functional Polymers Div.
Chem. Descrip.: Quat. ammonium (type 2) styrene-divinylbenzene copolymer
Category/Applications: Microporous strong-base anion-exchange resin with excellent regeneration efficiency; used for multiple-bed deionization, removal of weak acids from aq. solutions; used in chloride form to remove bicarbonate alkalinity; free of amine odors and FDA-approved for production of deionized water for use in food products; also used to deionize crude glycerol, to isolate and purify ingred. in fermentation broths, and to remove organic acids from nonaq. media; effectively operates at pH 0–14 and max. temps. of 85 C (salt form) and 40 C (OH form)

Form: Moist beads (mesh size: 16–50)
Gen'l. Prop.: Sp. Gr.: 1.11 (Cl); Density (bulk): 690 g/l BS&D
Std. Pkgs.: Shipped in chloride form
Ref.: Tech. sheet, data leaflet no. 2

Duolite A-104. Diamond Shamrock/ Functional Polymers Div.
Chem. Descrip.: Quat. ammonium styrene-divinylbenzene copolymer
Category/Applications: Microporous, strong-base anion exchange resin with outstanding operating capacity and excellent regeneration efficiency; used for multiple-bed deionization, dealkalization, and nitrate removal; operates effectively at pH 0–14 and max. temps. of 50 C (OH form) and 80 C (Cl form)
Form: Moist, opaque beads (20–50 mesh); Color: Ivory
Gen'l. Prop.: Sp. Gr.: 1.11 (Cl); Density: 720–750 g/l; Stability: Unaffected by dilute acids, alkalis, and most solvents; avoid severe oxidants
Std. Pkgs.: Shipped in chloride form in 5 ft^3 polyethylene-lined fiber drums
Ref.: Tech. sheet, data leaflet no. 2

Duolite A-109. Diamond Shamrock/ Functional Polymers Div.
Chem. Descrip.: Quat. ammonium (type 1) styrene-divinylbenzene copolymer
Category/Applications: Microporous, premium-grade, high capacity, strong-base anion exchange resin with exceptionally high crush resistance; used for condensate polishing, continuous counter-current deionization; operates at pH 0–14 and max. temps. of 60 C (OH) and 90 C (Cl)
Form: Moist, spherical beads (20–50 mesh); Color: Translucent, pale-yellow
Gen'l. Prop.: Sp. Gr.: 1.10 (Cl); Density: 705 g/l; Stability: Unaffected by acids, alkalis, and solvents; highly resistant to attrition
Std. Pkgs.: Shipped in chloride form in 5 ft^3 polyethylene-lined fiber drums
Ref.: Tech. sheet, data leaflet no. 2

Duolite A-113C. Diamond Shamrock/ Functional Polymers Div.
Chem. Descrip.: Quat. ammonium (type 1) crosslinked polystyrene resin
Category/Applications: Premium-grade, high capacity, strong-base anion-exchange resin with outstanding resistance to attrition; used in high flow rate condensate treatment or in continuous, moving bed processes; effective at pH 0–14 and max. temps. of 90 C (Cl) and 60 C (OH)
Form: Moist beads (16–40 mesh); Color: Light, translucent
Gen'l. Prop.: Density: 710–740 g/l; Stability: Unaffected by acids, alkalis, and solvents
Std. Pkgs.: Shipped in chloride form
Ref.: Tech. sheet

Duolite A-143. Diamond Shamrock/ Functional Polymers Div.
Chem. Descrip.: Quat. ammonium (type 1) polystyrene resin
Category/Applications: Microporous, strong-base resin specially formulated for adsorption of high m.w. colorants; used for color removal from sugar syrups and other food products; regeneration with salt brine; effective at pH 0–14 and max. temps. of 60 C (OH) and 90 C (Cl)
Form: Beads (20–50 mesh)
Gen'l. Prop.: Sp. Gr.: 1.09 (Cl)
Std. Pkgs.: Shipped in chloride form
Ref.: Data leaflet no. 2

Duolite A-161. Diamond Shamrock/ Functional Polymers Div.
Chem. Descrip.: Quat. ammonium (type 1) polystyrene resin
Category/Applications: Macroporous,

strong-base anion-exchange resin with outstanding resistance to osmotic shock; used for acid removal or anion interchange in nonaqueous media; catalysis; decolorizing/deionizing sugar syrups; effective at pH 0-14 and max. temps. of 60 C (OH) and 90 C (Cl)
Form: Beads (20–50 mesh)
Gen'l. Prop.: Sp. Gr.: 1.07 (Cl); Stability: Outstanding resistance to osmotic shock
Std. Pkgs.: Shipped in chloride form
Ref.: Data leaflet no. 2

Duolite A-162. Diamond Shamrock/ Functional Polymers Div.
Chem. Descrip.: Quat. ammonium (type 2) styrene-divinylbenzene copolymer (orthoporous)
Category/Applications: Macroporous strong-base anion-exchange resin with outstanding resistance to osmotic shock and good regeneration efficiency; used in deionization in a variety of chemical processes, esp. in solvents or solvent-water mixtures; may be used in food processing in conformance with FDA regulation 173.25; effective at pH 0–14 and max. temps. of 40 C (OH) and 80 C (Cl)
Form: Moist beads (20–50 mesh)
Gen'l. Prop.: Sp. Gr.: 1.09 (Cl); Density: 675 g/l; Stability: Unaffected by dilute acids, alkalis, and most solvents; avoid severe oxidants
Std. Pkgs.: Shipped in chloride form in 5 ft^3 polyethylene-lined fiber drums, and 1 and 2 ft^3 polyethylene-lined bags
Ref.: Tech. sheet, data leaflet no. 2

Duolite A-340. Diamond Shamrock/ Functional Polymers Div.
Chem. Descrip.: Polyfunctional amine epoxy-amine resin
Category/Applications: Microporous, weak-base, anion-exchange resin with exceptionally high capacity for strong acids; effective organics removal; used in multiple-bed deionization in water treatment and sugar purification;, and removal of some organic acids; effective at pH 0–9 and max. temp. of 50 C
Form: Moist beads (20–50 mesh); Color: Translucent
Gen'l. Prop.: Sp. Gr.: 1.15 (free-base); Density: 710 g/l (free-base); Stability: Unaffected by dilute acids and most solvents; avoid contact with conc. nitric acid and other oxidants
Std. Pkgs.: Shipped in free-base form in 2 ft^3 polyethylene-lined bags and 5 ft^3 polyethylene-lined fiber drums
Ref.: Tech. sheet, data leaflet no. 2

Duolite A-374. Diamond Shamrock/ Functional Polymers Div.
Chem. Descrip.: Tertiary amine polyacrylic resin
Category/Applications: Macroporous weak-base anion-exchange resin with high capacity and higher basicity than most weak-base resins; used in chemical processing, removal of organic acids, bisulfites, and in deionization; effective at pH 0–8 and max. temp. of 60 C
Form: Beads (20–50 mesh)
Gen'l. Prop.: Sp. Gr.: 1.08 (free-base)
Std. Pkgs.: Shipped in free-base form
Ref.: Data leaflet no. 2

Duolite A-377. Diamond Shamrock/ Functional Polymers Div.
Chem. Descrip.: Tertiary amine crosslinked polystyrene resin
Category/Applications: Macroporous, weak-base anion-exchange resin with high operating capacity and efficiency for the removal of mineral acids; used for removal of formic acid from formaldehyde, deionization of metal finishing wastes; effective at pH 0–9 and max. temp. of 100 C

Form: Beads (20–50 mesh); Color: Ivory
Gen'l. Prop.: Sp. Gr.: 1.04 (free-base); Density (bulk): ≈ 700 g/l; Stability: Good physical stability; high resistance to oxidation, fouling by organic matter, and breakage due to osmotic shock
Std. Pkgs.: Shipped in free-base form
Ref.: Tech. sheet 377, data leaflet no. 2

Duolite A-378. Diamond Shamrock/ Functional Polymers Div.
Chem. Descrip.: Tertiary amine styrene-divinylbenzene copolymer
Category/Applications: Macroporous, weak-base, anion-exchange resin with high porosity and good kinetics; used in water deionization and sugar processing; effective at pH 0–9 and max. temp. of 100 C
Form: Moist beads (20–50 mesh)
Gen'l. Prop.: Sp. Gr.: 1.05 (free-base); Density: 42 lb/ft^3; Stability: Unaffected by dilute acids, alkalis, and most solvents; avoid severe oxidants
Std. Pkgs.: Shipped in free-base form in 5 ft^3 polyethylene-lined fiber drums
Ref.: Tech. sheet, data leaflet no. 2

Duolite A-561. Diamond Shamrock/ Functional Polymers Div.
Chem. Descrip.: Polyfunctional amine crosslinked phenol-formaldehyde condensate
Category/Applications: Weak-base anion-exchange resin with rigid macroporous structure and very high resistance to attrition; used in removal of acids in chemical processing, deionization of starch hydrolysates; effective at pH 0–6 and max. temp. 60 C
Form: Granular (20–50 mesh)
Gen'l. Prop.: Sol.: Insol. in all common acids, alkalis, and solvents; Sp. Gr.: 1.12 (free-base); Density (bulk, fully hydrated): 640–720 g/l; Stability: Outstanding resistance to chemical and physical degradation
Std. Pkgs.: Shipped in partially dried, free-base form in 5 ft^3 polyethylene-lined fiber drums
Ref.: Tech. sheet, data leaflet no. 2

Duolite ARA-9366. Diamond Shamrock/ Functional Polymers Div.
Chem. Descrip.: Quat. ammonium (type 1) polystyrene resin
Category/Applications: High purity, analytical grade, strong-base ion-exchange resin with high capacity and basicity; used for deborating primary reactor coolant; effective at max. temp. of 60 C
Form: Beads (20–50 mesh)
Gen'l. Prop.: Density: 42 lb/ft^3
Std. Pkgs.: Shipped in OH form
Ref.: Data leaflet no. 3

Duolite ARA-9371. Diamond Shamrock/ Functional Polymers Div.
Chem. Descrip.: Polyfunctional amine epoxy-amine resin
Category/Applications: Analytical grade, high purity, weak-base ion-exchange resin for removal of Iodine-131 from borated reactor shims; effective at max. temp. of 80 C
Form: Beads (20–50 mesh)
Gen'l. Prop.: Density: 44 lb/ft^3
Std. Pkgs.: Shipped in free-base form
Ref.: Data leaflet no. 3

Duolite ARC-9323. Diamond Shamrock/ Functional Polymers Div.
Chem. Descrip.: Quat. ammonium (type 1) polystyrene resin
Category/Applications: Analytical grade, premium regenerated strong-base anion-exchange resin with high capacity and low organic Cl content; effective at max. temp. of 60 C
Form: Beads (20–50 mesh)
Gen'l. Prop.: Density: 43 lb/ft^3; Stabili-

ty: Exceptional resistance to breakage
Std. Pkgs.: Shipped in OH form
Ref.: Data leaflet no. 2

Duolite ARC-9351. Diamond Shamrock/ Functional Polymers Div.
Chem. Descrip.: Nuclear sulfonic polystyrene resin
Category/Applications: Highest purity analytical grade, strong-acid cation-exchange resin; full regeneration; for catalysis, epoxidation; max. temp. of 150 C
Form: Beads (20–50 mesh)
Gen'l. Prop.: Density: 50 lb/ft^3
Std. Pkgs.: Shipped in H form
Ref.: Data leaflet no. 3

Duolite ARC-9359. Diamond Shamrock/ Functional Polymers Div.
Chem. Descrip.: Methylene sulfonic phenolic resin
Category/Applications: Analytical grade, strong-acid, ion-exchange resin for selective removal of Cesium-137 from alkaline low-level radioactive wastes; effective at max. temp. of 60 C
Form: Beads (16–50 mesh)
Gen'l. Prop.: Density: 20 lb/ft^3
Std. Pkgs.: Shipped in H or Na form
Ref.: Data leaflet no. 3

Duolite ARM-9380. Diamond Shamrock/ Functional Polymers Div.
Category/Applications: Analytical grade, high purity, mixed bed ion-exchange resin (1:1 volume ratio of cation to anion resin); effective at max. temp. of 60 C
Form: Beads (20–50 mesh)
Std. Pkgs.: Shipped in H/OH form
Ref.: Data leaflet no. 3

Duolite ARM-9381. Diamond Shamrock/ Functional Polymers Div.
Category/Applications: Analytical grade, high purity, mixed-bed (1:1) ion-exchange resin for highest quality deionized water; effective at max. temp. of 60 C
Form: Beads (20–50 mesh)
Gen'l. Prop.: Density: 43 lb/ft^3
Std. Pkgs.: Shipped in H/OH form
Ref.: Data leaflet no. 3

Duolite ARM-9384. Diamond Shamrock/ Functional Polymers Div.
Category/Applications: Analytical grade, high purity, mixed bed (1:1) ion-exchange resin for clean-up and pH correction in ammoniated condensate treatment; effective at max. temp. of 60 C
Form: Beads (20–50 mesh)
Gen'l. Prop.: Density: 43 lb/ft^3
Std. Pkgs.: Shipped in NH_4/OH form
Ref.: Data leaflet no. 3

Duolite ARM-9386. Diamond Shamrock/ Functional Polymers Div.
Category/Applications: Analytical grade, high purity, mixed bed (1:1) ion-exchange resin for clean-up and pH correction in nuclear loops; effective at max. temp. of 60 C
Form: Beads (20–50 mesh)
Gen'l. Prop.: Density: 43 lb/ft^3
Std. Pkgs.: Shipped in Li/OH form
Ref.: Data leaflet no. 3

Duolite ARM-9390. Diamond Shamrock/ Functional Polymers Div.
Category/Applications: Analytical grade, high purity, mixed bed (1:1) ion-exchange resin for Cesium-137 and Iodine-131 removal from borated reactor shims at max. temp. of 80 C
Form: Beads (20–50 mesh)
Gen'l. Prop.: Density: 47 lb/ft^3
Std. Pkgs.: Shipped in H/free-base form
Ref.: Data leaflet no. 3

Duolite ARM-9395. Diamond Shamrock/ Functional Polymers Div.
Category/Applications: Analytical grade, high purity, mixed bed (1:1) ion-exchange resin with high capacity and low organic chloride content; used for applications requiring extremely high quality deionized water; effective at max. temp. of 60 C
Form: Beads (20–50 mesh)
Gen'l. Prop.: Density: 43 lb/ft^3
Std. Pkgs.: Shipped in H/OH form
Ref.: Data leaflet no. 3

Duolite C-3. Diamond Shamrock/Functional Polymers Div.
Chem. Descrip.: Methylene sulfonic acid phenolic resin
Category/Applications: Macroporous, strong-acid, cation-exchange resin with hydrophilic structure and excellent acid efficiency; used in deionization of starch hydrolysates, amino acid separations, selective removal of cesium; effective at pH 0–14 and max. temp. of 60 C
Form: Dry granules (16–50 mesh); Color: Black
Gen'l. Prop.: Sp. Gr.: 1.24 (Na); Density (dry): 17–23 lb/ft^3; Stability: Stable through wide range of pH and temps.; resistant to acids and bases; subject to oxidation by chlorine, esp. at high pH
Std. Pkgs.: Shipped in hydrogen form
Ref.: Tech. sheet, data leaflet no. 2

Duolite C-20. Diamond Shamrock/ Functional Polymers Div.
Chem. Descrip.: Aryl sulfonic cross-linked polystyrene resin
Category/Applications: Microporous, strong-acid cation-exchange resin with optimum particle size and resistance to attrition; used in household or industrial water softening and deionization, and a wide variety of chemical process operations; operates at pH 0–14 and max. temp. of 150 C
Form: Beads (20–50 mesh)
Gen'l. Prop.: Sp. Gr.: 1.32 (Na); Density (bulk): 800–832 g/l (BS&D); Stability: Unaffected by acids, alkalis, or solvents
Std. Pkgs.: Shipped in sodium form
Ref.: Tech. sheet, data leaflet no. 2

Duolite C-20C. Diamond Shamrock/ Functional Polymers Div.
Chem. Descrip.: Nuclear sulfonic polystyrene resin
Category/Applications: Microporous, strong-acid cation-exchange resin with high physical stability, excellent bead integrity, high capacity, and low pressure drop; used in condensate polishing, high flow rate mixed-bed deionization, continuous counter current water softening and deionization, and layered-bed cation exchange (Duo-Cat); effective at pH 0–14 and max. temp. of 150 C
Form: Beads (20–40 mesh)
Gen'l. Prop.: Sp. Gr.: 1.28 (H)
Std. Pkgs.: Shipped in hydrogen form
Ref.: Data leaflet no. 2

Duolite C-25D. Diamond Shamrock/ Functional Polymers Div.
Chem. Descrip.: Nuclear sulfonic polystyrene resin
Category/Applications: Macroporous, strong-acid, cation-exchange resin with good acid efficiency and resistance to osmotic shock; used in deionization of natural products, sugars, high sodium waters, isolation of amino acids, and stabilization of wines, effective at 0–14 and max. temp. of 120 C
Form: Beads (20–50 mesh)
Gen'l. Prop.: Sp. Gr.: 1.30 (Na)
Std. Pkgs.: Shipped in sodium form
Ref.: Data leaflet no. 2

Duolite C-26. Diamond Shamrock/Functional Polymers Div.
Chem. Descrip.: Nuclear sulfonic styrene-divinylbenzene copolymer
Category/Applications: Macroporous, strong-acid, cation-exchange resin with excellent chemical and physical stability, and resistance to oxidation; used in deionization of metal finishing baths, rinse waters, catalysis in nonaqueous systems
Form: Beads (20–50 mesh); Color: Gray opaque
Gen'l. Prop.: Sp. Gr.: 1.28 (Na); Density: 50 lb/ft^3; Stability: Excellent chemial and physical stability; resistant to oxidation
Std. Pkgs.: Shipped in sodium form
Ref.: Tech. sheet, data leaflet no. 2

Duolite C-225. Diamond Shamrock/Functional Polymers Div.
Chem. Descrip.: Arylsulfonate styrene-divinylbenzene copolymer
Category/Applications: Premium grade cation-exchange resin of high capacity, excellent bead integrity, and good resistance to mechanical attrition and chemical attack; used for condensate polishing, in deionization and chemical processing; effective at pH 0–14 and max. temps. of 120 C (H^+) and 160 C (Na^+)
Form: Moist beads (16–40 mesh); Color: Transparent pale yellow
Gen'l. Prop.: Density: 840 g/l (Na^+); Stability: Unaffected by acids, alkalis, or solvents; prolonged contact with vigorous oxidizing agents may result in gradual swelling
Std. Pkgs.: Shipped in sodium form in 5 ft^3 polyethylene-lined fiber drums
Ref.: Tech. sheet

Duolite C-225X10. Diamond Shamrock/Functional Polymers Div.
Chem. Descrip.: Aryl sulfonate crosslinked polystyrene resin
Category/Applications: Premium grade, high capacity, cation-exchange resin with optimum particle size for high flow rate and excellent resistance to mechanical attrition and chemical attack; used in condensate polishing and "Duo-Cat" (layered cation) applications; effective at pH 0–14 and max. temp. of 120 C (H^+ cycle)
Form: Bead (16–40 mesh); Color: Translucent
Gen'l. Prop.: Density: 800 g/l (H form); Stability: Unaffected by acids, alkalis, or solvents; gradual swelling by strong oxidizing agents
Std. Pkgs.: Shipped in hydrogen form
Ref.: Tech. sheet

Duolite C-433. Diamond Shamrock/Functional Polymers Div.
Chem. Descrip.: Carboxylic crosslinked polyacrylic acid resin
Category/Applications: Microporous, weak-acid, cation-exchange resin with very high capacity and exceptional physical strength; used in water deionization, softening, dealkalization, and selective heavy metal removal; effective at pH 5–14 and max. temp. of 120 C
Form: Clear beads (20–50 mesh); Color: Amber
Gen'l. Prop.: Sp. Gr.: 1.18 (H); Density (bulk): 760 g/l
Std. Pkgs.: Shipped in hydrogen form in 5 ft^3 polyethylene-lined fiber drums
Ref.: Tech. sheet, data leaflet no. 2

Duolite C-464. Diamond Shamrock/Functional Polymers Div.
Chem. Descrip.: Carboxylic crosslinked polyacrylic acid resin
Category/Applications: Macroporous, weak-acid, cation-exchange resin with excellent resistance to physical degradation and high capacity; used in ammonia removal, dealkalization, selective heavy

metal removal; effective at pH 5–14 and max. temp. of 120 C
Form: Beads (20–50 mesh); Color: Opaque beige
Gen'l. Prop.: Sp. Gr.: 1.13 (H); Density: 750–800 g/l; Stability: Outstanding resistance to osmotic attrition
Std. Pkgs.: Shipped in hydrogen form
Ref.: Tech. sheet 464, data leaflet no. 2

Duolite CS-100. Diamond Shamrock/ Functional Polymers Div.
Chem. Descrip.: Carboxylic and phenolic phenol-formaldehyde condensate
Category/Applications: Rigid macroporous, weak-acid, cation-exchange resin with unusually high selectivity for cesium and multivalent cations which makes it esp. effective for the treatment of alkaline low level radioactive waste solutions; also used for the adsorption of organic compounds; effective at pH 6–14 and max. temp. of 80 C
Form: Porous granules (16–50 mesh)
Gen'l. Prop.: Density (bulk): 600–700 g/l (moist, BS&D); Stability: Excellent resistance to most acids, good resistance to alkalis; structure is resistant to attrition
Std. Pkgs.: Shipped in hydrogen form
Ref.: Tech. sheet 100

Duolite CS-346. Diamond Shamrock/ Functional Polymers Div.
Chem. Descrip.: Amidoxime (some hydroxamic acid groups) crosslinked macroporous polymer
Category/Applications: Macroporous chelating resin used in removal of unwanted ions (e.g., copper, iron) from water or other polar solvents, recovery of trace materials (e.g., noble metals) from metal finishing wastes, isolation of valuable metals from ore extracts at low pH, recovery of uranium and other radioisotopes from radioactive solutions; effective at pH 1–11 and max. temp. of 50 C
Form: Beads (–30 to 70 mesh); Color: Opaque, beige
Gen'l. Prop.: Sol.: Insol. in all common organic and aq. solvent systems; Sp. Gr.: 1.15; Stability: Unaffected by moderately strong acids, alkalis, or solvents
Ref.: Tech. sheet

Duolite ES-131. Diamond Shamrock/ Functional Polymers Div.
Chem. Descrip.: Quat. ammonium (type 1) polystyrene resin
Category/Applications: Microporous strong-base, anion-exchange resin with optimized particle-size distribution and composition for selected applications; used in hydrometallurgical separations, esp. uranium; effective at pH 0–14 and max. temps. of 60 C (OH) and 90 C (Cl)
Form: Beads (20–30 mesh)
Gen'l. Prop.: Sp. Gr.: 1.10 (Cl)
Std. Pkgs.: Shipped in chloride form
Ref.: Data leaflet no. 2

Duolite ES-132. Diamond Shamrock/ Functional Polymers Div.
Chem. Descrip.: Quat. ammonium crosslinked acrylic (orthoporous) resin
Category/Applications: Strong-base anion-exchange resin with superior resistance to organic fouling; used in water deionization, sugar decolorization; effective at pH 0–14 and max. temps. of 35 C (OH) and 70 C (Cl)
Form: Moist beads (16–50 mesh)
Gen'l. Prop.: Sp. Gr.: 1.09 (OH); Density: 0.7 kg/l; Stability: Resistant to osmotic shock and physical attrition
Std. Pkgs.: Shipped in chloride form
Ref.: Tech. sheet

Duolite GPC-9301. Diamond Shamrock/Functional Polymers Div.
Chem. Descrip.: Nuclear sulfonic polystyrene resin
Category/Applications: General purpose, strong-acid, high purity, cation-exchange resin with high capacity; effective at max. temp. of 150 C
Form: Beads (20–50 mesh)
Gen'l. Prop.: Density: 50 lb/ft^3
Std. Pkgs.: Shipped in H form
Ref.: Data leaflet no. 3

Duolite GPA-9316. Diamond Shamrock/Functional Polymers Div.
Chem. Descrip.: Quat. ammonium (type 1) polystyrene resin
Category/Applications: General purpose, strong-base, high purity, anion-exchange resin with high porosity, highest basicity, silica removal; effective at max. temp. of 60 C
Form: Beads (20–50 mesh)
Gen'l. Prop.: Density: 42 lb/ft^3
Std. Pkgs.: Shipped in OH form
Ref.: Data leaflet no. 3

Duolite GPA-9319. Diamond Shamrock/Functional Polymers Div.
Chem. Descrip.: Quat. ammonium polystyrene resin
Category/Applications: General purpose, high capacity, high purity, anion-exchange regenerated type 2 resin for removal of radioactive materials from waste streams; effective at max. temp. of 60 C
Form: Beads (20–50 mesh)
Gen'l. Prop.: Density: 43 lb/ft^3
Std. Pkgs.: Shipped in OH form
Ref.: Data leaflet no. 3

Duolite GPA-9320. Diamond Shamrock/Functional Polymers Div.
Chem. Descrip.: Polyfunctional amine epoxy-amine resin
Category/Applications: General purpose, weak-base, high purity ion-exchange resin with high capacity, full regeneration; FMA removal; cartridge; effective at max. temp. of 80 C
Form: Beads (20–50 mesh)
Gen'l. Prop.: Density: 44 lb/ft^3
Std. Pkgs.: Shipped in free-base form
Ref.: Data leaflet no. 3

Duolite GPM-9331. Diamond Shamrock/Functional Polymers Div.
Category/Applications: General purpose, high purity, mixed-bed resin [1:1 (chemically equivalent)–GPC-9301/ GPA-9316] used for water deionization at max. temp. of 60 C
Form: Beads (20–50 mesh)
Gen'l. Prop.: Density: 43 lb/ft^3
Std. Pkgs.: Shipped in H/OH form
Ref.: Data leaflet no. 3

Duolite S-37. Diamond Shamrock/Functional Polymers Div.
Chem. Descrip.: Amine crosslinked phenol-formaldehyde resin
Category/Applications: Porous, very weak-base, adsorbent resin; reversible adsorption of aromatic acids; used in removal of organics, colorants from waters, sugar solutions, waste products; effective at pH 0–8 and max. temp. of 40 C
Form: Granules (20–50 mesh); Color: Ecru
Gen'l. Prop.: Sp. Gr.: 1.12 (free-base); Density (bulk): 40 lb/ft^3 (fully hydrated)
Std. Pkgs.: Shipped in salt form
Ref.: Data leaflet

Duolite S-761. Diamond Shamrock/Functional Polymers Div.
Chem. Descrip.: Hydroxyl crosslinked phenol-formaldehyde resin
Category/Applications: Macroporous,

very weak-acid adsorbent resin with good resistance to attrition; regeneration with alkali; used in removal of proteins and high m.w. colorants from sugar solutions, wines, fermentation broths, etc.; effective at pH 0–8 and max. temp. of 40 C
Form: Moist granules (20–50 mesh)
Gen'l. Prop.: Sp. Gr.: 1.11; Density: 477 g/l; Stability: Highly resistant to attrition; stable up to 80 C in neutral or acid, nonoxidizing media
Std. Pkgs.: 5 ft^3 polyethylene-lined fiber drums
Ref.: Tech. sheet, data leaflet no. 2

Duowet EHS, EHS Conc.. GAF Corp.
Chem. Descrip.: Sulfonated aliphatic polyester (anionic)
Category/Applications: Wetting agent used in padding, long-liquor dyeing, sizing, desizing, raw stock processing, package and beam dyeing; may be used with acid, direct, and disperse dyes, as well as printing pastes and lubricants
Form: Liquid
Ref.: Catalogue 2303-036

Durel 400. Hooker Chem. & Plastics Corp.
Chem. Descrip.: Polyarylate virgin resin
Category/Applications: Engineering thermoplastic with good weatherability, flame resistance, high impact strength, good flexural recovery, and good processability; used for glazing and lighting applications in the solar energy and transportation industries, UL-listed applications in the electrical, electronic, business-machine, and appliance industries, etc.
Form: Pellets; Color: Transparent
Gen'l. Prop.: Sp. Gr.: 1.2 (23 C, D792); Stability: Good chemical, heat (to 350 F), and flame resistance; good UV stability; Ref. Index: 1.6
Mech. Prop.: Tens. Str.: 68.9 Mpa (D-638); Tens. Elong.: 8% (@ yield, 5.1 mm/min, D638); Flex. Str.: 103.4 MPa (D790); Flex. Mod.: 2206 MPa (D790); Impact Str. (Izod): 294 J/m notched (23 C, D256)
Therm. Prop.: Distort. Temp.: 160 C (1/8 in. bar, 1.8 MPa, D648); Coeff. of Linear Exp.: 6.3×10^{-5} cm/cm/C (D-696); Flamm.: V-O (1.6 mm bar, UL-94)
Elec. Prop.: Dielec. Str.: 16.5 kV/mm (short time, D149); Dielec. Const.: 3.08 (60 Hz, D149); Vol. Resist.: 2×10^{16} ohm-cm (D257); Arc Resist.: 124 s (D-495); Dissip. Factor: 0.001 (60 Hz, D-150)
Ref.: Data sheets

Durel P400. Hooker Chem. & Plastics Corp.
Chem. Descrip.: Polyarylate resin (developmental grade)
Category/Applications: See Durel 400
Form: Pellets; Color: Opaque
Gen'l. Prop.: Sp. Gr.: 1.22; Stability: See Durel 400
Mech. Prop.: Tens. Str.: 10,100 psi; Tens. Elong.: 7.5%; Tens. Mod.: 354,000 psi; Impact Str. (Izod): 3.6 ft lb/in. notched (73 F)
Therm. Prop.: Distort. Temp.: 154 C (@ 254 psi); Flamm. V-O (1/16 in., UL94)
Elec. Prop.: Dielec. Str.: 613 V/mil; Dielec. Const.: 3.06 (60 Hz); Vol. Resist.: 3×10^{16} ohm-cm; Arc Resist.: 126 s; Dissip. Factor: 0.017 (10^6 Hz)
Ref.: Data sheets

Durel P410. Hooker Chem. & Plastics Corp.
Chem. Descrip.: Polyarylate resin (developmental grade)
Category/Applications: See Durel 400
Form: Pellets; Color: Opaque

Gen'l. Prop.: Sp. Gr.: 1.29; Stability: See Durel 400
Mech. Prop.: Tens. Str.: 13,100 psi; Tens. Elong.: 3.9%; Tens. Mod.: 626,000 psi; Impact Str. (Izod): 1.5 ft lb/in. notched (73 F)
Therm. Prop.: Distort. Temp.: 171 C (@ 264 psi), Flamm.: V-O (1/16 in., UL94)
Elec. Prop.: Dielec. Str.: 383 V/mil; Dielec. Const.: 3.18 (60 Hz); Vol. Resist.: 5×10^{16} ohm-cm; Arc Resist.: 98 s; Dissip. Factor: 0.016 (10^6 Hz)
Ref.: Data sheets

Durel P430. Hooker Chem. & Plastics Corp.
Chem. Descrip.: Polyarylate resin (developmental grade)
Category/Applications: See Durel 400
Form: Pellets; Color: Opaque
Gen'l. Prop.: Sp. Gr.: 1.43; Stability: See Durel 400
Mech. Prop.: Tens. Str.: 19,700 psi; Tens. Elong.: 2.2%; Tens. Mod.: 1.5×10^6 psi, Impact Str. (Izod): 2.0 ft lb/in. notched (73 F)
Therm. Prop.: Distort. Temp.: 175 C (@ 264 psi); Flamm.: V-O (1/16 in., UL94)
Elec. Prop.: Dielec. Str.: 409 V/mil; Dielec. Const.: 3.46 (60 Hz); Vol. Resist.: 3×10^{16} ohm-cm; Arc Resist.: 124 s; Dissip. Factor: 0.015 (10^6 Hz)
Ref.: Data sheets

Durethan A 30 S. Bayer AG
Chem. Descrip.: Nylon 6 and 66
Category/Applications: Engineering plastic for electrical engineering, mechanical and precision engineering, automotive engineering, household appliances and domestic ware, building construction and furniture manufacture, and packaging applications; A 30 S is a general-purpose grade for injection moldings with fast cycles and easy release
Form: Cylindrical granules; Color: Natural or opaque
Gen'l. Prop.: Density: 1.14 g/cc (DIN-53479, ISO/R 1183); M.P. (Kofler): 255 C; Stability: Good thermal stability and resistance to high dynamic loads, abrasion, and wear; resistant to hydrocarbons, oils and greases, esters, ketones, ethers, and to organic and inorganic bases up to med. concs.; largely resistant to many chlorinated or halogenated hydrocarbons; generally stable under outdoor conditions
Mech. Prop. (as molded; @ 23 C, 50% r.h.): Tens. Str.: 90 MPa (@ break, DIN 53455, ISO/R 527); Tens. Elong.: 40% (@ break, DIN 53455, ISO/R 527); Tens. Mod.: 3.4 GPa (DIN 53457); Flex. Mod.: 130 MPa (DIN 53452); Compr. Str.: 100 MPa (DIN 53454, ISO/R 604); Impact Str.: 4 kJ/m^2 notched (23 C, DIN 53453, ISO/R 179); Hardness (Rockwell): L 110 (D785)
Therm. Prop.: Conduct.: 0.23 W/Km (VDE 0304/1); Distort. Temp.: 100–110 C (@ 1.81 MPa, DIN 53461, ISO/R 75); Coeff. of Linear Exp.: 60–80 10^{-6} K^{-1} (VDE 0304/1); Sp. Heat: 1.7 kJ/kg K (20 C); Flamm.: V-2 (UL94)
Elec. Prop.: Dissip. Factor: 0.02 (dry, 1 MHz, VDE 0303/4, DIN 53483); Dielec. Str.: 100 kV/mm (as molded, 50 Hz, 0.5 kV/s, VDE 0303/2, DIN 53481); Dielec. Const.: 3.6 (dry, 1 MHz, VDE 0303/4, DIN 53483); Vol. Resist.: 10^{15} ohm-cm (as molded, VDE 0303/3, DIN 53482); Surf. Resist.: 10^{13} ohm (as molded, VDE 0303/3, DIN 53482)
Ref.: Product pamphlet KL 40360

Durethan AKV 30. Bayer AG
Chem. Descrip.: Nylon 6 and 66 with 30% glass fiber
Category/Applications: See Durethan A

30 S; AKV 30 is a general-purpose grade for injection moldings, with excellent stiffness and hardness
Form: Cylindrical granules; Color: Natural or opaque
Gen'l. Prop.: Density: 1.35 g/cc (DIN 53479, ISO/R 1183); M.P. (Kofler): 255 C; Stability: See Durethan A 30 S
Mech. Prop. (as molded; @ 23 C, 50% r.h.): Tens. Str.: 180 MPa (@ break, DIN 53455, ISO/R 527); Tens. Elong.: 4% (@ break, DIN 53455, ISO/R 527); Tens. Mod.: 9.3 GPa (DIN 53457); Flex. Str.: 270 MPa (DIN 53452, ISO 178); Compr. Str.: 190 MPa (DIN 53454, ISO/R 604); Impact Str.: 10 kJ/m^2 notched (23 C, DIN 53453, ISO/R 179); Hardness (Rockwell): L 109 (D785)
Therm. Prop.: Conduct.: 0.20 W/Km (VDE 0304/1); Distort. Temp.: ≈ 250 C (@ 1.81 MPa, DIN 53461, ISO/R 75); Coeff. of Linear Exp.: 20–30 10^{-5} K^{-1} (VDE 0304/1); Sp. Heat: 1.5 kJ/kg K (20 C); Flamm.: HB (UL94)
Elec. Prop.: Dissip. Factor: 0.02 (dry, 1 MHz, VDE 0303/4, DIN 53483); Dielec. Str.: 70 kV/mm (as molded, 50 Hz, 0.5 kV/s, VDE 0303/2, DIN 53481); Dielec. Const.: 3.8 (dry, @ 1 MHz, VDE 0303/4, DIN 53483); Vol. Resist.: 10^{15} ohm-cm (as molded, VDE 0303/3, DIN 53482); Surf. Resist.: 10^{14} ohm (as molded, VDE 0303/3, DIN 53482)
Ref.: Product pamphlet KL 40360

Durethan AKV 30 H. Bayer AG
Chem. Descrip.: Nylon 6 and 66 with 30% glass fiber and heat stabilizer
Category/Applications: See Durethan A 30 S; AKV 30 H is suitable for injection moldings exposed to elevated temp.; excellent stiffness and hardness
Form: Cylindrical granules; Color: Natural or opaque
Gen'l. Prop.: Density: 1.36–1.60 g/cc
Ref.: Product pamphet KL 40360

Durethan B 25 T. Bayer AG
Chem. Descrip.: Nylon 6 and 66
Category/Applications: See Durethan A 30 S; B 25 T is suitable for thin-section injection moldings, and has excellent transparency and easy release
Form: Cylindrical granules; Color: Transparent
Gen'l. Prop.: Density: 1.14 g/cc
Ref.: Product pamphelt KL 40360

Durethan B 30 P. Bayer AG
Chem. Descrip.: Nylon 6 and 66 with plasticizer
Category/Applications: See Durethan A 30 S; B 30 P is used for flexible injection moldings, and has good flexibility and easy release
Form: Cylindrical granules, Color: Natural or opaque
Gen'l. Prop.: Density: 1.14 g/cc
Ref.: Product pamphlet KL 40360

Durethan B 30 S, B 31 SK. Bayer AG
Chem. Descrip.: Nylon 6 and 66
Category/Applications: See Durethan A 30 S; B 30 S and B 31 SK are used for injection moldings with fast cycles; B 31 SK has higher impact strength
Form: Cylindrical granules; Color: Natural or opaque
Gen'l. Prop.: Density: 1.14 g/cc (DIN 53479, ISO/R 1183); M.P. (Kofler): 217–221 C; Stability: See Durethan A 30 S
Mech. Prop. (as molded; @ 23 C, 50% r.h.): Tens. Str.: 60 MPa (@ break, DIN 53455, ISO/R 527); Tens. Elong.: 50% (@ break, DIN 53455, ISO/R 527); Tens.

Mod.: 3.2 GPa (DIN 53457); Flex. Mod.: 120 MPa (DIN 53452); Compr. Str.: 90 MPa (DIN 53454, ISO/R 604); Impact Str. (Izod): 50 J/m (D256, ISO/R 180); Hardness (Rockwell): R 120 (D785)
Therm. Prop.: Conduct.: 0.25 W/Km (VDE 0304/1); Distort. Temp.: 75–80 C (@ 1.81 MPa, DIN 53461, ISO/R 75); Coeff. of Linear Exp.: 80–100 10^{-6} K^{-1} (VDE 0304/1); Sp. Heat: 1.6 kJ/kg K (20 C); Flamm.: V-2 (UL94)
Elec. Prop.: Dissip. Factor: 0.007 (dry, 50 Hz, VDE 0303/4, DIN 53483); Dielec. Str.: > 80 kV/mm (as molded, 50 Hz, 0.5 kV/s, VDE 0303/2, DIN 53481); Dielec. Const.: 3.8 (dry, 50 Hz, VDE 0303/4, DIN 53483); Vol. Resist.: 10^{15} ohm-cm (as molded, VDE 0303/3, DIN 53482); Surf. Resist.: 5 X 10^{12} ohm (as molded, VDE 0303/3, DIN 53482)
Ref.: Product pamphlet KL 40360

Durethan B 31 F. Bayer AG
Chem. Descrip.: Nylon 6 and 66
Category/Applications: See Durethan A 30 S; B 31 F is an extrusion grade used for films
Form: Cylindrical granules; Color: Natural or opaque
Gen'l. Prop.: Density: 1.14 g/cc
Ref.: Product pamphlet KL 40360

Durethan B 40 E. Bayer AG
Chem. Descrip.: Nylon 6 and 66
Category/Applications: See Durethan A 30 S; B 40 E is used for thick-section injection moldings, semifinished ware; also for extrusion
Form: Cylindrical granules; Color: Natural or opaque
Gen'l. Prop.: Density: 1.14 g/cc
Ref.: Product pamphlet KL 40360

Durethan B 40 F. Bayer AG
Chem. Descrip.: Nylon 6 and 66
Category/Applications: See Durethan A 30 S; B 40 F is an extrusion grade for use in flat and blown film
Form: Cylindrical granules; Color: Natural or opaque
Gen'l. Prop.: Density: 1.14 g/cc
Ref.: Product pamphlet KL 40360

Durethan B 40 SK. Bayer AG
Chem. Descrip.: Nylon 6 and 66
Category/Applications: See Durethan A 30 S; B 40 SK has very high impact strength, and is used for heavy duty injection moldings
Form: Cylindrical granules; Color: Natural or opaque
Gen'l. Prop.: See Durethan B 30 S, B 31 SK
Mech. Prop. (as molded; @ 23 C, 50% r.h.): Tens. Str.: 60 MPa (@ break, DIN 53455, ISO/R 527); Tens. Elong.: 70% (@ break, DIN 53455, ISO/R 527); Tens. Mod.: 3.0 GPa (DIN 53457); Flex. Mod.: 110 MPa (DIN 53452); Compr. Str.: 85 MPa (DIN 53454, ISO/R 604); Impact Str.: 4 kJ/m^2 notched (23 C, DIN 53453, ISO/R 179); Hardness (Rockwell): R 120 (D785)
Therm. & Elec. Prop.: See Durethan B 30 S, B 31 SK
Ref.: Product pamphlet KL 40360

Durethan B 50 E. Bayer AG
Chem. Descrip.: Nylon 6 and 66
Category/Applications: See Durethan A 30 S; B 50 E is used for small blown containers
Form: Cylindrical granules; Color: Natural or opaque
Gen'l. Prop.: See Durethan B 30 S, B 31 SK
Mech. Prop. (as molded; @ 23 C, 50%

r.h.): Tens. Str.: 50 MPa (@ break, DIN 53455, ISO/R 527); Tens. Elong.: 90% (@ break, DIN 53455, ISO/R 527); Tens. Mod.: 2.8 GPa (DIN 53457); Flex. Mod.: 100 MPa (DIN 53452); Compr. Str.: 80 MPa (DIN 53454, ISO/R 604); Impact Str.: 6 kJ/m^2 notched (23 C, DIN 53453, ISO/R 179); Hardness (Rockwell): R 115 (D785)
Therm. & Elec. Prop.: See Durethan B 30 S, B 31 SK
Ref.: Product pamphlet KL 40360

Durethan BC 30. Bayer AG
Chem. Descrip.: Nylon 6 and 66 with olefin polymer
Category/Applications: See Durethan A 3 S; BC 30 has high notched impact strenth (even freshly molded parts), and fast cyles; for use in injection moldings with good impact strength
Form: Cylindrical granules; Color: Natural or opaque
Gen'l. Prop.: Density: 1.14 g/cc
Ref.: Product pamphlet KL 40360

Durethan BC 40. Bayer AG
Chem. Descrip.: Nylon 6 and 66 with olefin polymer
Category/Applications: See Durethan A 30 S; BC 40 has very high notched impact strength (even at low temps.); used for heavy duty injection moldings, profiles, and blown containers
Form: Cylindrical granules; Color: Natural or opaque
Gen'l. Prop.: Density: 1.14 g/cc
Ref.: Product pamphlet KL 40360

Durethan BG 30 X. Bayer AG
Chem. Descrip.: Nylon 6 and 66 filled with glass fiber/glass beads
Category/Applications: See Durethan A 30 S; BG 30 X is a special grade noted for little warping
Form: Cylindrical granules; Color: Natural or opaque
Ref.: Product pamphlet KL 40360

Durethan BKV 30. Bayer AG
Chem. Descrip.: Nylon 6 and 66 with 30% glass fiber
Category/Applications: See Durethan A 30 S; BKV 30 is a general-purpose injection molding grade with excellent stiffness and hardness
Form: Cylindrical granules; Color: Natural or opaque
Gen'l. Prop.: Density: 1.36 g/cc; M.P. (Kofler): 217–221 C; Stability: See Durethan A 30 S
Therm. Prop.: Conduct.: 0.33 W/Km (VDE 0304/1); Distort. Temp.: ≈ 200 C (@ 1.81 MPa, DIN 53461, ISO/R 75); Coeff. of Linear Exp.: ≈ 25 10^{-6} K^{-1} (VDE 0304/1); Sp. Heat: 1.3 (20 C); Flamm.: HB (UL94)
Elec. Prop.: Dissip. Factor: 0.006 (dry, 50 Hz, VDE 0303/4, DIN 53483); Dielec. Str.: > 80 kV/mm (as molded, 50 Hz, 0.5 kV/s, VDE 0303/2, DIN 53481); Dielec. Const.: 4.1 (dry, 50 Hz, VDE 0303/4, DIN 53483); Vol. Resist.: 2 × 10^{15} ohm-cm (as molded, VDE 0303/3, DIN 53482); Surf. Resist.: 10^{14} ohm (as molded, VDE 0303/3, DIN 53482)
Ref.: Product pamphlet KL 40360

Durethan BKV 30 H. Bayer AG
Chem. Descrip.: Nylon 6 and 66 with 30% glass fiber and heat stabilizer
Category/Applications: See Durethan A 30 S; BKV 30 H is used for injection moldings exposed to elevated temps.; has excellent stiffness and hardness
Form: Cylindrical granules; Color: Natural or opaque
Gen'l. Prop.: Density: 1.36 g/cc (DIN 53479, ISO/R 1183); M.P. (Kofler):

217–221 C; Stability: See Durethan A 30 S
Mech. Prop. (as molded; @ 23 C, 50% r.h.): Tens. Str.: 160 MPa (@ break, DIN 53455, ISO/R 527); Tens. Elong.: 2.5% (@ break, DIN 53455, ISO/R 527); Tens. Mod.: 9.0 GPa (DIN 53457); Flex. Str.: 220 MPa (DIN 53452, ISO 178); Compr. Str.: 180 MPa (DIN 53454, ISO/R 604); Impact Str. (Izod): 100 J/m (D256, ISO/R 180); Hardness (Rockwell): R 120 (D785)
Therm. Prop.: Conduct.: 0.33 W/Km (VDE 0304/1); Distort. Temp.: ≈ 200 C (@ 1.81 MPa, DIN 53461, ISO/R 75); Coeff. of Linear Exp.: ≈ 25 10^{-6} K^{-1} (VDE 0304/1); Sp. Heat: 1.3 kJ/kg K (20 C); Flamm.: HB (UL94)
Elec. Prop.: Dissip. Factor: 0.007 (dry, 50 Hz, VDE 0303/4, DIN 53483); Dielec. Str.: > 80 kV/mm (as molded, 50 Hz, 0.5 kV/s, VDE 0303/2, DIN 53481); Dielec. Const.: 4.2 (dry, 50 Hz, VDE 0303/4, DIN 53483); Vol. Resist.: 10^{15} ohm-cm (as molded, VDE 0303/3, DIN 53482); Surf. Resist.: 10^{13} ohm (as molded, VDE 0303/3, DIN 53482)
Ref.: Product pamphlet KL 40360

Durethan BKV 30 N, BKV 30N1. Bayer AG
Chem. Descrip.: Nylon 6 and 66 with 30% glass fiber and flame retardant
Category/Applications: See Durethan A 30 S; BKV 30 N has excellent stiffness and hardness; used in injection moldings with flame retardant properties
Form: Cylindrical granules; Color: Natural or opaque
Gen'l. Prop.: Density: 1.36–1.60 g/cc; Stability: See Durethan A 30 S
Ref.: Product pamphlet KL 40360

Durethan BKV 35. Bayer AG
Chem. Descrip.: Nylon 6 and 66 with 35% glass fiber
Category/Applications: See Durethan A 30 S; BKV 35 has excellent stiffness and hardness, and is used in injection moldings
Form: Cylindrical granules; Color: Natural or opaque
Gen'l. Prop.: Density: 1.36–1.60 g/cc; Stability: See Durethan A 30 S
Ref.: Product pamphlet KL 40360

Durethan BKV 50 H. Bayer AG
Chem. Descrip.: Nylon 6 and 66 with 50% glass fiber, heat stabilizer
Category/Applications: See Durethan A 30 S; BKV 50 H has very high stiffness and hardness; used in injection moldings
Form: Cylindrical granules; Color: Natural or opaque
Gen'l. Prop.: Density: 1.58 g/cc (DIN 53479, ISO/R 1183); M.P. (Kofler): 217–221 C; Stability: See Durethan A 30 S
Mech. Prop. (as molded; @ 23 C, 50% r.h.): Tens. Str.: 180 MPa (@ break, DIN 53455, ISO/R 527); Tens. Elong.: 2.1% (@ break, DIN 53455, ISO/R 527); Tens. Mod.: 15 GPa (DIN 53457); Flex. Str.: 280 MPa (DIN 53452, ISO 178); Compr. Str.: 180 MPa (DIN 53454, ISO/R 604); Impact Str. (Izod): 120 J/m (D256, ISO/R 180); Hardness (Rockwell): R 120 (D785)
Therm. Prop.: Conduct.: 0.38 W/Km (VDE 0304/1); Distort. Temp.: ≈ 200 C (@ 1.81 MPa, DIN 53461, ISO/R 75); Coeff. of Linear Exp.: ≈ 16 10^{-6} K^{-1} (VDE 0304/1); Sp. Heat: 1.1 kJ/kg K (20 C)
Elec. Prop.: Dissip. Factor: 0.009 (dry, 50 Hz, VDE 0303/4, DIN 53483); Dielec. Str.: > 80 kV/mm (as molded, 50 Hz, 0.5 kV/s, VDE 0303/2, DIN 53481); Dielec. Const.: 4.6 (dry, 50 Hz, VDE 0303/4, DIN 53483); Vol. Resist.: 5×10^{14} ohm-cm (as molded, VDE 0303/3, DIN 53482); Surf. Resist.: 10^{14}

ohm (as molded, VDE 0303/3, DIN 53482)
Ref.: Product pamphlet KL 40360

Durethan C 38 F. Bayer AG
Chem. Descrip.: Nylon 6 and 66 with copolyamide
Category/Applications: See Durethan A 30 S; C 38 F is an extrusion grade with excellent transparency; used in film
Form: Cylindrical granules; Color: Transparent
Ref.: Product pamphlet KL 40360

Durez 111. Durez Div./Hooker Chem. & Plastics Corp.
Chem. Descrip.: Phenolic resin
Category/Applications: General purpose thermoset material for compression, transfer, and injection molding applications
Form: Granular; Color: Black, brown
Gen'l. Prop. (compression molded @ 340 F): Sp. Gr.: 1.38 (D792); Density: 0.60 g/cc (D1895)
Mech. Prop. (compression molded @ 340 F): Tens. Str.: 48 MPa (D638); Tens. Mod.: 8.3 GPa (D638); Flex. Str.: 69 MPa (D790); Compr. Str.: 207 MPa (D695); Impact Str. (Izod): 15 J/m (D-256)
Therm. Prop. (compression molded @ 340 F): Distort. Temp.: 182 C (@ 264 psi, D648); Flamm.: V-O (0.24 in., UL94)
Elec. Prop. (compression molded @ 340 F): Dielec. Str.: 14.8 MV/m (short time, D149); Dielec. Const.: 5.9 (60 Hz, D150); Vol. Resist.: 1×10^{12} ohm-cm (D257); Dissip. Factor: 0.12 (60 Hz, D150)
Ref.: Bulletin 328 MP5M379

Durez 118. Durez Div./Hooker Chem. & Plastics Corp.
Chem. Descrip.: Phenolic resin
Category/Applications: See Durez 111
Form: Granular; Color: Black, brown
Gen'l. Prop. (compression molded @ 340 F): Sp. Gr.: 1.40 (D792); Density: 0.58 g/cc (D1895)
Mech. Prop. (compression molded @ 340 F): Tens. Str.: 55 MPa (D638); Tens. Mod.: 9.7 GPa (D638); Flex. Str.: 76 MPa (D790); Compr. Str.: 207 MPa (D695); Impact Str. (Izod): 18 J/m (D-256)
Therm. Prop. (compression molded @ 340 F): Distort. Temp.: 171 C (@ 264 psi, D648); Flamm.: V-O (0.240 in., UL94)
Elec. Prop. (compression molded @ 340 F): Dielec. Str.: 15.7 MV/m (short time, D149); Dielec. Const.: 5.9 (60 Hz, D-150); Vol. Resist.: 1×10^{13} ohm-cm (D257); Dissip. Factor: 0.10 (60 Hz, D-150)
Ref.: Bulletin 328 MP5M379

Durez 123. Durez Div./Hooker Chem. & Plastics Corp.
Chem. Descrip.: Phenolic resin
Category/Applications: Impact-grade resin for molded materials requiring high shock resistance; used in compression, transfer, and injection molding processes
Form: Granular; Color: Black, brown
Gen'l. Prop. (compression molded @ 340 F): Sp. Gr.: 1.40 (D792); Density: 0.50 g/cc (D1895)
Mech. Prop. (compression molded @ 340 F): Tens. Str.: 48 MPa (D638); Tens. Mod.: 9.0 GPa (D638); Flex. Str.: 76 MPa (D790); Compr. Str.: 207 MPa (D695); Impact Str. (Izod): 20 J/m (D-256)

Therm. Prop. (compression molded @ 340 F): Distort. Temp.: 177 C (@ 264 psi, D648); Flamm.: V-O (0.240 in., UL94)
Elec. Prop. (compression molded @ 340 F): Dielec. Str.: 13.8 MV/m (short time, D149); Dielec. Const.: 6.2 (60 Hz, D-150); Vol. Resist.: 1 $\times$ 10^{12} ohm-cm (D257); Dissip. Factor: 0.10 (60 Hz, D-150)
Ref.: Bulletin 328 MP5M379

Durez 145. Durez Div./Hooker Chem. & Plastics Corp.
Chem. Descrip.: Phenolic resin
Category/Applications: Electrical grade resin with good dielectric strength and insulation resistance for use in automotive and communications products; used in compression, transfer, and injection molding processes
Form: Granular; Color: Black, brown
Gen'l. Prop. (compression molded @ 340 F): Sp. Gr.: 1.68 (D792); Density: 0.70 g/cc (D1895)
Mech. Prop. (compression molded @ 340 F): Tens. Str.: 38 MPa (D638); Tens. Mod.: 10.3 GPa (D638); Flex. Str.: 62 MPa (D790); Compr. Str.: 179 MPa (D695); Impact Str. (Izod): 19 J/m (D-256)
Therm. Prop. (compression molded @ 340 F): Distort. Temp.: 191 C (@ 264 psi, D648); Flamm.: V-O (0.020 in., UL94)
Elec. Prop. (compression molded @ 340 F): Dielec. Str.: 14.8 MV/m (short time, D149); Dielec. Const.: 6.0 (60 Hz, D-150); Vol. Resist.: 1 $\times$ 10^{12} ohm-cm (D257); Arc Resist.: 184 s (D495); Dissip. Factor: 0.08 (60 Hz, D150)
Ref.: Bulletin 328 MP5M379

Durez 152. Durez Div./Hooker Chem. & Plastics Corp.
Chem. Descrip.: Phenolic resin
Category/Applications: Heat-resistant, electrical asbestos-free resin for molded products requiring heat resistance and dimensional stability; used in compression, transfer, and injection molding processes
Form: Granular; Color: Black, brown
Gen'l. Prop. (compression molded @ 340 F): Sp. Gr.: 1.48 (D792); Density: 0.68 g/cc (D1895)
Mech. Prop. (compression molded @ 340 F): Tens. Str.: 55 MPa (D638); Tens. Mod.: 9.7 GPa (D638); Flex. Str.: 79 MPa (D790); Compr. Str.: 207 MPa (D695); Impact Str. (Izod): 15 J/m (D-256)
Therm. Prop. (compression molded @ 340 F): Distort. Temp.: 191 C (@ 264 psi, D648); Flamm.: V-O (0.120 in., UL94), V-1 (0.040 in., UL94)
Elec. Prop. (compression molded @ 340 F): Dielec. Str.: 16.7 MV/m (short time, D149); Dielec. Const.: 6.0 (60 Hz, D-150); Vol. Resist.: 1 $\times$ 10^{12} ohm-cm (D257); Dissip. Factor: 0.08 (60 Hz, D150)
Ref.: Bulletin 328 MP5M379

Durez 153. Durez Div./Hooker Chem. & Plastics Corp.
Chem. Descrip.: Phenolic resin
Category/Applications: See Durez 152
Form: Granular; Color: Black, brown
Gen'l. Prop. (compression molded @ 340 F): Sp. Gr.: 1.53 (D792); Density: 0.60 g/cc (D1895)
Mech. Prop. (compression molded @ 340 F): Tens. Str.: 55 MPa (D638); Tens. Mod.: 10.3 GPa (D638); Flex. Str.: 76 MPa (D790); Compr. Str.: 179 MPa (D695); Impact Str. (Izod): 18 J/m (D256)
Therm. Prop. (compression molded @ 340 F): Distort. Temp.: 182 C (@ 264 psi, D648); Flamm.: V-O (0.058 in., UL94), V-1 (0.040 in., UL94)

Elec. Prop. (compression molded @ 340 F): Dielec. Str.: 14.8 MV/m (short time, D149); Dielec. Const.: 6.0 (60 Hz, D-150); Vol. Resist.: 1 X 10^{12} ohm-cm (D257); Dissip. Factor: 0.07 (60 Hz, D-150)
Ref.: Bulletin 328 MP5M379

Durez 156. Durez Div./Hooker Chem. & Plastics Corp.
Chem. Descrip.: Phenolic resin
Category/Applications: See Durez 152
Form: Granular; Color: Black
Gen'l. Prop. (compression molded @ 340 F): Sp. Gr.: 1.58 (D792); Density: 0.59 g/cc (D1895)
Mech. Prop. (compression molded @ 340 F): Tens. Str.: 41 MPa (D638); Tens. Mod.: 10.3 GPa (D638); Flex. Str.: 65 MPa (D790); Compr. Str.: 186 MPa (D695); Impact Str. (Izod): 18 J/m (D256)
Therm. Prop. (compression molded @ 340 F): Distort. Temp.: 191 C (@ 264 psi, D648); Flamm.: V-O (0.040 in., UL94), V-1 (0.020 in., UL94)
Elec. Prop. (compression molded @ 340 F): Dielec. Str.: 11.8 MV/m (short time, D149); Dielec. Const.: 7.5 (60 Hz, D-150); Vol. Resist.: 1 X 10^{12} ohm-cm (D257); Arc Resist.: 180 s (D495); Dissip. Factor: 0.12 (60 Hz, D150)
Ref.: Bulletin 328 MP5M379

Durez 161. Durez Div./Hooker Chem. & Plastics Corp.
Chem. Descrip.: Phenolic resin
Category/Applications: Specialty molding materials for use in compression, transfer, and injection molding processes
Form: Granular; Color: Black, brown
Gen'l. Prop. (compression molded @ 340 F): Sp. Gr.: 1.39 (D792); Density: 0.61 g/cc (D1895)
Mech. Prop. (compression molded @ 340 F): Tens. Str.: 45 MPa (D638); Tens. Mod.: 6.9 GPa (D638); Flex. Str.: 62 MPa (D790); Compr. Str.: 172 MPa (D695); Impact Str. (Izod): 14 J/m (D-256)
Therm. Prop. (compression molded @ 340 F): Distort. Temp.: 191 C (@ 264 psi, D648); Flamm.: V-O (0.240 in., UL94)
Elec. Prop. (compression molded @ 340 F): Dielec. Str.: 11.8 MV/m (short time, D149); Dielec. Const.: 8.0 (60 Hz, D-150); Vol. Resist.: 1 X 10^{13} ohm-cm (D257); Dissip. Factor: 0.20 (60 Hz, D150)
Ref.: Bulletin 328 MP5M379

Durez 165. Durez Div./Hooker Chem. & Plastics Corp.
Chem. Descrip.: Phenolic resin
Category/Applications: See Durez 161
Form: Granular; Color: Black
Gen'l. Prop. (compression molded @ 340 F): Sp. Gr.: 1.52 (D792); Density: 0.50 g/cc (D1895)
Mech. Prop. (compression molded @ 340 F): Tens. Str.: 41 MPa (D638); Tens. Mod.: 9.7 GPa (D638); Flex. Str.: 59 MPa (D790); Compr. Str.: 172 MPa (D695); Impact Str. (Izod): 19 J/m (D-256)
Therm. Prop. (compression molded @ 340 F): Distort. Temp.: 232 C (@ 264 psi, D648); Flamm.: V-O (0.120 in., UL94)
Elec. Prop. (compression molded @ 340 F): Dielec. Str.: 9.8 MV/m (short time, D149); Dielec. Const.: 10.0 (60 Hz, D-150); Vol. Resist.: 1 X 10^{13} ohm-cm (D257); Dissip. Factor: 0.30 (60 Hz, D150)
Ref.: Bulletin 328 MP5M379

Durez 791. Durez Div./Hooker Chem. & Plastics Corp.
Chem. Descrip.: Phenolic resin
Category/Applications: General purpose

thermoset material for compression and transfer molding applications
Form: Granular; Color: Black (also avail. in scarlet and brown)
Gen'l. Prop. (compression molded @ 340 F): Sp. Gr.: 1.36 (D792); Density: 0.58 g/cc (D1895)
Mech. Prop. (compression molded @ 340 F): Tens. Str.: 48 MPa (D638); Tens. Mod.: 9.0 GPa (D638); Flex. Str.: 76 MPa (D790); Compr. Str.: 221 MPa (D695); Impact Str. (Izod): 16 J/m (D-256)
Therm. Prop. (compression molded @ 340 F): Distort. Temp.: 166 C (@ 264 psi, D648); Flamm.: V-1 (0.240 in., UL94)
Elec. Prop. (compression molded @ 340 F): Dielec. Str.: 13.8 MV/m (short time, D149); Dielec. Const.: 5.6 (60 Hz, D-150); Vol. Resist.: 1×10^{12} ohm-cm (D257); Dissip. Factor: 0.06 (60 Hz, D-150)
Ref.: Bulletin 328 MP5M379

Durez 3948. Durez Div./Hooker Chem. & Plastics Corp.
Chem. Descrip.: Phenolic resin
Category/Applications: Specialty molding material for use in compression molding
Form: Granular; Color: Brown
Gen'l. Prop. (compression molded @ 340 F): Sp. Gr.: 1.39 (D792); Density: 0.55 g/cc (D1895)
Mech. Prop. (compression molded @ 340 F); Tens. Str.: 52 MPa (D638); Tens. Mod.: 6.9 GPa (D638); Flex. Str.: 72 MPa (D790); Compr. Str.: 207 MPa (D695); Impact Str. (Izod): 16 J/m (D-256)
Therm. Prop. (compression molded @ 340 F): Distort. Temp.: 188 C (@ 264 psi, D648)
Elec. Prop. (compression molded @ 340 F): Dielec. Str.: 11.8 MV/m (short time, D149); Dielec. Const.: 6.5 (60 Hz, D150); Vol. Resist.: 1×10^{12} ohm-cm (D257); Dissip. Factor: 0.05 (60 Hz, D150)
Ref.: Bulletin 328 MP5M379

Durez 13124. Durez Div./Hooker Chem. & Plastics Corp.
Chem. Descrip.: Phenolic resin
Category/Applications: Impact-grade resin for molded materials requiring high shock resistance; used in compression and transfer molding processes
Form: Nodular; Color: Black
Gen'l. Prop. (compression molded @ 340 F): Sp. Gr.: 1.38 (D792); Density: 0.48 g/cc (D1895)
Mech. Prop. (compression molded @ 340 F): Tens. Str.: 48 MPa (D638); Tens. Mod.: 8.3 GPa (D638); Flex. Str.: 76 MPa (D790); Compr. Str.: 200 MPa (D695); Impact Str. (Izod): 29 J/m (D-256)
Therm. Prop. (compression molded @ 340 F): Distort. Temp.: 177 C (@ 264 psi, D648)
Elec. Prop. (compression molded @ 340 F): Dielec. Str.: 15.7 MV/m (short time, D149); Dielec. Const.: 5.8 (60 Hz, D-150); Vol. Resist.: 1×10^{12} ohm-cm (D257); Dissip. Factor: 0.07 (60 Hz, D-150)
Ref.: Bulletin 328 MP5M379

Durez 13856. Durez Div./Hooker Chem. & Plastics Corp.
Chem. Descrip.: Phenolic resin
Category/Applications: See Durez 161
Form: Granular; Color: Black
Gen'l. Prop. (compression molded @ 340 F): Sp. Gr.: 1.43 (D792); Density: 0.49 g/cc (D1895)
Mech. Prop. (compression molded @ 340 F): Tens. Str.: 48 MPa (D638); Tens. Mod.: 6.9 GPa (D638); Flex. Str.: 65 MPa (D790); Compr. Str.: 186 MPa

(D695); Impact Str. (Izod): 25 J/m (D-256)
Therm. Prop. (compression molded @ 340 F): Distort. Temp.: 204 C (@ 264 psi, D648); Flamm.: V-O (0.240 in., UL94)
Elec. Prop. (compression molded @ 340 F): Dielec. Str.: 6.9 MV/m (short time, D149); Dielec. Const.: 11.0 (60 Hz, D-150); Vol. Resist.: 1 X 10^{12} ohm-cm (D257); Dissip. Factor: 0.25 (60 Hz, D150)
Ref.: Bulletin 328 MP5M379

Durez 16378. Durez Div./Hooker Chem. & Plastics Corp.
Chem. Descrip.: Phenolic resin
Category/Applications: Specialty molding material for use in compression and transfer molding processes
Form: Granular; Color: Black
Gen'l. Prop. (compression molded @ 340 F): Sp. Gr.: 1.37 (D792); Density: 0.50 g/cc (D1895)
Mech. Prop. (compression molded @ 340 F): Tens. Str.: 48 MPa (D638); Tens. Mod.: 7.6 GPa (D638); Flex. Str.: 69 MPa (D790); Compr. Str.: 207 MPa (D695); Impact Str. (Izod): 20 J/m (D-256)
Therm. Prop. (compression molded @ 340 F): Distort. Temp.: 182 C (@ 264 psi, D648)
Elec. Prop. (compression molded @ 340 F): Dielec. Str.: 10.8 MV/m (short time, D149); Dielec. Const.: 7.4 (60 Hz, D-150); Vol. Resist.: 1 X 10^{11} ohm-cm (D257); Dissip. Factor: 0.16 (60 Hz, D150)
Ref.: Bulletin 328 MP5M379

Durez 17080. Durez Div./Hooker Chem. & Plastics Corp.
Chem. Descrip.: Phenolic resin
Category/Applications: See Durez 13124
Form: Nodular; Color: Black
Gen'l. Prop. (compression molded @ 340 F): Sp. Gr.: 1.40 (D792); Density: 0.41 g/cc (D1895)
Mech. Prop. (compression molded @ 340 F): Tens. Str.: 48 MPa (D638); Tens. Mod.: 10.3 GPa (D638); Flex. Str.: 69 MPa (D790); Compr. Str.: 193 MPa (D695); Impact Str. (Izod): 32 J/m (D256)
Therm. Prop. (compression molded @ 340 F): Distort. Temp.: 182 C (@ 264 psi, D648)
Elec. Prop. (compression molded @ 340 F): Dielec. Str.: 15.7 MV/m (short time, D149); Dielec. Const.: 6.3 (60 Hz, D-150); Vol. Resist.: 1 X 10^{12} ohm-cm (D257); Dissip. Factor: 0.15 (60 Hz, D-150)
Ref.: Bulletin 328 MP5M379

Durez 17156. Durez Div./Hooker Chem. & Plastics Corp.
Chem. Descrip.: Phenolic resin
Category/Applications: See Durez 13124
Form: Nodular; Color: Black
Gen'l. Prop. (compression molded @ 340 F): Sp. Gr.: 1.38 (D792); Density: 0.40 g/cc (D1895)
Mech. Prop. (compression molded @ 340 F): Tens. Str.: 45 MPa (D638); Tens. Mod.: 9.7 GPa (D638); Flex. Str.: 69 MPa (D790); Compr. Str.: 138 MPa (D695); Impact Str. (Izod): 59 J/m (D-256)
Therm. Prop. (compression molded @ 340 F): Distort. Temp.: 177 C (@ 264 psi, D648)
Elec. Prop. (compression molded @ 340 F): Dielec. Str.: 13.8 MV/m (short time, D149); Dielec. Const.: 6.0 (60 Hz, D-150); Vol. Resist.: 1 X 10^{13} ohm-cm (D257); Dissip. Factor: 0.16 (60 Hz, D-150)
Ref.: Bulletin 328 MP5M379

Durez 18420. Durez Div./Hooker Chem. & Plastics Corp.
Chem. Descrip.: Phenolic resin
Category/Applications: Specialty molding material for use in compression molding
Form: Granular; Color: Black
Gen'l. Prop. (compression molded @ 340 F): Sp. Gr.: 1.37 (D792); Density: 0.56 g/cc (D1895)
Mech. Prop. (compression molded @ 340 F): Tens. Str.: 48 MPa (D638); Tens. Mod.: 8.3 GPa (D638); Flex. Str.: 69 MPa (D790); Compr. Str.: 207 MPa (D695); Impact Str. (Izod): 15 J/m (D-256)
Therm. Prop. (compression molded @ 340 F): Distort. Temp.: 188 C (@ 264 psi, D648)
Elec. Prop. (compression molded @ 340 F): Dielec. Str.: 7.9 MV/m (short time, D149); Dielec. Const.: 10.0 (60 Hz, D-150); Vol. Resist.: 1×10^{12} ohm-cm (D257); Dissip. Factor: 0.15 (60 Hz, D150)
Ref.: Bulletin 328 MP5M379

Durez 18441. Durez Div./Hooker Chem. & Plastics Corp.
Chem. Descrip.: Phenolic resin
Category/Applications: See Durez 791
Form: Granular; Color: Black, brown
Gen'l. Prop. (compression molded @ 340 F): Sp. Gr.: 1.35 (D792); Density: 0.59 g/cc (D1895)
Mech. Prop. (compression molded @ 340 F): Tens. Str.: 52 MPa (D638); Tens. Mod.: 7.6 GPa (D638); Flex. Str.: 69 MPa (D790); Compr. Str.: 193 MPa (D695); Impact Str. (Izod): 14 J/m (D-256)
Therm. Prop. (compression molded @ 340 F): Distort. Temp.: 177 C (@ 264 psi, D648); Flamm.: V-O (0.240 in., UL94)
Elec. Prop. (compression molded @ 340 F): Dielec. Str.: 12.8 MV/m (short time, D149); Dielec. Const.: 6.2 (60 Hz, D150); Vol. Resist.: 1×10^{13} ohm-cm (D257); Dissip. Factor: 0.05 (60 Hz, D-150)
Ref.: Bulletin 328 MP5M379

Durez 18683. Durez Div./Hooker Chem. & Plastics Div.
Chem. Descrip.: Phenolic resin
Category/Applications: See Durez 13124
Form: Nodular; Color: Black
Gen'l. Prop. (compression molded @ 340 F): Sp. Gr.: 1.40 (D792); Density: 0.40 g/cc (D1895)
Mech. Prop. (compression molded @ 340 F): Tens. Str.: 38 MPa (D638); Tens. Mod.: 10.3 GPa (D638); Flex. Str.: 59 MPa (D790); Compr. Str.: 159 MPa (D695); Impact Str. (Izod): 48 J/m (D256)
Therm. Prop. (compression molded @ 340 F): Distort. Temp.: 160 C (@ 264 psi, D648)
Elec. Prop. (compression molded @ 340 F): Dielec. Str.: 12.8 MV/m (short time, D149); Dielec. Const.: 6.0 (60 Hz, D-150); Vol. Resist.: 1×10^{13} ohm-cm (D257); Dissip. Factor: 0.08 (60 Hz, D-150)
Ref.: Bulletin 328 MP5M379

Durez 21028. Durez Div./Hooker Chem. & Plastics Corp.
Chem. Descrip.: Phenolic resin
Category/Applications: See Durez 791
Form: Granular; Color: Black
Gen'l. Prop. (compression molded @ 340 F): Sp. Gr.: 1.39 (D792); Density: 0.60 g/cc (D1895)
Mech. Prop. (compression molded @ 340 F): Tens. Str.: 45 MPa (D638); Tens. Mod.: 7.6 GPa (D638); Flex. Str.: 69 MPa (D790); Compr. Str.: 207 MPa (D695); Impact Str. (Izod): 14 J/m (D-256)

Therm. Prop. (compression molded @ 340 F): Distort. Temp.: 171 C (@ 264 psi, D648); Flamm.: V-O (0.240 in., UL94)
Elec. Prop. (compression molded @ 340 F): Dielec. Str.: 13.8 MV/m (short time, D149); Dielec. Const.: 5.7 (60 Hz, D-150); Vol. Resist.: 1 X 10^{12} ohm-cm (D257); Dissip. Factor: 0.08 (60 Hz, D-150)
Ref.: Bulletin 328 MP5M379

Durez 21210. Durez Div./Hooker Chem. & Plastics Corp.
Chem. Descrip.: Phenolic resin
Category/Applications: Electrical grade resin with good dielectric strength and insulation resistance for use in automotive and communications products; used in compression and transfer molding processes
Form: Granular: Color: Black
Gen'l. Prop. (compression molded @ 340 F): Sp. Gr.: 1.36 (D792); Density: 0.60 g/cc (D1895)
Mech. Prop. (compression molded @ 340 F): Tens. Str.: 48 MPa (D638); Tens. Mod.: 8.3 GPa (D638); Flex. Str.: 69 MPa (D790); Compr. Str.: 214 MPa (D695); Impact Str. (Izod): 15 J/m (D-256)
Therm. Prop. (compression molded @ 340 F): Distort. Temp.: 160 C (@ 264 psi, D648)
Elec. Prop. (compression molded @ 340 F): Dielec. Str.: 14.8 MV/m (short time, D149); Dielec. Const.: 5.4 (60 Hz, D-150); Vol. Resist.: 1 X 10^{13} ohm-cm (D257); Dissip. Factor: 0.04 (60 Hz, D150)
Ref.: Bulletin 328 MP5M379

Durez 21372. Durez Div./Hooker Chem. & Plastics Corp.
Chem. Descrip.: Phenolic resin
Category/Applications: See Durez 21210
Form: Granular; Color: Black
Gen'l. Prop. (compression molded @ 340 F): Sp. Gr.: 1.38 (D792); Density: 0.62 g/cc (D1895)
Mech. Prop. (compression molded @ 340 F): Tens. Str.: 52 MPa (D638); Tens. Mod.: 9.7 GPa (D638); Flex. Str.: 65 MPa (D790); Compr. Str.: 186 MPa (D695); Impact Str. (Izod): 14 J/m (D-256)
Therm. Prop. (compression molded @ 340 F): Distort. Temp.: 171 C (@ 264 psi, D648)
Elec. Prop. (compression molded @ 340 F): Dielec. Str.: 12.8 MV/m (short time, D149); Dielec. Const.: 5,4 (60 Hz, D-150); Vol. Resist.: 1 X 10^{14} ohm-cm (D257); Dissip. Factor: 0.01 (60 Hz, D150)
Ref.: Bulletin 328 MP5M379

Durez 22257. Durez Div./Hooker Chem. & Plastics Corp.
Chem. Descrip.: Phenolic resin
Category/Applications: See Durez 13124
Form: Granular; Color: Black, brown
Gen'l. Prop. (compression molded @ 340 F): Sp. Gr.: 1.37 (D792); Density: 0.48 g/cc (D1895)
Mech. Prop. (compression molded @ 340 F): Tens. Str.: 52 MPa (D638); Tens. Mod.: 7.6 GPa (D638); Flex. Str.: 69 MPa (D790); Compr. Str.: 207 MPa (D695); Impact Str. (Izod): 23 J/m (D-256)
Therm. Prop. (compression molded @ 340 F): Distort. Temp.: 171 C (@ 264 psi, D648); Flamm.: V-O (0.240 in., UL94)
Elec. Prop. (compression molded @ 340 F): Dielec. Str.: 8.9 MV/m (short time, D149); Dielec. Const.: 6.8 (60 Hz, D-150); Vol. Resist.: 1 X 10^{12} ohm-cm (D257); Dissip. Factor: 0.10 (60 Hz, D-150)
Ref.: Bulletin 328 MP5M379

Durez 22486. Durez Div./Hooker Chem. & Plastics Corp.
Chem. Descrip.: Phenolic resin
Category/Applications: See Durez 791
Form: Granular; Color: Black (also avail. in brown)
Gen'l. Prop. (compression molded @ 340 F): Sp. Gr.: 1.36 (D792); Density: 0.60 g/cc (D1895)
Mech. Prop. (compression molded @ 340 F): Tens. Str.: 48 MPa (D638); Tens. Mod.: 8.3 GPa (D638); Flex. Str.: 69 MPa (D790); Compr. Str.: 207 MPa (D695); Impact Str. (Izod): 15 J/m (D-256)
Therm. Prop. (compression molded @ 340 F): Distort. Temp.: 177 C (@ 264 psi, D648); Flamm.: V-1 (0.240 in., UL94)
Elec. Prop. (compression molded @ 340 F): Dielec. Str.: 13.8 MV/m (short time, D149); Dielec. Const.: 5.8 (60 Hz, D-150); Vol. Resist.: 1×10^{12} ohm-cm (D257); Dissip. Factor: 0.08 (60 Hz, D-150)
Ref.: Bulletin 328 MP5M379

Durez 22761. Durez Div./Hooker Chem. & Plastics Corp.
Chem. Descrip.: Diallyl phthalate resin
Category/Applications: Molding material which retains its electrical properties when used in environments subject to high humidity and temp.; used in compression, transfer, and injection molding processes
Form: Granular; Color: Black, blue
Gen'l. Prop. (compression molded @ 340 F): Sp. Gr.: 1.71 (D792); Density: 0.74 g/cc (D1895)
Mech. Prop. (compression molded @ 340 F): Tens. Str.: 55 MPa (D638); Tens. Mod.: 9.6 GPa (D638); Flex. Str.: 96 MPa (D790); Compr. Str.: 214 MPa (D695); Impact Str. (Izod): 27 J/m (D256)
Therm. Prop. (compression molded @ 340 F): Distort. Temp.: 165 C (@ 264 psi, D648); Flamm.: V-O (0.240 in., UL94)
Elec. Prop. (compression molded @ 340 F): Dielec. Str.: 15.7 MV/m (short time, D149); Dielec. Const.: 4.6 (60 Hz, D-150); Vol. Resist.: 1×10^{14} ohm-cm (D257); Dissip. Factor: 0.009 (60 Hz, D150)
Ref.: Bulletin 328 MP5M379

Durez 23410. Durez Div./Hooker Chem. & Plastics Corp.
Chem. Descrip.: Alkyd resin
Category/Applications: Molding resin for arc-track resistance and to retain electrical properties at elevated temps.; used in compression, transfer, and injection molding processes
Form: Granular; Color: Black
Gen'l. Prop. (compression molded @ 340 F): Sp. Gr.: 1.92 (D792); Density: 0.76 g/cc (D1895)
Mech. Prop. (compression molded @ 340 F): Tens. Str.: 38 MPa (D638); Tens. Mod.: 13.8 GPa (D638); Flex. Str.: 69 MPa (D790); Compr. Str.: 152 MPa (D695); Impact Str. (Izod): 17 J/m (D256)
Therm. Prop. (compression molded @ 340 F): Distort. Temp.: 191 C (@ 264 psi, D648)
Elec. Prop. (compression molded @ 340 F): Dielec. Str.: 13.8 MV/m (short time, D149); Dielec. Const.: 6.4 (60 Hz, D-150); Vol. Resist.: 1×10^{14} ohm-cm (D257); Dissip. Factor: 0.03 (60 Hz, D150)
Ref.: Bulletin 328 MP5M379

Durez 23570. Durez Div./Hooker Chem. & Plastics Corp.
Chem. Descrip.: Glass-filled phenolic resin
Category / Applications: Glass-filled

molding material for a wide variety of high-technology applications; used in compression, transfer, and injection molding processes
Form: Granular; Color: Black, green
Gen'l. Prop. (compression molded @ 340 F): Sp. Gr.: 1.70 (D792); Density: 0.80 g/cc (D1895)
Mech. Prop. (compression molded @ 340 F): Tens. Str.: 59 MPa (D638); Tens. Mod.: 13.1 GPa (D638); Flex. Str.: 96 MPa (D790); Compr. Str.: 241 MPa (D695); Impact Str. (Izod): 24 J/m (D256)
Therm. Prop. (compression molded @ 340 F): Distort. Temp.: 185 C (@ 264 psi, D648); Flamm.: V-O (0.058 in., UL94)
Elec. Prop. (compression molded @ 340 F): Dielec. Str.: 13.8 MV/m (short time, D149); Dielec. Const.: 5.8 (60 Hz, D-150); Vol. Resist.: 1 X 10^{12} ohm-cm (D257); Dissip. Factor: 0.09 (60 Hz, D150)
Ref.: Bulletin 328 MP5M379

Durez 24150. Durez Div./Hooker Chem. & Plastics Corp.
Chem. Descrip.: Alkyd resin
Category/Applications: Molding resin for arc-track resistance and to retain electrical properties at elevated temps.; used in transfer and injection molding
Form: Granular; Color: Black
Gen'l. Prop. (compression molded @ 340 F): Sp. Gr.: 2.2 (D792); Density: 1.05 g/cc (D1895)
Mech. Prop. (compression molded @ 340 F): Tens. Str.: 62 MPa (D638); Tens. Mod.: 19.3 GPa (D638); Flex. Str.: 103 MPa (D790); Compr. Str.: 241 MPa (D695); Impact Str. (Izod): 19 J/m (D256)
Therm. Prop. (compression molded @ 340 F): Distort. Temp.: 232 C (@ 264 psi, D648); Flamm.: V-O (0.020 in., UL94)
Elec. Prop. (compression molded @ 340 F): Dielec. Str.: 13.8 MV/m (short time, D149); Dielec. Const.: 5.2 (60 Hz, D-150); Vol. Resist.: 1 X 10^{15} ohm-cm (D257); Arc Resist.: 184 s (D495); Dissip. Factor: 0.011 (60 Hz, D150)
Ref.: Bulletin 328 MP5M379

Durez 24337. Durez Div./Hooker Chem. & Plastics Corp.
Chem. Descrip.: Alkyd resin
Category/Applications: See Durez 23410
Form: Granular; Color: Black
Gen'l. Prop. (compression molded @ 340 F): Sp. Gr.: 1.98 (D792); Density: 0.83 g/cc (D1895)
Mech. Prop. (compression molded @ 340 F): Tens. Str.: 41 MPa (D638); Tens. Mod.: 13.1 GPa (D638); Flex. Str.: 62 MPa (D790); Compr. Str.: 172 MPa (D695); Impact Str. (Izod): 16 J/m (D256)
Therm. Prop. (compression molded @ 340 F): Distort. Temp.: 191 C (@ 264 psi, D648); Flamm.: HB (0.240 in., UL94)
Elec. Prop. (compression molded @ 340 F): Dielec. Str.: 13.8 MV/m (short time, D149); Dielec. Const.: 6.7 (60 Hz, D-150); Vol. Resist.: 1 X 10^{14} ohm-cm (D257); Arc Resist.: 180 s (D495); Dissip. Factor: 0.02 (60 Hz, D150)
Ref.: Bulletin 328 MP5M379

Durez 24668. Durez Div./Hooker Chem. & Plastics Corp.
Chem. Descrip.: Alkyd resin
Category/Applications: See Durez 23410
Form: Granular; Color: Black
Gen'l. Prop. (compression molded @ 340 F): Sp. Gr.: 2.08 (D792); Density: 0.80 g/cc (D1895)

Mech. Prop. (compression molded @ 340 F): Tens. Str.: 48 MPa (D638); Tens. Mod.: 20.7 GPa (D638); Flex. Str.: 83 MPa (D790); Compr. Str.: 193 MPa (D695); Impact Str. (Izod): 18 J/m (D256)
Therm. Prop. (compression molded @ 340 F): Distort. Temp.: 218 C (@ 264 psi, D648)
Elec. Prop. (compression molded @ 340 F): Dielec. Str.: 13.8 MV/m (short time, D149); Dielec. Const.: 5.1 (60 Hz, D150); Vol. Resist.: 1 × 10^{15} ohm-cm (D257); Dissip. Factor: 0.013 (60 Hz, D150)
Ref.: Bulletin 328 MP5M379

Durez 24776. Durez Div./Hooker Chem. & Plastics Corp.
Chem. Descrip.: Diallyl phthalate resin
Category/Applications: See Durez 22761
Form: Granular: Color: Green, gray
Gen'l. Prop. (compression molded @ 340 F): Sp. Gr.: 1.78 (D792); Density: 0.74 g/cc (D1895)
Mech. Prop. (compression molded @ 340 F): Tens. Str.: 62 MPa (D638); Tens. Mod.: 9 GPa (D638); Flex. Str.: 90 MPa (D790); Compr. Str.: 193 MPa (D695); Impact Str. (Izod): 27 J/m (D-256)
Therm. Prop. (compression molded @ 340 F): Distort. Temp.: 232 C (@ 264 psi, D648); Flamm.: V-O (0.240 in., UL94)
Elec. Prop. (compression molded @ 340 F): Dielec. Str.: 15.0 MV/m (short time, D149); Dielec. Const.: 4.6 (60 Hz, D-150); Vol. Resist.: 1 × 10^{15} ohm-cm (D257); Dissip. Factor: 0.010 (60 Hz, D150)
Ref.: Bulletin 328 MP5M379

Durez 27962. Durez Div./Hooker Chem. & Plastics Corp.
Chem. Descrip.: Alkyd resin
Category/Applications: See Durez 23410
Form: Granular; Color: Brown
Gen'l. Prop. (compression molded @ 340 F): Sp. Gr.: 2.00 (D792); Density: 0.80 g/cc (D1895)
Mech. Prop. (compression molded @ 340 F): Tens. Str.: 34 MPa (D638); Tens. Mod.: 17.9 GPa (D638); Flex. Str.: 59 MPa (D790); Compr. Str.: 138 MPa (D695); Impact Str. (Izod): 21 J/m (D256)
Therm. Prop. (compression molded @ 340 F): Distort. Temp.: 218 C (@ 264 psi, D648); Flamm.: V-O (0.020 in., UL94)
Elec. Prop. (compression molded @ 340 F): Dielec. Str.: 15.7 MV/m (short time, D149); Dielec. Const.: 5.4 (60 Hz, D-150); Vol. Resist.: 1 × 10^{14} ohm-cm (D257); Arc Resist.: 242 s (D495); Dissip. Factor: 0.01 (60 Hz, D150)
Ref.: Bulletin 328 MP5M379

Durez 28615. Durez Div./Hooker Chem. & Plastics Corp.
Chem. Descrip.: Phenolic resin
Category/Applications: See Durez 21210
Form: Granular; Color: Black
Gen'l. Prop. (compression molded @ 340 F): Sp. Gr.: 1.44 (D792); Density: 0.62 g/cc (D1895)
Mech. Prop. (compression molded @ 340 F): Tens. Str.: 41 MPa (D638); Tens. Mod.: 9.0 GPa (D638); Flex. Str.: 65 MPa (D790); Compr. Str.: 186 MPa (D695); Impact Str. (Izod): 14 J/m (D-256)
Therm. Prop. (compression molded @ 340 F): Distort. Temp.: 171 C (@ 264 psi, D648)
Elec. Prop. (compression molded @ 340 F): Dielec. Str.: 13.8 MV/m (short time, D149); Dielec. Const.: 7.0 (60 Hz, D-150); Vol. Resist.: 1 × 10^{13} ohm-cm

(D257); Dissip. Factor: 0.03 (60 Hz, D150)
Ref.: Bulletin 328 MP5M379

Durez 29053. Durez Div./Hooker Chem. & Plastics Corp.
Chem. Descrip.: Glass-filled phenolic resin
Category/Applications: See Durez 23570
Form: Granular; Color: Black
Gen'l. Prop. (compression molded @ 340 F): Sp. Gr.: 1.90 (D792); Density: 0.95 g/cc (D1895)
Mech. Prop. (compression molded @ 340 F): Tens. Str.: 41 MPa (D638); Tens. Mod.: 15.2 GPa (D638); Flex. Str.: 83 MPa (D790); Compr. Str.: 207 MPa (D695); Impact Str. (Izod): 21 J/m (D256)
Therm. Prop. (compression molded @ 340 F): Distort. Temp.: 177 C (@ 264 psi, D648); Flamm.: V-O (0.058 in., UL94)
Elec. Prop. (compression molded @ 340 F): Dielec. Str.: 14.8 MV/m (short time, D149); Dielec. Const.: 6.1 (60 Hz, D-150); Vol. Resist.: 1 $\times$ 10^{12} ohm-cm (D257); Dissip. Factor: 0.06 (60 Hz, D-150)
Ref.: Bulletin 328 MP5M379

Durez 29237. Durez Div./Hooker Chem. & Plastics Corp.
Chem. Descrip.: Phenolic resin
Category/Applications: See Durez 123
Form: Granular; Color: Black
Gen'l. Prop. (compression molded @ 340 F): Sp. Gr.: 1.36 (D792); Density: 0.52 g/cc (D1895)
Mech. Prop. (compression molded @ 340 F): Tens. Str.: 41 MPa (D638); Tens. Mod.: 9.0 GPa (D638); Flex. Str.: 62 MPa (D790); Compr. Str.: 172 MPa (D695); Impact Str. (Izod): 29 J/m (D-256)
Therm. Prop. (compression molded @ 340 F): Distort. Temp.: 166 C (@ 264 psi, D648); Flamm.: V-O (0.240 in., UL94)
Elec. Prop. (compression molded @ 340 F): Dielec. Str.: 13.8 MV/m (short time, D149); Dielec. Const.: 5.8 (60 Hz, D-150); Vol. Resist.: 1 $\times$ 10^{12} ohm-cm (D257); Dissip. Factor: 0.07 (60 Hz, D150)
Ref.: Bulletin 328 MP5M379

Durez 29276. Durez Div./Hooker Chem. & Plastics Corp.
Chem. Descrip.: Alkyd resin
Category/Applications: See Durez 23410
Form: Granular; Color: Black
Gen'l. Prop. (compression molded @ 340 F): Sp. Gr.: 1.90 (D792); Density: 0.80 g/cc (D1895)
Mech. Prop. (compression molded @ 340 F): Tens. Str.: 34 MPa (D638); Tens. Mod.: 8.9 GPa (D638); Flex. Str.: 69 MPa (D790); Compr. Str.: 152 MPa (D695); Impact Str. (Izod): 24 J/m (D-256)
Therm. Prop. (compression molded @ 340 F): Distort. Temp.: 204 C (@ 264 psi, D648); Flamm.: V-1 (0.058 in., UL94)
Elec. Prop. (compression molded @ 340 F): Dielec. Str.: 13.8 MV/m (short time, D149); Dielec. Const.: 5.1 (60 Hz, D-150); Vol. Resist.: 1 $\times$ 10^{14} ohm-cm (D257); Arc Resist.: 185 s (D495); Dissip. Factor: 0.02 (60 Hz, D150)
Ref.: Bulletin 328 MP5M379

Durez 29502. Durez Div./Hooker Chem. & Plastics Corp.
Chem. Descrip.: Glass-filled phenolic resin
Category / Applications: Glass - filled molding material for a wide variety of high-technology applications; used in

compression molding process
Form: Granular; Color: Black
Gen'l. Prop. (compression molded @ 340 F): Sp. Gr.: 2.13 (D792); Density: 1.02 g/cc (D1895)
Mech. Prop. (compression molded @ 340 F): Tens. Str.: 48 MPa (D638); Tens. Mod.: 20.7 GPa (D638); Flex. Str.: 83 MPa (D790); Compr. Str.: 193 MPa (D695); Impact Str. (Izod): 20 J/m (D256)
Therm. Prop. (compression molded @ 340 F): Distort. Temp.: 182 C (@ 264 psi, D648)
Elec. Prop. (compression molded @ 340 F): Dielec. Str.: 14.8 MV/m (short time, D149); Dielec. Const.: 6.4 (60 Hz, D-150); Vol. Resist.: 1×10^{12} ohm-cm (D257); Dissip. Factor: 0.07 (60 Hz, D150)
Ref.: Bulletin KA 30536 e

Durez 30245. Durez Div./Hooker Chem. & Plastics Corp.
Chem. Descrip.: Phenolic resin
Category/Applications: See Durez 145
Form: Nodular; Color: Black
Gen'l. Prop. (compression molded @ 340 F): Sp. Gr.: 1.68 (D792); Density: 0.60 g/cc (D1895)
Mech. Prop. (compression molded @ 340 F): Tens. Str.: 34 MPa (D638); Tens. Mod.: 13.7 GPa (D638); Flex. Str.: 58 MPa (D790); Compr. Str.: 145 MPa (D695); Impact Str. (Izod): 29 J/m (D256)
Therm. Prop. (compression molded @ 340 F): Distort. Temp.: 177 C (@ 264 psi, D648); Flamm.: V-O (0.130 in., UL94)
Elec. Prop. (compression molded @ 340 F): Dielec. Str.: 14.8 MV/m (short time, D149); Dielec. Const.: 10.0 (60 Hz, D-150); Vol. Resist.: 1×10^{12} ohm-cm (D257); Arc Resist.: 185 s (D495); Dissip. Factor: 0.35 (60 Hz, D150)
Ref.: Bulletin 328 MP5M379

Durez 30614. Durez Div./Hooker Chem. & Plastics Corp.
Chem. Descrip.: Phenolic resin
Category/Applications: See Durez 152
Form: Granular; Color: Black, green
Gen'l. Prop. (compression molded @ 340 F): Sp. Gr.: 1.79 (D792); Density: 0.90 g/cc (D1895)
Mech. Prop. (compression molded @ 340 F): Tens. Str.: 41 MPa (D638); Tens. Mod.: 16.5 GPa (D638); Flex. Str.: 65 MPa (D790); Compr. Str.: 138 MPa (D695); Impact Str. (Izod): 18 J/m (D256)
Therm. Prop. (compression molded @ 340 F): Distort. Temp.: 182 C (@ 264 psi, D648); Flamm.: V-O (0.020 in., UL94)
Elec. Prop. (compression molded @ 340 F): Dielec. Str.: 16.7 MV/m (short time, D149); Dielec. Const.: 6.6 (60 Hz, D-150); Vol. Resist.: 1×10^{13} ohm-cm (D257); Arc Resist.: 184 s (D495); Dissip. Factor: 0.12 (60 Hz, D150)
Ref.: Bulletin 328 MP5M379

Durez 30645. Durez Div./Hooker Chem. & Plastics Corp.
Chem. Descrip.: Glass-filled phenolic resin
Category/Applications: See Durez 23570
Form: Nodular; Color: Black
Gen'l. Prop. (compression molded @ 340 F): Sp. Gr.: 1.69 (D792); Density: 0.65 g/cc (D1895)
Mech. Prop. (compression molded @ 340 F): Tens. Str.: 62 MPa (D638); Tens. Mod.: 10.3 GPa (D638); Flex. Str.: 96 MPa (D790); Compr. Str.: 241

MPa (D695); Impact Str. (Izod): 35 J/m (D256)
Therm. Prop. (compression molded @ 340 F): Distort. Temp.: 191 C (@ 264 psi, D648); Flamm.: V-O (0.120 in., UL94)
Elec. Prop. (compression molded @ 340 F): Dielec. Str.: 14.8 MV/m (short time, D149); Dielec. Const.: 6.5 (60 Hz, D-150); Vol. Resist.: 1 X 10^{13} ohm-cm (D257); Arc Resist.: 150 s (D495); Dissip. Factor: 0.15 (60 Hz, D150)
Ref.: Bulletin KA 30536 e

Durez 30725. Durez Div./Hooker Chem. & Plastics Corp.
Chem. Descrip.: Phenolic resin
Category/Applications: See Durez 791
Form: Granular; Color: Black, brown
Gen'l. Prop. (compression molded @ 340 F): Sp. Gr.: 1.44 (D792); Density: 0.60 g/cc (D1895)
Mech. Prop. (compression molded @ 340 F): Tens. Str.: 48 MPa (D638); Tens. Mod.: 8.3 GPa (D638); Flex. Str.: 76 MPa (D790); Compr. Str.: 172 MPa (D695); Impact Str. (Izod): 17 J/m (D-256)
Therm. Prop. (compression molded @ 340 F): Distort. Temp.: 160 C (@ 264 psi, D648)
Elec. Prop. (compression molded @ 340 F): Dielec. Str.: 13.8 MV/m (short time, D149); Dielec. Const.: 6.2 (60 Hz, D-150); Vol. Resist.: 1 X 10^{12} ohm-cm (D257); Dissip. Factor: 0.07 (60 Hz, D-150)
Ref.: Bulletin 328 MP5M379

Durez 30871. Durez Div./Hooker Chem. & Plastics Corp.
Chem. Descrip.: Glass-filled phenolic resin
Category/Applications: See Durez 23570
Form: Nodular; Color: Black
Gen'l. Prop. (compression molded @ 340 F): Tens. Str.: 48 MPa (D638); Tens. Mod.: 8.3 GPa (D638); Flex. Str.: 58 MPa (D790); Compr. Str.: 103 MPa (D695); Impact Str. (Izod): 34 J/m (D256)
Therm. Prop. (compression molded @ 340 F): Distort. Temp.: 190 C (@ 264 psi, D648); Flamm.: V-O (0.120 in., UL94)
Elec. Prop. (compression molded @ 340 F): Dielec. Str.: 11.8 MV/m (short time, D149); Dielec. Const.: 12.6 (60 Hz, D-150); Vol. Resist.: 1 X 10^{12} ohm-cm (D257); Arc Resist.: 185 s (D495); Dissip. Factor: 0.55 (60 Hz, D150)
Ref.: Bulletin 328 MP5M379

Durez 30928. Durez Div./Hooker Chem. & Plastics Corp.
Chem. Descrip.: Glass-filled phenolic resin
Category/Applications: See Durez 23570
Form: Nodular; Color: Black
Gen'l. Prop. (compression molded @ 340 F): Sp. Gr.: 1.70 (D792); Density: 0.55 g/cc (D1895)
Mech. Prop. (compression molded @ 340 F): Tens. Str.: 59 MPa (D638); Tens. Mod.: 13.8 GPa (D638); Flex. Str.: 110 MPa (D790); Compr. Str.: 241 MPa (D695); Impact Str. (Izod): 40 J/m (D256)
Therm. Prop. (compression molded @ 340 F): Distort. Temp.: 246 C (@ 264 psi, D648)
Elec. Prop. (compression molded @ 340 F): Dielec. Str.: 11.8 MV/m (short time, D149); Dielec. Const.: 6.3 (60 Hz, D-150); Vol. Resist.: 1 X 10^{12} ohm-cm (D257); Dissip. Factor: 0.08 (60 Hz, D150)
Ref.: Bulletin 328 MP5M379

Durez SI-75. Durez Div./Hooker Chem. & Plastics Corp.
Chem. Descrip.: Diallyl phthalate resin
Category/Applications: See Durez 22761
Form: Granular; Color: Green
Gen'l. Prop. (compression molded @ 340 F): Sp. Gr.: 1.85 (D792); Density: 0.82 g/cc (D1895)
Mech. Prop. (compression molded @ 340 F): Tens. Str.: 48 MPa (D638); Tens. Mod.: 10.3 GPa (D638); Flex. Str.: 76 MPa (D790); Compr. Str.: 172 MPa (D695); Impact Str. (Izod): 22 J/m (D256)
Therm. Prop. (compression molded @ 340 F): Distort. Temp.: 149 C (@ 264 psi, D648); Flamm.: V-O (0.120 in., UL94)
Elec. Prop. (compression molded @ 340 F): Dielec. Str.: 15.0 MV/m (short time, D149); Dielec. Const.: 4.7 (60 Hz, D-150); Vol. Resist.: 1×10^{15} ohm-cm (D257); Dissip. Factor: 0.020 (60 Hz, D150)
Ref.: Bulletin 328 MP5M379

Dymerex. Hercules Inc.
Chem. Descrip.: Resin composed mainly of dimeric acids derived from rosin with lesser amounts of monomeric resin acids, and neutral materials of rosin origin

Category/Applications: Thermoplastic, high softening pt., oxidation-resistant, acidic resin for use in the manufacture of special finishes, protective coatings, spirit varnishes, epoxy ester resin vehicles, adhesives, and printing inks; also as a soldering flux; noncrystalline with excellent solvent-release properties

Form: Flakes; Color (USDA Rosin Scale): K max.

Gen'l. Prop.: M.W.: 502 (avg.); Sol.: Sol. in alcohols, esters, ketones, aliphatic and aromatic hydrocarbons, chlorinated solvents, and mineral oils; insol. in water; Density: 1.069 kg/l (20 C); Soften. Pt. (Hercules Drop): 150 C; Acid No.: 145; Sapon. No.: 145

Ref.: Data sheet 714-11

E

E-231. Fiberite Corp.
Chem. Descrip.: Epoxy
Category/Applications: Med.-impact, heat-resistant epoxy molding compound with excellent mechanical properties; suitable for applications where strength and performance at elevated temps. are required; recommended for continuous use @ 400 F
Form: Granular
Gen'l. Prop. (as molded): Sp. Gr.: 1.87
Mech. Prop. (as molded): Tens. Str.: 11,000 psi; Flex. Str.: 16,000 psi; Compr. Str.: 26,000 psi; Impact Str.: 0.50 ft lb/in. notch
Therm. Prop. (as molded): Conduct.: 12×10^{-4} Cal/s/cm^2/C/cm; Distort. Temp.: 500+ F (@ 264 psi); Coeff. of Linear Exp.: 1.8×10^{-5} in./in./C
Elec. Prop. (as molded): Power Factor: 0.012 (1 MC); Dielec. Str.: 400 (S/T); Dielec. Const.: 4.5 (1 MC); Vol. Resist.: 1×10^{14} ohm-cm; Arc Resist.: 180 s
Toxicity/Handling: Store in closed containers $\leqslant$ 65 F
Ref.: Product information

E-260. Fiberite Corp.
Chem. Descrip.: Epoxy
Category/Applications: High strength, epoxy glass material meeting MIL-P-46069; used for molding high performance structural parts for military and commercial use
Form: Chopped strands
Gen'l. Prop. (as molded): Sp. Gr.: 1.89; Stability: Good chemical resistance
Mech. Prop. (as molded): Tens. Str.: 22,000 psi; Flex. Str.: 50,000 psi; Compr. Str.: 30,000 psi; Impact Str.: 30 ft lb/in. notch
Therm. Prop. (as molded): Conduct.: 10×10^{-4} Cal/s/cm^2/C/cm; Distort. Temp.: 500 F (@ 264 psi); Coeff. of Linear Exp.: 1.4×10^{-5} in./in./C
Elec. Prop. (as molded): Power Factor: 0.017 (1 MC); Dielec. Str.: 400 (S/T); Dielec. Const.: 5.8 (1 MC); Vol. Resist.: 1×10^{16} ohm-cm; Arc Resist.: 180 s
Toxicity/Handling: See E-231
Ref.: Product information

E-260H. Fiberite Corp.
Chem. Descrip.: Epoxy
Category/Applications: See E-260
Form: ½ in. chopped strands
Gen'l. Prop. (as molded): Sp. Gr.: 1.89; Stability: Good chemical resistance
Mech. Prop. (as molded): Tens. Str.: 27,000 psi; Flex. Str.: 68,000 psi; Compr. Str.: 42,000 psi; Impact Str.: 30 ft lb/in. notch; Hardness (Rockwell): M 105
Therm. Prop. (as molded): Conduct.:

10 × 10^{-4} Cal/s/cm^2/C/cm; Distort. Temp.: 500 F (@ 264 psi); Coeff. of Linear Exp.: 1.4 × 10^{-5} in./in./C
Elec. Prop. (as molded): Power Factor: 0.017 (1 MC); Dielec. Str.: 375 V/mil (S/T); Dielec. Const.: 5.8 (1 MC); Vol. Resist.: 1 × 10^{16} ohm-cm; Arc Resist.: 120 s
Ref.: Product information

E-261H. Fiberite Corp.
Chem. Descrip.: Epoxy
Category/Applications: Highest strength version of a glass-reinforced epoxy impact compound; max. mechanical properties achieved with compression molding; transfer molding of parts produces a reduction in property values
Form: 1 in. chopped strands
Gen'l. Prop. (as molded): Sp. Gr.: 1.86–1.92; Stability: Good chemical and thermal resistance
Mech. Prop. (as molded): Tens. Str.: 45,100 psi; Flex. Str.: 110,600 psi; Compr. Str.: 31,000 psi; Impact Str.: 53 ft lb/in. notch;
Therm. Prop. (as molded): Conduct.: 10 × 10^{-4} Cal/s/cm^2/C/cm; Distort. Temp.: 482 F (@ 264 psi); Coeff. of Linear Exp.: 1.4 × 10^{-5} in./in./C
Elec. Prop. (as molded): Power Factor: 0.017 (1 MC); Dielec. Str.: 375 V/mil (S/T); Dielec. Const.: 5.8 (1 MC); Arc Resist.: 120 s
Ref.: Product information

E-264H. Fiberite Corp.
Chem. Descrip.: Epoxy
Category/Applications: Modified version of E-260H, having the capability of being screw in-lined injection molded; slight reduction of physical properties compared to E-260H
Form: ¼ in. chopped strands
Gen'l. Prop. (as molded): Sp. Gr.: 1.86–1.92
Mech. Prop. (as molded): Tens. Str.: 17,000 psi; Flex. Str.: 39,300 psi; Compr. Str.: 27,000 psi; Impact Str.: 14 ft lb/in. notch
Therm. Prop. (as molded): Conduct.: 10 × 10^{-4} Cal/s/cm^2/C/cm; Distort. Temp.: 482 F (@ 264 psi); Coeff. of Linear Exp.: 1.4 × 10^{-5} in./in./C
Elec. Prop. (as molded): Power Factor: 0.017 (1 MC); Dielec. Str.: 375 V/mil (S/T); Dielec. Const.: 5.8 (1 MC); Arc Resist.: 120 s
Toxicity/Handling: See E-231
Ref.: Product information

E-2068. Fiberite Corp.
Chem. Descrip.: Mineral-filled epoxy
Category/Applications: Molding compound with excellent moldability and release properties, for transfer and compression molding; suitable for encapsulation of coils, resistors, capacitors, and electronic components, terminal blocks, bobbins, connectors
Form: Granular; Color: Black, standard colors
Gen'l. Prop. (as molded): Sp. Gr.: 1.87–1.91 (D792); Stability: Good thermal shock, chemical, and heat resistance
Mech. Prop. (as molded): Tens. Str.: 10,000 psi (D638); Flex. Str.: 14,000 psi (D790); Flex. Mod.: 1.7 × 10^6 psi (D790); Compr. Str.: 28,000 psi (D695); Impact Str.: 0.35 ft lb/in. notch (D256); Hardness (Rockwell): M 110 (D785)
Therm. Prop. (as molded): Conduct.: 16 × 10^{-4} Cal/s/cm^2/C/cm (C177); Distort. Temp.: 141 C (@ 264 psi, D648); Coeff. of Linear Exp.: 2.8 × 10^{-5} in./in./C (−30 to 30 C, D696)
Elec. Prop. (as molded): Power Factor: 0.01 (dry, 1 MC, D150); Dielec. Str.: 420 V/mil (dry, S/T, D149); Dielec. Const.: 4.0 (dry, 1 MC, D150); Vol.

Resist.: 8 X 10^{14} ohm-cm (D257); Arc Resist.: 180 s (D495)
Ref.: Data sheet

E-2748. Fiberite Corp.
Chem. Descrip.: Glass-reinforced epoxy
Category/Applications: Med.-impact molding compound with good impact strength, electrical properties, and very soft flow; used in terminal blocks, connectors, pipe fittings, textile processing equipment, missile components, structural applications
Form: Granular; Color: Green, black
Gen'l. Prop. (as molded): Sp. Gr.: 1.70–1.74 (D792); Stability: Excellent heat and chemical resistance
Mech. Prop. (as molded): Tens. Str.: 11,000 psi (D638); Flex. Str.: 17,000 psi (D790); Flex. Mod.: 1.4 X 10^6 psi (D790); Compr. Str.: 28,000 psi (D-695); Impact Str.: 0.70 ft lb/in. notch (D256); Hardness (Rockwell): M 115 (D785)
Therm. Prop. (as molded): Conduct.: 8 X 10^{-4} Cal/s/cm^2/C/cm (C177); Distort. Temp.: 154 C (@ 264 psi, D-648); Coeff. of Linear Exp.: 1.8 X 10^{-5} in./in./C (–30 to 30 C, D696)
Elec. Prop. (as molded): Power Factor: 0.01 (dry, 1 MC, D150); Dielec. Str.: 400 V/mil (dry, S/T, D149); Dielec. Const.: 4.5 (dry, 1 MC, D150); Vol. Resist.: 9 X 10^{15} ohm-cm (D257); Arc Resist.: 180 s (D495)
Toxicity/Handling: See E-231
Std. Pkgs.: 50-lb drums
Ref.: Data sheet

E-3800 Series. Fiberite Corp.
Chem. Descrip.: Epoxy
Category/Applications: Epoxy molding compounds designed with the resin-hardener system first introduced with E-3824; excellent low pressure molding characteristics at a wide range of temps. and pressures for encapsulation of high performance devices, general purpose work with systems containing low ionic contamination and low hydrolyzable chloride, etc.; cure cycles can be as short as 30 s depending on part size and mold temp.

E-3800. Fiberite Corp.
Chem. Descrip.: Epoxy
Category/Applications: Lower cost version of E-3824 retaining excellent thermal shock resistance and electrical behavior; slightly higher melt viscosity; avail. in the same flow ranges as E-3824

E-3810. Fiberite Corp.
Chem. Descrip.: Epoxy
Category/Applications: Low-cost, general purpose encapsulant designed for applications requiring minimum thermal shock resistance but excellent electrical performance

E-3824. Fiberite Corp.
Chem. Descrip.: Glass-filled epoxy
Category/Applications: Low pressure molding compound for encapsulation in the electronics industry; good thermal shock resistance and excellent wet electrical properties; cure and molding characteristics from 250–350 F
Form: Granular
Gen'l. Prop. (as molded): Sp. Gr.: 1.63; Stability: Good thermal shock resistance
Mech. Prop. (as molded): Tens. Str.: 10,500 psi; Flex. Str.: 15,500 psi; Compr. Str.: 24,000 psi; Impact Str.: 0.40 ft lb/in. notch
Therm. Prop. (as molded): Conduct.: 9 X 10^{-4} Cal/s/cm^2/C/cm; Distort. Temp.: 320 F (@ 264 psi); Coeff. of Linear Exp.: 2.9 X 10^{-5} in./in./C
Elec. Prop. (as molded): Power Factor: 0.012 (1 MC); Dielec. Str.: 350 V/mil

(S/T); Dielec. Const.: 4.7 (1 MC); Vol. Resist.: 4.0×10^{15} ohm-cm; Arc Resist.: 140 s
Toxicity/Handling: See E-231
Ref.: Product information

E-3938. Fiberite Corp.
Chem. Descrip.: Epoxy
Category/Applications: Med.-impact molding material for use in connectors, rectifiers, and other applications where structural considerations are important; excellent moldability; good shelf life
Form: Granular
Gen'l. Prop. (as molded): Sp. Gr.: 1.85
Mech. Prop. (as molded): Tens. Str.: 10,500 psi; Flex. Str.: 18,000 psi; Compr. Str.: 30,000 psi; Impact Str.: 0.50 ft lb/in. notch
Therm. Prop. (as molded): Conduct.: 13×10^{-4} Cal/s/cm^2/C/cm; Distort. Temp.: 450 F (@ 264 psi); Coeff. of Linear Exp.: 1.7×10^{-5} in./in./C
Elec. Prop. (as molded): Power Factor: 0.017 (1 MC); Dielec. Str.: 400 V/mil (S/T); Dielec. Const.: 5.0 (1 MC); Vol. Resist.: 2.3×10^{15} ohm-cm; Arc Resist.: 180 s
Toxicity/Handling: See E-231
Ref.: Product information

E-8353. Fiberite Corp.
Chem. Descrip.: Glass-filled epoxy
Category/Applications: General purpose epoxy molding compound providing an excellent balance of mechanical and electrical properties, and adequate thermal shock resistance for most applications; ideal for coil encapsulation
Form: Granular; Color: Black, standard colors
Gen'l. Prop. (as molded): Sp. Gr.: 1.92–1.95 (D792)
Mech. Prop. (as molded): Tens. Str.: 10,000 psi (D638); Flex. Str.: 16,000 psi (D790); Flex. Mod.: 2.0×10^6 psi (D790); Compr. Str.: 31,000 psi (D695); Impact Str.: 0.35 ft lb/in. notch (D256)
Therm. Prop. (as molded): Conduct.: 17×10^{-4} Cal/s/cm^2/C/cm (C177); Distort. Temp.: 425 F (@ 264 psi, D648); Coeff. of Linear Exp.: 2.1×10^{-5} in./in. (−30 to 30 C, D696)
Elec. Prop. (as molded): Power Factor: 0.013 (dry, 1 MC, D150); Dielec. Str.: 410 V/mil (dry, S/T, D149); Dielec. Const.: 4.9 (dry, 1 MC, D150); Arc Resist.: 180 s (D495)
Toxicity/Handling: See E-231
Std. Pkgs.: 50-lb drums
Ref.: Data sheet

E-8354. Fiberite Corp.
Chem. Descrip.: Glass-filled epoxy
Category/Applications: General purpose epoxy molding compound with improved thermal shock resistance and superior mechanical properties; encapsulation flow material for use in transfer and injection molding equipment
Form: Granular
Gen'l. Prop. (as molded): Sp. Gr.: 1.80–1.83 (D792)
Mech. Prop. (as molded): Tens. Str.: 10,500 psi (D638); Flex. Str.: 18,000 psi (D790); Flex. Mod.: 2.0×10^6 pis (D790); Compr. Str.: 30,000 psi (D-695); Impact Str.: 0.43 ft lb/in. notch (D256)
Therm. Prop. (as molded): Conduct.: 15×10^{-4} Cal/s/cm^2/C/cm (C177); Distort. Temp.: 425 F (@ 264 psi, D-648); Coeff. of Linear Exp.: 2.2×10^{-5} in./in. (−30 to 30 C, D696); Flamm.: Self-extinguishing
Elec. Prop. (as molded): Power Factor: 0.014 (dry, 1 MC, D150); Dielec. Str.: 425 V/mil (dry, S/T, D149); Dielec. Const.: 4.75 (dry, 1 MC, D150); Arc

Resist.: 180 s (D495)
(D495)
Toxicity/Handling: See E-231
Std. Pkgs.: 50-lb drums
Ref.: Data sheet

E-8354J. Fiberite Corp.
Chem. Descrip.: Epoxy
Category/Applications: Version of E-8354 designed for injection molding; meets injection molding parameters of long barrel life and fast cure cycles

E-8405. Fiberite Corp.
Chem. Descrip.: Epoxy
Category/Applications: Encapsulation grade epoxy designed to have the lowest hydrolyzable chloride content; excellent moldability with good mechanical and electrical properties; used for semiconductor encapsulation
Form: Granular
Gen'l. Prop. (as molded): Sp. Gr.: 1.80
Mech. Prop. (as molded): Tens. Str.: 8400 psi; Flex. Str.: 12,000 psi; Flex. Mod.: 1.5 $\times$ 10^6 psi; Compr. Str.: 27,000 psi; Impact Str.: 0.30 ft lb/in. notch; Hardness (Rockwell): M 110
Therm. Prop. (as molded): Conduct.: 16 $\times$ 10^{-4} Cal/s/cm^2/C/cm; Distort. Temp.: 315 F (@ 264 psi); Coeff. of Linear Exp.: 2.7 $\times$ 10^{-5} in./in./C (−30 to 30 C)
Elec. Prop. (as molded): Power Factor: 0.016 (dry, 1 MC); Dielec. Str.: 330 V/mil (dry, S/T); Dielec. Const.: 4.6 (dry, 1 MC); Vol. Resist.: 1 $\times$ 10^{14} ohm-cm; Arc Resist.: 180 s
Toxicity/Handling: See E-231
Ref.: Product information

E-9405. Fiberite Corp.
Chem. Descrip.: Glass fiber-filled epoxy
Category/Applications: Molding compound for the encapsulation of delicate electrical components as well as large coils; excellent thermal shock, moldability, and retention of electrical properties under humid conditions
Form: Granular; Color: Black, standard colors
Gen'l. Prop. (as molded): Sp. Gr.: 1.62–1.65 (D792)
Mech. Prop. (as molded): Tens. Str.: 10,000 psi (D638); Flex. Str.: 19,000 psi (D790); Flex. Mod.: 1.5 $\times$ 10^6 psi (D790); Compr. Str.: 25,000 psi (D695); Imapct Str.: 0.40 ft lb/in. notch (D-256); Hardness (Rockwell): M 115 (D-785)
Therm. Prop. (as molded): Conduct.: 9 $\times$ 10^{-4} Cal/s/cm^2/C/cm (C177); Distort. Temp.: 146 C (@ 264 psi, D648); Coeff. of Linear Exp.: 2.4 $\times$ 10^{-5} in./in./C (−30 to 30 C, D696)
Elec. Prop. (as molded): Power Factor: 0.012 (dry, 1 MC, D150); Dielec. Str.: 400 V/mil (dry, S/T, D149); Dielec. Const.: 3.6 (dry, 1 MC, D150); Vol. Resist.: 4 $\times$ 10^{15} ohm-cm (D257); Arc Resist.: 140 s (D495)
Toxicity/Handling: See E-231
Std. Pkgs.: 50-lb drums
Ref.: Data sheet

E-9451. Fiberite Corp.
Chem. Descrip.: Mineral-filled epoxy
Category/Applications: Molding compound with good balance of mechanical and electrical properties, low water absorption, and high thermal conductivity; for encapsulation of coils, resistors, capacitors, and electronic components, terminal blocks, bobbins, connectors
Form: Granular; Color: Black, standard colors
Gen'l. Prop. (as molded): Sp. Gr.: 1.91–1.95 (D792)
Mech. Prop. (as molded): Tens. Str.: 9000 psi (D638); Flex. Str.: 16,000 psi (D790); Flex. Mod.: 2.0 $\times$ 10^6 psi (D-790); Compr. Str.: 30,000 psi (D695);

Impact Str.: 0.30 ft lb/in. notch (D256); Hardness (Rockwell): M 115 (D785)

Therm. Prop. (as molded): Conduct.: 15 X 10^{-4} Cal/s/cm^2/C/cm (C177); Distort. Temp.: 166 C (@ 264 psi, D648); Coeff. of Linear Exp.: 2.3 X 10^{-5} in./in./C (–30 to 30 C, D696)

Elec. Prop. (as molded): Power Factor: 0.01 (dry, 1 MC, D150); Dielec. Str.: 380 V/mil (dry, S/T, D149); Dielec. Const.: 3.8 (dry, 1 MC, D150); Vol. Resist.: 6 X 10^{15} ohm-cm (D257); Arc Resist.: 180 s (D495)

Ref.: Data sheet

Eccofoam SIL. Emerson & Cuming/W.R. Grace & Co.
Chem. Descrip.: RTV silicone product
Category/Applications: Pourable synthetic foam for embedment of electronics subject to shock and vibration; useful in aerospace and hydrospace for weight saving; continous use temps. to 400 F
Gen'l. Prop.: Sp. Gr.: Avail. in 0.33 and 0.50 g/cc grades (soft and firm respectively); Visc.: 17,000 cps
Mech. Prop.: Tens. Str.: 125 psi; Tens. Elong.: 300%
Therm. Prop.: Conduct.: 0.5 (Btu)(in.)
Toxicity/Handling: Use with adequate ventilation; avoid contact with skin, eyes, and clothing
Ref.: Bulletin

Eccosil 1776. Emerson & Cuming/W.R. Grace & Co.
Chem. Descrip.: RTV silicone product
Category/Applications: One-part G/P adhesive, caulk, sealant; easy-to-use, nonflowing flexible, moisture-resistant electrical grade for glass ceramics, metal, and plastics; cartridge packed; continuous use temps. to 400 F
Gen'l. Prop.: Sp. Gr.: 1.04 g/cc; Visc.: Thixotropic
Mech. Prop.: Tens. Str.: 380 psi; Tens. Elong.: 400%; Hardness (Shore A): 30
Therm. Prop.: Conduct.: 1.7 (Btu)(in.)
Toxicity/Handling: See Eccosil 2CN
Ref.: Bulletin

Eccosil 4122. Emerson & Cuming/W.R. Grace & Co.
Chem. Descrip.: RTV silicone product
Category/Applications: Silicone for encapsulating densely-packed components, filling thin sections, making flexible molds, casting; offers ease of use, pourability, flexibility, and continuous use temps. to 400 F
Gen'l. Prop.: Sp. Gr.: 1.15 g/cc; Visc.: 10,000 cps
Mech. Prop.: Tens. Str.: 400 psi; Tens. Elong.: 180%; Hardness (Shore A): 45
Therm. Prop.: Conduct.: 2.3 (Btu)(in.)
Toxicity/Handling: See Eccosil 2CN
Ref.: Bulletin

Eccosil 4451. Emerson & Cuming/W.R. Grace & Co.
Chem. Descrip.: RTV silicone product
Category/Applications: See Eccosil 4122
Gen'l. Prop.: Sp. Gr.: 1.65 g/cc; Visc.: 25,000 cps
Mech. Prop.: Tens. Str.: 700 psi; Tens. Elong.: 100%; Hardness (Shore A): 60
Therm. Prop.: Conduct.: 2.8 (Btu)(in.)
Toxicity/Handling: See Eccosil 2CN
Ref.: Bulletin

Eccosil 4553. Emerson & Cuming/W.R. Grace & Co.
Chem. Descrip.: RTV silicone product
Category/Applications: See Eccosil 4122
Gen'l. Prop.: Sp. Gr.: 1.60 g/cc; Visc.: 36,000 cps
Mech. Prop.: Tens. Str.: 800 psi; Tens. Elong.: 100%; Hardness (Shore A): 72
Therm. Prop.: Conduct.: 2.5 (Btu)(in.)
Toxicity/Handling: See Eccosil 2CN
Ref.: Bulletin

Eccosil 4640. Emerson & Cuming/W.R. Grace & Co.
Chem. Descrip.: RTV silicone product
Category/Applications: Pourable synthetic foam for embedment of electronics subject to shock and vibration; useful in aerospace and hydrospace for weight saving; continuous use temps. to 400 F
Gen'l. Prop.: Sp. Gr.: 0.75 g/cc; Visc.: 45,000 cps
Mech. Prop.: Tens. Str.: 760 psi; Tens. Elong.: 135%; Hardness (Shore A): 55
Therm. Prop.: Conduct.: 1.0 (Btu)(in.)
Toxicity/Handling: See Eccosil 2CN
Ref.: Bulletin

Eccosil 4712. Emerson & Cuming/W.R. Grace & Co.
Chem. Descrip.: Combination of epoxy and silicone
Category/Applications: Combination product for adhesion and flexibility; used for potting cable connectors and electrical and electronic components in severe environments; continuous use temps. to 375 F
Gen'l. Prop.: Sp. Gr.: 1.20 g/cc; Visc.: 120,000 cps
Mech. Prop.: Tens. Str.: 500 psi; Tens. Elong.: 60%; Hardness (Shore A): 68
Therm. Prop.: Conduct.: 2.5 (Btu)(in.)
Toxicity/Handling: See Eccosil 2CN
Ref.: Bulletin

Eccosil 4850. Emerson & Cuming/W.R. Grace & Co.
Chem. Descrip.: RTV silicone product
Category/Applications: See Eccosil 4122
Gen'l. Prop.: Sp. Gr.: 1.70 g/cc; Visc.: 38,000 cps
Mech. Prop.: Tens. Str.: 800 psi; Tens. Elong.: 140%; Hardness (Shore A): 60
Therm. Prop.: Conduct.: 3.2 (Btu)(in.)
Toxicity/Handling: See Eccosil 2CN
Ref.: Bulletin

Eccosil 4952. Emerson & Cuming/W.R. Grace & Co.
Chem. Descrip.: RTV silicone product
Category/Applications: Silicone for encapsulation and sealing of heater elements, heat dissipation from modules, and components in computers, appliances, aircraft, and autos; solar panel seals; continuous use temps. to 500 F
Gen'l. Prop.: Sp. Gr.: 2.28 g/cc; Visc.: 30,000 cps
Mech. Prop.: Tens. Str.: 650 psi; Tens. Elong.: 100%; Hardness (Shore A): 70
Therm. Prop.: Conduct.: 7.5 (Btu)(in.)
Toxicity/Handling: See Eccosil 2CN
Ref.: Bulletin

Eccosil 4954. Emerson & Cuming/W.R. Grace & Co.
Chem. Descrip.: RTV silicone product
Category/Applications: See Eccosil 4952
Gen'l. Prop.: Sp. Gr.: 2.33 g/cc; Visc.: 40,000 cps
Mech. Prop.: Tens. Str.: 525 psi; Tens. Elong.: 85%; Hardness (Shore A): 78
Therm. Prop.: Conduct.: 15.0 (Btu)(in.)
Toxicity/Handling: See Eccosil 2CN
Ref.: Bulletin

Eccosil 4995. Emerson & Cuming/W.R. Grace & Co.
Chem. Descrip.: RTV silicone product
Category/Applications: See Eccosil 4952; 4995 offers continuous use temps. to 400 F
Gen'l. Prop.: Sp. Gr.: 1.87 g/cc; Visc.: 25,000 cps
Mech. Prop.: Tens. Str.: 525 psi; Tens. Elong.: 85%; Hardness (Shore A): 75
Therm. Prop.: Conduct.: 5.0 (Btu)(in.)
Toxicity/Handling: See Eccosil 2CN
Ref.: Bulletin

Eccosil SC71. Emerson & Cuming/W.R. Grace & Co.
Chem. Descrip.: RTV silicone product
Category/Applications: Used in coating to cushion components prior to potting; continuous use temps. to 550 F
Gen'l. Prop.: Sp. Gr.: 1.55 g/cc; Visc.: 300 cps
Mech. Prop.: Tens. Str.: 300 psi; Tens. Elong.: 140%; Hardness (Shore A): 60
Therm. Prop.: Conduct.: 2.2 (Btu)(in.)
Toxicity/Handling: See Eccosil 2CN
Ref.: Bulletin

Eccosil TC-4, TC-8. Emerson & Cuming/W.R. Grace & Co.
Chem. Descrip.: RTV silicone product
Category/Applications: Greases for thermal conductivity offering high thermal "K", high insulation resistance, high dielec. strength on solar panels; heat sinks on semiconductor devices, power resistors, thermostats, etc.; continuous use temps. to 350 F
Gen'l. Prop.: Sp. Gr.: 2.20 and 2.30 g/cc respectively; Visc.: Thixotropic
Therm. Prop.: Conduct.: 11.0 and 16.0 (Btu)(in.) respectively
Toxicity/Handling: See Eccosil 2CN
Ref.: Bulletin

Eccosil TP50. Emerson & Cuming/W.R. Grace & Co.
Chem. Descrip.: RTV silicone product
Category/Applications: Nonflow, RTV adhesive for caulking, bonding applications, with good thermal conductivity; continuous use temps. to 450 F
Gen'l. Prop.: Sp. Gr.: 1.15 g/cc; Visc.: Thixotropic
Mech. Prop.: Tens. Str.: 700 psi; Tens. Elong.: 150%; Hardness (Shore A): 56
Therm. Prop.: Conduct.: 5.0 (Btu)(in.)
Toxicity/Handling: See Eccosil 2CN
Ref.: Bulletin

Eccosil 2CN. Emerson & Cuming/W.R. Grace & Co.
Chem. Descrip.: RTV silicone product
Category/Applications: Industrial grade RTV silicone for water clear repairable potting, embedment of circuits; continuous use temps. to 400 F
Gen'l. Prop.: Sp. Gr.: 0.99 g/cc; Visc.: 1700 cps
Mech. Prop.: Tens. Str.: Low; Tens. Elong.: 60%; Hardness (Shore A): 22
Therm. Prop.: Conduct.: 1.2 (Btu)(in.)
Toxicity/Handling: Use with adequate ventilation; avoid contact with eyes, skin, and clothing
Ref.: Bulletin

E-Form 7-51. Acme Chem. & Insulation Co.
Chem. Descrip.: Mineral-filled, short glass-reinforced epoxy comp.
Category/Applications: Molding comp. exhibiting exceptional hot strength, low shrinkage on cure, and resistance to hot transformer oils; used for applications requiring a rigid molding comp. exhibiting full hardness on ejectrion from the mold, and exhibiting a low ionizable content when subject to long-term oil immersion; esp. suitable for coils and solenoids with thin wall sections and/or irregular or sharp, angular projections
Form: Granular
Gen'l. Prop.: Sp. Gr.: 1.95; Stability: Outstanding dimensional stability in molded parts; excellent storage stability
Mech. Prop.: Tens. Str.: 7800 psi; Flex. Str.: 13,800 psi; Compr. Str.: 25,000 psi
Therm. Prop.: Distort. Temp.: 168 C (@ 264 psi, as molded); Coeff. of Linear Exp.: 27×10^{-6} in./in./C; Flamm.: Self-extinguishing
Elec. Prop.: Dielec. Str.: 410 V/mil; Dielec. Const.: 5.0 (1 kHz); Vol. Resist.: 9×10^{15} ohm-cm

Curing Charact.: Cure Schedule: 2 min @ 300 F (1/8-in. section)
Ref.: Data sheet

E-Form 7-61. Acme Chem. & Insulation Co.
Chem. Descrip.: Mineral-filled, glass-reinforced epoxy comp.
Category/Applications: Soft flow molding comp. exhibiting a very low ionizable content and exceptional retention of elec. properties after rigorous pressure-humidity tests; suitable for transfer molding and encapsulation of electronic components, e.g., metal film and wire-wound resistors, R-C networks, tantalum capacitors, diodes, and other semiconductor devices; also displays excellent thermal shock characteristics
Form: Granular
Gen'l. Prop.: Sp. Gr.: 1.92; Stability: Excellent storage stability (12 mos. @ 77 F)
Mech. Prop.: Tens. Str.: 11,400 psi (ultimate); Tens. Elong.: 0.8%; Tens. Mod.: 2.15×10^6 psi; Flex. Str.: 19,000 psi; Compr. Str.: 30,000 psi; Impact Str. (Izod): 0.4 ft lb/in. notch; Hardness (Rockwell): M 112
Therm. Prop.: Glass Trans.: 143 C; Distort. Temp.: 146 C (@ 264 psi, as molded); Coeff. of Linear Exp.: 29×10^{-6} in./in./C; Flamm.: V-O (UL94)
Elec. Prop.: Dissip. Factor: 0.01 (1 kHz); Dielec. Str.: 410 V/mil; Dielec. Const.: 5.0 (1 kHz); Vol. Resist.: 9×10^{15} ohm-cm
Curing Charact.: Cure Schedule: 2 min @ 300 F (1/8-in. section)
Ref.: Data sheet

E-Form 7-61 M. Acme Chem. & Insulation Co.
Chem. Descrip.: Mineral-filled, short glass-reinforced epoxy comp.
Category/Applications: Soft flow molding comp. with a low coeff. of expansion, for use in encapsulation of delicate and complex components; recommended for transfer molding and encapsulation of electronic components, e.g., glass diodes, thick and thin film resistor networks, and other devices packaged in SIP or DIP configurations; exhibits a very low ionizable content and exceptional retention of elec. properties after rigorous pressure-humidity tests
Form: Granular
Gen'l. Prop.: Sp. Gr.: 2.0; Stability: Excellent storage stability (12 mos. @ 77 F)
Mech. Prop.: Tens. Str.: 11,000 psi (ultimate); Tens. Elong.: 0.7%; Tens. Mod.: 2.25×10^6 psi; Flex. Str.: 19,000 psi; Compr. Str.: 32,000 psi; Impact Str. (Izod): 0.4 ft lb/in. notch; Hardness (Rockwell): M 115
Therm. Prop.: Glass Trans.: 143 C; Distort. Temp.: 154 C (@ 264 psi, as molded); Coeff. of Linear Exp.: 23.5×10^{-6} in./in./C; Flamm.: V-O (UL94)
Elec. Prop.: Dissip. Factor: 0.01 (1 kHz); Dielec. Str.: 410 V/mil; Dielec. Const.: 5.0 (1 kHz); Vol. Resist.: 9×10^{15} ohm-cm
Curing Charact.: Cure Schedule: 2 min @ 300 F (1/8-in. section)
Ref.: Data sheet

E-Form 7-61 MM. Acme Chem. & Insulation Co.
Chem. Descrip.: Mineral-filled, short glass-reinforced epoxy comp.
Category/Applications: Molding comp. offering low coeff. of thermal expansion, high glass trans. temp., very low ionizable content, exceptional moisture resistance, and nonstaining and nonburning properties, for use in encapsulation of transistors, integrated circuits, and other semiconductors in discrete, DIP,

or SIP configurations
Form: Granular
Gen'l. Prop.: Sp. Gr.: 1.93
Mech. Prop.: Tens. Str.: 11,000 psi (ultimate); Tens. Elong.: 0.7%; Tens. Mod.: 2.25×10^6 psi; Flex. Str.: 19,000 psi; Compr. Str.: 32,000 psi; Impact Str. (Izod): 0.45 ft lb/in. notch; Hardness (Rockwell): M 115
Therm. Prop.: Glass Trans.: 157 C; Conduct.: 18×10^{-4} Cal/s/cm^2/C/cm; Distort. Temp.: 170 C (@ 264 psi, as molded); Coeff. of Linear Exp.: 21.5×10^{-6} in./in./C (25–154 C); Flamm.: V-O (UL-94)
Elec. Prop.: Dissip. Factor: 0.01 (1 kHz); Dielec. Str.: 410 V/mil (1/8-in. section); Dielec. Const.: 5.0 (1 kHz); Vol. Resist.: 9×10^{15} ohm-cm
Curing Charact.: Cure Schedule: 2 min @ 300 F (1/8-in. section)
Ref.: Data sheet

E-Form 7-67. Acme Chem. & Insulation Co.
Chem. Descrip.: Mineral-filled, short glass-reinforced epoxy comp.
Category/Applications: Molding comp. offering excellent retention of insulation characteristics after prolonged exposure to high temps., short molding cycles which produce dimensionally stable parts with full hardness on ejection for easy removal from mold; recommended for encapsulation of electronic devices subjected to high operating temps., e.g., resistors, networks, capacitors, and semiconductors, and for low cost devices
Form: Granular
Gen'l. Prop.: Sp. Gr.: 1.95; Stability: Excellent storage stability
Mech. Prop.: Tens. Str.: 7800 psi; Flex. Str.: 13,800 psi; Compr. Str.: 25,000 psi; Impact Str. (Izod): 0.4 ft lb/in. notch; Hardness (Rockwell): M 110
Therm. Prop.: Distort. Temp.: 177 C (@ 264 psi, as molded); Coeff. of Linear Exp.: 27×10^{-6} in./in./C; Flamm.: Self-extinguishing
Elec. Prop.: Dielec. Str.: 410 V/mil; Dielec. Const.: 5.0 (1 kHz); Vol. Resist.: 10^{15} ohm-cm
Curing Charact.: Cure Schedule: 2 min @ 300 F
Ref.: Data sheet

E-Form 7-72. Acme Chem. & Insulation Co.
Chem. Descrip.: Mineral-filled epoxy comp.
Category/Applications: Soft flow, low temp. curing molding comp. exhibiting a very low ionizable content and exceptional retention of elec. properties after rigorous pressure-humidity tests; used for encapsulation of electronic devices with minimal thermal shock requirements, e.g., wirewound resistors, small metal film resistors, tantalum capacitors, diodes, and varistors
Form: Granular
Gen'l. Prop.: Sp. Gr.: 1.92; Stability: Excellent storage stability
Mech. Prop.: Tens. Str.: 7500 psi; Flex. Str.: 11,000 psi; Compr. Str.: 24,000 psi; Impact Str. (Izod): 0.4 ft lb/in. notch; Hardness (Rockwell): M 106
Therm. Prop.: Distort. Temp.: 141 C (@ 264 psi, as molded); Coeff. of Linear Exp.: 29.0×10^{-6} in./in./C; Flamm.: Self-extinguishing
Elec. Prop.: Dissip. Factor: 0.015 (dry, 1 kHz); Dielec. Str.: 410 V/mil (1/8-in. section); Dielec. Const.: 5.0 (dry, 1 kHz); Vol. Resist.: 9×10^{15} ohm-cm
Curing Charact.: Cure Schedule: 2 min @ 300 F (1/8-in. section)
Ref.: Data sheet

EHEC. Hercules Inc.
Chem. Descrip.: Ethyl hydroxyethyl cellulose; mixed cellulose ether
Category/Applications: Film-forming polymer used in printing inks, lacquers, varnishes, and other specialty finishes; gives clarity, color stability, toughness, flexibility, water resistance, and fast dry to lacquers; cold-cut addition to varnishes; as a vehicle-phase flatting agent; contributes bodying and toughening to heat-set inks, etc.; avail. in three viscosity types, extra-low, low, and high
Form: Granular powder; Color: White
Gen'l. Prop.: Sol.: Sol. in hydrocarbons, esters, alcohols, ketones, chlorinated solvents; unusual tolerance for low-cost aliphatic hydrocarbons; Sp. Vol.: 24.5 $in.^3$/lb (in sol'n.); Density: 19–22 lb/ft^3 (bulk, granules), 1.12 g/cc (film); Visc.: 10–20 cps, 20–35 cps, 125–250 cps for extra-low, low, and high grades respectively (5% conc. in 80/20 toluene/ethanol by wt. on sample dried 30 min @ 100 C); Flow Temp.: > 175 C (unplasticized); Stability: Unaffected by visible or uv light; retains flexibility over wide temp. range
Mech. Prop. (of 3-mil film): Tens. Str.: 4000–6000 psi (@ break); Tens. Elong.: 6–10% (@ break); Flexibility: 500–900 (MIT double folds)
Therm. Prop.: Flamm.: Low order flammability; can be rendered flame retardant
Ref.: Data sheet 475

Elvace 1870. Du Pont Co.
Chem. Descrip.: Acetate/ethylene copolymer emulsion
Category/Applications: Adhesive base providing significant improvements in plasticizer response, quick tack, and adhesion when compared with homopolymer
Composition: 55% solids
Gen'l. Prop.: Density: 1078 g/l; Visc. (Brookfield): 1700 cps (LV, 60 rpm, #3 spindle); pH: 4.5
Ref.: Tech. information

Elvace 1872. Du Pont Co.
Chem. Descrip.: Vinyl acetate/ethylene copolymer emulsion
Category/Applications: Adhesive base for use in vinyl/wood adhesives for laminates such as wood paneling on stereo and TV cabinets, paper/wood adhesives, caulking comp. for forming bonds around doors and windows, joint cement for filling in the space between gypsum board wall panels, panel adhesive to adhere vinyl or wood panels directly to wall studs without nails, side seam adhesive for kraft shopping bags, cardboard adhesive for carriers of bottled beverages, laminating adhesive for foil/cardboard, paper/polyethylene, paper/aluminum foil or K cellophane/polyethylene for a variety of food and drug packaging applications; Elvace 1872 is a general-purpose adhesive with med. ethylene content; offers slightly hazy film clarity and 116 C heat seal temp.
Composition: 55% solids
Gen'l. Prop.: Density: 1070 kg/m^3 (emulsion), 1140 kg/m^3 (polymer); Visc. (Brookfield): 1500 cps (LVF, 60 rpm, # 3 spindle); pH: 5
Mech. Prop. (polymer): Tens. Str.: 8.3 MPa; Tens. Elong.: 700%
Ref.: Bulletin E-07716

Elvace 1873. Du Pont Co.
Chem. Descrip.: Vinyl acetate/ethylene copolymer emulsion
Category/Applications: See Elvace 1872; 1873 is a water-resistant adhesive with med. ethylene content, hazy film clarity, and heat seal temp. of 110 C
Composition: 55% solids
Gen'l. Prop.: Density: 1070 kg/m^3

(emulsion), 1140 kg/m^3 (polymer); Visc. (Brookfield): 1500 cps (LVF, 60 rpm, #3 spindle); pH: 6
Mech. Prop. (polymer): Tens. Str.: 9.0 MPa; Tens. Elong.: 710%
Ref.: Bulletin E-07716

Elvace 1874. Du Pont Co.
Chem. Descrip.: Vinyl acetate/ethylene copolymer emulsion
Category/Applications: See Elvace 1872; 1874 is a general-purpose adhesive with med. ethylene content, slightly hazy film clarity, and heat seal temp. of 104 C
Composition: 55% solids
Gen'l. Prop.: Density: 1070 kg/m^3 (emulsion), 1140 kg/m^3 (polymer); Visc. (Brookfield): 4000 cps (LVF, 60 rpm, #4 spindle); pH: 5
Mech. Prop. (polymer): Tens. Str.: 6.9 MPa; Tens. Elong.: 600%
Ref.: Bulletin E-07716

Elvace 1875. Du Pont Co.
Chem. Descrip.: Vinyl acetate/ethylene copolymer emulsion
Category/Applications: See Elvace 1872; 1875 is a general purpose adhesive with high ethylene content offering the best adhesivity, viscosity build, and flexibility in series, and with slightly hazy film clarity and heat seal temp. of 82 C
Composition: 55% solids
Gen'l. Prop.: Density: 1070 kg/m^3 (emulsion), 1110 kg/m^3 (polymer); Visc. (Brookfield): 1700 cps (LVF, 60 rpm, #3 spindle); pH: 4.5
Mech. Prop. (polymer): Tens. Str.: 4.3 MPa; Tens. Elong.: 650%
Ref.: Bulletin E-07716

Elvace 1961. Du Pont Co.
Chem. Descrip.: Vinyl acetate/ethylene copolymer emulsion
Category/Applications: See Elvace 1872; 1961 is a water-resistant adhesive with low ethylene content, slightly hazy film clarity, and heat seal temp. of 138 C
Composition: 50% solids
Gen'l. Prop.: Density: 1070 kg/m^3 (emulsion), 1160 kg/m^3 (polymer); Visc. (Brookfield): 300 cps (LVF, 60 rpm, #3 spindle); pH: 5.5
Mech. Prop. (polymer): Tens. Str.: 11.7 MPa; Tens. Elong.: 500%
Ref.: Bulletin E-07716

Elvace TF. Du Pont Co.
Chem. Descrip.: Acetate/ethylene emulsion
Category/Applications: Emulsion for textile finishing applications
Composition: 55% min. solids
Gen'l. Prop.: Density: 1070 kg/m^3; Visc. (Brookfield): 500–1000 cps (60 rpm, #3 spindle); pH: 4.0–5.5
Mech. Prop. (film): Tens. Str.: 5.3 MPa; Tens. Elong.: 530% (break); Mod.: 17.2 MPa (initial)
Ref.: Tech. information E-16944

Elvace 1875 HV. Du Pont Co.
Chem. Descrip.: Acetate/ethylene emulsion
Category/Applications: High visc. version of Elvace 1875 for adhesive and/or binder applications where higher performance acetate/ethylene emulsions are required; offers cost savings when used in equal visc. adhesives at slightly lower solids, and slightly faster bonding speed than Elvace 1875; FDA approved for use in articles contacting food
Composition: 55% solids
Gen'l. Prop.: Density: 1070 kg/m^3; Visc. (Brookfield): 2100–3000 cps (LV, 60 rpm, #4 spindle); pH: 4.0–5.0
Toxicity/Handling: Emulsions contain minor amts. ($<$ 1%) of vinyl acetate monomer; use with adequate ventilation; ignitable vapor may accumulate in

the head space of containers; keep away from heat, sparks, and open flames when opening containers
Ref.: Bulletin E-35733

Elvanol 71-30. Du Pont Co.
Chem. Descrip.: Fully hydrolyzed polyvinyl alcohol
Category/Applications: Med. visc. grade offering film forming and binding characteristics along with insol. in cold water and organic solvents; used in adhesives, paper and paperboard sizing and coatings, textiles, gastight, solvent-resistant films, building products, hoses, gaskets, emulsification in emulsions and latices, as an additive for strength in concrete and cements, etc.; FDA approved for indirect food additive applications; biodegradable
Form: Granular; Color: White
Composition: 5% max. volatiles; 1.0% max. ash
Gen'l. Prop.: Sol.: Sol. in hot water, ethanol/water mixtures containing up to 35% by wt. of ethanol; insol. in cold water; Sp. Gr.: 1.30; Sp. Vol.: 7.7×10^{-4} m^3/kg; Density (Bulk): 400–432 kg/m^3; Visc. (Hoeppler Falling Ball): 28–32 cps (4% aq. sol'n., 20 C); Sapon. No.: 3–12; Stability: Unaffected by animal and veg. oils, greases, and petroleum hydrocarbons; substantially unaffected by esters, ethers, ketones, aliphatic and aromatic hydrocarbons, chlorohydrocarbons, and monohydric alcohols; high impermeability to oxygen, nitrogen, carbon dioxide, hydrogen, helium, and hydrogen sulfide gases; excellent stability to sunlight and artificial light; Ref. Index: 1.54; pH: 5.0–7.0 (sol'n.)
Mech. Prop.: Tens. Str.: 117 MPa (compr. molded sheeting, unplasticized, 35% r.h.); Tens. Elong.: 10% (break, dry, unplasticized) to 400% (plasticized films conditioned @ 50% r.h.); Hardness (Shore): > 100 (unplasticized)
Therm. Prop.: Sp. Heat: 1674 J/kg•K
Toxicity/Handling: Nontoxic; avoid inhalation of dust particles; store sol'ns. in corrosion-resistant containers; sol'ns. containing > 10% solids may gel @ R.T.
Std. Pkgs.: 22.7-kg net multiwall paper bags and bulk
Ref.: Pamphlet E-27696

Elvanol 75-15. Du Pont Co.
Chem. Descrip.: Fully hydrolyzed polyvinyl alcohol
Category/Applications: Med.–low visc. grade which imparts improved visc. stability and gel resistance to aq. sol'ns.; used in adhesives (esp. wet adhesives that require long-term storage stability and which must withstand temp. variations), films, paper applications, as a binder in the manufacture of joint cements and spray dried ceramic powder, and as an emulsifier for improved water resistance; biodegradable; FDA-approved as food packaging adhesive
Form: Granular; Color: White
Composition: 5% max. volatiles by wt.; 1.0% max. ash by wt.
Gen'l. Prop.: Sol.: Slurries easily in cold water and dissolves readily on heating; Visc. (Hoeppler Falling Ball): 13–15 cps (4% aq. sol'n., 20 C); pH: 5.0–7.0 (sol'n.)
Mech. Prop.: Tens. Str.: 55.2–138 MPa
Toxicity/Handling: Nontoxic; avoid inhalation of dust
Std. Pkgs.: See Elvanol 71-30
Ref.: Pamphlets E-34139, E-25195-1

Elvanol 85-50. Du Pont Co.
Chem. Descrip.: Polyvinyl alcohol
Category/Applications: Med.-visc., gel-resistant grade for use in adhesvies for cellulosic substrates where a high degree of wet tack is required; adhesives esp.

suited for high-speed tube, core, and can winding operations; also used in solid fiber laminating; FDA approved for food packaging adhesives
Form: Granular
Gen'l. Prop.: Sol.: Sol. in water with heat and stirring; Visc. (Brookfield): 19–21 cps (4% solids, LVF, 60 rpm, #1 spindle); pH: 3.6–5.1 (sol'n.)
Toxicity/Handling: See Elvanol 75-15
Std. Pkgs.: See Elvanol 71-30
Ref.: Pamphlet E-43352

Elvanol 85-60. Du Pont Co.
Chem. Descrip.: Polyvinyl alcohol
Category/Applications: Med.-visc., tackified grade for use in adhesives requiring a high degree of wet tack, e.g., for paper tube, core, and can winding; also used in solid fiber laminating adhesives; FDA approved for use in food packaging adhesives
Form: Granular
Gen'l. Prop.: Sol.: Sol. in water with heat and stirring; Visc. (Brookfield): 25–57 cps (4% aq. sol'n., LVF, #1 spindle, 60 rpm); pH: 5.9–6.6 (sol'n.)
Toxicity/Handling: See Elvanol 75015
Std. Pkgs.: See Elvanol 71-30
Ref.: Pamphlet E-43351

Elvanol 85-80. Du Pont Co.
Chem. Descrip.: Polyvinyl alcohol
Category/Applications: High-visc., tackified grade for adhesives use in tube, core, and can winding, and solid fiber laminating; useful in adhesives where a high degree of wet tack is required; adhesives have good flow properties at high machine speeds and provide bonds with excellent water resistance; FDA approved for use in food packaging adhesives
Form: Granular
Gen'l. Prop.: Sol.: Sol. in water with heat and stirring; Visc. (Brookfield): 29–33 cps (4% solids, LVF, 60 rpm, #1 spindle); pH: 3.1–4.1 (sol'n.)
Toxicity/Handling: See Elvanol 75-15
Std. Pkgs.: See Elvanol 71-30
Ref.: Pamphlet E-43353

Elvanol 85-82. Du Pont Co.
Chem. Descrip.: Fully hydrolyzed polyvinyl alcohol
Category/Applications: Med.-visc. grade which combines high tensile and adhesive strength with good gel resistance; used for adhesives, films, and specialty applications; FDA approved for use in food packaging adhesives
Composition: 5% max. volatiles by wt.; 1.0% max. ash by wt.
Gen'l. Prop.: Sol.: Resistant to cold water, but dissolves readily on heating; Visc. (Hoeppler Falling Ball): 25–31 cps (4% aq. sol'n., 20 C); pH: 5.0–7.0 (sol'n.)
Toxicity/Handling: See Elvanol 75-15
Ref.: Pamphlet E-37675

Elvanol 90-50. Du Pont Co.
Chem. Descrip.: Fully hydrolyzed polyvinyl alcohol
Category/Applications: Med.–low visc. grade designed to provide high film strength and binding power in relatively low visc. systems; used in paper and paperboard coating and sizing, in adhesives, and as a pigment binder for highly filled systems; FDA approved for indirect food additive applications
Composition: 5% max. volatiles by wt.; 1.0% max. ash by wt.
Gen'l. Prop.: Sol.: Resistant to cold water, but dissolves readily on heating; Visc. (Hoeppler Falling Ball): 13–15 cps (4% aq. sol'n., 20 C); Stability: Good oil, grease, and film resistance; pH: 5.0–7.0 (sol'n.)
Toxicity/Handling: See Elvanol 75-15
Ref.: Pamphlet E-39968

Elvanol HV. Du Pont Co.
Chem. Descrip.: Fully hydrolyzed polyvinyl alcohol
Category/Applications: High-visc. grade providing exceptionally high tensile and adhesive strength with strong resistance to grease and hydrocarbons; used in adhesive, textile, and paper applications; FDA approved for use in indirect food additive applications
Form: Granular; Color: White
Composition: 4% max. volatiles by wt.; 1% max. ash, by wt.
Gen'l. Prop.: Sol.: Sol. in hot water but insol. in cold water and common organic solvents; Visc. (Hoeppler Falling Ball): 55–65 cps (4% aq. sol'n., 20 C); pH: 5–7 (sol'n.)
Toxicity/Handling: See Elvanol 75-15
Std. Pkgs.: See Elvanol 71-30
Ref.: Pamphlet E-43358

Elvanol T-25. Du Pont Co.
Chem. Descrip.: Polyvinyl alcohol
Category/Applications: Unique grade for textile warp sizing, esp. for use with polyester/cotton blends and other spun yarns; provides excellent weaving performance and desizies readily even after heat-setting sized fabric in the greige; FDA approved for food packaging adhesive applications
Form: Granular; Color: White
Composition: 5% max. volatiles by wt.; 1% max. ash by wt.
Gen'l. Prop.: Sol.: Slurries easily in cold water without lumping and dissolves readily on heating; Density (Bulk): 512 kg/m^3 (packed); Visc. (Hoeppler Falling Ball): 25–31 cps (4% aq. sol'n., 20 C); pH: 5.0–7.0 (sol'n.)
Toxicity/Handling: See Elvanol 75-15
Std. Pkgs.: See Elvanol 71-30
Ref.: Pamphlet E-37657

Elvanol T-66. Du Pont Co.
Chem. Descrip.: Polyvinyl alcohol
Category/Applications: Unique grade for textile warp sizing for use on polyester and polyester blend spun yarns; designed for smooth slasher operation and easy desize, while still providing excellent weaving performance; FDA approved for food packaging adhesive applications; biodegradable
Form: Granular; Color: White
Composition: 3% max. volatiles by wt.; 0.5% ash by wt.
Gen'l. Prop.: Sol.: Slurries easily in cold water without lumping and dissolves readily on heating to 85 C or higher for 20–30 min; Density (Bulk): 513 kg/m^3 (packed); Visc. (Hoeppler Falling Ball): 13–15 cps (4% aq. sol'n., 20 C); Stability: Sol'ns. are stable in visc. and can be stored for long periods; pH: 5.0–7.0 (sol'n.)
Toxicity/Handling: See Elvanol 75-15
Std. Pkgs.: See Elvanol 71-30
Ref.: Pamphlet E-25194

Elvax 150. Du Pont Co.
Chem. Descrip.: Ethylene/vinyl acetate copolymer resin
Category/Applications: See Elvax 40-P
Form: Pellets; Color: Translucent; Odor: Slight
Composition: 33% vinyl acetate by wt.
Gen'l. Prop.: Sol.: Sol. in many organic solvents; Density: 957 kg/m^3 (23 C, D-1505); Visc. (Inherent): 0.78 (30 C, 0.25 g/100 ml toluene); Soften. Pt. (R&B): 110 C (E28); Melt Index: 43 g/10 min (D1238); Ref. Index: 1.482
Mech. Prop.: Tens. Str.: 5.9 MPa (D-1708); Tens. Elong.: 1050% (break, D-1708); Tens. Mod.: 4.8 MPa (D638); Hardness (Shore A-2): 65 (10 s, D2240)
Toxicity/Handling & Std. Pkgs.: See Elvax 40-P
Ref.: Bulletin E-37658

Elvax 210. Du Pont Co.
Chem. Descrip.: Ethylene/vinyl acetate copolymer resin
Category/Applications: Lowest m.w. and melt visc. of the Elvax 200 series; less flexible in blends with wax than other 200-series resins, but superior in gloss retention and equal in grease resistance
Composition: 27.2–28.8% vinyl acetate
Gen'l. Prop.: Melt Index: 365–440 dg/min (D1238, modified)
Ref.: Bulletin E-37658

Elvax 220, 230, 240. Du Pont Co.
Chem. Descrip.: Ethylene/vinyl acetate copolymer resins
Category/Applications: Similar to Elvax 210; used in paraffin wax systems where visc. limitations are less stringent or where higher viscosities are required
Composition: 27.2–28.8% vinyl acetate
Gen'l. Prop.: Melt Index: 134–168, 100–120, and 38.0–48.0 dg/min respectively (D1238, modified)
Ref.: Bulletin E-37658

Elvax 250. Du Pont Co.
Chem. Descrip.: Ethylene/vinyl acetate copolymer resins
Category/Applications: See Elvax 210, 220; Elvax 250 also used in footwear applications, incl. foamed midsoles for athletic and leisure footwear, foamed sandals, canvas box toes, and flocked or fabric-laminated counters;
Form: Pellets; Color: Nearly transparent
Composition: 28% vinyl acetate by wt.
Gen'l. Prop.: Density: 951 kg/m^3 (D-792); Melt Index: 25 dg/min (D1238)
Mech. Prop.: Tens. Str.: 9.3 MPa (23 C, D638); Tens. Elong.: 800% (23 C, D-638); Tens. Impact: 685 kJ/m^2 (23 C, D1822); Flex. Mod.: 10 MPa (23 C, D-790); Compr. Set: 62% (10 days @ 25 C, D395/B); Hardness (Shore D): 30 (D-2240)
Toxicity/Handling: Fumes may evolve at temps. above 220 C–use with proper ventilation
Ref.: Bulletins E-37672, E-37658

Elvax 260. Du Pont Co.
Chem. Descrip.: Ethylene/vinyl acetate copolymer resin
Category/Applications: High-visc. resin providing excellent toughness, flexibility, adhesion, and barrier properties for many coating and adhesive applications; also suitable for footwear applications, incl. foamed midsoles for athletic and leisure footwear, foamed sandals, canvas box toes, and flocked or fabric-laminated counters
Form: Pellets; Color: Nearly transparent
Composition: 28% vinyl acetate by wt.
Gen'l. Prop.: Density: 955 kg/m^3 (D-792); Melt Index: 6.0 dg/min (D1238)
Mech. Prop.: Tens. Str.: 15 MPa (23 C, D638); Tens. Elong.: 750% (23 C, D-638); Tens. Impact: 695 kJ/m^2 (23 C, D1822); Flex. Mod.: 14 MPa (23 C, D-790); Compr. Set: 54% (10 days, 25 C, D395/B); Hardness (Shore D): 34 (D-2240)
Toxicity/Handling: See Elvax 250
Ref.: Bulletins E-37658, E-37672

Elvax 265. Du Pont Co.
Chem. Descrip.: Ethylene/vinyl acetate copolymer resin
Category/Applications: Highest m.w. of Elvax 200 series imparting max. toughness, flexibility, and seal strength; provides excellent "hot-tack" and high visc. in wax blends
Composition: 27.2–28.8% vinyl acetate
Gen'l. Prop.: Melt Index: 2.6–3.4 dg/min (D1238, modified)
Ref.: Bulletin E-37658

Elvax 310. Du Pont Co.
Chem. Descrip.: Ethylene/vinyl acetate copolymer resin
Category/Applications: Wax-compatible resin developed to enhance the properties of hot melt blends containing microcrystalline waxes; recommended for use in heat-sealable barrier coatings and hot melt adhesives where optimum compatibility with microcrystalline waxes is essential; also used for blending with waxes and modifying resins for formulations giving fiber- or film-tearing bonds between paper substrates and such nonporous packaging materials as aluminum foil, polyethylene and polypropylene films, K cellophane, and Mylar polyester film; FDA approved for food packaging applications
Form: Pellets; Color: Translucent; Odor: Slight
Composition: 25% vinyl acetate by wt.
Gen'l. Prop.: Density: 0.948 g/cm^3 (23 C, D1505); Visc. (Inherent): 0.54 (30 C, 0.25 g/100 ml toluene); Soften. Pt. (R&B): 88 C (E28); Melt Index: 400 g/10 min (D1238, modified); Ref. Index: 1.486
Mech. Prop.: Tens. Str.: 3.3 MPa (D-1708 modified); Tens. Elong.: 800–1000% (break, D1708); Tens. Mod.: 16 MPa (D638); Hardness (Shore A-2): 70 (10 s, D2240)
Toxicity/Handling & Std. Pkgs.: See Elvax 40-P
Ref.: Bulletin E-35739

Elvax 350, 360. Du Pont Co.
Chem. Descrip.: Ethylene/vinyl acetate copolymer resins
Category/Applications: See Elvax 310; also suitable for footwear applications
Form: Pellets; Color: Translucent; Odor: Slight
Composition: 25% vinyl acetate by wt.
Gen'l. Prop.: Density: 0.948 and 0.950 g/cm^3 respectively (23 C, D1505); Visc. (Inherent): 0.84 and 1.05 respectively (30 C, 0.25 g/100 ml toluene); Soften. Pt. (R&B): 132 and 188 C respectively (E28); Melt Index: 19 and 2 g/10 min respectively (D1238 modified); Ref. Index: 1.489 and 1.491 respectively
Mech. Prop.: Tens. Str.: 14 and 26 MPa respectively (D1708 modified); Tens. Elong.: 800–1000% (both, @ break, D-1708 modified); Tens. Mod.: 25 and 35 MPa respectively (D638); Tens. Impact: 670 and 880 kJ/m^2 respectivley (23 C, D1822); Flex. Mod.: 18 and 26 MPa respectively (23 C, D790); Compr. Set: 59 and 49% respectively (10 days @ 25 C, D395/B); Hardness (Shore A-2): 80 and 85 respectively (10 s, D2240)
Toxicity/Handling & Std. Pkgs.: See Elvax 40-P
Ref.: Bulletin E-35739

Elvax 410. Du Pont Co.
Chem. Descrip.: Ethylene/vinyl acetate copolymer resin
Category/Applications: Resin imparts high gloss, good gloss retention, scuff resistance, hardness, and resistance to blocking in hot melt coatings (e.g., decorative coatings, grease-resistant coatings for bakery board, heat-sealable coatings for folding cartons, etc.), hot melt adhesives (e.g., low-cost, heat-sealable labels, packaging closures for multiwall bags, forming and sealing folding cartons, Kraft laminations, etc.), high-visc. industrial assembly adhesives (e.g., for shoe manufacturing, bonding molded plastic articles, etc.), and in specialty inks and lacquers
Form: Pellets; Color: Translucent; Odor: Slight
Composition: 18% vinyl acetate by wt.
Gen'l. Prop.: Density: 0.934 g/cm^3 (23 C, D1505); Visc. (Inherent): 0.52 (30 C,

0.25 g/100 ml toluene); Soften. Pt. (R&B): 88 C (E28); Melt Index: 500 g/10 min (D1238 modified); Ref. Index: 1.484
Mech. Prop.: Tens. Str.: 3.8 MPa (D-1708 modified); Tens. Elong.: 650% (@ break, D1708 modified); Tens. Mod.: 14 MPa (D638); Hardness (Shore A-2): 80 (10 s, D2240)
Toxicity/Handling & Std. Pkgs.: See Elvax 40-P
Ref.: Bulletin E-23431

Elvax 420. Du Pont Co.
Chem. Descrip.: Ethylene/vinyl acetate copolymer resin
Category/Applications: See Elvax 410
Form: Pellets; Color: Translucent; Odor: Slight
Composition: 18% vinyl acetate by wt.
Gen'l. Prop.: Density: 0.937 g/cc (23 C, D1505); Visc. (Inherent): 0.61 (30 C, 0.25 g/100 ml toluene); Soften. Pt. (R&B): 99 C (E28); Melt Index: 150 g/10 min (D1238 modified); Ref. Index: 1.492
Mech. Prop.: Tens. Str.: 6.6 MPa (D-1708 modified); Tens. Elong.: 700% (@ break, D1708 modified); Tens. Mod.: 19 MPa (D638); Hardness (Shore A-2): 84 (10 s, D2240)
Toxicity/Handling & Std. Pkgs.: See Elvax 40-P
Ref.: Bulletin E-23431

Elvax 450, 470. Du Pont Co.
Chem. Descrip.: Ethylene/vinyl acetate copolymer
Category/Applications: Lower melt visc. and higher melt visc. respectively than Elvax 460; similar in properties to other grades in 400 series
Composition: 18% vinyl acetate
Gen'l. Prop.: Density: 941 kg/m^3 (both, D792); Melt Index: 8.0 and 0.7 dg/min respectively (D1238)
Mech. Prop.: Tens. Str.: 15 and 28 MPa respectively (23 C, D638); Tens. Elong.: 800% and 750% respectively (23 C, D638); Tens. Impact: 525 and 610 kJ/m^2 respectively (23 C, D1822); Flex. Mod.: 40 and 52 MPa respectively (23 C, D790); Compr. Set: 49% and 56% respectively (10 days @ 25 C, D395/B); Hardness (Shore D): 38 and 41 respectively (D2240)
Ref.: Bulletins E-37658, E-37672

Elvax 460. Du Pont Co.
Chem. Descrip.: Ethylene/vinyl acetate copolymer resin
Category/Applications: See Elvax 410
Form: Pellets; Color: Translucent; Odor: Slight
Composition: 18% vinyl acetate by wt.
Gen'l. Prop.: Density: 0.941 g/cc (23 C, D1505); Visc. (Inherent): 1.02 (30 C, 0.25 g/100 ml toluene); Soften. Pt. (R&B): 199 C (E28); Melt Index: 2.5 g/10 min (D1238 modified); Ref. Index: 1.493
Mech. Prop.: Tens.Str.: 19 MPa (D-1708 modified); Tens. Elong.: 850% (@ break, D1708 modified); Tens. Mod.: 24 MPa (D638); Hardness (Shore A-2): 90 (10 s, D2240)
Toxicity/Handling & Std. Pkgs.: See Elvax 40-P
Ref.: Bulletin E-23431

Elvax 550, 560. Du Pont Co.
Chem. Descrip.: Ethylene/vinyl acetate copolymer resins
Category/Applications: High m.w. and high melt visc. grades used in paper roll wrap and carpet seaming tape
Composition: 15% vinyl acetate by wt.
Gen'l. Prop.: Density: 935 kg/m^3 (both, D792); Melt Index: 8.0 and 2.5 dg/min respectively (D1238)
Mech. Prop.: Tens. Str.: 14 and 20 MPa respectively (23 C, D638); Tens. Elong.:

700% and 750% respectively (23 C, D-638); Tens. Impact: 505 and 545 kJ/m^2 respectively (23 C, D1822); Flex. Mod.: 51 and 63 MPa respectively (23 C, D-790); Compr. Set: 54% and 56% respectively (10 days @ 25 C, D395/B); Hardness (Shore D): 41 and 43 respectively (D2240)
Toxicity/Handling: See Elvax 250
Ref.: Bulletin E-37672

Elvax 565. Du Pont Co.
Chem. Descrip.: Ethylene/vinyl acetate copolymer resin
Category/Applications: Used in footwear applications
Composition: 15% vinyl acetate by wt.
Gen'l. Prop.: Density: 935 kg/m^3 (D-792); Melt Index: 1.5 dg/min (D1238)
Mech. Prop.: Tens. Str.: 22 MPa (23 C, D638); Tens. Elong.: 750% (23 C, D-638); Tens. Impact: 770 kJ/m^2 (23 C, D1822); Flex. Mod.: 56 MPa (23 C, D-790); Hardness (Shore D): 40 (D2240)
Toxicity/Handling: See Elvax 250
Ref.: Bulletin E-37672

Elvax 650, 660, 670. Du Pont Co.
Chem. Descrip.: Ethylene/vinyl acetate copolymer resins
Category/Applications: High m.w. and high melt visc. grades for use in adhesive formulations where higher temp. performance is required
Composition: 11.0–13.0% vinyl acetate
Gen'l. Prop.: Melt Index: 6.7–9.3, 2.1–2.9, and 0.2–0.4 dg/min respectively (D1238 modified)
Ref.: Bulletin E-37658

Elvax 750, 760, 770. Du Pont Co.
Chem. Descrip.: Ethylene/vinyl acetate copolymer resins
Category/Applications: See Elvax 650, 660, 670
Composition: 8.0–10.0%, 8,8–9.8%, and 8.5–10.5% vinyl acetate respectively
Gen'l. Prop.: Melt Index: 6.3–7.7, 1.8–2.2, and 0.6–1.0 dg/min respectively (D-1238, modified)
Ref.: Bulletin E-37658

Elvax 3120. Du Pont Co.
Chem. Descrip.: Ethylene/vinyl acetate copolymer resin containing slip and antiblock additives
Category/Applications: Designed to provide a low-temp. heat seal to itself or many other materials commonly used in flexible packaging applications; can be processed on blown film equipment over a wide range of blow-up ratios and calipers; can also be coextruded with a variety of other polymers; FDA approved for use in contact with food
Form: Pellets
Composition: 7.5% vinyl acetate
Gen'l. Prop.: Density: 0.93 g/cc (D-792); Melt Index: 1.2 dg/min (D1238)
Toxicity/Handling: Fumes may evolve at temps. above 230 C–use with adequate ventilation
Ref.: Bulletin E-29193

Elvax 3128. Du Pont Co.
Chem. Descrip.: Ethylene/vinyl acetate copolymer resin
Category/Applications: See Elvax 3120
Form: Pellets
Composition: 8.9% vinyl acetate by wt.
Gen'l. Prop.: Density: 0.93 g/cc (D-792); Melt Index: 2.0 dg/min (D1238)
Toxicity/Handling: See Elvax 3120
Ref.: Bulletin E-37409

Elvax 4260. Du Pont Co.
Chem. Descrip.: Ethylene/vinyl acetate/acid terpolymer
Category/Applications: High m.w. resin for use in hot melt systems where improved adhesion to polar, nonporous

substrates is required; in coatings, provides superior hot tack, improved grease resistance, and optimum barrier properties
Composition: 27.0–29.0% vinyl acetate
Gen'l. Prop.: Melt Index: 5.0–7.0 dg/min (D1238 modified); Acid No.: 4–8
Ref.: Bulletin E-37658

Elvax 4310. Du Pont Co.
Chem. Descrip.: Ethylene/vinyl acetate/acid terpolymer
Category/Applications: Low m.w. resin designed to provide improved grease resistance and adhesion in low-visc. systems; permits max. Elvax content at a given visc. in solvent or hot melt systems
Composition: 24.0–26.0% vinyl acetate
Gen'l. Prop.: Melt Index: 420–580 dg/min (D1238 modified); Acid No.: 4–8
Ref.: Bulletin E-37658

Elvax 4320. Du Pont Co.
Chem. Descrip.: Ethylene/vinyl acetate/acid terpolymer
Category/Applications: Intermediate m.w. resin higher in visc. than Elvax 4310 and intermediate in performance between Elvax 4310 and 4355; can be combined (as can Elvax 4310) with Elvax 4355 or 4260 to optimize performance at a desired visc. level
Composition: 24.0–26.0% vinyl acetate
Gen'l. Prop.: Melt Index: 125–175 dg/min (D1238 modified); Acid No.: 4–8
Ref.: Bulletin E-37658

Elvax 4355. Du Pont Co.
Chem. Descrip.: Ethylene/vinyl acetate/acid terpolymer
Category/Applications: High m.w. resin preferred in high-hot-tack systems; most effective of the 4300 series resins in imparting toughness, flexibility, and seal strength in blends with wax
Composition: 24.0–26.0% vinyl acetate
Gen'l. Prop.: Melt Index: 5.0–7.0 dg/min (D1238 modified); Acid No.: 4–8
Ref.: Bulletin E-37658

Elvax EP 170-PS. Du Pont Co.
Chem. Descrip.: Segmented copolymer of ethylene and vinyl acetate
Category/Applications: Resin for use as base resin in hot melt pressure sensitive adhesives, esp. permanent label adhesives, and in other adhesives, coatings, and sealants; offers good adhesion, high melt strength, tensile strength, and elasticity; improves toughness, flexibility, and adhesion in blends with tackifying resins and other modifiers; FDA approved for use in contact with food
Form: Pellets; Color: Translucent; Odor: Slight
Composition: 36% vinyl acetate by wt.
Gen'l. Prop.: Density: 0.962 g/cc (23 C, D1505); Visc. (Inherent): 1.38 (30 C, 0.25 g/100 ml toluene); Soften. Pt. (R&B): 220 C (E28); Melt Index: 0.7 dg/min (190 C, D1238 modified); Stability: Excellent thermal and oxidative stability and resistance to uv light
Mech. Prop.: Tens. Str.: 11 MPa (D-1708); Tens. Elong.: 1400% (break, D-1708); Tens. Mod.: 2.8 MPa (D638); Hardness (Shore A-2): 63 (10 s, D2240)
Toxicity/Handling: See Elvax 40-P
Std. Pkgs.: 500-kg net boxes, and 22.7-kg net multiwall paper bags on 1134-kg shrink-wrapped pallets
Ref.: Bulletin E-41498

Elvax 40-P. Du Pont Co.
Chem. Descrip.: Ethylene/vinyl acetate copolymer resin
Category/Applications: High vinyl acetate grade resin useful as base polymers in hot melt and solvent-thinned coatings and adhesives (e.g., lacquers for wood,

metal, metal foils, paper; fabric and decal coatings; heat-seal coatings; wood primers and sealers; laminating adhesives; tape and label adhesives; etc.), and as modifiers in a variety of other applications to improve properties such as processability, thermoplasticity, adhesion, and abrasion resistance (e.g., in natural and synthetic elastomers, caulks and sealants, asphalt road repairing compositions, traffic paint formulations, solvent- and wax-based inks); FDA approved for use in food packaging applications
Form: Pellets; Color: Translucent
Composition: 40% vinyl acetate by wt.
Gen'l. Prop.: Sol.: Sol. in many organic solvents; Density: 965 kg/m^3 (23 C, D-1505); Visc. (Inherent): 0.70 (30 C, 0.25 g/100 ml toluene); Soften. Pt. (R (R&B): 104 C (E28); Melt Index: 57 g/10 min (D1238); Ref. Index: 1.476
Mech. Prop.: Tens. Str.: 4.5 MPa (D-1708); Tens. Elong.: 1450% (break, D-1708); Tens. Mod.: 2.1 MPa (D638); Hardness (Shore A-2): 40 (10 s, D2240)
Toxicity/Handling: Not considered hazardous with respect to inhalation, eye contact, or skin absorption; may generate acetic acid, and carbon monoxide if ignited
Std. Pkgs.: 22.7-kg net multiwall paper bags on 1134-kg shrink-wrapped pallets
Ref.: Bulletin E-37658

Elvax 3135SB. Du Pont Co.
Chem. Descrip.: Ethylene/vinyl acetate copolymer resin with slip and antiblock additives
Category/Applications: See Elvax 3120
Form: Pellets
Composition: 12% vinyl acetate by wt.
Gen'l. Prop.: Density: 0.94 g/cc (D792); Melt Index: 0.35 dg/min (D1238)
Toxicity/Handling: See Elvax 3120
Ref.: Bulletin E-37410

Elvax EVA 3135X. Du Pont Co.
Chem. Descrip.: Ethylene/vinyl acetate copolymer resin
Category/Applications: Resin providing a low-temp. heat seal to itself or many other materials commonly used in flexible packaging applications; its melt properties allow it to be processed on blown film equipment over a wide range of blow-up ratios and calipers; can also be coextruded with a variety of other polymers; FDA approved for use in contact with food
Form: Pellets
Composition: 12% vinyl acetate by wt.
Gen'l. Prop.: Density: 0.94 g/cc (D792); Melt Index: 0.35 dg/min (D1238)
Toxicity/Handling: Fumes may evolve at temps. above 230 C–use with adequate ventilation
Ref.: Bulletin E-20779-1

Emerez 1500. Emery Industries, Inc.
Chem. Descrip.: Reactive polyamide resin synthesized from polyfunctional fatty acids and alkylamines
Category/Applications: Resin used as modifier for adhesive systems, esp. polyamide systems, and for reaction with other chemicals to form solid materials, e.g., polyamide-epoxy thermoset systems (with good solvent resistance, adhesion to substrates, and low shrinkage–used in casting, encapsulation, laminating, concrete patching and coating, etc.), polyamide-phenolics (with good chemical resistance–used as corrosion-resistant coatings), etc.
Form: Viscous liquid; Color: Amber
Composition: 59–61% solids in toluene/isopropanol (1/1)
Gen'l. Prop.: Visc.: 5–9 poise (40 C); Amine Value: 51–57; Stabilty: Highly stable; not degraded by contact with steel

Toxicity/Handling: Reduced skin irritation potential over straight polyamines; plastic gloves suggested for prolonged contact; noncorrosive to most types of steel
Std. Pkgs.: 55 gal (400 lb net) tighthead unlined drums
Ref.: Data sheet 915

Emerez 1501. Emery Industries, Inc.
Chem. Descrip. & Category/Applications: See Emerez 1500
Form: Viscous liquid; Color: Amber
Composition: 59–61% solids in xylene/ethylcellosolve (9/1)
Gen'l. Prop.: Visc.: 10–15 poise (40 C); Amine Value: 51–57; Stability: Highly stable; not degraded by contact with steel
Toxicity/Handling & Std. Pkgs.: See Emerez 1500
Ref.: Data sheet 915

Emerez 1502. Emery Industries, Inc.
Chem. Descrip. & Category/Applications: See Emerez 1500
Form: Viscous liquid; Color: Amber
Composition: 69–71% solids in xylene
Gen'l. Prop.: Visc.: 3–8 poise (40 C); Amine Value: 161–173; Stability: Highly stable; not degraded by contact with steel
Toxicity/Handling & Std. Pkgs.: See Emerez 1500
Ref.: Data sheet 915

Emerez 1511. Emery Industries, Inc.
Chem. Descrip. & Category/Applications: See Emerez 1500
Form: Viscous liquid; Color: Amber
Composition: 100% solids
Gen'l. Prop.: Visc.: 31–38 poise (75 C); Amine Value: 230–246; Stability: Highly stable; not degraded by contact with steel
Toxicity/Handling: See Emerez 1500
Std. Pkgs.: 55 gal (400 lb net) openhead unlined drums
Ref.: Data sheet 915

Emerez 1512. Emery Industries, Inc.
Chem. Descrip. & Category/Applications: See Emerez 1500
Form: Viscous liquid; Color: Amber
Composition: 100% solids
Gen'l. Prop.: Visc.: 7–9 poise (75 C); Amine Value: 330–360; Stability: Highly stable; not degraded by contact with steel
Toxicity/Handling: See Emerez 1500
Std. Pkgs.: See Emerez 1511
Ref.: Data sheet 915

Emerez 1514. Emery Industries, Inc.
Chem. Descrip. & Category/Applications: See Emerez 1500
Form: Viscous liquid; Color: Amber
Composiiton: 100% solids
Gen'l. Prop.: Visc.: 2–6 poise (75 C); Amine Value: 350–400; Stability: Highly stable; not degraded by contact with steel
Toxicity/Handling & Std. Pkgs.: See Emerez 1500
Ref.: Data sheet 915

Emerez 1530. Emery Industries, Inc.
Chem. Descrip.: Polyamide resin
Category/Applications: Relatively high m.w. cosolvent resin requiring a mixed alcohol/hydrocarbon solvent system to achieve max. solubility; used in flexographic inks, esp. highly pigmented ink applications; provides excellent adhesion to polyethylene, polymer-coated cellophane, aluminum foil, etc.
Form: Beads; Color: Light; Odor: Low
Gen'l. Prop.: Sol. (30% solids @ 72 F) in

alcohols, amides, amines, and chlorinated solvents; Visc.: 21–27 poise (160 C); Soften. Pt. : 105–115 C
Std. Pkgs.: 50 lb multiwall bags
Ref.: Tech. bulletin 105E

Emerez 1533. Emery Industries, Inc.
Chem. Descrip.: Polyamide resin
Category/Applications: High m.w. cosolvent resin for flexographic inks with outstanding pigment wetting, good ethanol tolerance, excellent adhesion properties, and rapid gel recovery; provides superior brilliance and transparency to dye inks used for printing aluminum foil, etc.
Form: Beads; Color: Light; Odor: Low
Gen'l. Prop.: Sol.: See Emerez 1530; Visc.: 26–34 poise (160 C); Soften. Pt.: 99–108 C
Std. Pkgs.: 50 lb multiwall bags
Ref.: Tech. bulletin 105E

Emerez 1540. Emery Industries, Inc.
Chem. Descrip.: Polyamide resin
Category/Applications: Lower m.w. version of Emerez 1530, but providing lower solution viscosities and better compatibility with other resins
Form: Beads; Color: Light; Odor: Low
Gen'l. Prop.: Sol.: See Emerez 1530; Visc.: 12–18 poise (160 C); Soften. Pt.: 105–115 C
Std. Pkgs.: 50 lb multiwall bags
Ref.: Tech. bulletin 105E

Emerez 1548. Emery Industries, Inc.
Chem. Descrip.: Polyamide resin
Category/Applications: Low m.w. alcohol-soluble resin for use in flexographic inks; exhibits exceptional solvent release and high temp. blocking resistance; compatible with nitrocellulose and provides excellent gloss in pigment dispersions
Form: Beads; Color: Light; Odor: Low
Gen'l. Prop.: Sol.: See Emerez 1530; Visc.: 6–10 poise (160 C); Soften. Pt.: 115–125 C
Std. Pkgs.: 50 lb multiwall bags
Ref.: Tech. bulletin 105E

Emerez 1549. Emery Industries, Inc.
Chem. Descrip.: Polyamide resin
Category/Applications: Similar to Emerez 1548 but with better gel recovery
Form: Beads; Color: Light; Odor: Low
Gen'l. Prop.: Sol.: See Emerez 1530; Visc.: 10–18 poise (160 C); Soften. Pt.: 115–125 C
Std. Pkgs.: 50 lb multiwall bags
Ref.: Tech. bulletin 105E

Emerez 1552. Emery Industries, Inc.
Chem. Descrip.: Polyamide resin
Category/Applications: General purpose, low melting pt., thermoplastic resin providing excellent adhesion at a moderate viscosity;
Form: Bead
Gen'l. Prop.: Visc.: 60 poise (190 C); Soften. Pt.: 137 C
Curing Charact.: Speed of Set: 6 s
Ref.: Tech. bulletin 107C

Emerez 1553. Emery Industries, Inc.
Chem. Descrip.: Polyamide resin
Category/Applications: General purpose, low melting, tough thermoplastic resin for adhesives applications, with a low soften. pt. and excellent flexibility
Form: Bead
Gen'l. Prop.: Visc.: 100 poise (190 C); Soften. Pt.: 110 C
Curing Charact.: Speed of Set: 9 s
Ref.: Tech. bulletin 107C

Emerez 1556. Emery Industries, Inc.
Chem. Descrip.: Polyamide resin
Category/Applications: General purpose, low melting, low viscosity resin with

moderate flexibility; ideally suited for solution applications in the adhesives industry
Form: Tube
Gen'l. Prop.: Visc.: 88 poise (190 C); Soften. Pt.: 110 C
Curing Charact.: Speed of Set: 10 s
Ref.: Tech. bulletin 107C

Emerez 1565. Emery Industries, Inc.
Chem. Descrip.: Polyamide resin
Category/Applications: Specialty resin which can be used as a hot melt adhesive or as a reactive polyamide resin with epoxies; its moderately high amine value contributes to excellent surface wetting
Form: Solid
Gen'l. Prop.: Visc.: 5 poise (190 C); Soften. Pt.: 95 C
Curing Charact.: Speed of Set: 64 s
Std. Pkgs.: Release-coated breakaway boxes of ≈ 35 lb net wt; also avail. in 3 in. × 3 in. × 4 in. cubes
Ref.: Tech. bulletin 107C

Emerez 1566. Emery Industries, Inc.
Chem. Descrip.: Polyamide resin
Category/Applications: General purpose, high melting, high performance thermoplastic resin designed for demanding adhesive applications; outstanding thermal stability
Form: Bead and tube
Gen'l. Prop.: Visc.: 80 poise (190 C); Soften. Pt.: 165 C
Curing Charact.: Speed of Set: 5 s
Ref.: Tech. bulletin 107C

Emerez 1567. Emery Industries, Inc.
Chem. Descrip.: Polyamide resin
Category/Applications: General purpose, high melting resin similar to Emerez 1566, but with less thermal stability
Form: Bead and tube
Gen'l. Prop.: Visc.: 103 poise (190 C); Soften. Pt.: 165 C
Curing Charact.: Speed of Set;: 10 s
Ref.: Tech. bulletin 107C

Envex 1000. Rogers Corp.
Chem. Descrip.: Polyimide
Category/Applications: High-strength, thermoplastic polyimide which can be used at temps. up to 288 C in various structural or insulating applications, e.g., seals, gaskets, valve seats, insulators; can be modified with lubricating and/or structural fillers; avail. in assorted stock shapes and as molded and/or machined parts
Form: Solid
Gen'l. Prop.: Sp. Gr.: 1.33 (D792)
Mech. Prop.: Tens. Str.: 88.3 MPa (D-638-72); Tens. Elong.: 5.1% (@ break, D638-72); Tens. Mod.: 2593 MPa (D-638-72); Flex. Str.: 164.8 MPa (R.T., D790-71); Flex. Mod.: 3055 MPa (R.T., D790-71); Compr. Str.: 204 MPa (D-695-69); Compr. Mod.: 3227 MPa (D-695-69); Compr. Strain at Rupture: 19.3% (D695-69); Hardness (Rockwell): M 122
Therm. Prop.: Distort. Temp.: 288 C (D648-72); Coeff. of Linear Exp.: 54 × 10^{-6} in./in./C (50–200 C, TMA)
Elec. Prop.: Dielec. Str.: 12.3 MV/m (R.T., D149); Vol. Resist.: 1.4 × 10^{16} ohm-cm (R.T., D257)
Ref.: Product pamphlet

Envex 1115. Rogers Corp.
Chem. Descrip.: Polyimide
Category/Applications: Polyimide reinforced and lubricated with MoS_2; used at temps. up to 288 C in structural and/or self-lubricating applications, e.g., vacuum bearings, seals, high load applications, nuclear components
Form: Solid
Gen'l. Prop.: Sp. Gr.: 1.5 (D792)

Mech. Prop.: Tens. Str.: 55.3 MPa (D-638-72); Tens. Elong.: 2.4% (@ break, D638-72); Tens. Mod.: 3448 MPa (D-638-72); Flex. Str.: 110 MPa (R.T., D-790-71); Flex. Mod.: 4634 MPa (R.T., D790-71); Compr. Str.: 193 MPa (D-695-69); Compr. Mod.: 3647 MPa (D-695-69); Compr. Strain at Rupture: 13.4% (D695-69); Hardness (Rockwell): M 111
Therm. Prop.: Distort. Temp.: 288 C (D648-72); Coeff. of Linear Exp.: 64.8 $\times 10^{-6}$ in./in./C (50–200 C, TMA)
Ref.: Product pamphlet

Envex 1228. Rogers Corp.
Chem. Descrip.: Polyimide
Category/Applications: Polyimide lubricated with PTFE capable of operating from cryogenic temps. up to 288 C; used in unlubricated bearing applications, esp. when the motion is oscillating and reciprocating; used in carriage bearings, bushings, bearings, piston rings, sealing rings; has low wear and low coefficient of friction
Form: Solid
Gen'l. Prop.: Sp. Gr.: 1.52 (D792)
Mech. Prop.: Tens. Str.: 22.8 MPa (D-638-72); Tens. Elong.: 1.7% (@ break, D638-72); Tens. Mod.: 2069 MPa (D-638-72); Flex. Str.: 51.7 MPa (R.T., D-790-71); Flex. Mod.: 2965 MPa (R.T., D790-71); Compr. Str.: 90.3 MPa (D-695-69); Compr. Mod.: 2400 MPa (D-695-69); Compr. Strain at Rupture: 5.0% (D695-69); Hardness (Rockwell): M 93
Therm. Prop.: Distort. Temp.: 288 C (D648-72); Coeff. of Linear Exp.: 59.4 $\times 10^{-6}$ in./in./C (50–200 C, TMA)
Elec. Prop.: Dielec. Str.: 12.9 V/mil (R.T., D149); Vol. Resist.: 1.5 $\times 10^{16}$ ohm-cm (R.T., D257)
Ref.: Product pamphlet

Envex 1315. Rogers Corp.
Chem. Descrip.: Polyimide
Category/Applications: Polyimide lubricated with graphite; designed to have a unique balance of self-lubricating and structural properties up to 288 C; also an excellent bearing material; used in bushings, bearings, seals, structural parts
Form: Solid
Gen'l. Prop.: Sp. Gr.: 1.4 (D792)
Mech. Prop.: Tens. Str.: 65.5 MPa (D-638-72); Tens. Elong.: 3.0% (@ break, D638-72); Tens. Mod.: 3276 MPa (D-638-72); Flex. Str.: 118.6 MPa (R.T., D-790-71); Flex. Mod.: 4000 MPa (R.T., D790-71); Compr. Str.: 206.9 MPa (D-695-69); Compr. Mod.: 3751 MPa (D-695-69); Compr. Strain at Rupture: 8.2% (D695-69); Hardness (Rockwell): M 111
Therm. Prop.: Distort. Temp.: 288 C (D648-72); Coeff. of Linear Exp.: 46.8 $\times 10^{-6}$ in./in./C (50–200 C, TMA)
Ref.: Product pamphlet

Envex 1330. Rogers Corp.
Chem. Descrip.: Polyimide
Category/Applications: Polyimide lubricated with graphite, designed for unlubricated bearing applications, seals; has low coefficient of thermal expansion
Form: Solid
Gen'l. Prop.: Sp. Gr.: 1.51 (D792)
Mech. Prop.: Tens. Str.: 55.2 MPa (D-638-72); Tens. Elong.: 3.6% (@ break, D638-72); Tens. Mod.: 2855 MPa (D-638-72); Flex. Str.: 101.4 MPa (R.T., D790-71); Flex. Mod.: 4448 MPa (R.T., D790-71); Compr. Str.: 165.5 MPa (D-695-69); Compr. Mod.: 3310 MPa (D-695-69); Compr. Strain at Rupture: 12.4% (D695-69); Hardness (Rockwell): M 103
Therm. Prop.: Distort. Temp.: 288 C (D648-72); Coeff. of Linear Exp.: 45 $\times 10^{-6}$ in./in./C (50–200 C, TMA)
Ref.: Product pamphlet

Envex 1620. Rogers Corp.
Chem. Descrip.: Polyimide
Category/Applications: Random glass fiber-reinforced polyimide for use in structural applications over a broad temp. range

Envex 3540. Rogers Corp.
Chem. Descrip.: Polyimide
Category/Applications: High modulus, graphite fiber-reinforced polyimide for structural applications; has very low coefficient of thermal expansion and high flex modulus; can withstand continuous operation to 250 C
Form: Solid
Mech. Prop.: Tens. Str.: 69.1 MPa (D-638-72); Tens. Elong.: 1.3% (@ break, D638-72); Tens. Mod.: 13,861 MPa (D-538-72); Flex. Str.: 124 MPa (R.T., D-790-71); Flex. Mod.: 14,895 MPa (R.T., D790-71); Compr. Str.: 244.1 MPa (D-695-69); Compr. Mod.: 7034 MPa (D-695-69); Compr. Strain at Rupture: 7.2% (D695-69)
Therm. Prop.: Distort. Temp.: 260 C (D648-72); Coeff. of Linear Exp.: 13 $\times 10^{-6}$ in./in./C (50–200 C, TMA)
Ref.: Product pamphlet

EP 1760 LV. Hardman Inc.
Chem. Descrip.: One-part epoxy comp.
Category/Applications: Elevated temp. curing comp. for electrical/electronics applications
Gen'l. Prop.: Visc.: 2000 cps
Mech. Prop.: Hardness (Shore D): 88
Therm. Prop.: Coeff. of Linear Exp.: 140×10^{-6} in./in./C
Elec. Prop.: Dissip. Factor: 0.006 (100 kHz); Dielec. Const.: 2.8 (100 kHz); Vol. Resist.: 6×10^{15} ohm-cm
Curing Charact.: Pot Life: 18 min @ 150 C (200 g); Cure Schedule: 5 h @ 150 C
Ref.: Data sheet

EP 1765 TS. Hardman Inc.
Chem. Descrip.: One-part epoxy comp.
Category/Applications: Elevated temp. curing comp. for electrical/electronic applications
Gen'l. Prop.: Visc.: 56,000 cps
Mech. Prop.: Hardness (Shore D): 90
Therm. Prop.: Conduct.: 7.8×10^{-4} Cal/cm/s/cm^2/C; Distort. Temp.: 64 C; Coeff. of Linear Exp.: 105×10^{-6} in./in./C
Elec. Prop.: Dissip. Factor: 0.026 (100 kHz); Dielec. Const.: 4.3 (100 kHz); Vol. Resist.: 10^{15} ohm-cm
Curing Charact.: Pot Life: 40 min @ 150 C (200 g); Cure Schedule: 5 h @ 150 C
Ref.: Data sheet

EP 2220 TC. Hardman Inc.
Chem. Descrip.: Two-part epoxy comp.
Category/Applications: R.T. curing, epoxy comp. for electrical/electronic applications; mix ratio: 100A/5.5B by wt.
Gen'l. Prop.: Visc.: 3300 cps (mixed)
Mech. Prop.: Hardness (Shore D): 95
Therm. Prop.: Conduct.: 8.2×10^{-4} Cal/cm/s/cm^2/C; Distort. Temp.: 86 C; Coeff. of Linear Exp.: 43×10^{-6} in./in./C
Elec. Prop.: Dissip. Factor: 0.020 (100 kHz); Dielec. Const.: 4.2 (100 kHz); Vol. Resist.: 4×10^{13} ohm-cm
Curing Charact.: Pot Life: 61 min @ R.T. (200 g); Cure Schedule: 24 h @ R.T.
Ref.: Data sheet

EP 2225 MV. Hardman Inc.
Chem. Descrip.: Two-part epoxy comp.
Category/Applications: R.T. curing epoxy comp. for electrical/electronic applications; mix ratio: 100A/6B by wt.
Gen'l. Prop.: Visc.: 7100 cps (mixed)
Mech. Prop.: Hardness (Shore D): 96

Therm. Prop.: Distort. Temp.: 116 C; Coeff. of Linear Exp.: 44 X 10^{-6} in./in./C
Elec. Prop.: Dissip. Factor: 0.020 (100 kHz); Dielec. Const.: 4.3 (100 kHz); Vol. Resist.: 3 X 10^{15} ohm-cm
Curing Charact.: Pot Life: 48 min @ R.T. (200 g); Cure Schedule: 24 h @ R.T.
Ref.: Data sheet

EP 2230 TS. Hardman Inc.
Chem. Descrip.: Two-part epoxy comp.
Category/Applications: R.T. curing epoxy comp. for electrical/electronic applications; mix ratio 100A/50B by wt.
Gen'l. Prop.: Visc.: 9200 cps (mixed)
Mech. Prop.: Hardness (Shore D): 88
Therm. Prop.: Conduct.: 5.1 X 10^{-4} Cal/cm/s/cm^2/C; Distort. Temp.: 51 C; Coeff. of Linear Exp.: 152 X 10^{-6} in./in./C
Elec. Prop.: Dissip. Factor: 0.021 (100 kHz); Dielec. Const.: 3.6 (100 kHz); Vol. Resist.: 4 X 10^{15} ohm-cm
Curing Charact.: Pot Life: 30 min @ R.T. (200 g); Cure Schedule: 24 h @ R.T.
Ref.: Data sheet

EP 2305 LK. Hardman Inc.
Chem. Descrip.: Two-part epoxy comp.
Category/Applications: R.T. curing epoxy comp. for electrical/electronic applications; mix ratio: 100A/9B by wt.
Gen'l. Prop.: Visc.: 8000 cps (mixed)
Mech. Prop.: Hardness (Shore D): 88
Therm. Prop.: Conduct.: 6.9 X 10^{-4} Cal/cm/s/cm^2/C; Distort. Temp.: 67 C; Coeff. of Linear Exp.: 44 X 10^{-6} in./in./C
Elec. Prop.: Dissip. Factor: 0.001 (100 kHz); Dielec. Const.: 4.0 (100 kHz); Vol. Resist.: 3 X 10^{15} ohm-cm
Curing Charact.: Pot Life: 50 min @ R.T. (200 g); Cure Schedule: 24 h @ R.T.
Ref.: Data sheet

EP 2306 LK. Hardman Inc.
Chem. Descrip.: Two-part epoxy comp.
Category/Applications: R.T. curing epoxy comp. for electrical/electronic applications; mix ratio: 100A/11B by wt.
Gen'l. Prop.: Visc.: 6000 cps (mixed)
Mech. Prop.: Hardness (Shore D): 90
Therm. Prop.: Conduct.: 5.2 X 10^{-4} Cal/cm/s/cm^2/C; Distort. Temp.: 85 C; Coeff. of Linear Exp.: 107 X 10^{-6} in./in./C
Elec. Prop.: Dissip. Factor: 0.018 (100 kHz); Dielec. Const.: 3.8 (100 kHz); Vol. Resist.: 5 X 10^{15} ohm-cm
Curing Charact.: Pot Life: 37 min @ R.T. (200 g); Cure Schedule: 24 h @ R.T.
Ref.: Data sheet

EP 2400 TC. Hardman Inc.
Chem. Descrip.: Two-part epoxy comp.
Category/Applications: R.T. curing epoxy comp. for electrical/electronic applications; mix ratio: 100A/9B by wt.
Gen'l. Prop.: Visc.: 10,000 cps (mixed)
Mech. Prop.: Hardness (Shore D): 96
Therm. Prop.: Conduct.: 10.3 X 10^{-4} Cal/cm/s/cm^2/C; Distort. Temp.: 78 C; Coeff. of Linear Exp.: 78 X 10^{-6} in./in./C
Elec. Prop.: Dissip. Factor: 0.030 (100 kHz); Dielec. Const.: 4.2 (100 kHz); Vol. Resist.: 4 X 10^{14} ohm-cm
Curing Charact.: Pot Life: 17 min @ R.T. (200 g); Cure Schedule: 24 h @ R.T.
Ref.: Data sheet

EP 2404 GE. Hardman Inc.
Chem. Descrip.: Two-part epoxy comp.
Category/Applications: R.T. curing epoxy comp. for electrical/electronic

applications; mix ratio: 100A/6B by wt.
Gen'l. Prop.: Visc.: 5200 cps (mixed)
Mech. Prop.: Hardness (Shore D): 97
Therm. Prop.: Conduct.: 7.2 X 10^{-4} Cal/cm/s/cm^2/C; Distort. Temp.: 84 C; Coeff. of Linear Exp.: 68 X 10^{-6} in./in./C
Elec. Prop.: Dissip. Factor: 0.027 (100 kHz); Dielec. Const.: 4.2 (100 kHz); Vol. Resist.: 6 X 10^{14} ohm-cm
Curing Charact.: Pot Life: 45 min @ R.T. (200 g); Cure Schedule: 24 h @ R.T.
Ref.: Data sheet

EP 2405 FR. Hardman Inc.
Chem. Descrip.: Two-part epoxy comp.
Category/Applications: R.T. curing epoxy comp. for electrical/electronic applications; mix ratio: 100A/8.5B by wt.
Gen'l. Prop.: Visc.: 10,000 cps (mixed)
Mech. Prop.: Hardness (Shore D): 95
Therm. Prop.: Conduct.: 12.0 X 10^{-4} Cal/cm/s/cm^2/C; Distort. Temp.: 77 C; Coeff. of Linear Exp.: 62 X 10^{-6} in./in./C
Elec. Prop.: Dissip. Factor: 0.010 (100 kHz); Dielec. Const.: 4.2 (100 kHz); Vol. Resist.: 2 X 10^{15} ohm-cm
Curing Charact.: Pot Life: 15 min @ R.T. (200 g); Cure Schedule: 24 h @ R.T.
Ref.: Data sheet

EP 2408 TS. Hardman Inc.
Chem. Descrip.: Two-part epoxy comp.
Category/Applications: R.T. curing epoxy comp. for electrical/electronic applications; mix ratio: 100A/83B by wt.
Gen'l. Prop.: Visc.: 1300 cps (mixed)
Mech. Prop.: Hardness (Shore D): 53
Elec. Prop.: Dissip. Factor: 0.050 (100 kHz); Dielec. Const.: 4.0 (100 kHz); Vol. Resist.: 4 X 10^{15} ohm-cm
Curing Charact.: Pot Life: 105 min @ R.T. (200 g); Cure Schedule: 48 h @ R.T.
Ref.: Data sheet

EP 2415 TC. Hardman Inc.
Chem. Descrip.: Two-part epoxy comp.
Category/Applications: R.T. curing epoxy comp. for electrical/electronic applications; mix ratio: 100A/15B by wt.
Gen'l. Prop.: Visc.: 2000 cps (mixed)
Mech. Prop.: Hardness (Shore D): 90
Therm. Prop.: Conduct.: 10.3 X 10^{-4} Cal/cm/s/cm^2/C; Distort. Temp.: 60 C; Coeff. of Linear Exp.: 115 X 10^{-6} in./in./C
Elec. Prop.: Dissip. Factor: 0.036 (100 kHz); Dielec. Const.: 5.2 (100 kHz); Vol. Resist.: 10^{12} ohm-cm
Curing Charact.: Pot Life: 45 min @ R.T. (200 g); Cure Schedule: 24 h @ R.T.
Ref.: Data sheet

EP 2420 TC. Hardman Inc.
Chem. Descrip.: Two-part epoxy comp.
Category/Applications: R.T. curing epoxy comp. for electrical/electronic applications; mix ratio: 100A/7B by wt.
Gen'l. Prop.: Visc.: 12,000 cps (mixed)
Mech. Prop.: Hardness (Shore D): 86
Therm. Prop.: Conduct.: 13.6 X 10^{-4} Cal/cm/s/cm^2/C; Distort. Temp.: 44 C; Coeff. of Linear Exp.: 135 X 10^{-6} in./in./C
Elec. Prop.: Dissip. Factor: 0.020 (100 kHz); Dielec. Const.: 5.4 (100 kHz); Vol. Resist.: 2 X 10^{14} ohm-cm
Curing Charact.: Pot Life: 130 min @ R.T. (200 g); Cure Schedule: 24 h @ R.T.
Ref.: Data sheet

EP 2740 HT. Hardman Inc.
Chem. Descrip.: Two-part epoxy comp.
Category/Applications: Elevated temp.

curing comp. for electrical/electronic applications; mix ratio: 100A/60B by wt.
Gen'l. Prop.: Visc.: 13,600 cps (mixed)
Mech. Prop.: Hardness (Shore D): 93
Therm. Prop.: Conduct.: 6.8 $\times$ 10^{-4} Cal/cm/s/cm^2/C; Distort. Temp.: 100 C; Coeff. of Linear Exp.: 62 $\times$ 10^{-6} in./in./C
Elec. Prop.: Dissip. Factor: 0.008 (100 kHz); Dielec. Const.: 3.8 (100 kHz); Vol. Resist.: 7 $\times$ 10^{15} ohm-cm
Curing Charact.: Pot Life: 63 min @ 110 C (200 g); Cure Schedule: 16 h @ 110 C
Ref.: Data sheet

EP 2770 TC. Hardman Inc.
Chem. Descrip.: Two-part epoxy comp.
Category/Applications: Elevated temp. curing comp. for electrical/electronic applications; mix ratio: 100A/100B by wt.
Gen'l. Prop.: Visc.: 10,000 cps (mixed)
Mech. Prop.: Hardness (Shore D): 98
Therm. Prop.: Conduct.: 7.3 $\times$ 10^{-4} Cal/cm/s/cm^2/C; Distort. Temp.: 103 C; Coeff. of Linear Exp.: 50 $\times$ 10^{-6} in./in./C
Elec. Prop.: Dissip. Factor: 0.012 (100 kHz); Dielec. Const.: 3.7 (100 kHz); Vol. Resist.: 10^{15} ohm-cm
Curing Charact.: Pot Life: 120 min @ 121 C (200 g); Cure Schedule: 2 h @ 121 C
Ref.: Data sheet

EP 2800 GU. Hardman Inc.
Chem. Descrip.: Two-part epoxy comp.
Category/Applications: Elevated temp. curing comp. for electrical/electronic applications; mix ratio: 100A/10B by wt.
Gen'l. Prop.: Visc.: 1700 cps (mixed)
Mech. Prop.: Hardness (Shore D): 91
Therm. Prop.: Distort. Temp.: 79 C; Coeff. of Linear Exp.: 97 $\times$ 10^{-6} in./in./C
Elec. Prop.: Dissip. Factor: 0.020 (100 kHz); Dielec. Const.: 3.5 (100 kHz); Vol. Resist.: 3 $\times$ 10^{15} ohm-cm
Curing Charact.: Pot Life: 11 min @ 100 C (200 g); Cure Schedule: 4 h @ 100 C
Ref.: Data sheet

EP 2805 TC. Hardman Inc.
Chem. Descrip.: Two-part epoxy comp.
Category/Applications: Elevated temp. curing comp. for electrical/electronic applications; mix ratio: 100A/5B by wt.
Gen'l. Prop.: Visc. : 10,600 cps (mixed)
Mech. Prop.: Hardness (Shore D): 98
Therm. Prop.: Conduct.: 8.8 $\times$ 10^{-4} Cal/cm/s/cm^2/C; Distort. Temp.: 87 C; Coeff. of Linear Exp.: 51 $\times$ 10^{-6} in./in./C
Elec. Prop.: Dissip. Factor: 0.020 (100 kHz); Dielec. Const.: 3.8 (100 kHz); Vol. Resist.: 6 $\times$ 10^{14} ohm-cm
Curing Charact.: Pot Life: 12 min @ 100 C (200 g); Cure Schedule: 4 h @ 100 C
Ref.: Data sheet

EP 2810 TS. Hardman Inc.
Chem. Descrip.: Two-part epoxy comp.
Category/Applications: Elevated temp. curing comp. for electrical/electronic applications; mix ratio: 100A/135B by wt.
Gen'l. Prop.: Visc.: 4500 cps (mixed)
Mech. Prop.: Hardness (Shore D): 82
Therm. Prop.: Coeff. of Linear Exp.: 156 $\times$ 10^{-6} in./in./C
Elec. Prop.: Dissip. Factor: 0.004 (100 kHz); Dielec. Const.: 2.9 (100 kHz); Vol. Resist.: 9 $\times$ 10^{15} ohm-cm
Curing Charact.: Pot Life: 25 min @ 121 C (200 g); Cure Schedule: 16 h @ 121 C
Ref.: Data sheet

EP 2820 TS. Hardman Inc.
Chem. Descrip.: Two-part epoxy comp.
Category/Applications: Elevated temp. curing comp. for electrical/electronic applications; mix ratio: 100A/86B by wt.
Gen'l. Prop.: Visc.: 2800 cps (mixed)
Mech. Prop.: Hardness (Shore A): 85
Therm. Prop.: Conduct.: 4.0×10^{-4} $Cal/cm/s/cm^2/C$
Elec. Prop.: Dissip. Factor: 0.063 (100 kHz); Dielec. Const.: 4.4 (100 kHz); Vol. Resist.: 6×10^{12} ohm-cm
Curing Charact.: Pot Life: 32 min @ 90 C (200 g); Cure Schedule: 2¼ h @ 90 C
Ref.: Data sheet

EP 2825 TS. Hardman Inc.
Chem. Descrip.: Two-part epoxy comp.
Category/Applications: Elevated temp. curing comp. for electrical/electronic applications; mix ratio: 100A/100B by wt.
Gen'l. Prop.: Visc.: 1900 cps (mixed)
Mech. Prop.: Hardness (Shore D): 26
Therm. Prop.: Conduct.: 3.8×10^{-4} $Cal/cm/s/cm^2/C$
Elec. Prop.: Dissip. Factor: 0.054 (100 kHz); Dielec. Const.: 4.2 (100 kHz); Vol. Resist.: 10^{11} ohm-cm
Curing Charact.: Pot Life: 30 min @ 121 C (200 g); Cure Schedule: 1 h @ 121 C
Ref.: Data sheet

EP 2840 TS. Hardman Inc.
Chem. Descrip.: Two-part epoxy comp.
Category/Applications: Elevated temp. curing comp. for electrical/electronic applications; mix ratio: 100A/200B by wt.
Gen'l. Prop.: Visc.: 4800 cps (mixed)
Mech. Prop.: Hardness (Shore D): 67
Therm. Prop.: Conduct.: 3.8×10^{-4} $Cal/cm/s/cm^2/C$
Elec. Prop.: Dissip. Factor: 0.029 (100 kHz); Dielec. Const.: 3.3 (100 kHz); Vol. Resist.: 8×10^{14} ohm-cm
Curing Charact.: Pot Life: 50 min @ 121 C (200 g); Cure Schedule: 3 h @ 121 C
Ref.: Data sheet

Epcar EPDM, EPM. BF Goodrich Co.
Chem. Descrip.: Ethylene-propylene rubbers
Category/Applications: Rubbers with improved resistance to ozone and weathering; good heat resistance and electrical properties; accept higher loadings of fillers and extender oils; excellent mixing, molding, and extruding characteristics; used for molded and extruded products, wire insulation and jacketing, molded O-rings, brake components, tire flaps, white sidewalls, hose tube, hose covers, matting, and automobile weatherstripping
Gen'l. Prop.: Stability: Good dimensional stability; good resistance to heat, ozone cracking, and weathering
Ref.: Product information

Epiceram. Delta Plastics Co.
Chem. Descrip.: Two-component epoxy resin polymeric systems
Category/Applications: Polymeric systems used as self-extinguishing closed cell foams in structural repair applications of damaged buildings as may occur in earthquakes

Epolene C-10. Eastman Chem. Products, Inc.
Chem. Descrip.: Low m.w. polyolefin resin
Category/Applications: Nonemulsifiable low-density wax used in hot melt coatings for papers and packaging materials, as paraffin modifiers, in slush and cast molding, in hot melt adhesives, rubber compounding, and in rigid and flexible

vinyl compounds; its coatings exhibit high gloss, low water-vapor-transmission rates, and good grease- and blocking-resistance
Form: Color (Gardner): 1
Gen'l. Prop.: M.W.: 8000; Density: 0.906 g/cc; Visc. (Brookfield Thermosel): 7800 cps (150 C); Cloud Pt.: 77 C (2% in 130 F paraffin); Soften. Pt. (Ring & Ball): 104 C; Acid No.: < 0.05
Ref.: Bulletin F-165G

Epolene C-13. Eastman Chem. Products, Inc.
Chem. Descrip.: Low m.w. polyolefin resin
Category/Applications: Nonemulsifiable wax designed for use with other Epolene waxes or blends containing lower m.w. materials; used as petroleum-wax modifiers to increase blend viscosity, improve grease resistance, blocking temp., scuff resistance, and gloss of paraffin; also as additives for inks and as compounding resins for hot melt adhesives
Form: Color (Gardner): 1
Gen'l. Prop.: M.W.: 12,000; Density: 0.913 g/cc; Cloud Pt.: 81 C (2% in 130 F paraffin); Soften. Pt. (Ring & Ball): 110 C; Acid No.: < 0.05
Ref.: Bulletin F-165G

Epolene C-14. Eastman Chem. Products, Inc.
Chem. Descrip.: Polyolefin resin
Category/Applications: See Epolene C-13
Form: Solid; Color (Gardner): 1
Gen'l. Prop.: M.W.: 23,000; Density: 0.918 g/cc; Cloud Pt.: 84 C (2% in 130 F paraffin); Soften. Pt. (Ring & Ball): > 133 C; Acid No.: < 0.05
Ref.: Bulletin F-165G

Epolene C-15. Eastman Chem. Products, Inc.
Chem. Descrip.: Polyolefin resin
Category/Applications: Nonemulsifiable low-density wax used in hot melt coatings for papers and packaging materials, as paraffin modifiers, in slush and cast molding, in hot melt adhesives, rubber compounding, and in rigid and flexible vinyl compounds; also as an additive in candle wax to provide gloss, sheen, opacity, and mold release
Form: Color (Gardner): 1
Gen'l. Prop.: M.W.: 4000; Density: 0.906 g/cc; Visc. (Brookfield Thermosel): 3900 cps (150 C); Cloud Pt.: 75 C (2% in 130 F paraffin); Soften. Pt. (Ring & Ball): 102 C; Acid No.: < 0.05
Ref.: Bulletin F-165G

Epolene C-16. Eastman Chem. Products, Inc.
Chem. Descrip.: Polyolefin resin
Category/Applications: Nonemulsifiable wax used as hot melt coatings for paper to provide glossy barrier coatings which may be readily heat sealed to most paper products, metal foils, and plastic films; also in petroleum-wax coatings, as additives in wax-copolymer coatings providing increased scuff resistance, gloss stabilization, and good hot tack; tolerates high levels of inorganic fillers without drastic increases in melt viscosity
Form: Color (Gardner): 1
Gen'l. Prop.: M.W.: 8000; Density: 0.908 g/cc; Visc. (Brookfield Thermosel): 8500 cps (150 C); Cloud Pt.: 78 C (2% in 130 F paraffin); Soften. Pt. (Ring & Ball): 106 C; Sapon. No.: 5
Ref.: Bulletin F-165G

Epolene C-17. Eastman Chem. Products, Inc.
Chem. Descrip.: Polyolefin resin
Category/Applications: See Epolene C-13

Form: Solid; Color (Gardner): 1
Gen'l. Prop.: M.W.: 19,000; Density: 0.917 g/cc; Cloud Pt.: 81 C (2% in 130 F paraffin); Soften. Pt. (Ring & Ball): 133 C; Acid No.: < 0.05
Ref.: Bulletin F-165G

Epolene C-18. Eastman Chem. Products, Inc.
Chem. Descrip.: Polyolefin resin
Category/Applications: See Epolene C-16
Form: Color (Gardner): 1
Gen'l. Prop.: M.W.: 4000; Density: 0.905 g/cc; Visc. (Brookfield, Thermosel): 4000 cps (150 C); Cloud Pt.: 71 C (2% in 130 F paraffin); Soften. Pt. (Ring & Ball): 102 C; Sapon. No.: 5
Ref.: Bulletin F-165G

Epolene E-10. Eastman Chem. Products, Inc.
Chem. Descrip.: Polyolefin resin
Category/Applications: Emulsifiable wax for water-emulsion floor polishes, imparting excellent slip resistance, toughness, and durability to polish films; also as finishing agents for cotton and rayon fabrics, as textile softeners in conjunction with wash-and-wear finishing waxes, and as lubricants in clay coatings for paper to reduce dusting during calendering
Form: Color (Gardner): 2
Gen'l. Prop.: M.W.: ≈ 3000; Density: 0.942 g/cc; Visc. (Brookfield Thermosel): 900 cps (125 C); Soften. Pt. (Ring & Ball): 106 C; Acid No.: 15
Ref.: Bulletin F-165G

Epolene E-11. Eastman Chem. Products, Inc.
Chem. Descrip.: Polyolefin resin
Category/Applications: Low-density emulsifiable wax used in emulsion floor polishes where good slip resistance is required in formulations containing < 20% emulsifiable waxes (lower m.w. and melt visc. than Epolene E-10); also as textile finishing agent
Form: Color (Gardner): 2
Gen'l. Prop.: M.W.: ≈ 2200; Density: 0.941 g/cc; Visc. (Brookfield Thermosel): 350 cps (125 C); Soften. Pt. (Ring & Ball): 106 C; Acid No.:15
Ref.: Bulletin F-165G

Epolene E-12. Eastman Chem. Products, Inc.
Chem. Descrip.: Polyolefin resin
Category/Applications: High-density emulsifiable wax esp. for use as a textile lubricant (used with wash-and-wear finishing resins to impart better tensile strength, abrasion resistance, sewability, and hand); also as softeners for synthetic fabrics, floor polishes, and emulsions for coating citrus fruit
Form: Color (Gardner): 1
Gen'l. Prop.: M.W.: ≈ 2300; Density: 0.955 g/cc; Visc. (Brookfield, Thermosel): 250 cps (125 C); Soften. Pt. (Ring & Ball): 112 C; Acid No.: 16
Ref.: Bulletin F-165G

Epolene E-14. Eastman Chem. Products, Inc.
Chem. Descrip.: Polyolefin resin
Category/Applications: Low-density and low soften. pt. emulsifiable wax imparting excellent slip resistance to floor polish films
Form: Powder; Color (Gardner): 2
Gen'l. Prop.: M.W.: ≈ 1800; Density: 0.939 g/cc; Visc. (Brookfield Thermosel): 250 cps (125 C); Soften. Pt. (Ring & Ball): 104 C; Acid No.: 16
Ref.: Bulletin F-165G

Epolene E-15. Eastman Chem. Products, Inc.
Chem. Descrip.: Polyolefin resin
Category/Applications: See Epolene E-14
Form: Color (Gardner): 2
Gen'l. Prop.: M.W.: ≈ 3400; Density: 0.925 g/cc; Visc. (Brookfield Thermosel): 350 cps (125 C); Soften. Pt. (Ring & Ball): 100 C; Acid No.: 16
Ref.: Bulletin F-165G

Epolene E-43. Eastman Chem. Products, Inc.
Chem. Descrip.: Polyolefin resin
Category/Applications: Emulsifiable wax with greatest hardness and highest melting pt. of the emulsifiable Epolene waxes; imparts outstanding slip resistance to floor polishes
Form: Solid; Color (Gardner): 11
Gen'l. Prop.: M.W.: ≈ 4500; Density: 0.934 g/cc; Visc. (Brookfield Thermosel): 400 cps (190 C); Soften. Pt. (Ring & Ball): 157 C; Acid No.: 47
Ref.: Bulletin F-165G

Epolene N-10. Eastman Chem. Products, Inc.
Chem. Descrip.: Polyolefin resin
Category/Applications: Nonemulsifiable wax which is easily blended with natural or synthetic waxes to improve tensile strength, abrasion resistance, and adhesion to fibrous substrates; used in paper-coating applications, e.g., folding cartons (increases coverage, provides a glossy, flexible, scuff-resistant finish), and in printing inks to improve resistance to scuffing and rub-off
Form: Color (Gardner): 1
Gen'l. Prop.: M.W.: 3000; Density: 0.925 g/cc; Visc. (Brookfield Thermosel): 1500 cps (125 C); Cloud Pt.: 85 C (2% in 130 F paraffin); Soften. Pt. (Ring & Ball): 111 C; Acid No.: < 0.05
Ref.: Bulletin F-165G

Epolene N-11. Eastman Chem. Products, Inc.
Chem. Descrip.: Polyolefin resin
Category/Applications: Similar to Epolene N-10 but with lower m.w. and density; also used as a solid lubricant in the manufacture of corrugated board
Form: Color (Gardner): 1
Gen'l. Prop.: M.W.: 2200; Density: 0.921 g/cc; Visc. (Brookfield Thermosel): 350 cps (125 C); Cloud Pt.: 79 C (2% in 130 F paraffin); Soften. Pt. (Ring & Ball): 108 C; Acid No.: < 0.05
Ref.: Bulletin F-165G

Epolene N-12. Eastman Chem. Products, Inc.
Chem. Descrip.: Polyolefin resin
Category/Applications: High-density, high-melting point, hard polyethylene wax used to raise the blocking temp. of wax blends and to upgrade low-melting paraffins; also as a mold release additive in rubber molding and as an extrusion aid for vinyl processing
Form: Color (Gardner): 1
Gen'l. Prop.: M.W.: 2300; Density: 0.938 g/cc; Visc. (Brookfield Thermosel): 450 cps (125 C); Cloud Pt.: 87 C (2% in 130 F paraffin); Soften. Pt. (Ring & Ball): 117 C; Acid No.: < 0.05
Ref.: Bulletin F-165G

Epolene N-14. Eastman Chem. Products, Inc.
Chem. Descrip.: Polyolefin resin
Category/Applications: Similar to Epolene N-10 but with lower m.w. and density; also used as a lubricant in processing vinyl plastics
Form: Powder; Color (Gardner): 1
Gen'l. Prop.: M.W.: 1800; Density: 0.920 g/cc; Visc. (Brookfield Thermosel): 150 cps (125 C); Cloud Pt.: 77 C (2% in 130 F paraffin); Soften. Pt. (Ring & Ball): 106 C; Acid No.: < 0.05
Ref.: Bulletin F-165G

Epolene N-15. Eastman Chem. Products, Inc.
Chem. Descrip.: Polyolefin resin
Category/Applications: Low-density, low-visc. polypropylene resin with a very high melting pt. and hardness; used as a modifier for petroleum waxes to increase resistance to blocking, scuffing, and abrasion, and as a compounding resin for hot melt adhesives
Form: Solid; Color (Gardner): 1
Gen'l. Prop.: M.W.: 11,000; Density: 0.860 g/cc; Visc. (Brookfield Thermosel): 600 cps (190 C); Cloud Pt.: 104 C (2% in 130 F paraffin); Soften. Pt. (Ring & Ball): 163 C; Acid No.: < 0.05
Ref.: Bulletin F-165G

Epolene N-34. Eastman Chem. Products, Inc.
Chem. Descrip.: Polyolefin resin
Category/Applications: See Epolene N-11
Form: Color (Gardner): 1
Gen'l. Prop.: M.W.: 2900; Density: 0.910 g/cc; Visc. (Brookfield, Thermosel): 450 cps (125 C); Cloud Pt.: 69 C (2% in 130 F paraffin); Soften. Pt. (Ring & Ball): 103 C; Acid No.: < 0.05
Ref.: Bulletin F-165G

Epolite 1301 (formerly S/916AD). Hexcel Corp./Rezolin Div.
Chem. Descrip.: Two-part epoxy resin
Category/Applications: General purpose, chip-resistant, epoxy surface coat with fast tack free time; for use in models, masters, strectch dies, checking and drill fixtures, patterns; can be used with Epolite 2300 and 2302 laminating systems; excellent nonsag properties; cures readily under high humidity conditions; mix ratio: 100 parts resin/10 parts hardener by weight
Form: Paste; Color: White
Gen'l. Prop.: Density: 1.384 g/cc (D792-66); Visc. (mixed): Thixotropic (D2393-71)
Mech. Prop.: Tens. Str.: 329 kg/cm^2 (D-638-68); Flex. Str.: 770 kg/cm^2 (D790-70); Compr. Str.: 1.050 kg/cm^2 (D695-69); Hardness (Shore D): 86 (D2240-68)
Curing Charact. (R.T. cure): Gel Time: 20 min (D2471-71)
Toxicity/Handling: Resin: harmful if swallowed; may cause skin irritation, sensitization, or dermatitis; Hardener: harmful if swallowed; prolonged or repeated skin contact may cause burns, sensitization, or dermatitis; use with adequate ventilation; for industrial use only; store in a cool, dry area (65-80 F); 1 year min. shelf life in unopened containers
Std. Pkgs.: 12 qt. Pack containing 12 qt. cans (resin) and 12 ½-pt. cans (hardener), 29¼ lb net wt.; Pail Pack containing 5-gal pail (resin) and 1-gal can (hardener), 26 lb net wt.
Ref.: Tech. data bulletin

Epolite 1302 (formerly F/916/AD). Hexcel Corp./Rezolin Div.
Chem. Descrip.: Two-part epoxy resin and hardener
Category/Applications: Abrasion- and impact-resistant epoxy, metallic black surface coat with rapid cure; used for automotive gates, router and trim fixtures, foundry patterns and core boxes, check and inspection fixtures, draw and stretch dies; can be used with Epolite 2300, 2301, 2302, and 2350 laminating systems; mix ratio: 100 parts resin/7 parts hardener by wt.
Form: Paste; Color: Black
Gen'l. Prop.: Density: 1.605 g/cc (D792-66); Visc. (mixed): 20,000 cps (D2393-71)
Mech. Prop.: Tens. Str.: 290.5 kg/cm^2 (D638-68); Flex. Str.: 392 kg/cm^2 (D-

790-70); Hardness (Shore D): 88 (D-2240-68)
Therm. Prop. (casting): Distort. Temp.: 50 C (D648-56)
Curing Charact. (R.T. cure): Gel Time: 23 min (D2471-71)
Toxicity/Handling: See Epolite 1301
Std. Pkgs.: 12 qt. Pack containing 12 qt. cans (resin) and 12 ½-pt. cans (hardener), 34 lb 2 oz net wt.; Pail Pack containing 5-gal pail (resin) and ¼-gal can (hardener), 53½ lb net wt.
Ref.: Tech. data bulletin

Epolite 1350 (formerly 111/111). Hexcel/Rezolin Div.
Chem. Descrip.: Two-part epoxy
Category/Applications: Epoxy surface coat with med. heat resistance, max. dimensional stability, and capability of bonding readily to wet plaster or concrete; used in plastic-faced plasters, Paraplast 8100 and 8101 molds, thermoplastic sheet forming molds, spotting racks, Keller patterns and blue blocks; can be used with Epolite 2300 and 2302 laminating systems; mix ratio: 100 parts resin/16 parts hardener, by wt.
Form: Paste; Color: White
Gen'l. Prop.: Density: 1.218 g/cc (D-792-66); Visc. (mixed): Thixotropic
Mech. Prop.: Flex. Str.: 490 kg/cm^2 (D-790-70); Compr. Str.: 770 kg/cm^2 (D-695-69); Hardness (Shore D): 88 (D-2240-68)
Therm. Prop.: Distort. Temp.: 196.7 C (short period, D648-56)
Curing Charact. (R.T. cure): Gel Time: 15 min (D2471-71)
Toxicity/Handling: See Epolite 1301
Std. Pkgs.: 12 Qt. Pack containing 12 qt. cans (resin) and 12 ½-pt. cans (hardener), 28 lb 2 oz net wt.; Pail Pack containing 5 gal pail (resin) and 1 gal can (hardener), 58 lb net wt.
Ref.: Tech. data bulletin

Epolite 1353 (formerly 57C/953A). Hexcel Corp./Rezolin Div.
Chem. Descrip.: Two-part epoxy
Category/Applications: Epoxy surface coat with high strength retention in high temp. applications to 400 F; used in high temp. laminating molds, vacuum forming molds, autoclave and oven bonding fixtures, spray masks and metal spray molds; can be used with Epolite 2353 laminating system; mix ratio: 100 parts resin/18 parts hardener by wt.
Form: Paste; Color: Black
Gen'l. Prop.: Density: 1.522 g/cc (D-792-66); Visc. (mixed): 20,000 cps (D-2393-71)
Mech. Prop.: Flex. Str.: 546 kg/cm^2 (D-790-70); Compr. Str.: 1484 kg/cm^2; Hardness (Shored D): 88 (D2240-68)
Therm. Prop.: Distort. Temp.: 173 C (casting, D648-56)
Curing Charact.: Gel Time: 4 h (D2471-71)
Toxicity/Handling: See Epolite 1301
Std. Pkgs.: 12 Qt. Pack containing 12 qt. cans (resin) and 12 ½-pt. cans (hardener), 26 lb 13 oz net wt.; Pail Pack containing 5 gal pail (resin) and 1 gal can (hardener), 60 lb 12 oz net wt.
Ref.: Tech. data bulletin

Epolite 1354 (formerly 57C/954). Hexcel Corp./Rezolin Div.
Chem. Descrip.: Two-part epoxy
Category/Applications: Epoxy surface coat designed to give high temp. properties with R.T. setting; may be buffed to a glossy finish; can be used with Epolite 2354 laminating system; used for high temp. molds, fixtures used in autoclaves or ovens, vacuum form molds, and tooling aids; mix ratio: 100 parts resin/17 parts hardener by wt.
Form: Paste; Color: Black
Gen'l. Prop.: Density: 1.584 g/cc (D792-66); Visc. (mixed): 15,000 cps (D2393-71)

Mech. Prop.: Flex. Str.: 651 kg/cm^2 (D-790-70); Compr. Str.: 1330 kg/cm^2 (D-695-69); Hardness (Shore D): 88 (D-2240-68)
Therm. Prop.: Distort. Temp.: 157 C (casting, D648-56)
Curing Charact. (R.T. and post cure): Gel Time: 45 min (D2471-71)
Toxicity/Handling: See Epolite 1301
Std. Pkgs.: 12 Qt. Pack containing 12 qt. cans (resin) and 12 ½-pt. cans (hardener), 26 lb 10 oz net wt.; Pail Pack containing 5 gal pail (resin) and 1 gal can (hardener), 60 lb 4 oz net wt.
Ref.: Tech. data bulletin

Epolite 2300 (formerly 90EF/921). Hexcel Corp./Rezolin Div.
Chem. Descrip.: Two-part epoxy resin laminating system
Category/Applications: Laminating system with max. wettability, low visc., and high mechanical strength; used in checking and inspection fixtures and gauges, Keller models and spotting racks, router and trim fixtures, foundry patterns and core boxes, assembly and drill fixtures, and stretch dies and die models; mix ratio: 8 parts resin/1 part hardener
Form: Paste; Color: White
Gen'l. Prop.: Density: 1.273 g/cc (D-792-66); Visc. (mixed): 3000 cps (D-2393-71)
Mech. Prop.: Tens. Str.: 1750 kg/cm^2 (D638-68); Flex. Str.: 2450 kg/cm^2 (D-790-70); Hardness (Shore D): 89 (D-2240-68)
Therm. Prop.: Distort. Temp.: 57.2 C (casting, D648-56)
Curing Charact. (R.T. cure): Gel Time: 22 min (D2471-71)
Toxicity/Handling: See Epolite 1301
Std. Pkgs.: 12 Qt. Pack containing 12 qt. cans (resin) and 12 ½-pt. cans (hardener), 27-3/8 lb net wt.; also avail. in Pail Pack, net wt. 56¼ lb, and Drum Pack, net wt. 562½ lb
Ref.: Tech. data bulletin

Epolite 2302 (formerly 90S/924). Hexcel Corp./Rezolin Div.
Chem. Descrip.: Two-part epoxy resin laminating system
Category/Applications: Laminating system with excellent wettability without run-out, high strength, and very low toxicity; ideal for R.T. wet lay-up tooling and production applications, spotting racks, Keller models, router and trim fixtures, checking and inspection fixtures and gauges, stretch dies, die models, assembly and drill fixtures; mix ratio: 100 parts resin/16 parts hardener
Form: Paste; Color: White; Odor: Low
Gen'l. Prop.: Density: 1.433 g/cc (D-792-66); Visc. (mixed): 2600 cps (D-2393-71)
Mech. Prop.: Tens. Str.: 1519 kg/cm^2 (D638-68); Flex. Str.: 2170 kg/cm^2 (D-790-70); Hardness (Shore D): 89 (D-2240-68)
Therm. Prop.: Distort. Temp.: 56.1 C (casting, D648-56)
Curing Charact. (R.T. cure): Gel Time: 20 min (D2471-71)
Toxicity/Handling: See Epolite 1301
Std. Pkgs.: 12 Qt. Pack containing 12 qt. cans (resin), 12 ½-pt. cans (hardener), 30¾ lb net wt.; also avail. in Pail Pack, 58 lb net wt. and Drum Pack, 585 lb net wt.
Ref.: Tech. data bulletin

Epolite 2353 (formerly 52B/953A). Hexcel Corp./Rezolin Div.
Chem. Descrip.: Two-part epoxy resin laminating system
Category/Applications: Laminating system for high temp. tooling applications, bonding fixtures, laminating molds; its long pot life make it ideal for vacuum

bag-oven cure processes; mix ratio: 100 parts resin/22 parts hardener
Form: Paste; Color: Black
Gen'l. Prop.: Density: 1.384 g/cc (D-792-66); Visc. (mixed): 11,800 cps (D-2393-71); Stability: High heat resistance
Mech. Prop.: Flex. Str.: 5565 kg/cm^2 (D-790-70); Hardness (Shore D): 86 (D-2240-68)
Therm. Prop.: Distort. Temp.: 182 C (casting, D648-56)
Curing Charact.: Gel Time: 5 h (25 C, D2471-71)
Toxicity/Handling: See Epolite 1301
Std. Pkgs.: 12 Qt. Pack containing 12 qt. cans (resin), 12 ½-pt. cans (hardener), 27 lb 12 oz net wt.; Pail Pack containing 5 gal pail (resin), 1 gal can (hardener), 67 lb 2 oz net wt.
Ref.: Tech. data bulletin

Epolite 2354 (formerly 52B/954). Hexcel Corp./Rezolin Div.
Chem. Descrip.: Two-part epoxy resin laminating system
Category/Applications: Laminating system offering R.T. hardening with high temp. properties for most tooling applications; used in spray masks, vacuum forming molds, high temp. laminating molds, fixtures used in ovens and autoclaves; mix ratio: 100 parts resin/22 parts hardener by wt.
Form: Paste; Color: Black
Gen'l. Prop.: Density: 1.430 g/cc (D-792-66); Visc. (mixed): 5500 cps (D-2393-71)
Mech. Prop.: Flex. Str.: 3605 kg/cm^2 (D790-70); Hardness (Shore D): 86 (D2240-68)
Therm. Prop.: Distort. Temp.: 158.9 C (casting, D648-56)
Curing Charact.: Gel Time: 40 min (25 C, D2471-71); Hardening Time: 16 h (25 C); Post Cure: 2 h @ 200 F, 3 h @ 325 F
Toxicity/Handling: See Epolite 1301
Std. Pkgs.: 12 Qt. Pack containing 12 qt. cans (resin), 12 ½-pt. cans (hardener), 27¾ lb net wt.; Pail Pack containing 5 gal pail (resin), 1 gal can (hardener), 67 lb, 2 oz net wt.
Ref.: Tech. data bulletin

Epolite 2360. Hexcel Corp./Rezolin Div.
Chem. Descrip.: Two-part epoxy resin laminating system
Category/Applications: Two-part laminating system formulated for high temp. tooling applications using Kevlar fabrics, structures to be used in high temp. autoclave applications, high temp. laminating molds and bonding fixtures; long pot life for vacuum bagging and rub-out; mix ratio: 1 part resin/1 part hardener by wt.
Form: Liquid; Color: Dark brown
Gen'l. Prop.: Density: 1.217 g/cc (D-792-66); Visc. (mixed): 1000 cps (D-2393-71)
Mech. Prop.: Tens. Str.: 5180 kg/cm^2 (D-638-68); Flex. Str.: 343 kg/cm^2 (D-790-70); Hardness (Shored D): 90 (D-2240-68)
Therm. Prop.: Distort. Temp.: 198 C (D648-56)
Curing Charact.: Recommended Cure Cycle: 2–4 h gel @ 150 F; 2 h @ 150 F; 2 h @ 225 F; 2 h @ 325 F; 2 h @ 400 F
Toxicity/Handling: See Epolite 1301
Std. Pkgs.: Pail Pack containing 5 gal pail (resin), 5 gal pail (hardener), 80 lb net wt.
Ref.: Tech. data bulletin

Epolite 3300 (formerly 30/960). Hexcel Corp./Rezolin Div.
Chem. Descrip.: Two-part epoxy compound
Category/Applications: Mass casting epoxy compound with variable hardness

and high impact resistance; used in hammer forms, stretch dies, drop hammer dies, close tolerance masters or models; mix ratio: 100 parts resin/10–30 parts hardener by wt.
Form: Paste; Color: Black
Gen'l. Prop.: Density: 1.688–1.938 g/cc (D792-66); Visc. (mixed): 25,000 cps (D2393-71)
Mech. Prop.: Tens. Str.: 182 kg/cm^2 (D-638-68); Flex. Str.: 630 kg/cm^2 (D790-70); Compr. Str.: 777 kg/cm^2 (D695-69); Hardness (Shore D): 38–85 (D2240-68)
Therm. Prop.: Distort. Temp.: 52.8 C (casting, D648-56)
Curing Charact. (R.T. cure): Gel Time: 70 min (D2471-71)
Toxicity/Handling: See Epolite 1301
Std. Pkgs.: 12 Qt. Pail containing 12 qt. cans (resin), 12 ½-pt. cans (hardener), 46½ lb net wt.; Pail Pack with 5 gal pail (resin), 1 gal can (hardener), 82½ lb net wt.; Drum Pack with 55 gal drum (resin), 15 gal drum (hardener), 676 lb net wt.
Ref.: Tech. data bulletin

Epolite 3301 (formerly 30/932B). Hexcel Corp./Rezolin Div.
Chem. Descrip.: Two-part epoxy casting compound
Category/Applications: Casting system with excellent dimensional stability, long pot life, high strength; used for stretch dies, draw dies, impact dies, foundry patterns, casting molds; mix ratio: 100 parts resin/10 parts hardener
Form: Paste; Color: Black
Gen'l. Prop.: Density: 2.24 g/cc (D792-66); Visc. (mixed): 20,000 cps (D2393-71)
Mech. Prop.: Tens. Str.: 525 kg/cm^2 (D-638-68); Flex. Str.: 518 kg/cm^2 (D790-70); Compr. Str.: 980 kg/cm^2 (D695-69); Hardness (Shore D): 88 (D2240-68)
Curing Charact. (R.T. cure): Gel Time: 70 min (D2471-71)
Toxicity/Handling: See Epolite 1301
Std. Pkgs.: 12 Qt. Pack containing 12 qt. cans (resin), 12 ½-pt. cans (hardener), 46½ lb net wt.; Pail Pack with 5 gal pail (resin), 1 gal can (hardener), 82½ lb net wt.; Drum Pack with 55 gal drum (resin), 15 gal drum (hardener), 660 lb net wt.
Ref.: Tech. data bulletin

Epolite 3302 (formerly F/933A). Hexcel Corp./Rezolin Div.
Chem. Descrip.: Two-part epoxy casting compound
Category/Applications: Abrasion-resistant metallic black casting compound which is easy to pour and provides accurate reproduction of details and excellent dimensional stability; used for foundry patterns and core boxes, match plates and inserts, hydropress blocks, and vise and chuck jaws; mix ratio: 100 parts resin/10 parts hardener by wt.
Form: Liquid; Color: Metallic black
Gen'l. Prop.: Density: 1.550 g/cc (D-792-66); Visc. (mixed): 10,000 cps (D-2393-71)
Mech. Prop.: Tens. Str.: 385 kg/cm^2 (D-638-68); Flex. Str.: 630 kg/cm^2 (D790-70); Compr. Str.: 875 kg/cm^2 (D695-69); Hardness (Shore D): 88 (D2240-68)
Curing Charact. (R.T. cure): Gel Time: 30 min (D2471-71)
Toxicity/Handling: See Epolite 1301
Std. Pkgs.: 12 Qt. Pack containing 12 qt. cans (resin), 12 ½-pt. cans (hardener), 35 lb 4 oz net wt.; Pail Pack with 5 gal pail (resin), 1 gal can (hardener), 55 lb net wt.
Ref.: Tech. data bulletin

Epolite 3306 (formerly 33/G-3). Hexcel Corp./Rezolin Div.
Chem. Descrip.: Two-part epoxy casting compound

Category/Applications: General purpose casting compound giving excellent detail reproduction, high impact resistance, and bonding readily to metal substructures; used in hammer forms, foundry patterns and core boxes, chuck jaws, production parts, surface table tops; mix ratio: 100 parts resin/10 parts hardener
Form: Liquid; Color: Metallic gray
Gen'l. Prop.: Density: 1.495 g/cc (D792-66); Visc. (mixed): 6000 cps (D2393-71)
Mech. Prop.: Tens. Str.: 441 kg/cm^2 (D-638-68); Flex. Str.: 574 kg/cm^2 (D790-70); Compr. Str.: 980 kg/cm^2 (D695-69); Hardness (Shore D): 86 (D2240-68)
Curing Charact. (R.T. cure): Gel Time: 45 min (D2471-71)
Toxicity/Handling: See Epolite 1301
Std. Pkgs.: 12 Qt. Pack containing 12 qt. cans (resin), 12 ½-pt. cans (hardener), 33 lb 6 oz net wt.; Pail Pack with 5 gal pail (resin), 1 gal can (hardener), 55 lb net wt.; Drum Pack with 55 gal drum (resin), 15 gal drum (hardener), 660 lb net wt.
Ref.: Tech. data bulletin

Epolite 3353 (formerly 56A/953A). Hexcel Corp./Rezolin Div.
Chem. Descrip.: Two-part epoxy casting compound
Category/Applications: Heat cured casting compound with high thermal conductivity, excellent machineability and surface reproduction; used in oven and autoclave tooling, vacuum forming molds, high temp. laminating molds, and honeycomb bonding fixtures; mix ratio: 100 parts resin/12 parts hardener by wt.
Form: Paste; Color: Metallic gray
Gen'l. Prop.: Density: 1.522 g/cc (D792-66); Visc. (mixed): 137,000 cps (D2393-71)
Mech. Prop.: Hardness (Shore D): 89 (D2240-68)
Therm. Prop.: Distort. Temp.: 176.7 C (D648-56)
Curing Charact.: Gel Time: 2–3 h @ 25 C (D2471-71)
Toxicity/Handling: See Epolite 1301
Std. Pkgs.: 12 Qt. Pack containing 12 qt. cans (resin), 12 ¼-pt. cans (hardener), 25 lb 11 oz net wt.; Pail Pack with 5 gal pail (resin), 1 gal can (hardener), 68 lb net wt.
Ref.: Tech. data bulletin

Epolite 3354 (formerly 56A/954). Hexcel Corp./Rezolin Div.
Chem. Descrip.: Two-part epoxy casting compound
Category/Applications: R.T. set/high temp. curing casting compound with good flowability, and excellent machineability and reproduction characteristics; used in high temp. casting and lay-up molds, honeycomb bonding fixtures, oven and autoclave tooling, and vacuum form molds; mix ratio: 100 parts resin/11 parts hardener by wt.
Form: Paste; Color: Metallic gray
Gen'l. Prop.: Density: 1.55 g/cc (D792-66); Visc. (mixed): 52,000 cps (D2393-71)
Therm. Prop.: Distort. Temp.: 154.4 C (D648-56)
Curing Charact. (R.T.): Gel Time: 45 min (D2471-71)
Toxicity/Handling: See Epolite 1301
Std. Pkgs.: 12 Qt. Pack containing 12 qt. cans (resin), 12 ¼-pt. cans (hardener), 25 lb 5 oz net wt.; Pail Pack with 5 gal pail (resin), 1 gal can (hardener), 66 lb 10 oz net wt.
Ref.: Tech. data bulletin

Epolite 3357. Hexcel Corp./Rezolin Div.
Chem. Descrip.: Two-part epoxy casting compound
Category/Applications: High temp. curing, casting resin system with good flowability, reproduction characteristics, and machineability; exhibits low shrinkage for thick high temp. castings; used in high temp. casting and lay-up molds, oven and autoclave tooling, vacuum form molds, and thermoforming molds; mix ratio: 6 parts resin/1 part hardener by wt.
Form: Paste; Color: Black
Gen'l. Prop.: Density: 2.463 g/cc (D792-66); Visc. (mixed): 40,000 cps (D2393-71)
Mech. Prop.: Tens. Str.: 189 kg/cm^2 (D-638-68); Flex. Str.: 501 kg/cm^2 (D790-70); Compr. Str.: 1295 kg/cm^2 (D695-69); Hardness (Shore D): 90 (D2240-68)
Therm. Prop.: 141 C (casting per recommended cure cycle, D648-56)
Curing Charact.: Recommended Cure Cycle: 2–4 h gel @ 150 F, 2 h @ 150 F, 2 h @ 225 F, 2 h @ 325 F
Toxicity/Handling: See Epolite 1301
Std. Pkgs.: Pail Pack containing 5 gal pail (resin), 2 gal can (hardener), 87½ lb net wt.
Ref.: Tech. data bulletin

Epotuf Hardener 37-605. Reichhold Chem., Inc.
Chem. Descrip.: Modified polyamine epoxy hardener
Category/Applications: Low-visc. epoxy hardener with good pot life for electrical potting, encapsulating, and casting, hand lay-up laminating and tooling, flooring and surfacing, and adhesives
Form: Liquid
Gen'l. Prop.: Visc. (Brookfield): 700–900 cps
Ref.: Product pamphlet

Epotuf Hardener 37-610. Reichhold Chem., Inc.
Chem. Descrip.: Modified polyamine epoxy hardener
Category/Applications: Epoxy hardener imparting resiliency; used in electrical potting, encapsulating, and casting, hand lay-up laminating and tooling, flooring and surfacing, and adhesives
Form: Liquid
Gen'l. Prop.: Visc. (Brookfield): 200–300 cps; Stability: Resistant to mechanical and thermal shock
Ref.: Product pamphlet

Epotuf Hardener 37-611. Reichhold Chem., Inc.
Chem. Descrip.: Modified polyamine epoxy hardener
Category/Applications: Epoxy hardener with high flexibility; used in electrical potting, encapsulating, and casting, hand lay-up laminating and tooling, flooring and surfacing, and adhesives
Form: Liquid
Gen'l. Prop.: Visc. (Brookfield): 3500–4500 cps
Ref.: Product pamphlet

Epotuf Hardener 37-612. Reichhold Chem., Inc.
Chem. Descrip.: Polyamide epoxy hardener
Category/Applications: General purpose epoxy hardener with good resiliency and adhesion; used in electrical potting, encapsulating, and casting, flooring and surfacing, coatings, and adhesives
Form: Liquid
Gen'l. Prop.: Visc. (Brookfield): 10,000–19,000 cps; Amine Value: 330–350
Ref.: Product pamphlet

Epotuf Hardener 37-614. Reichhold Chem., Inc.
Chem. Descrip.: Modified polyamine epoxy hardener
Category/Applications: Epoxy hardener with exceptionally wide use range; used in electrical potting, encapsulating, and casting, hand lay-up laminating and tooling, flooring and surfacing, coatings, and adhesives
Form: Liquid
Gen'l. Prop.: Visc. (Brookfield): 3500–5500 cps
Ref.: Product pamphlet

Epotuf Hardener 37-617. Reichhold Chem., Inc.
Chem. Descrip.: Polyamide epoxy hardener
Category/Applications: Epoxy hardener for use in solution form in coatings
Form: Liquid
Gen'l. Prop.: Visc. (Brookfield): 31–38 poise (170 F); Amine Value: 210–220
Ref.: Product pamphlet

Epotuf Hardener 37-618. Reichhold Chem., Inc.
Chem. Descrip.: Polyamide
Category/Applications: Solution of Epotuf Hardener 37-617
Form: Liquid
Gen'l. Prop.: Visc. (Gardner-Holdt): V–Z
Ref.: Product pamphlet

Epotuf Hardener 37-619. Reichhold Chem., Inc.
Chem. Descrip.: Aromatic polyamine
Category/Applications: Nonstaining, heat-resistant epoxy hardener for use in electrical potting, encapsulating, and casting, high-pressure laminating, filament winding, coatings, and adhesives
Form: Liquid
Gen'l. Prop.: Visc. (Brookfield): 600–700 poise; Stability: Heat-resistant
Ref.: Product pamphlet

Epotuf Hardener 37-620. Reichhold Chem., Inc.
Chem. Descrip.: Polyamide epoxy hardener
Category/Applications: Low-visc. epoxy hardener with good adhesion and resiliency; used in electrical potting, encapsulating, and casting, hand lay-up laminating and tooling, filament winding, flooring and surfacing, coatings, and adhesives
Form: Liquid
Gen'l. Prop.: Visc. (Brookfield): 400–700 cps; Amine Value: 425–450
Ref.: Product pamphlet

Epotuf Hardener 37-621. Reichhold Chem., Inc.
Chem. Descrip.: Polyamide epoxy hardener
Category/Applications: Epoxy hardener in solution form; used in coatings
Form: Liquid
Gen'l. Prop.: Visc. (Gardner-Holdt): $W–Z_1$; Amine Value: 85–95
Ref.: Product pamphlet

Epotuf Hardener 37-622. Reichhold Chem., Inc.
Chem. Descrip.: Modified polyamine epoxy hardener
Category/Applications: Very versatile, general purpose epoxy hardener; used in electrical potting, encapsulating, and casting, hand lay-up laminating and tooling, flooring and surfacing, coatings, and adhesives
Form: Liquid
Gen'l. Prop.: Visc. (Brookfield): 80–150 cps
Ref.: Product pamphlet

Epotuf Hardener 37-623. Reichhold Chem., Inc.
Chem. Descrip.: Aromatic polyamine
Category/Applications: Epoxy hardener with outstanding high temp. strength and chemical resistance; used in electrical potting, encapsulating, and casting, high-pressure laminating, hand lay-up laminating and tooling, adhesives, and molding and fluidized bed compounds
Form: Solid
Gen'l. Prop.: M.P. (Durran): 140–150 F
Ref.: Product pamphlet

Epotuf Hardener 37-624. Reichhold Chem., Inc.
Chem. Descrip.: Anhydride
Category/Applications: Epoxy hardener with very low viscosity; used in electrical potting, encapsulating, and casting, and filament winding
Form: Liquid
Gen'l. Prop.: Visc. (Brookfield): 100–200 cps
Ref.: Product pamphlet

Epotuf Hardener 37-625. Reichhold Chem., Inc.
Chem. Descrip.: Polyamide epoxy hardener
Category/Applications: Faster-reacting epoxy hardener for use in coatings and adhesives
Form: Liquid
Gen'l. Prop.: Visc. (Brookfield): 25,000–38,000 cps; Amine Value: 335–365
Ref.: Product pamphlet

Epotuf Hardener 37-640. Reichhold Chem., Inc.
Chem. Descrip.: Polyamide epoxy hardener
Category/Applications: Versatile epoxy hardener with instant compatibility with resin; used in electrical potting, encapsulating, and casting, flooring and surfacing, coatings, and adhesives
Form: Liquid
Gen'l. Prop.: Visc. (Brookfield): 9000–15,000 cps; Amine Value: 370–410
Ref.: Product pamphlet

Epotuf Resin 37-127, 37-128. Reichhold Chem., Inc.
Chem. Descrip.: Epoxy resin
Category/Applications: 100% reactive, low-visc. resin for electrical potting, encapsulating, and casting, hand lay-up laminating and tooling, flooring and surfacing, coatings, and adhesives applications
Form: Liquid
Gen'l. Prop.: Visc. (Brookfield): 500–1000 cps
Ref.: Product pamphlet

Epotuf Resin 37-130. Reichhold Chem., Inc.
Chem. Descrip.: Epoxy resin
Category/Applications: See Epotuf 37-127, 37-128
Form: Liquid
Gen'l. Prop.: Visc. (Brookfield): 500–700 cps
Ref.: Product pamphlet

Epotuf Resin 37-134. Reichhold Chem., Inc.
Chem. Descrip.: Epoxy resin
Category/Applications: Resilient, med.-visc. resin for electrical potting, encapsulating, and casting, flooring and surfacing, coatings, and adhesives
Form: Liquid
Gen'l. Prop.: Visc. (Brookfield): 2100–3100 cps
Ref.: Product pamphlet

Epotuf Resin 37-135. Reichhold Chem., Inc.
Chem. Descrip.: Epoxy resin
Category/Applications: 100% reactive,

med.-visc. resin for electrical potting, encapsulating, and casting, hand lay-up laminating and tooling, flooring and surfacing, coatings, and adhesives
Form: Liquid
Gen'l. Prop.: Visc. (Brookfield): 5000–8500 cps
Ref.: Product pamphlet

Epotuf Resin 37-139. Reichhold Chem., Inc.
Chem. Descrip.: Epoxy resin
Category/Applications: Unmodified, med.-visc. resin for electrical potting, encapsulation, and casting, filament winding, and adhesives
Form: Liquid
Gen'l. Prop.: Visc. (Brookfield): 7000–10,000 cps
Ref.: Product pamphlet

Epotuf Resin 37-140. Reichhold Chem., Inc.
Chem. Descrip.: Epoxy resin
Category/Applications: Unmodified, med.-high visc. resin for electrical potting, encapsulation, and casting, high-pressure laminating, filament winding, coatings, and adhesives
Form: Liquid
Gen'l. Prop.: Visc. (Brookfield): 11,000–14,000 cps
Ref.: Product pamphlet

Epotuf Resin 37-141. Reichhold Chem., Inc.
Chem. Descrip.: Epoxy resin
Category/Applications: Unmodified, high-visc. resin for coatings and adhesives
Form: Liquid
Gen'l. Prop.: Visc. (Brookfield): 20,000–35,000 cps
Ref.: Product pamphlet

Epotuf Resin 37-151. Reichhold Chem., Inc.
Chem. Descrip.: Epoxy resin
Category/Applications: Resin with inherent flexiblity for electrical potting, encapsulation, and casting, flooring and surfacing, coatings, and adhesives
Form: Liquid
Gen'l. Prop.: Visc. (Brookfield): 30,000–70,000 cps
Ref.: Product pamphlet

Epotuf Resin 37-170. Reichhold Chem., Inc.
Chem. Descrip.: Epoxy novolac resin
Category/Applications: Resin used for electrical potting, encapsulation, and casting, high-pressure laminating, filament winding, coatings, and adhesives
Form: Liquid
Gen'l. Prop.: Visc. (Brookfield): 30,000–90,000 cps (125 F)
Ref.: Product pamphlet

Epotuf Resin 37-200. Reichhold Chem., Inc.
Chem. Descrip.: Brominated epoxy resin
Category/Applications: Flame-retardant, unmodified resin for electrical potting, encapsulation, and casting, high-pressure laminating, filament winding, and coatings
Form: Liquid
Gen'l. Prop.: Visc. (Brookfield): 5000–10,000 poise
Ref.: Product pamphlet

Epotuf Resin 37-250. Reichhold Chem., Inc.
Chem. Descrip.: Epoxy resin
Category/Applications: Resilient, low-visc. resin for flooring and surfacing, and adhesives applications
Form: Liquid
Gen'l. Prop.: Visc. (Brookfield): 1000–2000 cps
Ref.: Product pamphlet

Epoxsilrub 400. Isochem Resins Co.
Category/Applications: Flexible two-part potting or encapsulating resin; cuttable, silicone rubber replacement; stress free—will not crush delicate components during cure; mix ratio: 60 phr of Hardener 114
Form: Color: Tan
Gen'l. Prop.: Sp. Gr.: 1.08; Visc.: 14,000 cps (mixed, 27 C)
Mech. Prop.: Tens. Str.: 2200 psi; Compr. Str.: 4400 psi; Hardness (Shore D): 50
Therm. Prop.: Conduct.: 4.5×10^{-4} Cal/s cm^2 C/cm; Coeff. of Linear Exp.: 6.2×10^{-5}/C
Elec. Prop.: Dissip. Factor: 0.035 (100 KC, 27 C); Dielec. Str.: 310 V/mil; Dielec. Const.: 3.1 (100 KC, 27 C); Vol. Resist.: 10^{13} ohm-cm (27 C)
Curing Charact.: Pot Life: 3 h (1 lb mass @ 27 C); Suggested Cure: R.T. or mild heat
Ref.: Bulletin

Epoxsilrub 411. Isochem Resins Co.
Chem. Descrip.: Filled version of Epoxsilrub 400
Category/Applications: Excellent encapsulant offering resilience, good adhesion, and excellent electrical properties; mix ratio: 30 phr of Hardener 114
Form: Color: Tan
Gen'l. Prop.: Sp. Gr.: 1.42; Visc.: 32,000 cps (mixed, 27 C)
Mech. Prop.: Tens. Str.: 2400 psi; Compr. Str.: 6400 psi; Hardness (Shore D): 50
Therm. Prop.: Conduct.: 5.2×10^{-4} Cal/s cm^2 C/cm; Coeff. of Linear Exp.: 6.6×10^{-5}/C
Elec. Prop.: Dissip. Factor: 0.032 (100 KC, 27 C); Dielec. Str.: 410 V/mil; Dielec. Const.: 4.2 (100 KC, 27 C); Vol. Resist.: 10^{14} ohm-cm (27 C)
Ref.: Bulletin

Ertalon 6SA. Chemplast, Inc.
Chem. Descrip.: Nylon type 6
Category/Applications: General purpose extruded nylon for use in bearings, bushings, gears, rollers, cable wheels, roll covers, wear plates, couplings, conveyor rollers, impact plates
Form: Solid
Gen'l. Prop.: Sp. Gr.: 1.14 (D792); M.P.: 420 ± 5 F (D569); Stability: Resistant to common solvents, hydrocarbons, esters, ketones, alkalis, and dilute acids alkalis, and dilute acids
Mech. Prop. (@ 73 F): Tens. Str.: 10,000–11,000 psi (D638); Tens. Elong.: 320% (D638); Tens. Mod.: 425,000–485,000 psi (D638); Compr. Str.: 850 psi (@ 1% deformat'n., D695); Shear Str.: 7100 psi (D732); Hardness (Rockwell): R 87 (D785)
Therm. Prop.: Conduct.: 2 Btu/h/ft^2/F/in. (C177); Distort. Temp.: 168 F (@ 264 psi, D648); Coeff. of Linear Exp.: 47×10^{-6} in./in./F (D696); Flamm.: Self-extinguishing (D635)
Elec. Prop.: Dissip. Factor: 0.06 (60 cycles, D150); Dielec. Str.: 290 V/mil (S/T, D149); Dielec. Const.: 7.5 (60 cycles, D150); Vol. Resist.: 54×10^{12} ohm-cm (D257)
Ref.: Publication C80-20M879N

Ertalon 6SAU. Chemplast, Inc.
Chem. Descrip.: Nylon type 6
Category/Applications: Wear-resistant extruded nylon for bushings, gears, wear plates, cable wheels, etc.
Form: Solid
Gen'l. Prop.: Sp. Gr.: 1.14 (D792); M.P.: 420 ± 10 F (D569); Stability: Wear- and temp.-resistant; resistant to common solvents, hydrocarbons, esters, ketones, alkalis, and dilute acids
Mech. Prop. (@ 73 F): Tens. Str.: 9000 psi (D638); Tens. Elong.: 50–200% (D-638); Tens. Mod.: 220,000 psi (D638);

Compr. Str.: 1200 psi (@ 1% deformation, D695); Shear Str.: 7500 psi (D-732); Impact Str. (Izod): 1.4 ft lb/in. (D256); Hardness (Rockwell): R 90 (D-785)

Therm. Prop.: Conduct.: 1.92 Btu/h/ft^2/F/in. (C177); Distort. Temp.: 168 F (@ 264 psi, D648); Coeff. of Linear Exp.: 45 $\times$ 10^{-6} in./in./F (D696); Flamm.: Self-extinguishing (D635)

Elec. Prop.: Dissip. Factor: 0.06 (60 cycles, D150); Dielec. Str.: 295 V/mil (S/T, D149); Dielec. Const.: 7.5 (60 cycles, D150); Vol. Resist.: 47 $\times$ 10^{12} ohm-cm (D257)

Ref.: Publication C80-20M879N

Ertalon 11SA. Chemplast, Inc.

Chem. Descrip.: Nylon type 11

Category/Applications: Extruded nylon with best chemical, electrical, and dimensional properties in series; used in gears, bushings, valves, pumps, insulating parts

Form: Solid

Gen'l. Prop.: Sp. Gr.: 1.04 (D792); M.P.: 365 ± 5 F (D569); Stability: See Ertalon 6SA

Mech. Prop. (@ 73 F): Tens. Str.: 7700 psi (D638); Tens. Elong.: 300% (D638); Tens. Mod.: 150,000 psi (D638); Compr. Str.: 800 psi (@ 1% deformation, D695); Shear Str.: 6000 psi (D732); Hardness (Rockwell): R 108 (D785)

Therm. Prop.: Conduct.: 2.01 Btu/h/ft^2/F/in. (C177); Distort. Temp.: 131 F (@ 264 psi, D648); Coeff. of Linear Exp.: 83 $\times$ 10^{-6} in./in./F (D696); Flamm.: Self-extinguishing (D635)

Elec. Prop.: Dissip. Factor: 0.045 (60 cycles, D150); Dielec. Str.: 425 V/mil (S/T, D149); Dielec. Const.: 4.0 (60 cycles, D150); Vol. Resist.: 78 $\times$ 10^{12} ohm-cm (D257)

Ref.: Publication C80-20M879N

Ertalon 66SA. Chemplast, Inc.

Chem. Descrip.: Nylon type 66

Category/Applications: Extruded nylon with high rigidity for use in smaller bushings, sleeves, gaskets, clamps, etc.

Form: Solid

Gen'l. Prop.: Sp. Gr.: 1.14–1.15 (D792); M.P.: 500 ± 5 F (D569); Stability: See Ertalon 6SA

Mech. Prop. (@ 73 F): Tens. Str.: 11,500 psi (D638); Tens. Elong.: 240% (D538); Tens. Mod.: 480,000–540,000 psi (D638); Compr. Str.: 1400 psi (@ 1% deformation, D695); Shear Str.: 10,500 psi (D732); Impact Str. (Izod): 0.9 ft lb/in. (D256); Hardness (Rockwell): R 108 (D785)

Therm. Prop.: Conduct.: 2 Btu/h/ft^2/F/in. (C177); Distort. Temp.: 203 F (@ 264 psi, D648); Coeff. of Linear Exp.: 42 $\times$ 10^{-6} in./in./F (D696); Flamm.: Self-extinguishing (D635)

Elec. Prop.: Dissip. Factor: 0.06 (60 cycles, D150); Dielec. Str.: 295 V/mil (S/T, D149); Dielec. Const.: 7.6 (60 cycles, D150); Vol. Resist.: 45 $\times$ 10^{12} ohm-cm (D257)

Ref.: Publication C80-20M879N

Ertalon 66SAM. Chemplast, Inc.

Chem. Descrip.: Nylon type 66 filled with molybdenum disulfide

Category/Applications: See Ertalon 66SA

Form: Solid

Gen'l. Prop.: Sp. Gr.: 1.15 (D792); M.P.: 500 ± 5 F (D569); Stability: See Ertalon 6SA

Mech. Prop. (@ 73 F): Tens. Str.: 12,400–13,500 psi (D638); Tens. Elong.: 5–150% (D638); Tens. Mod.: 500,000–550,000 psi (D638); Compr. Str.: 9000 psi (@ 1% deformation, D695); Shear Str.: 10,500–11,500 psi (D732); Hardness (Rockwell): R 120 (D785)

Therm. Prop.: Conduct.: 2 Btu/h/ft^2/

F/in. (C177); Distort. Temp.: 212 F (@ 264 psi, D648); Coeff. of Linear Exp.: 40 × 10^{-6} in./in./F (D696); Flamm.: Self-extinguishing (D635)

Elec. Prop.: Dissip. Factor: 0.02 (60 cycles, D150); Dielec. Str.: 300–400 V/mil (S/T, D149); Dielec. Const.: 3.7 (60 cycles, D150); Vol. Resist.: 2.5 × 10^{13} ohm-cm (D257)

Ref.: Publication C80-20M879N

Estane 5701 F1. BF Goodrich Co.

Chem. Descrip.: Polyester-urethane polymer, calcium stearate-dusted

Category/Applications: Thermoplastic polymer with excellent flexibility, resistance to chemicals and abrasion, and compatibility with many additives and other resins; used in binders, coated fabrics, adhesive blending resins

Form: Granular; Color: Pale amber

Gen'l. Prop.: Visc. (Brookfield): 300–700 cps (15% in THF)

Mech. Prop.: Hardness (Shore): 88A

Ref.: Data sheet

Estane 5702, 5702 F1, 5702 F2. BF Goodrich Co.

Chem. Descrip.: Polyester-urethane polymer, calcium stearate-dusted

Category/Applications: Thermoplastic polymer which is readily soluble, easily processed, and broadly compatible with adhesives and other resins; used in fabric coating, "tie coat" for transfer coating process, blending resin for coating and adhesives

Form: Granular; Color: Pale amber

Gen'l. Prop.: Visc. (Brookfield): 300–500 cps, 500–800 cps, 800–1200 cps respectively (15% in THF)

Mech. Prop.: Hardness (Shore): 70A

Ref.: Data sheet

Estane 5703, 5703 F1, 5703 F2. BF Goodrich Co.

Chem. Descrip.: Polyester-urethane polymer, talc-dusted

Category/Applications: Thermoplastic polymer which is readily soluble, compatible with additives and resins, and has high adhesion; used in shoe construction adhesives, commercial adhesives systems, fabric lamination, tape and sheet adhesives, binders

Form: Granular; Color: Pale amber

Gen'l. Prop.: Visc. (Brookfield): 300–500 cps, 500–800 cps, 800–1200 cps respectively (15% in THF)

Mech. Prop.: Hardness (Shore): 70A

Ref.: Data sheet

Estane 5707 F1. BF Goodrich Co.

Chem. Descrip.: Polyester-urethane polymer, calcium stearate-dusted

Category/Applications: Thermoplastic polymer with most resistance to chemicals, abrasion, and tearing, and highest tensile strength in series; used in binders, outerwear, leather finishes by direct coating, gloss, matte, and leather finish outer wear by transfer coating for coats, boots, purses, and luggage; free film casting

Form: Granular; Color: Pale amber

Gen'l. Prop.: Visc. (Brookfield): 670–1290 cps (15% solids in DMF)

Mech. Prop.: Hardness (Shore): 50D

Ref.: Data sheet

Estane 5708 F1. BF Goodrich Co.

Chem. Descrip.: Polyester-urethane polymer, calcium stearate-dusted

Category/Applications: Thermoplastic polymer with best low-temp. flexibility in series; used in fabric coating

Form: Granular; Color: Pale amber

Gen'l. Prop.: Visc. (Brookfield): 420–780 cps (15% in DMF)

Mech. Prop.: Hardness (Shore): 87A

Ref.: Data sheet

Estane 5710 F1. BF Goodrich Co.
Chem. Descrip.: Polyester-urethane polymer, calcium stearate-dusted
Category/Applications: Soft, tough thermoplastic polymer which produces very flexible film with exceptional "hand"; blend with other resins to improve low-temp. flexibility; good chemical resistance; used in fabric coating by direct and transfer methods, binders, commercial coating systems
Form: Granular; Color: Pale amber
Gen'l. Prop.: Visc. (Brookfield): 400–800 cps (15% in THF)
Mech. Prop.: Hardness (Shore): 78A
Ref.: Data sheet

Estane 5711, 5712, 5713, 5716. BF Goodrich Co.
Chem. Descrip.: Polyester-urethane polymer, talc-dusted
Category/Applications: Thermoplastic "open tack" polymers; dries to non-tacky film; tack is easily restored by heat or solvent treatment; high strength, flexible bonds; resistant to plasticizers and other chemicals; used in high speed assembly of footwear multimaterial laminates, hot-melt sealing, heat activated "open tack" adhesives in sheet and tape form
Form: Granular; Color: Pale amber
Gen'l. Prop.: Visc. (Brookfield): 1800–3000 cps (20% in MEK), 2100–3800 cps (20% in MEK), 800–1300 cps (15% in THF), and 2200–4000 cps (20% in MEK) respectively
Mech. Prop.: Hardness (Shore): 42D, 42D, 42D, 30D respectively
Ref.: Data sheet

Estane 5714F1. BF Goodrich Co.
Chem. Descrip.: Polyether-based polyurethane polymer
Category/Applications: Thermoplastic polymer with excellent resistance to hydrolysis and fungus, exceptional low-temp. flexibility, good resistance to chemicals and abrasion; limited compatibility with other Estane polymers; used in fabric coating, lifejackets, tarpaulins
Form: Granular; Color: Pale amber
Gen'l. Prop.: Visc. (Brookfield): 600–1200 cps (15% in THF)
Mech. Prop.: Hardness (Shore): 80A
Ref.: Data sheet

Estane 5715. BF Goodrich Co.
Chem. Descrip.: Polyester-urethane polymer, calcium stearate-dusted
Category/Applications: Thermoplastic polymer readily sol. in MEK; produces hard, flexible, abrasion-resistant coating; used in binders, leather dressing, commercial coatings, paper and foil coatings
Form: Granular; Color: Pale amber
Gen'l. Prop.: Visc. (Brookfield): 100–200 cps (20% in MEK)
Mech. Prop.: Hardness (Shore): 63D
Ref.: Data sheet

Estane 58013. BF Goodrich Co.
Chem. Descrip.: Polyester-urethane compound
Category/Applications: Med. hardness, thermoplastic comp. with excellent resistance to cuts, tears, abrasion; used for extrusion of tubing, shapes, sheet
Gen'l. Prop.: Sp. Gr.: 1.21
Mech. Prop.: Hardness (Shore): 86A
Ref.: Data sheet

Estane 58091. BF Goodrich Co.
Chem. Descrip.: Polyester-urethane compound
Category/Applications: High hardness, semirigid, thermoplastic comp. with excellent oil and grease resistance; used

in injection molding, automotive parts, appliance gears
Gen'l. Prop.: Sp. Gr.: 1.15
Mech. Prop.: Hardness (Shore): 70D
Ref.: Data sheet

Estane 58092. BF Goodrich Co.
Chem. Descrip.: Polyester-urethane compound
Category/Applications: Firm, flexible, low-blocking, thermoplastic comp. with highest resistance to chemicals and abrasion, and highest properties in series; used in injection molding and extrusion of film, sheet, tubing, and shapes
Gen'l. Prop.: Sp. Gr.: 1.22
Mech. Prop.: Hardness (Shore): 45D
Ref.: Data sheet

Estane 58109. BF Goodrich Co.
Chem. Descrip.: Polyester-urethane compound
Category/Applications: Fast molding and release thermoplastic comp. with exceptional low-temp. resilience and flexibility, melt strength, abrasion resistance, and compression set; used in injection molding, O-rings, seals
Gen'l. Prop.: Sp. Gr.: 1.22
Mech. Prop.: Hardness (Shore): 90A
Ref.: Data sheet

Estane 58121. BF Goodrich Co.
Chem. Descrip.: Polyester-urethane compound
Category/Applications: Softest Estane comp. with excellent abrasion resistance and flow; used for shoe soles and heels and in applications similar to 58109 where softer material is desired; injection molding
Gen'l. Prop.: Sp. Gr.: 1.21
Mech. Prop.: Hardness (Shore): 75A
Ref.: Data sheet

Estane 58130. BF Goodrich Co.
Chem. Descrip.: Polyester-urethane compound
Category/Applications: Med.-hard thermoplastic comp. with excellent low-temp. impact strength and good heat resistance; used in injection molding, automotive parts
Gen'l. Prop.: Sp. Gr.: 1.21
Mech. Prop.: Hardness (Shore): 50D
Ref.: Data sheet

Estane 58133. BF Goodrich Co.
Chem. Descrip.: Polyester-urethane compound
Category/Applications: High mod. thermoplastic comp. with excellent heat resistance and fast molding cycles; used where some structural strength is required; used in injection molding, ice skates, automotive parts
Gen'l. Prop.: Sp. Gr.: 1.21
Mech. Prop.: Hardness (Shore): 55D
Ref.: Data sheet

Estane 58134. BF Goodrich Co.
Chem. Descrip.: Polyester-urethane compound
Category/Applications: Med.-soft comp. with excellent molding characteristics, fast cycle time; used in injection molding, all purpose applications, and for applications requiring good abrasion and cut resistance
Gen'l. Prop.: Sp. Gr.: 1.21
Mech. Prop.: Hardness (Shore): 45D
Ref.: Data sheet

Estane 58136. BF Goodrich Co.
Chem. Descrip.: Polyester-urethane compound
Category/Applications: High performance, injection molding comp. similar to 58134 but with excellent outdoor

weathering properties; used for cattle tags, etc.
Gen'l. Prop.: Sp. Gr.: 1.21
Mech. Prop.: Hardness (Shore): 45D
Ref.: Data sheet

Estane 58137. BF Goodrich Co.
Chem. Descrip.: Polyester-urethane compound
Category/Applications: Very high mod., thermoplastic comp. with excellent heat resistance and fast molding cycles; used where good structural strength is required; used in injection molding, automotive parts, ski boots
Gen'l. Prop.: Sp. Gr.: 1.21
Mech. Prop.: Hardness (Shore): 65D
Ref.: Data sheet

Estane 58271, 58277. BF Goodrich Co.
Chem. Descrip.: Polyester-urethane compound
Category/Applications: Extrusion compounds with USDA approval for use in processing or storage areas for direct contact with meat or poultry food products
Gen'l. Prop.: Sp. Gr.: 1.21
Mech. Prop.: Hardness (Shore): 87A and 93A respectively
Ref.: Data sheet

Estane 58300. BF Goodrich Co.
Chem. Descrip.: Polyether base urethane comp.
Category/Applications: Thermoplastic comp. with excellent resistance to hydrolysis, improved fungus resistance, and exceptional low-temp. flexibility; used in extrusion and injection molding for hose and wire jacketing, molded junctions
Gen'l. Prop.: Sp. Gr.: 1.13
Mech. Prop.: Hardness (Shore): 80A
Ref.: Data sheet

Estane 58309. BF Goodrich Co.
Chem. Descrip.: Polyether base urethane comp.
Category/Applications: Thermoplastic comp. with excellent hydrolysis resistance, good antiblock characteristics, low temp. properties, and tear properties; used in extrusion and injection molding for hose, film, and sheet
Gen'l. Prop.: Sp. Gr.: 1.14
Mech. Prop.: Hardness (Shore): 85A
Ref.: Data sheet

Estane 58311. BF Goodrich Co.
Chem. Descrip.: Polyether base urethane comp.
Category/Applications: Thermoplastic comp. with excellent resistance to hydrolysis, improved fungus resistance, good low-temp. properties, and good abrasion and high tear properties; used in extrusion and injection molding for hose, wire jacketing, film, and sheet
Gen'l. Prop.: Sp. Gr.: 1.14
Mech. Prop.: Hardness (Shore): 85A
Ref.: Data sheet

Estane 58360. BF Goodrich Co.
Chem. Descrip.: Polyester-based polyurethane comp.
Category/Applications: Thermoplastic comp. with characteristics similar to Estane 58600; used in flame-retarded extrusions, film/sheet, wire, cable jacketing; oxygen index 25.4% (D69)
Gen'l. Prop.: Sp. Gr.: 1.28
Mech. Prop.: Hardness (Shore): 93A
Ref.: Data sheet

Estane 58370. BF Goodrich Co.
Chem. Descrip.: Polyether-based polyurethane comp.
Category/Applications: Thermoplastic comp. with characteristics similar to Estane 58300; used in flame-retarded

extrusions, wire and cable jacketing; oxygen index: 26.5% (D69)
Gen'l. Prop.: Sp. Gr.: 1.18
Mech. Prop.: Hardness (Shore): 86A
Ref.: Data sheet

Estane 58630. BF Goodrich Co.
Chem. Descrip.: Polyether-based comp.
Category/Applications: Low hardness thermoplastic comp. with hydrolytic and fungus resistance, good low-temp. flexibility, and low soften. pt.; used in calenders, Zimmers, and Bema coaters; industrial coated fabrics and films; good adhesion
Gen'l. Prop.: Sp. Gr.: 1.14
Mech. Prop.: Hardness (Shore): 82A
Ref.: Data sheet

Estane Polyurethanes. BF Goodrich Co.
Chem. Descrip.: Polyester- and polyether-based polyurethanes
Category/Applications: Thermoplastic polyurethanes with good resistance to abrasion, cutting, tearing, and to grease, oils, and chemicals; used in lightweight fabric coatings for apparel or for industrial fabric coatings, pigment binders for magnetic tape, wire and cable jacketing, molded parts requiring strength and quiet operation, gears and cams for small appliances, sheet/film, and in adhesives
Mech. Prop.: High tensile strength, wide hardness range, good low-temp. flexibility
Ref.: Product information

Ethyl Cellulose NF Grade. Hercules Inc.
Chem. Descrip.: Ethyl ether of cellulose
Category/Applications: Organo-soluble, NF grade cleared by the FDA for pharmaceutical uses as protective coatings, and as binders and fillers for vitamins and mineral preparations for humans and animals
Form: Free-flowing powder
Composition: 47.5–49% ethoxyl (Hercules N-type)
Gen'l. Prop.: Visc.: 6 grades as follows (dissolved in 80/20 toluene/ethanol): N-7, 6–8 cps; N-10, 8–11 cps; N-14, 12–16 cps; N-22, 18–24 cps; N-50, 40–52 cps; N-200, 80–105 cps; Ref. Index: 1.47 (film); pH: Neutral to litmus (water slurry)
Ref.: Product data 466-2

EVA 3121. Du Pont Co.
Chem. Descrip.: Ethylene/vinyl acetate copolymer resin
Category/Applications: Resin for flexible packaging applications; for use in conventional extrusion equipment designed to process polyethylene resins; designed to provide a low-temp. heat seal to itself or many other materials used in flexible packaging applications; its melt properties allow it to be processed on blown film equipment over a wide range of blow-up ratios and calipers; can also be coextruded with a variety of other polymers; FDA approved for use in contact with food
Form: Pellets
Composition: 7.5% vinyl acetate
Gen'l. Prop.: Density: 0.93 g/cc (D792); Melt Index: 0.50 dg/min (D1238)
Toxicity/Handling: Fumes may evolve at temps. above 230 C—use with adequate ventilation
Ref.: Bulletin E-20775

EVA 3134. Du Pont Co.
Chem. Descrip.: Ethylene/vinyl acetate copolymer resin
Category/Applications: Resin providing a low-temp. heat seal to itself or many other materials commonly used in flexible packaging applications; its melt properties allow it to be processed on

cast film equipment over a wide range of line speeds and calipers; can also be coextruded with a variety of other polymers; FDA approved for use in contact with food
Form: Pellets
Composition: 12% vinyl acetate by wt.
Gen'l. Prop.: Density: 0.94 g/cc (D792); Melt Index: 8.0 dg/min (D1238)
Toxicity/Handling: See EVA 3121
Ref.: Bulletin E-20776

EVA 3152. Du Pont Co.
Chem. Descrip.: Ethylene/vinyl acetate copolymer resin
Category/Applications: See EVA 3134
Form: Pellets
Composition: 15% vinyl acetate by wt.
Gen'l. Prop.: Density: 0.94 g/cc (D792); Melt Index: 8.0 dg/min (D1238)
Toxicity/Handling: See EVA 3121
Ref.: Bulletin E-20780

EVA 3159. Du Pont Co.
Chem. Descrip.: Ethylene/vinyl acetate copolymer resin
Category/Applications: See EVA 3121
Form: Pellets
Composition: 15% vinyl acetate by wt.
Gen'l. Prop.: Density: 0.94 g/cc (D792); Melt Index: 0.5 dg/min (D1238)
Toxicity/Handling: See EVA 3121
Ref.: Bulletin E-20781

EVA 3167. Du Pont Co.
Chem. Descrip.: Ethylene/vinyl acetate copolymer resin
Category/Applications: See EVA 3121
Form: Pellets
Composition: 18% vinyl acetate
Gen'l. Prop.: Density: 0.94 g/cc (D792); Melt Index: 0.7 dg/min (D1238)
Toxicity/Handling: See EVA 3121
Ref.: Bulletin E-20782

EVA 3174. Du Pont Co.
Chem. Descrip.: Ethylene/vinyl acetate copolymer resin
Category/Applications: See EVA 3134
Form: Pellets
Composition: 18% vinyl acetate by wt.
Gen'l. Prop.: Density: 0.94 g/cc (D792); Melt Index: 8 dg/min (D1238)
Toxicity/Handling: See EVA 3121
Ref.: Bulletin E-20785

EVA 3200. Du Pont Co.
Chem. Descrip.: Ethylene/vinyl acetate copolymer resin with paraffin wax
Category/Applications: Resin providing a low-temp. heat seal to itself or many other materials commonly used in flexible packaging applications; its melt properties allow it to be processed on extrusion coating equipment over a wide range of line speeds and calipers; priming is usually required for best adhesion to transparent and smooth substrates; matte finish chill rolls are recommended for best release characteristics; can also be coextruded with a variety of other polymers; FDA approved for use in contact with food
Form: Pellets
Gen'l. Prop.: Density: 0.95 g/cc (D792); Melt Index: 32 dg/min (D1238)
Toxicity/Handling: See EVA 3121
Ref.: Bulletin E-20786

EVA PE-4928. Du Pont Co.
Chem. Descrip.: Ethylene/vinyl acetate copolymer resin
Category/Applications: See EVA 3121
Form: Pellets
Composition: 9.3% vinyl acetate by wt.
Gen'l. Prop.: Density: 0.93 g/cc (D792); Melt Index: 2.0 dg/min (D1238)
Toxicity/Handling: See EVA 3121
Ref.: Bulletin E-20787

EVA 3130SB. Du Pont Co.
Chem. Descrip.: Ethylene/vinyl acetate copolymer resin
Category/Applications: Resin for use in conventional plastics processing equipment, e.g., extruders, intensive mixers, and injection molding machines, for flexible packaging applications; provides an injection molded part with excellent toughness and environmental stress-crack resistance; can be used in dry blends with other extrudable resins; may be considered an extrudable adhesive inner layer in many coextrusion applications; FDA approved for use in contact with food
Form: Pellets
Composition: 12% vinyl acetate
Gen'l. Prop.: Density: 0.94 g/cc (D792); Melt Index: 2.5 dg/min (D1238)
Toxicity/Handling: See EVA 3121
Ref.: Bulletin E-21930

EVA 3135SB. Du Pont Co.
Chem. Descrip.: Ethylene/vinyl acetate copolymer resin
Category/Applications: See EVA 3121
Form: Pellets
Composition: 12% vinyl acetate
Gen'l. Prop.: Density: 0.94 g/cc (D792); Melt Index: 0.35 dg/min (D1238)
Toxicity/Handling: See EVA 3121
Ref.: Bulletin E-20778

EVA 3170SB. Du Pont Co.
Chem. Descrip.: Ethylene/vinyl acetate copolymer resin with slip and antiblock additives
Category/Applications: See EVA 3121; slip and antiblock provide ease of handling of the finished film on automatic packaging equipment
Form: Pellets
Composition: 18% vinyl acetate by wt.
Gen'l. Prop.: Density: 0.94 g/cc (D792); Melt Index: 2.5 dg/min (D1238)
Toxicity/Handling: See EVA 3121
Ref.: Bulletin E-20783

EVA 3170SHB. Du Pont Co.
Chem. Descrip.: Ethylene/vinyl acetate copolymer resin with slip and high antiblock additives
Category/Applications: See EVA 3121; slip and antiblock provide ease of handling of the finished film on automatic packaging equipment
Form: Pellets
Composition: 18% vinyl acetate by wt.
Gen'l. Prop.: Density: 0.94 g/cc (D792); Melt Index: 2.5 dg/min (D1238)
Toxicity/Handling: See EVA 3121
Ref.: Bulletin E-20784

Everflex BG. W.R. Grace & Co./Organic Chem. Div.
Chem. Descrip.: Vinyl copolymer emulsion
Category/Applications: Very flexible, high m.w. emulsion with excellent flow properties, low temp. coalescence, and high flexibility to provide outstanding performance in exterior coatings for wood and masonry; uses incl. latex paints for interior flats, primer sealers, roof coatings, caulks, sealants, and mastics
Form: Emulsion; particle size 1.5 μ; Color: White; Odor: Slight, characteristic; Films of the polymer are clear, transparent, and with no odor
Composition: 50–52% total solids
Gen'l. Prop.: Density: 9.57 lb/gal; Visc.: 200–500 cps; Stability: See Everflex 81L; pH: 4.0–6.5
Ref.: Bulletin

Everflex E. W.R. Grace & Co./Organic Chem. Div.
Chem. Descrip.: Vinyl acrylic emulsion
Category/Applications: General purpose

paint/coatings emulsion for interior and exterior paints with well-balanced properties; used for latex paints for interior flats, primer sealers, and exterior masonry
Form: Emulsion; particle size 0.2 μ; Color: White; Odor: Slight, characteristic; Films of the polymer are clear, transparent, with no odor
Composition: 54–56% total solids
Gen'l. Prop.: Density: 9.7 lb/gal; Visc.: 300–700 cps; Stability: See Everflex 81L; pH: 4.5–5.5
Ref.: Bulletin

Everflex G. W.R. Grace & Co./Organic Chem. Div.
Chem. Descrip.: Vinyl acetate copolymer emulsion
Category/Applications: High m.w., permanently flexible emulsion offering excellent flow, holdout, and machining properties; widely used in primers, sealers, specialty coating applications, latex paints for exterior masonry, exterior wood, and roof coatings, in caulks, sealants, and mastics, concrete admixtures
Form: Emulsion; particle size 1.5 μ; Color: White; Odor: Slight, characteristic; Films of the polymer are clear, transparent, with no odor
Composition: 51–53% total solids
Gen'l. Prop.: Density: 9.76 lb/gal; Visc.: 300–700 cps; Stability: See Everflex 81L; pH: 4.0–6.5
Ref.: Bulletin

Everflex GT. W.R. Grace & Co./Organic Chem. Div.
Chem. Descrip.: Vinyl copolymer emulsion
Category/Applications: Emulsion designed for high solids, mastic compounds, caulks, sealants, and joint cements; offers outstanding chemical and mechanical stability; can produce highly loaded formulations with excellent rheological characteristics; also for fire-retardant coatings, paper and felt saturation
Form: Emulsion; particle size 1.0 μ; Color: White; Odor: Slight, characteristic; Films of the polymer are clear, transparent, with no odor
Composition: 54–56% total solids
Gen'l. Prop.: Density: 9.76 lb/gal; Visc.: 1500–3000 cps; Stability: See Everflex 81L; pH: 4.0–6.5
Ref.: Bulletin

Everflex MA. W.R. Grace & Co./Organic Chem. Div.
Chem. Descrip.: Vinyl acrylic copolymer emulsion
Category/Applications: Colloid-free microparticle size emulsion with excellent chemical and mechanical stability; used as a vehicle for trade sales paints, factory finishes for wood and composition board, and in the coating and impregnation of paper and textiles
Form: Emulsion; particle size 0.1 μ; Color: White; Odor: Slight, characteristic; Films of the polymer are clear, transparent, with no odor
Composition: 50–52% total solids
Gen'l. Prop.: Density: 9.76 lb/gal; Visc.: 200–800 cps; Stability: See Everflex 81L; pH: 4.0–6.5
Ref.: Bulletin

Everflex MB. W.R. Grace & Co./Organic Chem. Div.
Chem. Descrip.: Vinyl acrylic copolymer emulsion
Category/Applications: Softer, more flexible version of Everflex MA; suitable for use in latex paints, factory finishes, stains, roof coatings, and paper and industrial coatings
Form: See Everflex MA
Composition: 54–56% total solids
Gen'l. Prop.: See Everflex MA
Ref.: Bulletin

Everflex SP-1084. W.R. Grace & Co./Organic Chem. Div.
Chem. Descrip.: Vinyl acetate copolymer emulsion
Category/Applications: High m.w. emulsion used as concrete additive to improve adhesion and and compressive strength
Form: Emulsion; particle size 1.5 μ; Color: White; Odor: Slight, characteristic; Films of the polymer are clear, transparent, with no odor
Composition: 52.5–54.5% total solids
Gen'l. Prop.: Density: 9.8 lb/gal; Visc.: 300–700 cps; Stability: See Everflex 81L; pH: 4.5–5.5
Ref.: Bulletin

Everflex SP-1087. W.R. Grace & Co./Organic Chem. Div.
Chem. Descrip.: Vinyl acrylic copolymer emulsion
Category/Applications: General purpose trade sales paint vehicle providing good flow and leveling, burnish resistance, low temp. coalescence, and wide compounding latitude; used in latex paints for interior flats, primer sealers, exterior masonry, interior and exterior gloss and semigloss finishes, and exterior wood
Form: Emulsion; particle size 0.2 μ; Color: White; Odor: Slight, characteristic; Films of the polymer are clear, transparent, with no odor
Composition: 54–56% total solids
Gen'l. Prop.: Density: 9.7 lb/gal; Visc.: 300–700 cps; Stability: See Everflex 81L; pH: 4.5–5.5
Ref.: Bulletin

Everflex T. W.R. Grace & Co./Organic Chem. Div.
Chem. Descrip.: Vinyl acrylic emulsion
Category/Applications: High solids emulsion binder for use in the production of both gloss and flat interior and exterior coatings prepared by pigment slurry techniques; used in latex paints for interior flats, primer sealers, exterior masonry, exterior wood, factory finishes, ceiling tile, wall board, stains, roof coatings
Form: Emulsion; particle size 0.25 μ; Color: White; Odor: Slight, characteristic; Films of the polymer are clear, transparent, with no odor
Composition: 64–66% total solids
Gen'l. Prop.: Density: 9.7 lb/gal; Visc.: 3000–4000 cps; Stability: See Everflex 81L; pH: 4.5–5.5
Ref.: Bulletin

Everflex TMF. W.R. Grace & Co./Organic Chem. Div.
Chem. Descrip.: Vinyl acrylic terpolymer emulsion
Category/Applications: Fine particle size, med. m.w. emulsion with high pigment binding capacity; well-balanced emulsifier-surfactant system permitting formulation of high quality gloss, semigloss, and flat paints
Form: Emulsion; particle size 0.2 μ; Color: White; Odor: Slight, characteristic; Films of the polymer are clear, transparent, with no odor
Composition: 54–56% total solids
Gen'l. Prop.: Density: 9.7 lb/gal; Visc.: 300–700 cps; Stability: See Everflex 81L; pH: 4.5–5.5
Ref.: Bulletin

Everflex 81L. W.R. Grace & Co./Organic Chem. Div.
Chem. Descrip.: Vinyl acetate copolymer emulsion
Category/Applications: Fine particle size paint/coating emulsion designed for use as a binder for high-quality clay coating of paper and paperboards; offers excellent color, brightness, and printability; also used in factory finishes, ceiling tile, wall board, and textile treatments

Form: Emulsion; particle size 0.15 μ; Color: White; Odor: Slight, characteristic; Films of the polymer are clear, transparent, with no odor
Composition: 49–51% total solids
Gen'l. Prop.: Density: 9.9 lb/gal; Visc.: 50–250 cps; Stability: Emulsions have 20+ min mechanical stability; films exhibit good water resistance; pH: 4.5–5.5
Ref.: Bulletin

F

FA-1. Bacon Industries Inc.
Chem. Descrip.: Unfilled gyro-grade epoxy resin adhesive
Category/Applications: Adhesive for bonding and sealing aluminum and other metals; mix ratio: 100 pbw adhesive/3.2 pbw of Activator BA-4
Form: Color: Amber
Gen'l. Prop.: Sp. Gr.: 1.1; Visc.: 100 poise (mixed, 75 F); Stability: Not softened by low-boiling alcohols and aromatics, and aliphatic solvents; softened by ketones and chlorinated solvents
Mech. Prop.: Bond Str.: 2250 psi (70 F); Hardness (Shore D): 85
Therm. Prop.: Coeff. of Linear Exp.: 40 $\times 10^{-6}$ in./in./F (–65 to 80 F)
Curing Charact.: Pot Life: 3 h @ 75 F; Recommended Cure: 2 h @ 200 F
Ref.: Data sheet 2001

FA-8. Bacon Industries Inc.
Chem. Descrip.: Unfilled gyro-grade epoxy resin adhesive
Category/Applications: Exceptionally fluid adhesive for bonding and sealing beryllium and other metals; the activator is colored to make it easy to tell when adhesive is properly mixed; mix ratio: 100 pbw adhesive/13.5 pbw of Activator BA-5
Form: Color: Blue-green
Gen'l. Prop.: Sp. Gr.: 1.1; Visc.: 60 poise (mixed, 75 F); Stability: Not softened by aliphatic solvents; softened by ketones, chlorinated solvents, and low-boiling alcohols and aromatics
Mech. Prop.: Bond Str.: 2250 psi (70 F); Hardness (Shore D): 85
Therm. Prop.: Coeff. of Linear Exp.: 35 $\times 10^{-6}$ in./in./F (–65 to 80 F)
Curing Charact.: Pot Life: 90 min @ 75 F; Recommended Cure: 2 h @ 200 F
Ref.: Data sheet 2001

FA-13. Bacon Industries Inc.
Chem. Descrip.: Unfilled gyro-grade eposy resin adhesive
Category/Applications: Semiflexible adhesive for sealing joints between dissimilar metals which are difficult or impossible to seal with regular gyro-grade adhesives; mix ratio: 100 pbw adhesive/26.5 pbw Activator BA-39
Form: Color: Light brown
Gen'l. Prop.: Sp. Gr.: 1.22; Visc.: 5 poise (mixed, 75 F); Stability: See FA-8
Mech. Prop.: Bond Str.: 2250 psi (70 F); Hardness (Shore D): 84
Therm. Prop.: Coeff. of Linear Exp.: 35 $\times 10^{-6}$ in./in./F (–65 to 80 F)
Curing Charact.: Pot Life: 24 h @ 75 F; Recommended Cure: 2 h @ 160 F plus 4 h @ 200 F
Ref.: Data sheet 2001

FA-14. Bacon Industries Inc.
Chem. Descrip.: Two-part modified epoxy comp.
Category/Applications: Adhesive coil impregnant for the impregnation of electronic components containing a high percentage of fine wires where complete penetration and freedom from voids is important; also for casting and coating applications where a low-visc. material is required; FA-14 offers very low visc. and excellent ability to penetrate cracks and crevices; also used for bonding components made from fused beryllium oxide and for other bonding applications requiring a low surface tension comp.; suitable for gyro use in contact with poly(bromotrifluoroethylene) oil
Form: Liquid
Gen'l. Prop. (activated impregnant–24.5 phr of Activator BA-45, @ R.T.): Sp. Gr.: 1.13 (cured); Visc.: 275 cps; Surface Tension: 24.9 dyne/cm
Mech. Prop. (cured): Flex. Str.: 8600 psi (yield); Flex. Mod. (Young's): 400,000 psi; Hardness (Shore D): 85
Therm. Prop. (cured): Distort. Temp.: 132 F (@ 264 psi); Coeff. of Linear Exp.: 34×10^{-6}/F (–65 to 80 F)
Elec. Prop. (cured): Dissip. Factor: 0.003 (1 kHz); Dielec. Const.: 3.6 (1 kHz); Vol. Resist.: $> 10^{14}$ ohm-cm
Curing Charact.: Pot Life: 8 h @ R.T.; Cure Schedule: 8 h @ 160 F
Toxicity/Handling: See CI-6
Std. Pkgs.: Pt containers
Ref.: Data sheet 1071

FA-21. Bacon Industries Inc.
Chem. Descrip.: Unfilled epoxy resin comp.
Category/Applications: Fluid adhesive with unusually high heat distortion temp.; offers ease of application and good pot life; mix ratio: 100 pbw adhesive/100 pbw Activator BA-2XM
Form: Liquid
Gen'l. Prop.: Sp. Gr.: 1.41 (cured); Visc.: 40 poise (mixed, @ 160 F)
Mech. Prop. (cured): Flex. Str.: 12,500 psi (yield); Flex. Mod. (Young's): 450,000 psi; Hardness (Shore D): 85
Therm. Prop.: Distort. Temp.: 348 F (@ 264 psi); Coeff. of Linear Exp.: 33×10^{-6}/F (75–200 F)
Curing Charact.: Pot Life: 2 h @ 160 F; Recommended Cure: 8 h @ 212 F plus 16 h @ 300 F
Toxicity/Handling: May cause injury to skin on prolonged or repeated contact; Activator BA-2XM, a powder, causes irritation to the skin, nose, and throat–avoid breathing dust and use with adequate ventilation; > 1 yr shelf life when stored in unopened containers below 85 F
Std. Pkgs.: Qt containers
Ref.: Data sheet 2321

FFA-2. Bacon Industries Inc.
Chem. Descrip.: Epoxy resin adhesive
Category/Applications: Flexible adhesive which distributes stresses developed in bonded joints because of impact or differences in the coefficients of expansion between dissimilar materials; FFA-2 offers more flexibility than FFA-5; flexibility is variable depending upon adhesive/activator ratio; recommended mix ratio: 100 pbw adhesive/70 pbw Activator BA-6A
Form: Color: Amber
Gen'l. Prop.: Sp. Gr.: 1.1; Visc.: 60 poise (mixed, 75 F); Stability: Not softened by aliphatic solvents; softened by ketones, chlorinated solvents, and low-boiling alcohols and aromatics
Mech. Prop.: Bond Str.: 825 psi (70 F); Hardness (Shore D): 50
Therm. Prop.: Coeff. of Linear Exp.: 60×10^{-6} in./in./F (–65 to 80 F)

Curing Charact.: Pot Life: 3 h @ 75 F; Recommended Cure: 3½ h @ 200 F
Ref.: Data sheet 2001

FFA-5. Bacon Industries Inc.
Chem. Descrip.: Epoxy resin adhesive
Category/Applications: See FFA-2; FFA-5 offers stronger bond than FFA-2; mix ratio: 100 pbw adhesive/150 pbw Activator BA-15
Form: Color: Amber
Gen'l. Prop.: Sp. Gr.: 1.1; Visc.: 60 poise (mixed, 75 F); Stability: See FFA-2
Mech. Prop.: Bond Str.: 2250 psi (70 F); Hardness (Shore D): 65
Therm. Prop.: Coeff. of Linear Exp.: 75 X 10^{-6} in./in./F (–65 to 80 F)
Curing Charact.: Pot Life: 2 h @ 75 F; Recommended Cure: 1 h @ 160 F
Ref.: Data sheet 2001

FFA-9. Bacon Industries Inc.
Chem. Descrip.: Epoxy resin adhesive
Category/Applications: See FFA-2; FFA-9 is a fast setting, R.T. curing adhesive; bond is more brittle than FFA-5; mix ratio: 100 pbw adhesive/100 pbw Activator BA-11
Form: Color: Amber
Gen'l. Prop.: Sp. Gr.: 1.1; Stability: See FFA-2
Mech. Prop.: Bond Str.: 2250 psi (70 F); Hardness (Shore D): 65
Therm. Prop.: Coeff. of Linear Exp.: 75 X 10^{-6} in./in./F (–65 to 80 F)
Curing Charact.: Pot Life: 15 min @ 75 F; Recommended Cure: 1 h @ 75 F
Ref.: Data sheet 2001

Firestone Nylon 200-001. Firestone Synthetic Fibers Co.
Chem. Descrip.: Nylon 6
Category/Applications: Med. m.w., general purpose thermoplastic resin for molding and extrusion applications
Form: Semicrystalline; Color: Natural
Gen'l. Prop. (dry, as molded): Sp. Gr.: 1.13 (D792); M.P.: 215 C (D789); Stability: Good thermal stability, and abrasion- and impact-resistance
Mech. Prop. (dry, as molded): Tens. Str.: 83 MPa (yield, D638); Tens. Elong.: 170% (ultimate, D638); Flex. Str.: 104 MPa (D790); Flex. Mod.: 2000 MPa (D-790); Impact Str. (Izod): 37 J/m notch (D256); Hardness (Rockwell): R 118 (D785)
Therm. Prop. (dry, as molded): Distort. Temp.: 63 C (@ 1820 kPa, D648)
Toxicity/Handling: Hygroscopic–avoid moisture
Std. Pkgs.: 50 lb net wt., hermetically-sealed bags
Ref.: Bulletin MTB-7903

Firestone Nylon 210-001L. Firestone Synthetic Fibers Co.
Chem. Descrip.: Nylon 6
Category/Applications: Heat-stabilized thermoplastic resin for wire jacketing applications
Form: Semicrystalline; Color: Natural
Gen'l. Prop. (dry, as molded): Sp. Gr.: 1.13 (D792); M.P.: 215 C (D789); Stability: See Firestone Nylon 200-001; 210-001L also has outstanding resistance to long term oxidative embrittlement
Mech. Prop. (dry, as molded): Tens. Str.: 86 MPa (yield, D638); Tens. Elong.: 170% (ultimate, D638); Flex. Str.: 104 MPa (D790); Flex. Mod.: 2070 MPa (D-790); Impact Str. (Izod): 32 J/m notch (D256); Hardness (Rockwell): R 118 (D785)
Therm. Prop. (dry, as molded): Distort. Temp.: 66 C (@ 1820 kPa, D648)
Toxicity/Handling & Std. Pkgs.: See Firestone Nylon 200-001
Ref.: Bulletin MTB-7903

Firestone Nylon 213-001. Firestone Synthetic Fibers Co.
Chem. Descrip.: Nylon 6
Category/Applications: Heat- and light-

stabilized, nucleated, thermoplastic resin designed to offer faster molding cycles, improved shrinkage control, and improved part dimensional stability; U.L. listed at 105 C
Form: Semicrystalline; Color: Natural
Gen'l. Prop. (dry, as molded): Sp. Gr.: 1.14 (D792); M.P.: 219 C (D789); Stability: See Firestone Nylon 200-001
Mech. Prop. (dry, as molded): Tens. Str.: 97 MPa (yield, D638); Tens. Elong.: 25% (ultimate, D638); Flex. Str.: 131 MPa (D790); Flex. Mod.: 3170 MPa (D-790); Impact Str. (Izod): 32 J/m notch (D256); Hardness (Rockwell): R 129 (D785)
Therm. Prop. (dry, as molded): Distort. Temp.: 76 C (@ 1820 kPa, D648)
Toxicity/Handling & Std. Pkgs.: See Firestone Nylon 200-001
Ref.: Bulletin MTB-7903

Firestone Nylon 228-001. Firestone Synthetic Fibers Co.
Chem. Descrip.: Nylon 6
Category/Applications: Low-visc. thermoplastic resin formulated for injection molding operations, esp. for molding complex or thin section parts; a heat-stabilized version, 228-001HS is avail. for applications requiring long-term heat stability
Form: Semicrystalline; Color: Natural
Gen'l. Prop. (dry, as molded): Sp. Gr.: 1.13 (D792); M.P.: 219 C (D789); Stability: See Firestone Nylon 200-001
Mech. Prop. (dry, as molded): Tens. Str.: 79 MPa (yield, D638); Tens. Elong.: 90% (ultimate, D638); Flex. Str.: 107 MPa (D790); Flex. Mod.: 2750 MPa (D-790); Impact Str. (Izod): 40 J/m notch (D256); Hardness (Rockwell): R 130 (D785)
Therm. Prop. (dry, as molded): Distort. Temp.: 71 C (@ 1820 kPa, D648)
Toxicity/Handling & Std. Pkgs.: See Firestone Nylon 200-001
Ref.: Bulletin MTB-7903

Firestone Nylon 415-001HS. Firestone Synthetic Fibers Co.
Chem. Descrip.: Glass-reinforced nylon 6 resin
Category/Applications: Heat-stabilized thermoplastic resin reinforced with 15% of specially treated, short glass fibers; designed for injection molding of nylon parts requiring excellent thermal and dimensional stability, and high strength; U.L. listed at 105 C
Form: Semicrystalline; Color: Natural
Gen'l. Prop. (dry, as molded): Sp. Gr.: 2.14 (D792); M.P.: 219 C (D789); Stability: See Firestone Nylon 200-001
Mech. Prop. (dry, as molded): Tens. Str.: 120 MPa (yield, D638); Tens. Elong.: 2.8% (ultimate, D638); Flex. Str.: 186 MPa (D790); Flex. Mod.: 5700 MPa (D-790); Impact Str. (Izod): 48 J/m notch (D256); Hardness (Rockwell): R 123 (D785)
Therm. Prop. (dry, as molded): Distort. Temp.: 202 C (@ 1820 kPa, D648)
Toxicity/Handling & Std. Pkgs.: See Firestone Nylon 200-001
Ref.: Bulletin MTB-7903

Firestone Nylon 430-001HS. Firestone Synthetic Fibers Co.
Chem. Descrip.: Glass-reinforced nylon 6 resin
Category/Applications: Heat-stabilized, thermoplastic resin reinforced with 30% of specially treated, short glass fibers; has high strength and rigidity; U.L. listed at 130 C
Form: Semicrystalline; Color: Natural
Gen'l. Prop. (dry, as molded): Sp. Gr.: 1.37 (D792); M.P.: 215 C (D789); Stability: See Firestone Nylon 200-001
Mech. Prop. (dry, as molded): Tens. Str.: 169 MPa (yield, D638); Tens. Elong.:

3.3% (ultimate, D638); Flex. Str.: 246 MPa (D790); Flex. Mod.: 9000 MPa (D-790); Impact Str. (Izod): 96 J/m notch (D256); Hardness (Rockwell): R 121 (D785)
Toxicity/Handling & Std. Pkgs.: See Firestone Nylon 200-001
Ref.: Bulletin MTB-7903

Firestone Nylon 540-001HS. Firestone Synthetic Fibers Co.
Chem. Descrip.: Mineral-reinforced nylon 6 resin
Category/Applications: Heat- and light-stabilized thermoplastic resin with 40% of a specially treated, particulate mineral reinforcement; combines the processability of thermoplastics with the strength and durability of "light weight" metals
Form: Semicrystalline; Color: Natural
Gen'l. Prop. (dry, as molded): Sp. Gr.: 1.50 (D792); M.P.: 219 C (D789); Stability: See Firestone Nylon 200-001
Mech. Prop. (dry, as molded): Tens. Str.: 95 MPa (yield, D638); Tens. Elong.: 8.0% (ultimate, D638); Flex. Str.: 150 MPa (D790); Flex. Mod.: 6000 MPa (D-790); Impact Str. (Izod): 44 J/m notch (D256); Hardness (Rockwell): R 122 (D785)
Therm. Prop. (dry, as molded): Distort. Temp.: 160 C (@ 1820 kPa, D648)
Toxicity/Handling & Std. Pkgs.: See Firestone Nylon 200-001
Ref.: Bulletin MTB-7903

Firestone Nylon 640-001HS. Firestone Synthetic Fibers Co.
Chem. Descrip.: Glass- and mineral-reinforced nylon 6 resin
Category/Applications: Heat- and light-stabilized thermoplastic resin with 25% of a specially treated, particulate mineral and 15% of specially treated, short glass fibers; combines the strength of glass reinforcement and the economics of mineral reinforcement; also has improved part dimensional stability
Form: Semicrystalline; Color: Natural
Gen'l. Prop. (dry, as molded): Sp. Gr.: 1.46 (D792); M.P.: 219 C (D789); Stability: See Firestone Nylon 200-001
Mech. Prop. (dry, as molded): Tens. Str.: 145 MPa (yield, D638); Tens. Elong.: 3.1% (ultimate, D638); Flex. Str.: 207 MPa (D790); Flex. Mod.: 8250 MPa (D-790); Impact Str. (Izod): 54 J/m notch (D256); Hardness (Rockwell): R 123 (D785)
Therm. Prop. (dry, as molded): Distort. Temp.: 208 C (@ 1820 kPa, D648)
Toxicity/Handling & Std. Pkgs.: See Firestone Nylon 200-001
Ref.: Bulletin MTB-7903

Fluorel 2145. 3M/Commercial Chem. Div.
Chem. Descrip.: Fluoroelastomer
Category/Applications: Fluoroelastomer gum without incorporated cure; low-visc. gum for blending with incorporated cure grades to lower viscosity and increase flow; FDA-approved
Gen'l. Prop.: Sp. Gr.: 1.81; Visc. (Mooney): 28 M_L (1 + 10) @ 250 F
Mech. Prop.: Tens. Str.: 2000 psi; Tens. Elong.: 200%; 100% Mod.: 800 psi; Hardness (Shore A): 73
Ref.: Bulletin Y-1FPCG(99.5)11

Fluorel 2170. 3M/Commercial Chem. Div.
Chem. Descrip.: Fluoroelastomer
Category/Applications: Fluoroelastomer with low compression set, fast cures for use in O-rings, O-ring cord, calendered sheet, rolls
Gen'l. Prop.: Sp. Gr.: 1.81; Visc. (Mooney): 30 M_L (1 + 10) @ 250 F
Mech. Prop.: Tens. Str.: 1950 psi; Tens. Elong.: 195%; 100% Mod.: 900 psi; Hardness (Shore A): 75
Ref.: Bulletin Y-1FPCG(99.5)11

Fluorel 2172. 3M/Commercial Chem. Div.
Chem. Descrip.: Masterbatch of cure accelerators
Category/Applications: Masterbatch of cure accelerators to be used with incorporated cure Fluorel fluoroelastomers; accelerates cure over wide range of temps. (esp. < 350 F); also helpful in formulations that contain fillers that have a retarding effect on the cure rate
Gen'l. Prop.: Sp. Gr.: 1.80; Visc. (Mooney): 30 M_L (1 + 10) @ 250 F
Mech. Prop.: Tens. Str.: 2025 psi; Tens. Elong.: 180%; 100% Mod.: 925 psi; Hardness (Shore A): 72
Ref.: Bulletin Y-1FPCG(99.5)11

Fluorel 2173. 3M/Commercial Chem. Div.
Chem. Descrip.: Specialty masterbatch of incorporated cure fluoroelastomer gum
Category/Applications: Best-flowing incorporated cure polymer with improved release over 2170 without sacrifice in physical properties; slightly higher shrinkage can be minimized by blending with other Fluorel fluoroelastomer incorporated cure gums; used in molded goods, sheet goods, O-rings, hose
Gen'l. Prop.: Sp. Gr.: 1.80; Visc. (Mooney): 28 M_L (1 + 10) @ 250 F
Mech. Prop.: Tens. Str.: 1950 psi; Tens. Elong.: 190%; 100% Mod.: 900 psi; Hardness (Shore A): 76
Ref.: Bulletin Y-1FPCG(99.5)11

Fluorel 2174. 3M/Commercial Chem. Div.
Chem. Descrip.: Fluoroelastomer
Category/Applications: Fluoroelastomer with excellent compression set resistance, ultrafast cure, improved mold release; used for O-rings, O-ring cord, calendered sheet, lathe cuts
Gen'l. Prop.: Sp. Gr.: 1.80; Visc. (Mooney): 38 M_L (1 + 10) @ 250 F
Mech. Prop.: Tens. Str.: 2100 psi; Tens. Elong.: 200%; 100% Mod.: 875 psi; Hardness (Shore A): 77
Ref.: Bulletin Y-1FPCG(99.5)11

Fluorel 2175. 3M/Commercial Chem. Div.
Chem. Descrip.: Fluoroelastomer
Category/Applications: Fluoroelastomer gum without incorporated cure; med-visc. gum blendable with incorporated cure grades; FDA-approved
Gen'l. Prop.: Sp. Gr.: 1.81; Visc. (Mooney: 37 M_L (1 + 10) @ 250 F
Mech. Prop.: Tens. Str.: 2150 psi; Tens. Elong.: 200%; 100% Mod.: 800 psi; Hardness (Shore A): 74
Ref.: Bulletin Y-1FPCG(99.5)11

Fluorel 2176. 3M/Commercial Chem. Div.
Chem. Descrip.: Fluoroelastomer
Category/Applications: Fluoroelastomer with good tear resistance, fast cures with excellent processing safety, all purpose grade for injection, compression, and transfer molding; used in O-rings, shaft seals, molded goods, hose
Gen'l. Prop.: Sp. Gr.: 1.80; Visc. (Mooney): 30 M_L (1 + 10) @ 250 F
Mech. Prop.: Tens. Str.: 1900 psi; Tens. Elong.: 280%; 100% Mod.: 525 psi; Hardness (Shore A): 73
Ref.: Bulletin Y-1FPCG(99.5)11

Fluorel 2177. 3M/Commercial Chem. Div.
Chem. Descrip.: Fluoroelastomer
Category/Applications: Fluoroelastomer with slower cure rate, higher tear str.

and elong. than 2170, but less resistance to compression set; used for molded goods, shaft seals, metal bonded composites
Gen'l. Prop.: Sp. Gr.: 1.81; Visc. (Mooney): 32 M_L (1 + 10) @ 250 F
Mech. Prop.: Tens. Str.: 2000 psi; Tens. Elong.: 225%; 100% Mod.: 780 psi; Hardness (Shore A): 74
Ref.: Bulletin Y-1FPCG(99.5)11

Fluorel 2178. 3M/Commercial Chem. Div.
Chem. Descrip.: Fluoroelastomer
Category/Applications: Fluoroelastomer gum without incorporated cure; high-visc. gum for blending with incorporated grades to increase viscosity and green strength; FDA-approved
Gen'l. Prop.: Sp. Gr.: 1.82; Visc. (Mooney): 120 M_L (1 + 10) @ 250 F
Mech. Prop.: Tens. Str.: 2540 psi; Tens. Elong.: 290%; 100% Mod.: 615 psi; Hardness (Shore A): 74
Ref.: Bulletin Y-1FPCG(99.5)11

Fluorel 2179. 3M/Commercial Chem. Div.
Chem. Descrip.: Fluoroelastomer
Category/Applications: Fluoroelastomer with high viscosity, excellent resistance to compression set and deformation at high pressures; used for O-rings, downhole packers, blow-out preventers
Gen'l. Prop.: Sp. Gr.: 1.81; Visc. (Mooney): 80 M_L (1 + 10) @ 250 F
Mech. Prop.: Tens. Str.: 2100 psi; Tens. Elong.: 200%; 100% Mod.: 950 psi; Hardness (Shore A): 75
Ref.: Bulletin Y-1FPCG(99.5)11

Fluorel 2181. 3M/Commercial Chem. Div.
Chem. Descrip.: Fluoroelastomer
Category/Applications: Fluoroelastomer combining high elongation with good compression set and tear resistance, ultrafast cure rate, improved mold release, and excellent processing safety; used in O-rings, molded goods, lathe cuts, hose
Gen'l. Prop.: Sp. Gr.: 1.80; Visc. (Mooney): 28 M_L (1 + 10) @ 250 F
Mech. Prop.: Tens. Str.: 2100 psi; Tens. Elong.: 265%; 100% Mod.: 625 psi; Hardness (Shore A): 74
Ref.: Bulletin Y-1FPCG(99.5)11

Fluorel 2182. 3M/Commercial Chem. Div.
Chem. Descrip.: Fluoroelastomer
Category/Applications: Fluoroelastomer optimized for scorch safety and low extrusion pressures; used for hose, tubing, profile extrusions, injection, and transfer molded goods
Gen'l. Prop.: Sp. Gr.: 1.80; Visc. (Mooney): 28 M_L (1 + 10) @ 250 F
Mech. Prop.: Tens. Str.: 2000 psi; Tens. Elong.: 290%; 100% Mod.: 500 psi; Hardness (Shore A): 72
Ref.: Bulletin Y-1FPCG(99.5)11

Fluorocomp 101. Liquid Nitrogen Processing Corp.
Chem. Descrip.: Glass-filled TFE comp.
Category/Applications: Used in bearings, gaskets, V-rings, O-rings, back-up rings, hydraulic ring seals, nonlubricated compressor rings, valve seats, valve liners, swivel liners, thrust washers, lip seals, standoff and feed-through insulators, and other applications in the chemical, electrical, and mechanical fields; used at service temps. from –450 to +500 F, in chemical service in liquid oxygen and fuming nitric acid, and in severe applications, e.g., wear pads for railway cars and nonlubricated compressor rings that can last 12,000 h
Composition: 95% TFE, 5% glass fiber

Gen'l. Prop.: Sp. Gr.: 2.14 g/cc (D1457-62)
Mech. Prop. (CD): Tens. Str.: 3700 psi (D1457-62); Tens. Elong.: 320% (D-1457-62); Flex. Str.: 1125 psi (@ 1% strain, D790-59T); Flex. Mod.: 120,000 psi (D790-59T); Compr. Str.: 810 psi (@ 1% strain, D695-54); Compr. Mod.: 83,000 psi (D695-54); Hardness (Shore D): 59 (initial, D785-62)
Therm. Prop.: Conduct.: 2.1 Btu/h/ft^2/F/in. (Cenco-Fitch); Coeff. of Linear Exp.: 6.0×10^{-5} in./in./F (CD, 78–200 F, D696-44)
Ref.: Product data 103-869

Fluorocomp 102. Liquid Nitrogen Processing Corp.
Chem. Descrip.: Glass-filled TFE comp.
Category/Applications: See Fluorocomp 101
Composition: 90% TFE, 10% glass fiber
Gen'l. Prop.: Sp. Gr.: 2.16 g/cc (D1457-62)
Mech. Prop. (CD): Tens. Str.: 3400 psi (D1457-62); Tens. Elong.: 310% (D-1457-62); Flex. Str.: 1175 psi (@ 1% strain, D790-59T); Flex. Mod.: 135,000 psi (D790-59T); Compr. Str.: 840 psi (@ 1% strain, D695-54); Compr. Mod.: 88,000 psi (D695-54); Hardness (Shore D): 60 (initial, D785-62)
Therm. Prop.: Conduct.: 2.4 Btu/h/ft^2/F/in. (Cenco-Fitch); Coeff. of Linear Exp.: 5.5×10^{-5} in./in./F (CD, 78–200 F, D696-44)
Ref.: Product data 103-869

Fluorocomp 103. Liquid Nitrogen Processing Corp.
Chem. Descrip.: Glass-filled TFE comp.
Category/Applications: See Fluorocomp 101
Composition: 85% TFE, 15% glass fiber
Gen'l. Prop.: Sp. Gr.: 2.19 g/cc (D1457-62)
Mech. Prop. (CD): Tens. Str.: 3100 psi (D1457-62); Tens. Elong.: 290% (D-1457-62); Flex. Str.: 1225 psi (@ 1% strain, D790-59T); Flex. Mod.: 155,000 psi (D790-59T); Compr. Str.: 880 psi (@ 1% strain, D695-54); Compr. Mod.: 94,000 psi (D695-54); Hardness (Shore D): 61 (initial, D785-62)
Therm. Prop.: Conduct.: 2.7 Btu/h/ft^2/F/in. (Cenco-Fitch); Coeff. of Linear Exp.: 5.0×10^{-5} in./in./F (CD, 78–100 F, D696-44)
Ref.: Product data 103-869

Fluorocomp 104. Liquid Nitrogen Processing Corp.
Chem. Descrip.: Glass-filled TFE comp.
Category/Applications: See Fluorocomp 101
Composition: 80% TFE, 20% glass fiber
Gen'l. Prop.: Sp. Gr.: 2.21 g/cc (D1457-62)
Mech. Prop. (CD): Tens. Str.: 2900 psi (D1457-62); Tens. Elong.: 270% (D-1457-62); Flex. Str.: 1250 psi (@ 1% strain, D790-59T); Flex. Mod.: 170,000 psi (D790-59T); Compr. Str.: 910 psi (@ 1% strain, D695-54); Compr. Mod.: 98,000 psi (D695-54); Hardness (Shore D): 62 (initial, D785-62)
Therm. Prop.: Conduct.: 2.9 Btu/h/ft^2/F/in. (Cenco-Fitch); Coeff. of Linear Exp.: 4.5×10^{-5} in./in./F (CD, 78–200 F, D696-44)
Ref.: Product data 103-869

Fluorocomp 105. Liquid Nitrogen Processing Corp.
Chem. Descrip.: Glass-filled TFE comp.
Category/Applications: See Fluorocomp 101
Composition: 75% TFE, 25% glass fiber
Gen'l. Prop.: Sp. Gr.: 2.22 g/cc (D1457-62)
Mech. Prop. (CD): Tens. Str.: 2700 psi (D1457-62); Tens. Elong.: 250% (D-

1457-62); Flex. Str.: 1275 psi (@ 1% strain, D790-59T); Flex. Mod.: 190,000 psi (D790-59T); Compr. Str.: 950 psi (@ 1% strain, D695-54); Compr. Mod.: 103,000 psi (D695-54); Hardness (Shore D): 63 (initial, D785-62)
Therm. Prop.: Conduct.: 3.1 Btu/h/ft^2/F/in. (Cenco-Fitch); Coeff. of Linear Exp.: 4.1 × 10^{-5} in./in./F (CD, 78–200 F, D696-44)
Ref.: Product data 103-869

Fluorocomp 106. Liquid Nitrogen Processing Corp.
Chem. Descrip.: Glass-filled TFE comp.
Category/Applications: See Fluorocomp 101
Composition: 70% TFE, 30% glass fiber
Gen'l. Prop.: Sp. Gr.: 2.21 g/cc (D1457-62)
Mech. Prop. (CD): Tens. Str.: 2300 psi (D1457-62); Tens. Elong.: 230% (D-1457-62); Flex. Str.: 1200 psi (@ 1% strain, D790-59T); Flex. Mod.: 190,000 psi (D790-59T); Compr. Str.: 980 psi (@ 1% strain, D695-54); Compr. Mod.: 109,000 psi (D695-54); Hardness (Shore D): 64 (initial, D785-62)
Therm. Prop.: Conduct.: 3.3 Btu/h/ft^2/F/in. (Cenco-Fitch); Coeff. of Linear Exp.: 3.8 × 10^{-5} in./in./F (CD, 78–200 F, D696-44)
Ref.: Product data 103-869

Fluorocomp 122. Liquid Nitrogen Processing Corp.
Chem. Descrip.: Graphite-filled TFE comp.
Category/Applications: Filled fluorocarbon comp. for long-lasting, nonlubricated bearing service; used as compressor rings, wear rings, packings, seals, etc.
Composition: 90% TFE, 10% graphite
Gen'l. Prop.: Sp. Gr.: 2.08 (D1457)
Mech. Prop.: Tens. Str.: 3300 psi (CD, D1457); Tens. Elong.: 240% (CD, D-1457); Flex. Str.: 850 psi (@ 1% strain, D790); Flex. Mod.: 120,000 psi (D790); Compr. Str.: 750 psi (CD, @ 1% strain, D695); Compr. Mod.: 76,000 psi (CD, D695); Hardness (Shore D): 57 (initial, CD, D785)
Therm. Prop.: Conduct.: 3.9 Btu/h/ft^2/F/in. (Cenco-Fitch); Coeff. of Linear Exp.: 6.7 × 10^{-5} in./in./F (CD, 78–200 F, D696)
Ref.: Product data 106-470

Fluorocomp 123. Liquid Nitrogen Processing Corp.
Chem. Descrip.: Graphite-filled TFE comp.
Category/Applications: See Fluorocomp 122
Composition: 85% TFE, 15% graphite
Gen'l. Prop.: Sp. Gr.: 2.08 (D1457)
Mech. Prop.: Tens. Str.: 2900 psi (CD, D1457); Tens. Elong.: 200% (CD, D-1457); Flex. Str.: 900 psi (@ 1% strain, D790); Flex. Mod.: 125,000 psi (D790); Compr. Str.: 875 psi (CD, @ 1% strain, D695); Compr. Mod.: 79,000 psi (CD, D695); Hardness (Shore D): 59 (CD, initial, D785)
Therm. Prop.: Conduct.: 4.8 Btu/h/ft^2/F/in. (Cenco-Fitch); Coeff. of Linear Exp.: 5.8 × 10^{-5} in./in./F (CD, 78–200 F, D696)
Ref.: Product data 106-470

Fluorocomp 124. Liquid Nitrogen Processing Corp.
Chem. Descrip.: Graphite-filled TFE comp.
Category/Applications: See Fluorocomp 122
Composition: 80% TFE, 20% graphite
Gen'l. Prop.: Sp. Gr.: 2.09 (D1457)
Mech. Prop.: Tens. Str.: 2600 psi (CD, D1457); Tens. Elong.: 170% (CD, D-1457); Flex. Str.: 950 psi (@ 1% strain, D790); Flex. Mod.: 125,000 psi (D790);

Compr. Str.: 1000 psi (CD, @ 1% strain, D695); Compr. Mod.: 82,000 psi (CD, D695); Hardness (Shore D): 60 (CD, initial, D785)
Therm. Prop.: Conduct.: 5.9 Btu/h/ft^2/F/in. (Cenco-Fitch); Coeff. of Linear Exp.: 4.8 × 10^{-5} in./in./F (CD, 78–200 F, D696)
Ref.: Product data 106-470

Fluorocomp 131. Liquid Nitrogen Processing Corp.
Chem. Descrip.: Coke flour-filled TFE comp.
Category/Applications: See Fluorocomp 122
Composition: 95% TFE, 5% coke flour
Gen'l. Prop.: Sp. Gr.: 2.12 (D1457)
Mech. Prop.: Tens. Str.: 3600 psi (CD, D1457); Tens. Elong.: 280% (CD, D-1457); Flex. Str.: 825 psi (@ 1% strain, D790); Flex. Mod.: 105,000 psi (D790); Compr. Str.: 800 psi (CD, @ 1% strain, D695); Compr. Mod.: 80,000 psi (CD, D695); Hardness (Shore D): 55 (CD, initial, D785)
Therm. Prop.: Conduct.: 3.1 Btu/h/ft^2/F/in. (Cenco-Fitch); Coeff. of Linear Exp.: 6.1 × 10^{-5} in./in./F (CD, 78–200 F, D696)
Ref.: Product data 106-470

Fluorocomp 133. Liquid Nitrogen Processing Corp.
Chem. Descrip.: Coke flour-filled TFE
Category/Applications: See Fluorocomp 122
Composition: 85% TFE, 15% coke flour
Gen'l. Prop.: Sp. Gr.: 2.10 (D1457)
Mech. Prop.: Tens. Str.: 3000 psi (CD, D1457); Tens. Elong.: 230% (CD, D-1457); Flex. Str.: 1100 psi (@ 1% strain, D790); Flex. Mod.: 140,000 psi (D790); Compr. Str.: 900 psi (CD, @ 1% strain, D695); Compr. Mod.: 90,000 psi (CD, D695); Hardness (Shore D): 59 (CD, initial, D785)
Therm. Prop.: Conduct.: 4.3 Btu/h/ft^2/F/in. (Cenco-Fitch); Coeff. of Linear Exp.: 4.7 × 10^{-5} in./in./F (CD, 78–200 F, D696)
Ref.: Product data 106-470

Fluorocomp 135. Liquid Nitrogen Processing Corp.
Chem. Descrip.: Coke flour-filled TFE comp.
Category/Applications: See Fluorocomp 122
Composition: 75% TFE, 25% coke flour
Gen'l. Prop.: Sp. Gr.: 2.04 (D1457)
Mech. Prop.: Tens. Str.: 2300 psi (CD, D1457); Tens. Elong.: 100% (CD, D-1457); Flex. Str.: 1175 psi (@ 1% strain, D790); Flex. Mod.: 160,000 psi (D790); Compr. Str.: 1000 psi (CD, @ 1% strain, D695); Compr. Mod.: 100,000 psi (CD, D695); Hardness (Shore D): 61 (CD, initial, D785)
Therm. Prop.: Conduct.: 4.5 Btu/h/ft^2/F/in. (Cenco-Fitch); Coeff. of Linear Exp.: 4.5 × 10^{-5} in./in./F (CD, 78–200 F, D696)
Ref.: Product data 106-470

Fluorocomp 144. Liquid Nitrogen Processing Corp.
Chem. Descrip.: Bronze-filled TFE comp.
Category/Applications: Long-wearing, self-lubricating filled TFE comp. for most mechanical and some chemical applications; used for self-lubricating bearings, rings, gaskets, seals, valve seats, and liners
Composition: 60% TFE, 40% bronze
Gen'l. Prop.: Sp. Gr.: 3.05 g/cc (D1457-62)
Mech. Prop.: Tens. Str.: 2850 psi (CD, D1457-62); Tens. Elong.: 220% (CD, D-1457-62); Flex. Str.: 1400 psi (@ 1% strain, D790-59T); Flex. Mod.: 140,000

psi; Compr. Str.: 1000 psi (CD, 1% strain, D695-54); Compr. Mod.: 10,000 psi (CD, D695-54); Hardness (Shore D): 60 (CD, initial)
Therm. Prop.: Conduct.: 4.3 Btu/h/ft^2/F/in. (Cenco-Fitch); Coeff. of Linear Exp.: 5.4 in./in./F (CD, 78–200 F, D-696-90)
Ref.: Product data 102-367

Fluorocomp 145. Liquid Nitrogen Processing Corp.
Chem. Descrip.: Bronze-filled TFE comp.
Category/Applications: See Fluorocomp 144
Composition: 50% TFE, 50% bronze
Gen'l. Prop.: Sp. Gr.: 3.35 g/cc (D1457-62)
Mech. Prop.: Tens. Str.: 2700 psi (CD, D1457-62); Tens. Elong.: 170% (CD, D-1457-62); Flex. Str.: 1700 psi (1% strain, D790-59T); Flex. Mod.: 170,000 psi; Compr. Str.: 1150 psi (CD, 1% strain, D695-54); Compr. Mod.: 11,000 psi (CD, D695-54); Hardness (Shore D): 64 (CD, initial)
Therm. Prop.: Conduct.: 5.0 Btu/h/ft^2/F/in. (Cenco-Fitch); Coeff. of Linear Exp.: 4.9 in./in./F (CD, 78–200 F, D-696-90)
Ref.: Product data 102-367

Fluorocomp 146. Liquid Nitrogen Processing Corp.
Chem. Descrip.: Bronze-filled TFE comp.
Category/Applications: See Fluorocomp 144
Composition: 40% TFE, 60% bronze
Gen'l. Prop.: Sp. Gr.: 3.78 g/cc (D1457-62)
Mech. Prop.: Tens. Str.: 2600 psi (CD, D1457-62); Tens. Elong.: 140% (CD, D-1457-62); Flex. Str.: 2000 psi (1% strain, D790-59T); Flex. Mod.: 200,000 psi; Compr. Str.: 1500 psi (CD, 1% strain, D695-54); Compr. Mod.: 16,000 psi (CD, D695-54); Hardness (Shore D): 68 (CD, initial)
Therm. Prop.: Conduct.: 5.8 Btu/h/ft^2/F/in. (Cenco-Fitch); Coeff. of Linear Exp.: 4.3 in./in./F (CD, 78–200 F, D-696-90)
Ref.: Product data 102-367

Fluorocomp 147. Liquid Nitrogen Processing Corp.
Chem. Descrip.: Bronze-filled TFE comp.
Category/Applications: See Fluorocomp 144
Composition: 30% TFE, 70% bronze
Gen'l. Prop.: Sp. Gr.: 4.25 g/cc (D1457-62)
Mech. Prop.: Tens. Str.: 1700 psi (CD, D1457-62); Tens. Elong.: 100% (CD, D-1457-62); Flex. Str.: 2400 psi (1% strain, D790-59T); Flex. Mod.: 240,000 psi; Compr. Str.: 1800 psi (CD, 1% strain, D695-54); Compr. Mod.: 20,000 psi (CD, D695-54); Hardness (Shore D): 73 (CD, initial)
Therm. Prop.: Conduct.: 6.6 Btu/h/ft^2/F/in. (Cenco-Fitch); Coeff. of Linear Exp.: 3.5 in./in./F (CD, 78–200 F, D-696-90)
Ref.: Product data 102-367

Fluorocomp 148. Liquid Nitrogen Processing Corp.
Chem. Descrip.: Bronze-filled TFE comp.
Category/Applications: See Fluorocomp 144
Composition: 20% TFE, 80% bronze
Gen'l. Prop.: Sp. Gr.: 4.70 g/cc (D1457-62)
Mech. Prop.: Tens. Str.: 1700 psi (CD, D1457-62); Tens. Elong.: 25% (CD, D-1457-62); Flex. Str.: 2700 psi (1% strain, D790-59T); Flex. Mod.: 270,000

psi; Compr. Str.: 2000 psi (CD, 1% strain, D695-54); Compr. Mod.: 22,000 psi (CD, D695-54); Hardness (Shore D): 76 (CD, initial)
Therm. Prop.: Conduct.: 7.5 Btu/h/ft^2/F/in. (Cenco-Fitch); Coeff. of Linear Exp.: 2.4 in./in./F (CD, 78–200 F, D-696-90)
Ref.: Product data 102-367

Fluorocomp 169. Liquid Nitrogen Processing Corp.
Chem. Descrip.: Glass- and graphite-filled TFE comp.
Category/Applications: See Fluorocomp 122
Composition: 80% TFE, 10% glass fiber, 10% graphite
Gen'l. Prop.: Sp. Gr.: 2.16 (D1457)
Mech. Prop.: Tens. Str.: 2700 psi (CD, D1457); Tens. Elong.: 240% (CD, D-1457); Flex. Str.: 875 psi (@ 1% strain, D790); Flex. Mod.: 140,000 psi (D790); Compr. Str.: 750 psi (CD, @ 1% strain, D695); Compr. Mod.: 73,000 psi (CD, D695); Hardness (Shore D): 57 (CD, initial, D785)
Therm. Prop.: Conduct.: 4.7 Btu/h/ft^2/F/in. (Cenco-Fitch); Coeff. of Linear Exp.: 4.7 $\times$ 10^{-5} in./in./F (CD, 78–200 F, D696)
Ref.: Product data 106-470

Fluorocomp 170. Liquid Nitrogen Processing Corp.
Chem. Descrip.: Glass- and graphite-filled TFE comp.
Category/Applications: See Fluorocomp 122
Composition: 90% TFE, 5% glass fiber, 5% graphite
Gen'l. Prop.: Sp. Gr.: 2.13 (D1457)
Mech. Prop.: Tens. Str.: 3400 psi (CD, D1457); Tens. Elong.: 270% (CD, D-1457); Flex. Str.: 800 psi (@ 1% strain, D790); Flex. Mod.: 115,000 psi (D790); Compr. Str.: 750 psi (CD, @ 1% strain, D695); Compr. Mod.: 70,000 psi (CD, D695); Hardness (Shore D): 56 (CD, initial, D785)
Therm. Prop.: Conduct.: 2.8 Btu/h/ft^2]F/in. (Cenco-Fitch); Coeff. of Linear Exp.: 5.8 $\times$ 10^{-5} in./in./F (CD, 78–200 F, D696)
Ref.: Product data 106-470

Fluorocomp 171. Liquid Nitrogen Processing Corp.
Chem. Descrip.: Glass- and graphite-filled TFE comp.
Category/Applications: See Fluorocomp 122
Composition: 75% TFE, 20% glass fiber, 5% graphite
Gen'l. Prop.: Sp. Gr.: 2.18 (D1457)
Mech. Prop.: Tens. Str.: 2500 psi (CD, D1457); Tens. Elong.: 230% (CD, D-1457); Flex. Str.: 1050 psi (@ 1% strain, D790); Flex. Mod.: 155,000 psi (D790); Compr. Str.: 850 psi (CD, @ 1% strain, D695); Compr. Mod.: 90,000 psi (CD, D695); Hardness (Shore D): 58 (CD, initial, D785)
Therm. Prop.: Conduct.: 3.8 Btu/h/ft^2/F/in. (Cenco-Fitch); Coeff. of Linear Exp.: 4.5 $\times$ 10^{-5} in./in./F (CD, 78–200 F, D696)
Ref.: Product data 106-470

Fluorocomp 174. Liquid Nitrogen Processing Corp.
Chem. Descrip.: Glass- and MoS_2-filled TFE comp.
Category/Applications: See Fluorocomp 101; MoS_2 added to glass-filled TFE extends usefulness to applications requiring greater hardness and stiffness, and increased wear resistance; MoS_2 has little effect on electrical and chemical

properties, but provides lower deformation under load, lower break-in wear, and higher PV values
Composition: 80% TFE, 15% glass fiber, 5% MoS_2
Gen'l. Prop.: Sp. Gr.: 2.22 g/cc (D1457-62)
Mech. Prop. (CD): Tens. Str.: 2800 psi (D1457-62); Tens. Elong.: 265% (D-1457-62); Flex. Str.: 1170 psi (@ 1% strain, D790-59T); Flex. Mod.: 132,000 psi (D790-59T); Compr. Str.: 990 psi (@ 1% strain, D695-54); Compr. Mod.: 101,000 psi (D695-54); Hardness (Shore D): 62 (initial, D785-62)
Therm. Prop.: Conduct.: 3.2 Btu/h/ft^2/F/in. (Cenco-Fitch); Coeff. of Linear Exp.: 4.6 × 10^{-5} in./in./F (CD, 78–200 F, D696-44)
Ref.: Product data 103-869

Fluorocomp 175. Liquid Nitrogen Processing Corp.
Chem. Descrip.: Glass- and MoS_2-filled TFE comp.
Category/Applications: See Fluorocomp 174
Composition: 60% TFE, 20% glass fiber, 20% MoS_2
Gen'l. Prop.: Sp. Gr.: 2.47 g/cc (D1457-62)
Mech. Prop. (CD): Tens. Str.: 2400 psi (D1457-62); Tens. Elong.: 70% (D1457-62); Flex. Str.: 1550 psi (@ 1% strain, D790-59T); Flex. Mod.: 210,000 psi (D-790-59T); Compr. Str.: 1350 psi (@ 1% strain, D695-54); Compr. Mod.: 141,000 psi (D695-54); Hardness (Shore D): 69 (initial, D785-62)
Therm. Prop.: Conduct.: 4.1 Btu/h/ft^2/F/in. (Cenco-Fitch); Coeff. of Linear Exp.: 2.9 × 10^{-5} in./in./F (CD, 78–200 F, D696-44)
Ref.: Product data 103-869

Fluorocomp 176. Liquid Nitrogen Processing Corp.
Chem. Descrip.: Glass- and MoS_2-filled TFE comp.
Category/Applications: See Fluorocomp 174
Composition: 75% TFE, 23% glass fiber, 2% MoS_2
Gen'l. Prop.: Sp. Gr.: 2.21 g/cc (D1457-62)
Mech. Prop. (CD): Tens. Str.: 2600 psi (D1457-62); Tens. Elong.: 245% (D-1457-62); Flex. Str.: 1350 psi (@ 1% strain, D790-59T); Flex. Mod.:172,000 psi (D790-59T); Compr. Str.: 1160 psi (@ 1% strain, D695-54); Compr. Mod.: 118,000 psi (D695-54); Hardness (Shore D): 65 (initial, D785-62)
Therm. Prop.: Conduct.: 3.4 Btu/h/ft^2/F/in. (Cenco-Fitch); Coeff. of Linear Exp.: 3.9 × 10^{-5} in./in./F (CD, 78–200 F, D696-44)
Ref.: Data bulletin 103-869

Fluorocomp 177. Liquid Nitrogen Processing Corp.
Chem. Descrip.: Glass- and MoS_2-filled TFE comp.
Category/Applications: See Fluorocomp 174
Composition: 75% TFE, 12.5% glass fiber, 12.5% MoS_2
Gen'l. Prop.: Sp. Gr.: 2.30 g/cc (D1457-62)
Mech. Prop. (CD): Tens. Str.: 2550 psi (D1457-62); Tens. Elong.: 240% (D-1457-62); Flex. Str.: 1350 psi (@ 1% strain, D790-59T); Flex. Mod.: 153,000 psi (D790-59T); Compr. Str.: 1170 psi (@ 1% strain, D695-54); Compr. Mod.: 120,000 psi (D695-54); Hardness (Shore D): 66 (initial, D785-62)
Therm. Prop.: Conduct.: 3.6 Btu/h/ft^2/F/in. (Cenco-Fitch); Coeff. of Linear

Exp.: 3.7 × 10^{-5} in./in./F (CD, 78–200 F, D696-44)
Ref.: Data bulletin 103-869

Fluorocomp 181. Liquid Nitrogen Processing Corp.
Chem. Descrip.: Bronze- and MoS_2-filled TFE comp.
Category/Applications: See Fluorocomp 144
Composition: 55% TFE, 40% bronze, 5% MoS_2
Gen'l. Prop.: Sp. Gr.: 3.14 g/cc (D1457-62)
Mech. Prop.: Tens. Str.: 2700 psi (CD, D1457-62); Tens. Elong.: 200% (CD, D-1457-62); Flex. Str.: 1600 psi (1% strain, D790-59T); Flex. Mod.: 160,000 psi; Compr. Str.: 1150 psi (CD, 1% strain, D695-54); Compr. Mod.: 11,000 psi (CD, D695-54); Hardness (Shore D): 60 (CD, initial)
Therm. Prop.: Conduct.: 4.3 Btu/h/ft^2/F/in. (Cenco-Fitch); Coeff. of Linear Exp.: 4.7 in./in./F (CD, 78–200 F, D-696-90)
Ref.: Product data 102-367

Fluorocomp 182. Liquid Nitrogen Processing Corp.
Chem. Descrip.: Bronze- and MoS_2-filled TFE comp.
Category/Applications: See Fluorocomp 144
Composition: 40% TFE, 55% bronze, 5% MoS_2
Gen'l. Prop.: Sp. Gr.: 3.67 g/cc (D1457-62)
Mech. Prop.: Tens. Str.: 2300 psi (CD, D1457-62); Tens. Elong.: 130% (CD, D-1457-62); Flex. Str.: 2100 psi (1% strain, D790-59T); Flex. Mod.: 210,000 psi; Compr. Str.: 1450 psi (CD, 1% strain, D695-54); Compr. Mod.: 15,000 psi (CD, D695-54); Hardness (Shore D): 66 (CD, initial)
Therm. Prop.: Conduct.: 5.0 Btu/h/ft^2/F/in. (Cenco-Fitch); Coeff. of Linear Exp.: 3.9 in./in./F (CD, 78–200 F, D-696-90)
Ref.: Product data 102-367

Fluorocomp 183. Liquid Nitrogen Processing Corp.
Chem. Descrip.: Bronze- and MoS_2-filled TFE comp.
Category/Applications: See Fluorocomp 144
Composition: 30% TFE, 65% bronze, 5% MoS_2
Gen'l. Prop.: Sp. Gr.: 4.16 g/cc (D1457-62)
Mech. Prop.: Tens. Str.: 2200 psi (CD, D1457-62); Tens. Elong.: 65% (CD, D-1457-62); Flex. Str.: 2500 psi (1% strain, D790-59T); Flex. Mod.: 250,000 psi; Compr. Str.: 1700 psi (CD, 1% strain, D695-54); Compr. Mod.: 18,000 psi (CD, D695-54); Hardness (Shore D): 70 (CD, initial)
Therm. Prop.: Conduct.: 6.0 Btu/h/ft^2/F/in. (Cenco-Fitch); Coeff. of Linear Exp.: 3.3 in./in./F (CD, 78–200 F, D-696-90)
Ref.: Product data 102-367

Fluorocomp 190. Liquid Nitrogen Processing Corp.
Chem. Descrip.: Carbon/graphite-filled TFE comp.
Category/Applications: See Fluorocomp 122
Composition: 90% TFE, 10% carbon/graphite
Gen'l. Prop.: Sp. Gr.: 2.10 (D1457)
Mech. Prop.: Tens. Str.: 3450 psi (CD, D1457); Tens. Elong.: 230% (CD, D-1457); Flex. Str.: 950 psi (@ 1% strain,

D790); Flex. Mod.: 130,000 psi (D790); Compr. Str.: 750 psi (CD, @ 1% strain, D695); Compr. Mod.: 74,000 psi (CD, D695); Hardness (Shore D): 57 (CD, intial, D785)
Therm. Prop.: Conduct.: 3.3 Btu/h/ft^2/F/in. (Cenco-Fitch); Coeff. of Linear Exp.: 6.3 X 10^{-5} in./in./F (CD, 78–200 F, D696)
Ref.: Product data 106-470

Fluorocomp 191. Liquid Nitrogen Processing Corp.
Chem. Descrip.: Carbon/graphite-filled TFE comp.
Category/Applications: See Fluorocomp 122
Composition: 75% TFE, 25% carbon/graphite
Gen'l. Prop.: Sp. Gr.: 2.08 (D1457)
Mech. Prop.: Tens. Str.: 2200 psi (CD, D1457); Tens. Elong.: 65% (CD, D-1457); Flex. Str.: 1375 psi (@ 1% strain, D790); Flex. Mod.: 160,000 psi (D790); Compr. Str.: 900 psi (CD, @ 1% strain, D695); Compr. Mod.: 96,000 psi (CD, D695); Hardness (Shore D): 60 (CD, initial, D785)
Therm. Prop.: Conduct.: 5.0 Btu/h/ft^2/F/in. (Cenco-Fitch); Coeff. of Linear Exp.: 4.7 X 10^{-5} in./in./F (CD, 78–200 F, D696)
Ref.: Product data 106-470

Fluorocomp 192. Liquid Nitrogen Processing Corp.
Chem. Descrip.: Carbon/graphite-filled TFE comp.
Category/Applications: See Fluorocomp 122
Composition: 65% TFE, 35% carbon/graphite; avail. in pressure sintered and free sintered grades
All properties–pressure sintered @ 720 F for 4 h @ 4000 psi, cooled @ 100 F/h
Gen'l. Prop.: Sp. Gr.: 2.05 (D1457)
Mech. Prop.: Tens. Str.: 2000 psi (CD, D1457); Tens. Elong.: 55% (CD, D-1457); Flex. Str.: 1425 psi (@ 1% strain, D790); Flex. Mod.: 180,000 psi (D790); Compr. Str.: 1000 psi (CD, @ 1% strain, D695); Compr. Mod.: 108,000 psi (CD, D695); Hardness (Shore D): 62 (CD, initial, D785)
Therm. Prop.: Conduct.: 6.3 Btu/h/ft^2/F/in. (Cenco-Fitch); Coeff. of Linear Exp.: 4.0 X 10^{-5} in./in./F (CD, 78–200 F, D696)
Ref.: Product data 106-470

Fluorocomp 193. Liquid Nitrogen Processing Corp.
Chem. Descrip.: Carbon-filled TFE comp.
Category/Applications: See Fluorocomp 122
Composition: 85% TFE, 15% carbon
Gen'l. Prop.: Sp. Gr.: 2.08 (D1457)
Mech. Prop.: Tens. Str.: 3000 psi (CD, D1457); Tens. Elong.: 250% (CD, D-1457); Flex. Str.: 1100 psi (@ 1% strain, D790); Flex. Mod.: 140,000 psi (D790); Compr. Str.: 700 psi (CD, @ 1% strain, D695); Compr. Mod.: 90,000 psi (CD, D695); Hardness (Shore D): 59 (CD, initial, D785)
Therm. Prop.: Conduct.: 4.3 Btu/h/ft^2/F/in. (Cenco-Fitch); Coeff. of Linear Exp.: 4.7 X 10^{-5} in./in./F (CD, 78–200 F, D696)
Ref.: Product data 106-470

Fluorocomp 195. Liquid Nitrogen Processing Corp.
Chem. Descrip.: Carbon-filled TFE comp.
Category/Applications: See Fluorocomp 122
Composition: 75% TFE, 25% carbon
Gen'l. Prop.: Sp. Gr.: 2.04 (D1457)

Mech. Prop.: Tens. Str.: 2300 psi (CD, D1457); Tens. Elong.: 100% (CD, D-1457); Flex. Str.: 1175 psi (@ 1% strain, D790); Flex. Mod.: 160,000 psi (D790); Compr. Str.: 800 psi (CD, @ 1% strain, D695); Compr. Mod.: 100,000 psi (CD, D695); Hardness (Shore D): 61 (CD, initial, D785)
Therm. Prop.: Conduct.: 4.5 $Btu/h/ft^2/$ F/in. (Cenco-Fitch); Coeff. of Linear Exp.: 4.5 $\times 10^{-5}$ in./in./F (CD, 78–200 F, D696)
Ref.: Product data 106-470

Fluorocomp 197. Liquid Nitrogen Processing Corp.
Chem. Descrip.: Glass- and carbon-filled TFE comp.
Category/Applications: See Fluorocomp 122
Composition: 85% TFE, 10% glass fiber, 5% carbon
Gen'l. Prop.: Sp. Gr.: 2.11 (D1457)
Mech. Prop.: Tens. Str.: 3000 psi (CD, D1457); Tens. Elong.: 240% (CD, D-1457); Flex. Str.: 1050 psi (@ 1% strain, D790); Flex. Mod.: 140,000 psi (D790); Compr. Str.: 750 psi (CD, @ 1% strain, D695); Compr. Mod.: 68,000 psi (CD, D695); Hardness (Shore D): 57 (CD, initial, D785)
Therm. Prop.: Conduct.: 4.0 $Btu/h/ft^2/$ F/in. (Cenco-Fitch); Coeff. of Linear Exp.: 5.2 $\times 10^{-5}$ in./in./F (CD, 78–200 F, D696)
Ref.: Product data 106-470

Fluorocomp 199. Liquid Nitrogen Processing Corp.
Chem. Descrip.: Glass- and carbon-filled TFE comp.
Category/Applications: See Fluorocomp 122
Composition: 85% TFE, 5% glass fiber, 10% carbon
Gen'l. Prop.: Sp. Gr.: 2.15 (D1457)
Mech. Prop.: Tens. Str.: 2900 psi (CD, D1457); Tens. Elong.: 235% (CD, D-1457); Flex. Str.: 1175 psi (@ 1% strain, D790); Flex. Mod.: 135,000 psi (D790); Compr. Str.: 750 psi (CD, @ 1% strain, D695); Compr. Mod.: 70,000 psi (CD, D695); Hardness (Shore D): 59 (CD, initial, D785)
Therm. Prop.: Conduct.: 4.2 $Btu/h/ft^2/$ F/in. (Cenco-Fitch); Coeff. of Linear Exp.: 4.8 $\times 10^{-5}$ in./in./F (CD, 78–200 F, D696)
Ref.: Product data 106-470

FM 510. Fiberite Corp.
Chem. Descrip.: Two-stage phenolic resin
Category/Applications: Molding comp. with fast cure and good molded finish; free-flowing and suitable for automatic equipment
Form: Nodular; Color: Black
All properties as molded:
Gen'l. Prop.: Sp. Gr.: 1.37–1.39 (D792)
Mech. Prop.: Tens. Str.: 6500 psi (D-638); Flex. Str.: 10,000 psi (D790); Flex. Mod.: 1.1 $\times 10^6$ psi (D790); Compr. Str.: 29,000 psi (D695); Impact Str.: 0.6 ft lb/in. notch (D256); Hardness (Rockwell): M 110
Therm. Prop.: Distort. Temp.: 320 F (@ 264 psi, D648);
Elec. Prop.: Dissip. Factor: 0.05 (1 MC); Dielec. Str.: 300 V/mil (S/T, dry, D149); Dielec. Const.: 4.2 (1 MC); Vol. Resist.: 2 $\times 10^{13}$ ohm-cm; Arc Resist.: 75 s
Ref.: Data sheet

FM 1132. Fiberite Corp.
Chem. Descrip.: Two-stage phenolic resin with macerated fabric as reinforcement
Category/Applications: Molding comp. with good shock resistance and dielec. properties for impact, compression, and transfer molding

Form: Macerated; Color: Black, natural, brown
All properties as molded:
Gen'l. Prop.: Sp. Gr.: 1.36–1.38 (D792)
Mech. Prop.: Tens. Str.: 7000 psi (D-651); Flex. Str.: 12,500 psi (D790); Flex. Mod.: 10^6 psi (D790); Compr. Str.: 28,000 psi (D695); Impact Str.: 1.6 ft lb/in. notch (D256); Hardness (Rockwell): M 115 (D785)
Therm. Prop.: Conduct.: 9×10^{-4} Cal/s/cm^2/C/cm (C177); Distort. Temp.: 163 C (@ 264 psi, D648); Coeff. of Linear Exp.: 1.8×10^{-5} in./in./C (−30 to 30 C, D696)
Elec. Prop.: Dissip. Factor: 0.06 (dry, 1 MC, D150); Dielec. Str.: 300 V/mil (dry, S/T, D149); Dielec. Const.: 4.4 (dry, 1 MC, D150); Vol. Resist.: 7×10^{12} ohm-cm (D257)
Ref.: Data sheet

FM 1132P. Fiberite Corp.
Chem. Descrip.: Two-stage phenolic resin with macerated fabric as reinforcement
Category/Applications: Pelleted form of FM 1132; pellets have the advantage of lower bulk factor and the ability to be automatically preformed
Form: Nodular; Color: Black, natural, green
All properties as molded
Gen'l. Prop.: Sp. Gr.: 1.36–1.38 (D792)
Mech. Prop.: Tens. Str.: 6500 psi (D-651); Flex. Str.: 10,000 psi (D790); Flex. Mod.: 1.2×10^6 psi (D790); Compr. Str.: 28,000 psi (D695); Impact Str.: 1.1 ft lb/in. notch (D256); Hardness (Rockwell): M 115 (D785)
Therm. Prop.: Conduct.: 9×10^{-4} Cal/s/cm^2/C/cm (C177); Distort. Temp.: 160 C (@ 264 psi, D648); Coeff. of Linear Exp.: 2.4×10^{-5} in./in./C (−30 to 30 C, D696)
Elec. Prop.: Dissip. Factor: 0.05 (dry, 1 MC, D150); Dielec. Str.: 300 V/mil (dry, S/T, D149); Dielec. Const.: 4.2 (dry, 1 MC, D150); Vol. Resist.: 3×10^{13} ohm-cm (D257)
Ref.: Data sheet

FM 1303. Fiberite Corp.
Chem. Descrip.: Two-stage nylon-reinforced phenolic resin
Category/Applications: Resin with good impact strength, resilience, low sp. gr., low water absorption; used in pulleys, gears, self-lubricating bearings, pump housings
Form: Granular; Color: Black, brown
All properties as molded:
Gen'l. Prop.: Sp. Gr.: 1.23–1.25, black, 1.27–1.29, brown (D792)
Mech. Prop.: Tens. Str.: 7500 psi (D-651); Flex. Str.: 12,000 psi (D790); Flex. Mod.: 500,000 psi (D790); Compr. Str.: 23,000 psi (D695); Impact Str.: 0.55 ft lb/in. notch (D256); Hardness (Rockwell): M 110 (D785)
Therm. Prop.: Conduct.: 7×10^{-4} Cal/s/cm^2/C/cm (C177); Distort. Temp.: 124 C (@ 264 psi, D648); Coeff. of Linear Exp.: 5.2×10^{-5} in./in./C (−30 to 30 C, D696)
Elec. Prop.: Dissip. Factor: 0.03 (dry, 1 MC, D150); Dielec. Str.: 340 V/mil (dry, S/T, D149); Dielec. Const.: 3.6 (dry, 1 MC, D150); Vol. Resist.: 10^{13} ohm-cm (D257)
Ref.: Data sheet

FM 1389. Fiberite Corp.
Chem. Descrip.: One-stage cellulose-filled phenolic resin
Category/Applications: Specially lubricated molding material of med. impact strength with good flow, surface luster; designed for wear resistance in heavy duty applications; used in trays, housings, bushings, pulleys, gears, handles, heavy duty machine components, cas-

ters, bearings, textile pulleys, and bobbins
Form: Macerated; Color: Green, black
All properties as molded:
Gen'l. Prop.: Sp. Gr.: 1.45–1.47 (D-792); Stability: Good resistance to chemicals and moisture
Mech. Prop.: Tens. Str.: 6500 psi (D-651); Flex. Str.: 11,000 psi (D790); Flex. Mod.: 1.1×10^6 psi (D790); Compr. Str.: 21,000 psi (D695); Impact Str.: 1.3 ft lb/in. notch (D256); Hardness (Rockwell): M 105 (D785)
Therm. Prop.: Conduct.: 12×10^{-4} Cal/s/cm^2/C/cm (C177); Distort. Temp.: 149 C (@ 264 psi, D648); Coeff. of Linear Exp.: 2.2×10^{-5} in./in./C (–30 to 30 C, D696)
Ref.: Data sheet

FM 3000. Fiberite Corp.
Chem. Descrip.: Two-stage cellulose-filled phenolic resin
Category/Applications: Molding comp. with good strength, electrical properties, wear resistance; for use in conduit bushings, timing gears, pulleys, textile spools, casters, conveyor wheels, rifle stock grips, self-lubricating bearings, insulators, instrument housings, switch gear
Form: Nodular; Color: Black
All properties as molded:
Gen'l. Prop.: Sp. Gr.: 1.44–1.46 (D792)
Mech. Prop.: Tens. Str.: 5500 psi (D-651); Flex. Str.: 8000 psi (D790); Flex. Mod.: 960,000 psi (D790); Compr. Str.: 15,000 psi (D695); Impact Str.: 0.50 ft lb/in. notch (D256); Hardness (Rockwell): M 114 (D785)
Therm. Prop.: Conduct.: 8.0×10^{-4} Cal/s/cm^2/C/cm (C177); Distort. Temp.: 350 F (@ 264 psi, D648); Coeff. of Linear Exp.: 2.6×10^{-5} in./in./C (78–212 F, D696)
Elec. Prop.: Dissip. Factor: 0.035 (dry, 1 MC, D150); Dielec. Str.: 380 V/mil (dry, S/T, D149); Dielec. Const.: 3.6 (dry, 1 MC, D150)
Ref.: Data sheet

FM 3001. Fiberite Corp.
Chem. Descrip.: Two-stage cellulose-filled phenolic resin
Category/Applications: See FM 3000
Form: Nodular; Color: Black
All properties as molded:
Gen'l. Prop.: Sp. Gr.: 1.40–1.42 (D792)
Mech. Prop.: Tens. Str.: 5800 psi (D-651); Flex. Str.: 8500 psi (D790); Flex. Mod.: 1.1×10^6 psi (D790); Compr. Str.: 22,800 psi (D695); Impact Str.: 0.6 ft lb/in. notch (D256); Hardness (Rockwell): M 113 (D785)
Therm. Prop.: Conduct.: 9.0×10^{-4} Cal/s/cm^2/C/cm (C177); Distort. Temp.: 365 F (@ 264 psi, D648); Coeff. of Linear Exp.: 2.4×10^{-5} in./in./C (78–212 F, D696)
Elec. Prop.: Dissip. Factor: 0.04 (dry, 1 MC, D150); Dielec. Str.: 300 V/mil (dry, S/T, D149); Dielec. Const.: 4.40 (dry, 1 MC, D150)
Ref.: Data sheet

FM 3002. Fiberite Corp.
Chem. Descrip.: Two-stage cellulose-filled phenolic resin
Category/Applications: See FM 3000
Form: Nodular; Color: Black
All properties as molded:
Gen'l. Prop.: Sp. Gr.: 1.38–1.40 (D792)
Mech. Prop.: Tens. Str.: 6000 psi (D-651); Flex. Str.: 8500 psi (D790); Flex. Mod.: 970,000 psi (D790); Compr. Str.: 22,000 psi (D695); Impact Str.: 0.7 ft lb/in. notch (D256); Hardness (Rockwell): M 115 (D785)
Therm. Prop.: Conduct.: 7.0×10^{-4} Cal/s/cm^2/C/cm (C177); Distort. Temp.: 350 F (@ 264 psi, D648); Coeff. of Linear Exp.: 26×10^{-5} in./in./C (78–212 F, D696)

Elec. Prop.: Dissip. Factor: 0.04 (dry, 1 MC, D150); Dielec. Str.: 380 V/mil (dry, S/T, D149); Dielec. Const.: 3.8 (dry, 1 MC, D150)
Ref.: Data sheet

FM 3510. Fiberite Corp.
Chem. Descrip.: Two-stage cotton fabric-filled phenolic resin
Category/Applications: Impact molding resin with good torque strength and molding surface
Form: Macerated; Color: Black, natural, brown
All properties as molded:
Gen'l. Prop.: Sp. Gr.: 1.36–1.38 (D792)
Mech. Prop.: Tens. Str.: 7000 psi (D-651); Flex. Str.: 14,000 psi (D790); Flex. Mod.: 1.2×10^6 psi (D790); Compr. Str.: 26,000 psi (D695); Impact Str.: 2.2 ft lb/in. notch (D256); Hardness (Rockwell): M 115 (D785)
Therm. Prop.: Conduct.: 8×10^{-4} Cal/s/cm^2/C/cm (C177); Distort. Temp.: 166 C (@ 264 psi, D648); Coeff. of Linear Exp.: 1.5×10^{-5} in./in./C (–30 to 30 C, D696)
Elec. Prop.: Dissip. Factor: 0.05 (dry, 1 MC, D150); Dielec. Str.: 300 V/mil (dry, S/T, D149); Dielec. Const.: 4.5 (dry, 1 MC, D150)
Ref.: Data sheet

FM 4002. Fiberite Corp.
Chem. Descrip.: Glass-filled phenolic resin
Category/Applications: 25% higher tens. and flex. strength properties and lower mold shrinkage than FM-4005; used where minimum molding shrinkage and strength is required
Ref.: Data sheet

FM 4003. Fiberite Corp.
Chem. Descrip.: Glass-filled phenolic resin
Category/Applications: Thermosetting comp. with excellent dimensional stability and a marginal decrease in mechanical properties from FM 4004; lowest cost in FM-4000 Series
Form: Color: Green, black
All properties as molded:
Gen'l. Prop.: Sp. Gr.: 1.92–1.97
Mech. Prop.: Tens. Str.: 6000 psi; Flex. Str.: 14,500 psi; Compr. Str.: 35,000 psi; Impact Str.: 0.40 ft lb/in. notch
Therm. Prop.: Distort. Temp.: 340 F (@ 264 psi)
Elec. Prop.: Dissip. Factor: 0.020 (dry, 1 MC); Dielec. Str.: 280 V/mil (dry, S/S); Dielec. Const.: 4.5 (dry, 1 MC); Insul. Resist.: 10^{13} ohm (dry)
Ref.: Data sheet

FM 4004. Fiberite Corp.
Chem. Descrip.: Two-stage glass-filled phenolic resin
Category/Applications: Thermosetting comp. with good dimensional stability, moldability, electrical properties, and heat resistance; used in computer frames, housings, connectors, coil bobbins, terminal strips, insulators, dimensionally stable mechanical components; 4004 is esp. for use in high speed injection molding machines
Form: Granular; Color: Green, black
All properties as molded:
Gen'l. Prop.: Sp. Gr.: 1.90–1.95 (D792)
Mech. Prop.: Tens. Str.: 8000 psi (D-651); Flex. Str.: 12,000 psi (D790); Flex. Mod.: 2.3×10^6 psi (D790); Compr. Str.: 38,000 psi (D695); Impact Str.: 0.45 ft lb/in. notch (D256); Hardness (Rockwell): M 110 (D785)
Therm. Prop.: Conduct.: 13×10^{-4} Cal/s/cm^2/C/cm (C177); Distort. Temp.: 177 C (@ 264 psi, D648); Coeff. of Linear Exp.: 1.4×10^{-5} in./in./C (–30 to 30 C, D696)
Elec. Prop.: Dissip. Factor: 0.020 (dry,

1 MC, D150); Dielec. Str.: 320 V/mil (dry, S/T, D149); Dielec. Const.: 4.5 (dry, 1 MC, D150); Vol. Resist.: 2 X 10^{13} ohm-cm (D257); Arc Resist.: 180 s (D495)
Ref.: Data sheet

FM 4004F. Fiberite Corp.
Chem. Descrip.: Two-stage glass-filled phenolic resin
Category/Applications: Resin suited for automatic preforming prior to transfer, and for compression or automatic molding; with improved tensile, flexural, and compressive strength, and exhibiting very low post mold shrinkage
Form: Color: Green, black
All properties as molded:
Gen'l. Prop.: Sp. Gr.: 1.78–1.81
Mech. Prop.: Tens. Str.: 12,000 psi; Flex. Str.: 16,000 psi; Compr. Str.: 45,000 psi; Impact Str.: 0.55 ft lb/in. notch
Therm. Prop.: Distort. Temp.: 445 F (@ 264 psi)
Elec. Prop.: Dissip. Factor: 4.6 (dry, 1 MC); Dielec. Str.: 430 V/mil (dry, S/S); Dielec. Const.: 4.6 (dry, 1 MC)
Ref.: Data sheet

FM 4005. Fiberite Corp.
Chem. Descrip.: Two-stage glass-filled phenolic resin
Category/Applications: See FM 4004; 4005 esp. suited for automatic preforming prior to transfer, compression, or automatic molding, and for highly intricate close tolerance parts
Form: Granular; Color: Green, black
All properties as molded:
Gen'l. Prop.: Sp. Gr.: 1.78–1.81 (D792)
Mech. Prop.: Tens. Str.: 10,000 psi (D-651); Flex. Str.: 15,000 psi (D790); Flex. Mod.: 2 X 10^6 psi (D790); Compr. Str.: 40,000 psi (D695); Shear Str.: 14,000 psi; Impact Str.: 0.60 ft lb/in. notch (D256); Hardness (Rockwell): M 120 (D785)
Therm. Prop.: Conduct.: 10 X 10^{-4} Cal/s/cm^2/C/cm (C177); Distort. Temp.: 191 C (@ 264 psi, D648); Coeff. of Linear Exp.: 1.9 X 10^{-5} in./in./C (–30 to 30 C, D696)
Elec. Prop.: Dissip. Factor: 0.012 (dry, 1 MC, D150); Dielec. Str.: 400 V/mil (dry, S/T, D149); Dielec. Const.: 4.6 (dry, 1 MC, D150); Vol. Resist.: 2 X 10^{13} ohm-cm (D257); Arc Resist.: 180 s (D495)
Ref.: Data sheet

FM 4005F. Fiberite Corp.
Chem. Descrip.: Two-stage glass-filled phenolic resin
Category/Applications: See FM-4004F
Form: Color: Green, black
All properties as molded:
Gen'l. Prop.: Sp. Gr.: 1.78–1.81
Mech. Prop.: Tens. Str.: 16,000 psi; Flex. Str.: 22,000 psi; Compr. Str.: 45,000 psi; Impact Str.: 0.60 ft lb/in. notch
Therm. Prop.: Distort. Temp.: 445 F (@ 264 psi)
Elec. Prop.: Dissip. Factor: 0.012 (dry, 1 MC); Dielec. Str.: 430 V/mil (dry, S/S); Dielec. Const.: 4.5 (dry, 1 MC)
Ref.: Data sheet

FM 4007. Fiberite Corp.
Chem. Descrip.: Two-stage glass-filled phenolic resin
Category/Applications: See FM 4004
Form: Granular; Color: Green
All properties as molded:
Gen'l. Prop.: Sp. Gr.: 1.81–1.83 (D792)
Mech. Prop.: Tens. Str.: 9000 psi (D-651); Flex. Str.: 14,000 psi (D790); Flex. Mod.: 2 X 10^6 psi (D790); Compr. Str.: 40,000 psi (D695); Impact Str.:

0.50 ft lb/in. notch (D256); Hardness (Rockwell): M 120 (D785)
Therm. Prop.: Conduct.: 11 X 10^{-4} Cal/s/cm^2/C/cm (C177); Distort. Temp.: 191 C (@ 264 psi, D648); Coeff. of Linear Exp.: 1.9 X 10^{-5} in./in./C (−30 to 30 C, D696)
Elec. Prop.: Dissip. Factor: 0.011 (dry, 1 MC, D150); Dielec. Str.: 380 V/mil (dry, S/T, D149); Dielec. Const.: 4.1 (dry, 1 MC, D150); Vol. Resist.: 2 X 10^{13} ohm-cm (D257); Arc Resist.: 180 s (D495)
Ref.: Data sheet

FM 4007F. Fiberite Corp.
Chem. Descrip.: Two-stage glass-filled phenolic resin
Category/Applications: See FM-4004F
Form: Color: Green, black
All properties as molded:
Gen'l. Prop.: Sp. Gr.: 1.78–1.81
Mech. Prop.: Tens. Str.: 14,000 psi; Flex. Str.: 20,000 psi; Compr. Str.: 45,000 psi; Impact Str.: 0.70 ft lb/in. notch
Therm. Prop.: Distort. Temp.: 445 F (@ 264 psi)
Elec. Prop.: Dissip. Factor: 0.012 (dry, 1 MC); Dielec. Str.: 430 V/mil (dry, S/S); Dielec. Const.: 4.6 (dry, 1 MC)
Ref.: Data sheet

FM 4008. Fiberite Corp.
Chem. Descrip.: Glass-filled phenolic resin
Category/Applications: Recommended for difficult, hermetically sealed electrical products
Form: Color: Green, black
All properties as molded:
Gen'l. Prop.: Sp. Gr.: 1.78–1.82
Mech. Prop.: Tens. Str.: 8000 psi; Flex. Str.: 14,500 psi; Compr. Str.: 35,000 psi; Impact Str.: 0.50 ft lb/in. notch
Therm. Prop.: Distort. Temp.: 340 F (@ 264 psi)
Elec. Prop.: Dissip. Factor: 0.025 (dry, 1 MC); Dielec. Str.: 280 V/mil (dry, S/S); Dielec. Const.: 5.5 (dry, 1 MC); Insul. Resist.: 10^{13} ohm (dry)
Ref.: Data sheet

FM 4009. Fiberite Corp.
Chem. Descrip.: Glass-filled phenolic comp.
Category/Applications: Thermosetting comp. esp. formulated to achieve low friction properties; excellent wear resistance properties for use where sliding wear is increased
Ref.: Data sheet

FM 4011. Fiberite Corp.
Chem. Descrip.: Glass-filled phenolic comp.
Category/Applications: Thermosetting comp. for structural applications where superior physical strength properties are required; exceeds FM 4002 and 4005 physical properties without sacrificing excellent molding characteristics
Ref.: Data sheet

FM 4017. Fiberite Corp.
Chem. Descrip.: Glass-filled phenolic comp.
Category/Applications: Thermosetting comp. with superior electrical properties, good dimensional stability, and easy injection molding processability; used for electrical connectors, sockets, tube bases, and other electronic/electrical components
Form: Color: Green, black
All properties as molded:
Gen'l. Prop.: Sp. Gr.: 1.78–1.82
Mech. Prop.: Tens. Str.: 9000 psi; Flex. Str.: 15,000 psi; Compr. Str.: 40,000 psi; Impact Str.: 0.60 ft lb/in. notch

Therm. Prop.: Distort. Temp.: 375 F (@ 264 psi)
Elec. Prop.: Dissip. Factor: 0.010 (dry, 1 MC); Dielec. Str.: 360 V/mil (dry, S/S); Dielec. Const.: 4.4 (dry, 1 MC); Insul. Resist.: 10^{13} ohm (dry)
Ref.: Data sheet

FM 4029B. Fiberite Corp.
Chem. Descrip.: Glass fiber- and mineral-reinforced phenolic comp.
Category/Applications: Thermoset molding comp. for high heat resistance applications; can resist short term temp. > 450 F
Form: Color: Blue
All properties as molded:
Gen'l. Prop.: Sp. Gr.: 2.1 + 0.05
Mech. Prop.: Tens. Str.: 6500 psi; Flex. Str.: 15,000 psi; Compr. Str.: 26,000 psi; Impact Str.: 0.4 ft lb/in. notch
Therm. Prop.: Distort. Temp.:482 F (@ 264 psi)
Elec. Prop.: Dissip. Factor: 0.013 (dry, 1 MC); Dielec. Str.: 410 V/mil (dry, S/S); Dielec. Const.: 5.9 (dry, 1 MC); Insul. Resist.: 10^{13} ohm (dry)
Ref.: Data sheet

FM 4030-190. Fiberite Corp.
Chem. Descrip.: Two-stage glass roving-reinforced phenolic comp.
Category/Applications: Molding comp. with high strength, electrical properties, heat resistance; used in missile components, switch gear, lantern and flashlight cases, motor gear housings, power tool insulating housings, rifle hand guard, conveyor wheels
Form: Chopped fiber; Color: Natural, black
All properties as molded:
Gen'l. Prop.: Sp. Gr.: 1.72–1.76 (D792)
Mech. Prop.: Tens. Str.: 7500 psi (D-651); Flex. Str.: 16,000 psi (D790); Flex. Mod.: 2.4 × 10^6 psi (D790); Compr. Str.: 27,000 psi (D695); Impact Str.: 12 ft lb/in. notch (D256); Hardness (Rockwell): M 115 (D785)
Therm. Prop.: Conduct.: 10 × 10^{-4} Cal/s/cm^2/C/cm (C177); Distort. Temp.: 260 C (@ 264 psi, D648); Coeff. of Linear Exp.: 1.5 × 10^{-5} in./in./C (–30 to 30 C, D696)
Elec. Prop.: Dissip. Factor: 0.02 (dry, 1 MC, D150); Dielec. Str.: 380 V/mil (dry, S/T, D149); Dielec. Const.: 5.0 (dry, 1 MC, D150); Vol. Resist.: 10^{13} ohm-cm (D257)
Ref.: Data sheet

FM 4036. Fiberite Corp.
Chem. Descrip.: Glass fiber- and mineral-reinforced phenolic comp.
Category/Applications: Thermoset molding comp. for high heat resistance applications; requires step postcuring to achieve short term temp. resistance > 450 F
Form: Color: Green, black
All properties as molded:
Gen'l. Prop.: Sp. Gr.: 1.78–1.82
Mech. Prop.: Tens. Str.: 10,000 psi; Flex. Str.: 16,000 psi; Compr. Str.: 35,000 psi; Impact Str.: 0.5 ft lb/in. notch
Therm. Prop.: Distort. Temp.: 446 F (@ 264 psi)
Elec. Prop.: Dissip. Factor: 0.030 (dry, 1 MC); Dielec. Str.: 230 V/mil (dry, S/S); Dielec. Const.: 6.0 (dry, 1 MC); Insul. Resit.: 10^{12} ohm (dry)
Ref.: Data sheet

FM 4201. Fiberite Corp.
Chem. Descrip.: Two-stage cellulose-reinforced phenolic comp.
Category/Applications: See FM 3000; 4201 esp. for use in automatic equipment with volumetric feed
Form: Nodular; Color: Black
All properties as molded:

Gen'l. Prop.: Sp. Gr.: 1.37–1.39 (D792)
Mech. Prop.: Tens. Str.: 6300 psi (D-651); Flex. Str.: 10,000 psi (D790); Flex. Mod.: 1.1 × 10^6 psi (D790); Compr. Str.: 31,000 psi (D695); Impact Str.: 0.70 ft lb/in. notch (D256); Hardness (Rockwell): M 115 (D785)
Therm. Prop.: Conduct.: 9 × 10^{-4} Cal/s/cm^2/C/cm (C177); Distort. Temp.: 174 C (@ 264 psi, D648); Coeff. of Linear Exp.: 2.6 × 10^{-5} in./in./C (−30 to 30 C, D696)
Elec. Prop.: Dissip. Factor: 0.05 (dry, 1 MC, D150); Dielec. Str.: 340 V/mil (dry, S/T, D149); Dielec. Const.: 4.3 (dry, 1 MC, D150); Vol. Resist.: 2 × 10^{13} ohm-cm (D257)
Ref.: Data sheet

FM 5064. Fiberite Corp.
Chem. Descrip.: Two-stage glass roving-reinforced phenolic comp.
Category/Applications: See FM 4030-190
Form: Chopped fiber (avail. in chop lengths from ½–1½ in.); Color: Black, yellow
All properties as molded (½ in. chop):
Gen'l. Prop.: Sp. Gr.: 1.72–1.75 (D792)
Mech. Prop.: Tens. Str.: 7500 psi (D-651); Flex. Str.: 17,000 psi (D790); Flex. Mod.: 2 × 10^6 psi (D790); Compr. Str.: 28,000 psi (D695); Impact Str.: 11 ft lb/in. notch (D256); Hardness (Rockwell): M 120 (D785)
Therm. Prop.: Conduct.: 9 × 10^{-4} Cal/s/cm^2/C/cm (C177); Distort. Temp.: 250 C (@ 264 psi, D648); Coeff. of Linear Exp.: 1.8 × 10^{-5} in./in./C (−30 to 30 C, D696)
Elec. Prop.: Dissip. Factor: 0.025 (dry, 1 MC, D150); Dielec. Str.: 320 V/mil (dry, S/T, D149); Dielec. Const.: 6.0 (dry, 1 MC, D150); Vol. Resist.: 10^{13} ohm-cm (D257)
Ref.: Data sheet

FM 5367. Fiberite Corp.
Chem. Descrip.: Two-stage phenolic comp. reinforced with cellulosic fabric and glass fibers
Category/Applications: Molding comp. with excellent combination of properties from a cost/in.^{3} basis
Form: Macerated; Color: Black
All properties as molded:
Gen'l. Prop.: Sp. Gr.: 1.44–1.46
Mech. Prop.: Tens. Str.: 10,000 psi; Flex. Str.: 21,000 psi; Flex. Mod.: 1.5 × 10^6 psi; Compr. Str.: 30,000 psi; Impact Str.: 4.5 ft lb/in. notch
Therm. Prop.: Distort. Temp.: 410 F (@ 264 psi)
Ref.: Data sheet

FM 6101. Fiberite Corp.
Chem. Descrip.: Two-stage asbestos yarn-reinforced phenolic comp.
Category/Applications: Molding comp. with high impact, heat resistance, good molding properties, arc resistance, wear resistance; used in terminal strips, switch gear, motor starters, circuit breakers, self-lubricating bearings, welding gun tips
Form: Chopped fiber; Color: Natural, black
All properties as molded (½ in. chop):
Gen'l. Prop.: Sp. Gr.: 1.82–1.85 (D792)
Mech. Prop.: Tens. Str.: 8000 psi (D-651); Flex. Str.: 15,000 psi (D790); Flex. Mod.: 2.2 × 10^6 psi (D790); Compr. Str.: 24,000 psi (D695); Impact Str.: 3.5 ft lb/in. notch (D256); Hardness (Rockwell): M 110 (D785)
Therm. Prop.: Conduct.: 15 × 10^{-5} Cal/s/cm^2/C/cm (C177); Distort. Temp.: 183 C (@ 264 psi, D648); Coeff. of Linear Exp.: 1.5 × 10^{-5} in./in./C (−30 to 30 C, D696)
Elec. Prop.: Dissip. Factor: 0.3 (dry, 1 MC, D150); Dielec. Str.: 80 V/mil (dry, S/T, D149); Dielec. Const.: 5.5

(dry, 1 MC, D150); Vol. Resist.: 8 × 10^{12} ohm-cm (D257); Arc Resist.: 180+ s (D495)
Ref.: Data sheet

FM 7676. Fiberite Corp.
Chem. Descrip.: Two-stage cellulose fiber-reinforced phenolic comp.
Category/Applications: See FM 3000
Form: Nodular; Color: Black
All properties as molded:
Gen'l. Prop.: Sp. Gr.: 1.37–1.39 (D792)
Mech. Prop.: Tens. Str.: 6500 psi (D-651); Flex. Str.: 11,000 psi (D790); Flex. Mod.: 10^6 psi (D790); Compr. Str.: 28,000 psi (D695); Impact Str.: 1.1 ft lb/in. notch (D256); Hardness (Rockwell): M 105 (D785)
Therm. Prop.: Conduct.: 9 × 10^{-4} Cal/s/cm^2/C/cm (C177); Distort. Temp.: 160 C (@ 264 psi, D648); Coeff. of Linear Exp.: 2.4 × 10^{-5} in./in./C (−30 to 30 C, D696)
Elec. Prop.: Dissip. Factor: 0.05 (dry, 1 MC, D150); Dielec. Str.: 300 V/mil (dry, S/T, D149); Dielec. Const.: 4.5 (dry, 1 MC, D150); Vol. Resist.: 6 × 10^{12} ohm-cm (D257)
Ref.: Data sheet

FM 7700. Fiberite Corp.
Chem. Descrip.: Two-stage phenolic comp. reinforced with cotton fabric
Category/Applications: See FM 3000; the pellet form permits the use of automatic preforming and molding techniques
Form: Nodular; Color: Black
All properties as molded:
Gen'l. Prop.: Sp. Gr.: 1.37–1.39 (D792)
Mech. Prop.: Tens. Str.: 6100 psi (D-651); Flex. Str.: 12,500 psi (D790); Flex. Mod.: 1.3 × 10^6 psi (D790); Compr. Str.: 24,000 psi (D695); Impact Str.: 1.9 ft lb/in. notch (D256); Hardness (Rockwell): M 105 (D785)
Therm. Prop.: Conduct.: 9 × 10^{-4} Cal/s/cm^2/C/cm (C177); Distort. Temp.: 166 C (@ 264 psi, D648); Coeff. of Linear Exp.: 2 × 10^{-5} in./in./C (−30 to 30 C, D696)
Elec. Prop.: Dissip. Factor: 0.05 (dry, 1 MC, D150); Dielec. Str.: 300 V/mil (dry, S/T, D149); Dielec. Const.: 4.2 (dry, 1 MC, D150); Vol. Resist.: 2 × 10^{13} ohm-cm
Ref.: Data sheet

FM 8130. Fiberite Corp.
Chem. Descrip.: Two-stage glass roving-reinforced phenolic comp.
Category/Applications: See FM 4030-190
Form: Chopped fiber (avail. in chop lengths from ½–1½ in.); Color: Black, natural
All properties as molded (½ in. chop):
Gen'l. Prop.: Sp. Gr.: 1.76–1.80 (D792)
Mech. Prop.: Tens. Str.: 7200 psi (D-651); Flex. Str.: 15,800 psi (D790); Flex. Mod.: 2.3 × 10^6 psi (D790); Compr. Str.: 27,000 psi (D695); Impact Str.: 12 ft lb/in. notch (D256); Hardness (Rockwell): M 115 (D785)
Therm. Prop.: Conduct.: 10 × 10^{-4} Cal/s/cm^2/C/cm (C177); Distort. Temp.: 260 C (@ 264 psi, D648); Coeff. of Linear Exp.: 1.2 × 10^{-5} in./in./C (−30 to 30 C, D696)
Elec. Prop.: Dissip. Factor: 0.02 (dry, 1 MC, D150); Dielec. Str.: 400 V/mil (dry, S/T, D149); Dielec. Const.: 5.0 (dry, 1 MC, D150); Vol. Resist.: 2 × 10^{13} ohm-cm (D257)
Ref.: Data sheet

FM 9294. Fiberite Corp.
Chem. Descrip.: Two-stage cotton cord-reinforced phenolic comp.
Category/Applications: See FM 3000

Form: Chopped strands; Color: Natural, black
All properties as molded:
Gen'l. Prop.: Sp. Gr.: 1.37–1.39 (D792)
Mech. Prop.: Tens. Str.: 9000 psi (D-651); Flex. Str.: 16,000 psi (D790); Flex. Mod.: 1.4×10^6 psi (D790); Compr. Str.: 22,000 psi (D695); Impact Str.: 4.5 ft lb/in. notch (D256); Hardness (Rockwell): M 110 (D785)
Therm. Prop.: Conduct.: 9×10^{-4} Cal/s/cm^2/C/cm (C177); Distort. Temp.: 149 C (@ 264 psi, D648); Coeff. of Linear Exp.: 1.3×10^{-5} in./in./C (−30 to 30 C, D696)
Elec. Prop.: Dissip. Factor: 0.05 (dry, 1 MC, D150); Dielec. Str.: 320 V/mil (dry, S/T, D149); Dielec. Const.: 4.5 (dry, 1 MC, D150); Vol. Resist.: 2×10^{13} ohm-cm (D257)
Ref.: Data sheet

FM 10365. Fiberite Corp.
Chem. Descrip.: Two-stage cellulose-reinforced phenolic comp.
Category/Applications: High impact molding comp.
Form: Macerated; Color: Black, tan
All properties as molded:
Gen'l. Prop.: Sp. Gr.: 1.37–1.39 (D792)
Mech. Prop.: Tens. Str.: 8000 psi (D-638); Flex. Str.: 14,000 psi (D790); Flex. Mod.: 10^6 psi (D790); Compr. Str.: 20,000 psi (D695); Impact Str.: 3.5 ft lb/in. notch (D256)
Therm. Prop.: Distort. Temp.: 300 F (@ 264 psi, D648)
Elec. Prop.: Dielec. Str.: 300 V/mil (dry, S/T, D149)
Ref.: Data sheet

FM 11547, FM 17610. Fiberite Corp.
Chem. Descrip.: Two-stage asbestos yarn-reinforced phenolic comp.
Category/Applications: See FM 6101
Form: Chopped fiber strands; Color: Black and natural, respectively
All properties as molded (½ in. chop):
Gen'l. Prop.: Sp. Gr.: 1.64–1.68 (D792)
Mech. Prop.: Tens. Str.: 9000 psi (D-651); Flex. Str.: 16,000 psi (D790); Flex. Mod.: 2×10^6 psi (D790); Compr. Str.: 23,000 psi (D695); Impact Str.: 3.5 ft lb/in. notch (D256); Hardness (Rockwell): M 110 (D785)
Therm. Prop.: Conduct.: 12×10^{-4} Cal/s/cm^2/C/cm (C177); Distort. Temp.: 177 C (@ 264 psi, D648); Coeff. of Linear Exp.: 1.1×10^{-5} in./in./C (−30 to 30 C, D696)
Elec. Prop.: Dissip. Factor: 0.3 (dry, 1 MC, D150); Dielec. Str.: 60 V/mil (dry, S/T, D149); Dielec. Const.: 4.6 (dry, 1 MC, D150); Vol. Resist.: 8×10^{12} ohm-cm (D257)
Ref.: Data sheet

FM 14111. Fiberite Corp.
Chem. Descrip.: One-stage cellulose-filled phenolic comp.
Category/Applications: Molding comp. with good strength, no ammonia odor or corrosion, chemical and moisture resistance, and exhibiting high surface luster; used in trays, housings, bushings, pulleys, gears, handles, heavy duty machine components, casters, bearings, textile pulleys, and bobbins
Form: Macerated; Color: Tan
All properties as molded:
Gen'l. Prop.: Sp. Gr.: 1.35–1.37 (D792)
Mech. Prop.: Tens. Str.: 6000 psi (D-651); Flex. Str.: 11,000 psi (D790); Flex. Mod.: 900,000 psi (D790); Compr. Str.: 25,000 psi (D695); Impact Str.: 1.5 ft lb/in. notch (D256); Hardness (Rockwell): M 105 (D785)
Therm. Prop.: Conduct.: 8×10^{-5} Cal/s/cm^2/C/cm (C177); Distort. Temp.: 200 C (@ 264 psi, D648); Coeff. of Linear Exp.: 2.1×10^{-5} in./in./C (−30 to 30 C, D696)

Elec. Prop.: Dissip. Factor: 0.06 (dry, 1 MC, D150); Dielec. Str.: 200 V/mil (dry, S/T, D149); Dielec. Const.: 4.8 (dry, 1 MC, D150); Vol. Resist.: 2 × 10^{12} ohm-cm (D257)
Ref.: Data sheet

FM 16771. Fiberite Corp.
Chem. Descrip.: One-stage glass roving-reinforced phenolic comp.
Category/Applications: High impact molding comp.
Form: Avail. in chop lengths from ¼ to 1½ in.
All properties as molded (½ in. chop):
Gen'l. Prop.: Sp. Gr.: 1.83–1.87 (D792)
Mech. Prop.: Tens. Str.: 5000 psi (D-638); Flex. Str.: 18,000 psi (D790); Compr. Str.: 16,000 psi (D695); Impact Str.: 9.0 ft lb/in. notch (D296); Hardness (Rockwell): E 54 (D785)
Therm. Prop.: Distort. Temp.: 600 F (@ 264 psi, D648)
Elec. Prop.: Dissip. Factor: 0.02 (1 MC, D150); Dielec. Str.: 400 V/mil (S/T, D-149); Dielec. Const.: 6.0 (1 MC, D150); Vol. Resist.: 10^{12} ohm-cm (D257)
Ref.: Data sheet

FM 21288. Fiberite Corp.
Chem. Descrip.: One-stage glass roving-reinforced phenolic comp.
Category/Applications: High impact molding material
Form: Avail. in chop lengths from ¼ to 1½ in.
All properties as molded (½ in. chop, partial orientation):
Gen'l. Prop.: Sp. Gr.: 1.85–1.87 (D792)
Mech. Prop.: Tens. Str.: 5000 psi (D-638); Flex. Str.: 15,000 psi (D790); Compr. Str.: 16,000 psi (D695); Impact Str.: 9.0 ft lb/in. notch (D256); Hardness (Rockwell): E 64 (D785)
Therm. Prop.: Distort. Temp.: 600 F (@ 264 psi, D648)
Elec. Prop.: Dissip. Factor: 0.04 (1 MC, D150); Dielec. Str.: 275 V/mil (S/T, D-149); Dielec. Const.: 6.3 (1 MC, D150); Vol. Resist.: 10^{13} ohm-cm (D257)
Ref.: Data sheet

Foam-Kon 20. Liquid Nitrogen Processing Corp.
Category/Applications: Thermoplastic structural foam concentrate for high pressure foam systems; used in injection molding compounds (polyethylene, polypropylene, ABS, SAN, polystyrene, acetal, etc.) processed at melt temps. < 500 F
Ref.: Product data 226-979

Foam-Kon 28. Liquid Nitrogen Processing Corp.
Category/Applications: Thermoplastic structural foam concentrate for high pressure foam systems; for use in nylon, polyphenylene sulfide, polysulfone, and polycarbonate based compounds
Form: Pellet
All properties with Thermocomp PF-1003:
Gen'l. Prop.: Sp. Gr.: 1.08 (D792)
Mech. Prop.: Tens. Str.: 12,000 psi (D-638); Tens. Elong.: 2–3% (D638); Flex. Str.: 18,000 psi (D790); Flex. Mod.: 700,000 psi (D790); Impact Str. (Izod): 0.6 ft lb/in. notch (¼ in. bar, D256)
Therm. Prop.: Distort. Temp.: 390 F (@ 264 psi, D648)
Ref.: Product data 226-979

Foam-Kon 29. Liquid Nitrogen Processing Corp.
Category/Applications: Thermoplastic structural foam concentrate for high pressure foam systems; used in modified polyphenylene oxide and thermoplastic polyester based compounds
Form: Pellets
Ref.: Product data 226-979

Foral 85. Hercules Inc.
Chem. Descrip.: Glycerol ester of highly hydrogenated rosin
Category/Applications: Thermoplastic synthetic resin used as a tackifier resin in adhesives, and in protective and barrier-type coatings that require an exceptional degree of color retention and oxidation resistance
Form: Solid, flake; Color: Pale; Odor: Low
Gen'l. Prop.: Sol.: Sol. in esters, ketones, higher alcohols, glycol ethers, and aliphatic, aromatic, and chlorinated hydrocarbons; insol. in ethanol and water; Sp. Gr.: 1.07; Density: 1.07 kg/l; Soften. Pt. (Drop): 82 C; Acid No.: 9; Stability: Outstanding resistance to oxidation and to discoloration caused by heat and aging
Toxicity/Handling: Flaked form may be prone to lumping or to gradual oxidation; may result in darkening or adverse effect on solubility in organic solvents; use proper storage conditions and strict inventory control
Std. Pkgs.: 223-kg net lightweight metal drums (solid), 22.7-kg net multiwall paper bags (flakes)
Ref.: Product data 7115-10

Foral 105. Hercules Inc.
Chem. Descrip.: Pentaerythritol ester of highly hydrogenated rosin
Category/Applications: See Foral 85; Foral 105 has a substantially higher softening pt. and is indicated where a harder resin may be desirable
Form: Solid, flakes; Color: Pale; Odor: Low
Gen'l. Prop.: Sol.: See Foral 85; Sp. Gr.: 1.07; Density: 1.07 kg/l; Visc. (Gardner-Holdt): C–D (60% sol'n. in min. spirits); Soften. Pt. (Drop): 104 C; Acid No.: 12; Stability: See Foral 85
Toxicity/Handling: See Foral 85
Std. Pkgs.: 224-kg net lightweight metal drums (solid), 22.7-kg net multiwall paper bags (flakes)
Ref.: Product data 7205-8

Foral AX. Hercules Inc.
Chem. Descrip.: Highly hydrogenated wood rosin
Category/Applications: Thermoplastic, acidic resin with excellent resistance to oxidation and exceptionally pale color; used as the tackifier or modifying resin in adhesives and in hot-melt-applied decorative, pressure sensitive, and heat-sealable coatings
Form: Solid; Color: Very pale; Odor: Low
Gen'l. Prop.: Sol.: Sol. in alcohols, esters, ketones, hydrocarbons, chlorinated solvents, and mineral oils; insol. in water; Soften. Pt. (Drop): 75 C; Acid No.: 160; Ref. Index: 1.4955 (100 C); Stability: Excellent heat stability and color retention
Std. Pkgs.: 227-kg net lightweight metal drums
Ref.: Product data 7225-7

Fosta Nylon 438. Amer. Hoechst Corp./Plastics Div.
Chem. Descrip.: Nylon 6
Category/Applications: High flow, fast-setting, injection molding grade resin for operation at very fast cycles in difficult-to-fill molds; NSF-approved for use in molded pipe fittings for potable water
Form: Solid
Gen'l. Prop.: Sp. Gr.: 1.14 (23 C, D792); M.P.: 220 C (D789)
Mech. Prop.: Tens. Str.: 83 MPa (@ yield, D638); Tens. Elong.: 15% (@ fail, D638); Flex. Str.: 110 MPa (@ yield, D-790); Impact Str. (Izod): 37 J/m (D-256); Hardness (Rockwell): R 119 (D-785)
Ref.: Bulletin No. N-20-78

Fosta Nylon 446. Amer. Hoechst Corp./ Plastics Div.
Chem. Descrip.: Nucleated nylon 6
Category/Applications: High viscosity, nucleated nylon for extrusion of tubing, rods, sheet, and film with fast set-up time
Form: Solid
Gen'l. Prop.: Sp. Gr.: 1.13 (23 C, D-792); M.P.: 220 C (D789)
Mech. Prop.: Tens. Str.: 83 MPa (@ yield, D638); Tens. Elong.: 130% (@ fail, D638); Flex. Str.: 114 MPa (@ yield, D790); Impact Str. (Izod): 59 J/m (D256); Hardness (Rockwell): R 119 (D785)
Ref.: Bulletin No. N-21-78

Fosta Nylon 471. Amer. Hoechst Corp./ Plastics Div.
Chem. Descrip.: Nylon 6
Category/Applications: Low viscosity extrusion resin suitable for the production of cast film and for extrusion coating and coextrusion processes; FDA-approved for use in food packaging applications (esp. frozen food and boil-in-bag applications) with proper processing; offers barrier properties to organic vapors and most gases
Form: Solid
Gen'l. Prop.: Sp. Gr.: 1.13 (23 C, D792); M.P.: 220 C (D789); Stability: Outstanding resistance to oils and grease
Mech. Prop.: Tens. Str.: 76 MPa (@ yield, D638); Tens. Elong.: 150% (@ fail, D638); Flex. Str.: 103 MPa (@ yield, D790); Flex. Mod.: 2586 MPa (D-790); Impact Str. (Izod): 48 J/m (D-256); Hardness (Rockwell): R 115 (D-785)
Ref.: Bulletin No. N-22-78

Fosta Nylon 512. Amer. Hoechst Corp./ Plastics Div.
Chem. Descrip.: Nylon 6
Category/Applications: Med. viscosity, highly crystalline, general purpose injection molding resin with good stiffness, and excellent mold release; NSF-approved for pipe fittings used in potable water applications with proper molding techniques
Form: Crystalline
Gen'l. Prop.: Sp. Gr.: 1.14 (23 C, D-792); M.P.: 220 C (D789)
Mech. Prop.: Tens. Str.: 86 MPa (@ yield, D638); Tens. Elong.: 20% (@ fail, D638); Flex. Str.: 117 MPa (@ yield, D790); Impact Str. (Izod): 43 J/m (D256)
Ref.: Bulletin No. N-23-78

Fosta Nylon 523. Amer. Hoechst Corp./ Plastics Div.
Chem. Descrip.: Nylon 6
Category/Applications: Med. viscosity, general purpose, highly crystalline nylon molding resin with excellent flow, fast-setting, and quick mold release properties; used in combs, bearings, bushings, valves, door slides, hardware items, and gears
Form: Small uniform spherulites
Gen'l. Prop.: Sp. Gr.: 1.14 (D792); M.P.: 220 C (D789)
Mech. Prop.: Tens. Str.: 86 MPa (@ yield, D638); Tens. Elong.: 20% (@ fail, D638); Flex. Str.: 110 MPa (@ yield, D-790); Impact Str. (Izod): 43 J/m (D256)
Ref.: Bulletin No. N-35-79

Fosta Nylon 567. Amer. Hoechst Corp./ Plastics Div.
Chem. Descrip.: Plasticized nylon 6
Category/Applications: High impact strength extrusion and molding resin with superior flexibility and toughness; used in diversified fields, incl. injection molded shoe heel lifts, extruded soft-hand fishing line and monofilament
Form: Solid

Gen'l. Prop.: Sp. Gr.: 1.13 (D792); M.P.: 220 C (D789)
Mech. Prop.: Tens. Str.: 48 MPa (@ yield, D638); Tens. Elong.: 300% (@ fail, D638); Flex. Str.: 48 MPa (@ yield, D790); Impact Str. (Izod): 117 J/m (D256)
Ref.: Bulletin No. N-19-78

Fosta Nylon 589. Amer. Hoechst Corp./ Plastics Div.
Chem. Descrip.: Nylon 6 (pure linear polymer)
Category/Applications: Very high melt viscosity and high purity extrusion grade resin; used for the extrusion of heavy cross-sectional shapes, for blown film, tubing, and blow molded articles such as bottles and containers
Form: Solid
Gen'l. Prop.: Sp. Gr.: 1.14 (23 C, D-792); M.P.: 220 C; Stability: High chemical resistance and high resistance to aging
Mech. Prop.: Tens. Str.: 76 MPa (@ yield, D638); Tens. Elong.: 200% (@ fail, D638); Flex. Str.: 103 MPa (@ yield, D-790); Impact Str. (Izod): 64 J/m (D256)
Ref.: Bulletin No. N-25-78

Fosta Nylon 714. Amer. Hoechst Corp./ Plastics Div.
Chem. Descrip.: Plasticized nylon 6 copolymer
Category/Applications: Nylon with outstanding softness and flexibility for extrusion and molding into a variety of specialty products, e.g., molded athletic shoe soles, fishing line, etc.
Form: Solid
Gen'l. Prop.: Sp. Gr.: 1.14 (23 C, D-792); M.P.: 185 C (D789)
Mech. Prop.: Tens. Str.: 41.3 MPa (D-638); Flex. Str.: 18.6 MPa (@ yield, D-790); Flex. Mod.: 345 MPa (D790); Impact Str. (Izod): No break (D256); Hardness (Rockwell): R 56 (D785)
Ref.: Bulletin No. N-38-79

Fosta Nylon 870. Amer. Hoechst Corp./ Plastics Div.
Chem. Descrip.: Heat-stabilized nylon 6
Category/Applications: Heat-stabilized extrusion resin for jacketing wire and cable, incl. thermoplastic-insulated building wires and gasoline-resistant types
Form: Solid
Gen'l. Prop.: Sp. Gr.: 1.13 (23 C, D-792); M.P.: 220 C (D789); Stability: Excellent abrasion resistance and heat stability
Mech. Prop.: Tens. Str.: 79 MPa (@ yield, D638); Tens. Elong.: 150% (@ fail, D638); Flex. Str.: 110 MPa (@ yield, D-790); Impact Str. (Izod): 48 J/m (D256)
Ref.: Bulletin No. N-26-78

Fosta Nylon 1047. Amer. Hoechst Corp./ Plastics Div.
Chem. Descrip.: Heat-stabilized nylon 6
Category/Applications: Heat-stabilized, med. viscosity injection molding resin with excellent toughness and processability; used in problem applications where general purpose molding nylons do not have sufficient strength or stability, and special copolymers are too difficult to process
Form: Solid
Gen'l. Prop.: Sp. Gr.: 1.13 (23 C, D792); M.P.: 220 C (D789)
Mech. Prop.: Tens. Str.: 79 MPa (@ yield, D638); Tens. Elong.: 100% (@ fail, D638); Flex. Str.: 110 MPa (@ yield, D-790); Flex. Mod.: 2586 MPa (D790); Impact Str. (Izod): 48 J/m (D256); Hardness (Rockwell): R 116 (D785)
Ref.: Bulletin No. N-28-78

Fosta Nylon 1210. Amer. Hoechst Corp./ Plastics Div.
Chem. Descrip.: Plasticized nylon 6 homopolymer
Category/Applications: Nylon for use in extrusion in the monofilament fishing line market to produce a good quality fishing line with intermediate softness; also for specialty molding in selected applications
Form: Solid
Gen'l. Prop.: Sp. Gr.: 1.15 (23 C, D-792); M.P.: 218 C (D792)
Mech. Prop.: Tens. Str.: 58.4 MPa (@ yield, D638); Tens. Elong.: 110% (@ fail, D638); Flex. Str.: 51 MPa (@ yield, D790); Flex. Mod.: 1308 MPa (D790); Impact Str. (Izod): 54 J/m (D256); Hardness (Rockwell): R 100 (D785)
Therm. Prop.: Distort. Temp.: 51.5 C (@ 264 psi, D648)
Ref.: Bulletin No. N-37-79

Fosta Nylon 1379. Amer. Hoechst Corp./ Plastics Div.
Chem. Descrip.: Plasticized, heat-stabilized nylon 6
Category/Applications: High impact strength, heat-stabilized extrusion resin used by the wire and cable industry for protective jacketing (e.g., thermoplastic-insulated wire), esp. where exceptional flexibility is required; used in applications involving exposure to elevated temps.; also as molding resin
Form: Solid
Gen'l. Prop.: Sp. Gr.: 1.13 (D792); M.P.: 220 C (D789)
Mech. Prop.: Tens. Str.: 48 MPa (@ yield, D638); Tens. Elong.: 300% (@ fail, D638); Flex. Str.: 48 MPa (@ yield, D-790); Flex. Mod.: 1034 MPa (D790); Impact Str. (Izod): 117 J/m (D256)
Ref.: Bulletin No. N-31-78

Fosta Nylon 1417. Amer. Hoechst Corp./ Plastics Div.
Chem. Descrip.: Impact-modified nylon 6 resin
Category/Applications: Med. high impact resin designed for applications where notch sensitivity of standard nylons produces inconsistent part performance; avail. in colored and heat stabilized versions
Form: Solid
Gen'l. Prop.: Tens. Str.: 62 MPa (@ yield, D638); Tens. Elong.: 150% (@ fail, D638); Flex. Str.: 69 MPa (@ yield, D-790); Flex. Mod.: 1896 MPa (D790); Impact Str. (Izod): 160 J/m (D256); Hardness (Rockwell): R 113 (D785)
Therm. Prop.: Distort. Temp.: 56 C (@ 264 psi, D648)
Ref.: Bulletin No. N-34-79

Fosta Nylon 1525. Amer. Hoeshst Corp./ Plastics Div.
Chem. Descrip.: Plasticized, heat-stabilized nylon 6
Category/Applications: High viscosity, heat-stabilized extrusion nylon for wire and cable industry; used for protective jacketing requiring exceptional clarity and flexibility (e.g., thermoplastic-insulated wire, oil resistant at 75 C, and gasoline-resistant wire); applications involving exposure to elevated temps. and outdoor environments for extended periods; designed to facilitate high speed coating
Form: Solid
Gen'l. Prop.: Sp. Gr.: 1.13 (23 C, D792); M.P.: 220 C (D789)
Mech. Prop.: Tens. Str.: 70.3 MPa (D-638); Tens. Elong.: 300+% (@ fail, D-638); Flex. Str.: 41.3 MPa (@ yield, D-790); Impact Str. (Izod): 162 J/m (D-256); Hardness (Rockwell): R 90 (D-785)

Therm. Prop.: Distort. Temp.: 51.5 C (@ 264 psi, D648)
Ref.: Bulletin No. N-36-79

Fosta Nylon 1693. Amer. Hoechst Corp./ Plastics Div.
Chem. Descrip.: High impact-modified nylon 6 resin
Category/Applications: High impact strength molding resin; also avail. in colored and heat stabilized versions
Form: Solid
Gen'l. Prop.: Sp. Gr.: 1.11 (23 C, D792); M.P.: 220 C (D792)
Mech. Prop.: Tens. Str.: 48 MPa (@ yield, D638); Tens. Elong.: 200% (@ fail, D638); Flex. Str.: 55 MPa (@ yield, D-790); Flex. Mod.: 1480 MPa (D790); Impact Str. (Izod): 1080 J/m (D256); Hardness (Rockwell): R 102 (D785)
Therm. Prop.: Distort. Temp.: 50 C (@ 264 psi, D648)
Ref.: Bulletin No. N-33-79

Fosta Nylon 1722. Amer. Hoechst Corp./ Plastics Div.
Chem. Descrip.: High impact-modified nylon 6 resin
Category/Applications: See Fosta Nylon 1693
Form: Solid
Gen'l. Prop.: Sp. Gr.: 1.11 (23 C, D-792); M.P.: 220 C (D792)
Mech. Prop.: Tens. Str.: 62 MPa (@ yield, D638); Tens. Elong.: 200% (@ fail, D638); Flex. Str.: 71.6 MPa (@ yield, D-790); Flex. Mod.: 1893 MPa (D790); Impact Str. (Izod): 1080 J/m (D256); Hardness (Rockwell): R 110 (D785)
Therm. Prop.: Distort. Temp.: 57 C (@ 264 psi, D648)
Ref.: Bulletin No. N-32-79

Fosta Tuf-Flex 240. Amer. Hoechst Corp./Plastics Div.
Chem. Descrip.: Rubber-modified high impact polystyrene
Category/Applications: High-impact resin with outstanding shock resistance at ambient and low temps., good gloss characteristics, and good thermal properties; used for various refrigerator parts and other appliances
Form: Solid
Gen'l. Prop.: Sp. Gr.: 1.04 (D792); Soften. Pt. (Vicat): 101 C (D1525)
Mech. Prop.: Tens. Str.: 28 MPa (D-638); Tens. Elong.: 40% (D638); Tens. Mod.: 2140 MPa (D638); Impact Str.: 107 J/m notch (injection molded, D-256); Hardness (Rockwell): L 55 (D-785)
Therm. Prop.: Distort. Temp.: 82 C (@ 264 psi, D648)
Ref.: Bulletin No. I-38-79

Fosta Tuf-Flex 271. Amer. Hoechst Corp./Plastics Div.
Chem. Descrip.: Rubber-modified polystyrene
Category/Applications: Med. impact polystyrene with outstanding thermal resistance and high gloss for easy processing on ram and screw injection molding equipment; used for designing furniture, parts, utensils, and other items in a variety of shapes where very fast set up times are required
Form: Solid
Gen'l. Prop.: Sp. Gr.: 1.04 (D792); Soften. Pt. (Vicat): 104 C (D1525)
Mech. Prop.: Tens. Str.: 41 MPa (D638); Tens. Elong.: 20% (D638); Tens. Mod.: 2415 MPa (D638); Impact Str.: 59 J/m notch (injection molded, D256); Hardness (Rockwell): M 45 (D785)
Therm. Prop.: Distort. Temp.: 93 C (@ 264 psi, D648)
Ref.: Bulletin No. I-25-78

Fosta Tuf-Flex 283. Amer. Hoechst Corp./Plastics Div.
Chem. Descrip.: Rubber-modified polystyrene
Category/Applications: High heat, high impact polymer for production of parts with exceptional gloss characteristics, e.g., illuminated signs and displays, small appliance housings, and applications requiring vacuum metalizing
Form: Solid
Gen'l. Prop.: Sp. Gr.: 1.04 (D792); Soften. Pt. (Vicat): 100 C (D1525)
Mech. Prop.: Tens. Str.: 31 MPa (D638); Tens. Elong.: 35% (D638); Tens. Mod.: 2280 MPa (D638); Impact Str.: 78 J/m notch (injection molded, D256); Hardness (Rockwell): L 75 (D785)
Therm. Prop.: Distort. Temp.: 81 C (@ 264 psi, D648)
Ref.: Bulletin No. I-39-79

Fosta Tuf-Flex 326. Amer. Hoechst Corp./Plastics Div.
Chem. Descrip.: Polystyrene
Category/Applications: Med. impact, high-flow polystyrene with high speed moldability, good strength, and good gloss characteristics; used primarily for the food packaging industry, and in cameras, housewares, etc. where toughness is required
Form: Solid
Gen'l. Prop.: Sp. Gr.: 1.04 (D792); Soften. Pt. (Vicat): 91 C (D1525)
Mech. Prop.: Tens. Str.: 35 MPa (D-638); Tens. Elong.: 25% (D638); Tens. Mod.: 2415 MPa (D638); Impact Str.: 43 J/m notch (injection molded, D256); Hardness (Rockwell): M 45 (D785)
Therm. Prop.: Distort. Temp.: 74 C (@ 264 psi, D648)
Ref.: Bulletin No. I-27-78

Fosta Tuf-Flex 329. Amer. Hoechst Corp./Plastics Div.
Chem. Descrip.: Polystyrene
Category/Applications: High impact extrusion or injection molding material with high elongation, deep draw characteristics, superior low temp. properties; used for packaging applications, and for furniture and toys due to its excellent nail and screw retention and ability to take normal wood finishing operations
Form: Solid
Gen'l. Prop.: Sp. Gr.: 1.04 (D792); Soften. Pt. (Vicat): 100 C (D1525)
Mech. Prop.: Tens. Str.: 26 MPa (D-638); Tens. Elong.: 50% (D638); Tens. Mod.: 2000 MPa (D638); Impact Str.: 107 J/m notch (injection molded, D-256); Hardness (Rockwell): L 65 (D785)
Therm. Prop.: Distort. Temp.: 85 C (@ 264 psi, D648)
Ref.: Bulletin No. I-28-78

Fosta Tuf-Flex 474. Amer. Hoechst Corp./Plastics Div.
Chem. Descrip.: Rubber-modified polystyrene
Category/Applications: High-heat, high-impact polymer with high thermal properties, outstanding shock resistance at ambient and low temps., and good moldability; used for small appliances, household items, radio and TV consoles and accessories, and interior parts for the automotive industry
Form: Solid
Gen'l. Prop.: Sp. Gr.: 1.04 (D792); Soften. Pt. (Vicat): 103 C (D1525)
Mech. Prop.: Tens. Str.: 35 MPa (D638); Tens. Elong.: 35% (D638); Tens. Mod.: 2140 MPa (D638); Impact Str.: 80 J/m notch (injection molded, D256); Hardness (Rockwell): L 75 (D785)

Therm. Prop.: Distort. Temp.: 91 C (@ 264 psi, D648)
Ref.: Bulletin No. I-29-78

Fosta Tuf-Flex 702. Amer. Hoechst Corp./Plastics Div.
Chem. Descrip.: Polystyrene containing flame retardant additives
Category/Applications: High impact polystyrene containing chemicals to increase resistance to ignition in laboratory flame tests; used in TV cabinets, appliances, business machine housings, and various electrical applications
Form: Solid
Gen'l. Prop.: Soften. Pt. (Vicat): 99 C (D1525); Stability: May discolor when exposed to uv light
Mech. Prop.: Tens. Str.: 4000 psi (@ yield, D638); Tens. Elong.: 30% (@ break, D638); Tens. Mod.: 300,000 psi (D638); Flex. Str.: 5900 psi (D790/B); Impact Str. (Izod): 1.9 ft lb/in. (D256); Hardness (Rockwell): L 55 (D785)
Therm. Prop.: Distort. Temp.: 83 C (@ 264 psi, D648); Flamm.: V-O (0.058 in., UL94)
Toxicity/Handling: May release hazardous or toxic vapors on prolonged exposure to high heat—use with adequate ventilation; avoid melt temps. > 500 F to prevent resin degradation and possible machine damage through corrosion and/or high internal pressure
Ref.: Bulletin No. I-41-79

Fosta Tuf-Flex 707. Amer. Hoechst Corp./Plastics Div.
Chem. Descrip.: Polystyrene containing flame retardant additives
Category/Applications: See Fosta Tuf-Flex 702
Form: Solid
Gen'l. Prop.: Soften. Pt. (Vicat): 95 C (D1525); Stability: May discolor when exposed to uv light
Mech. Prop.: Tens. Str.: 3100 psi (@ yield, D638); Tens. Elong.: 50% (@ break, D638); Tens. Mod.: 240,000 psi (D638); Impact Str. (Izod): 3.2 ft lb/in. (D256); Hardness (Rockwell): L 38 (D-785)
Therm. Prop.: Distort. Temp.: 80 C (@ 264 psi, D648)
Toxicity/Handling: May release hazardous or toxic vapors on prolonged exposure to high heat—use with adequate ventilation; avoid melt temps. > 475 F to prevent resin degradation and possible machine damage through corrosion and/or high internal pressure
Ref.: Bulletin No. I-42-79

Fosta Tuf-Flex 717. Amer. Hoechst Corp./Plastics Div.
Chem. Descrip.: Polystyrene
Category/Applications: High flow, high impact polystyrene for injection molding applications, esp. for molding thin walled items; also for structural foam molding
Form: Solid
Gen'l. Prop.: Sp. Gr.: 1.04 (D792); Soften. Pt. (Vicat): 97 C (D1525)
Mech. Prop.: Tens. Str.: 24 MPa (D-638); Tens. Elong.: 40% (D638); Tens. Mod.: 1960 MPa (D638); Impact Str.: 75 J/m notch (injection molded, D256); Hardness (Rockwell): L 65 (D785)
Therm. Prop.: Distort. Temp.: 85 C (@ 264 psi, D648)
Ref.: Bulletin I-34-78

Fosta Tuf-Flex 721-M. Amer. Hoechst Corp./Plastics Div.
Chem. Descrip.: Rubber-modified polystyrene
Category/Applications: High impact polymer with good thermal properties,

shock resistance at ambient and low temps., and good moldability; used in furniture industry, automotive industry, and for small appliance housings
Form: Solid
Gen'l. Prop.: Sp. Gr.: 1.04 (D792); Soften. Pt. (Vicat): 101 C (D1525)
Mech. Prop.: Tens. Str.: 35 MPa (D-638); Tens. Elong.: 35% (D638); Tens. Mod.: 2140 MPa (D638); Impact Str.: 80 J/m notch (injection molded, D256); Hardness (Rockwell): L 75 (D785)
Therm. Prop.: Distort. Temp.: 88 C (@ 264 psi, D648)
Ref.: Bulletin No. I-30-78

Fosta Tuf-Flex 730. Amer. Hoechst Corp./Plastics Div.
Chem. Descrip.: Low monomer content polystyrene
Category/Applications: High impact resin for extrusion and thermoforming applications; used in food and dairy packaging, profiles, lamp shades, containers, refrigerator door liners and trays, hot and cold drink cups and lids, and other packaging applications
Form: Solid
Gen'l. Prop.: Sp. Gr.: 1.04 (D792); Soften. Pt. (Vicat): 101 C (D1525)
Mech. Prop.: Tens. Str.: 28 MPa (D-638); Tens. Elong.: 50% (D638); Tens. Mod.: 1800 MPa (D638); Impact Str.: 92 J/m notch (injection molded, D256); Hardness (Rockwell): L 56 (D785)
Therm. Prop.: Distort. Temp.: 84 C (@ 264 psi, D648)
Ref.: Bulletin No. I-36-79

Fosta Tuf-Flex 742. Amer. Hoechst Corp./Plastics Div.
Chem. Descrip.: Rubber-modified polystyrene
Category/Applications: High impact polystyrene with good gloss, toughness, and easy processability; used in toys, air conditioner and fan grills, etc.
Form: Solid
Gen'l. Prop.: Sp. Gr.: 1.04 (D792); Soften. Pt. (Vicat): 96 C (D1525)
Mech. Prop.: Tens. Str.: 26 MPa (D-638); Tens. Elong.: 50% (D638); Tens. Mod.: 2070 MPa (D638); Impact Str.: 107 J/m notch (injection molded, D-256); Hardness (Rockwell): L 60 (D785)
Therm. Prop.: Distort. Temp.: 81 C (@ 264 psi, D648)
Ref.: Bulletin No. I-40-79

Fosta Tuf-Flex 782. Amer. Hoechst Corp./Plastics Div.
Chem. Descrip.: Polystyrene
Category/Applications: High impact resin with toughness, good flowability, gloss, excellent nail and screw retention, and the ability to take normal wood finishing operations; used for furniture applications
Form: Solid
Gen'l. Prop.: Sp. Gr.: 1.04 (D792); Soften. Pt. (Vicat): 97 C (D1525)
Mech. Prop.: Tens. Str.: 29 MPa (D-638); Tens. Elong.: 45% (D638); Tens. Mod.: 2140 MPa (D638); Impact Str.: 69 J/m notch (injection molded, D256); Hardness (Rockwell): L 75 (D785)
Therm. Prop.: Distort. Temp.: 85 C (@ 264 psi, D648)
Ref.: Bulletin No. I-37-79

Fosta Tuf-Flex 840. Amer. Hoechst Corp./Plastics Div.
Chem. Descrip.: Polystyrene
Category/Applications: Extra high impact polystyrene for sheet extrusion applications where the thermoformed items require deep draw, excellent surface appearance, and extra strength
Form: Solid
Gen'l. Prop.: Sp. Gr.: 1.04 (D792); Soften. Pt. (Vicat): 101 C (D1525)
Mech. Prop.: Tens. Str.: 25 MPa (D-

638); Tens. Elong.: 50% (@ break, D-638); Tens. Mod.: 1930 MPa (D638); Impact Str.: 161 J/m (injection molded, D256); Hardness (Rockwell): L 50 (D-785)
Therm. Prop.: Distort. Temp.: 88 C (@ 264 psi, D648)
Ref.: Bulletin No. I-35-79

Fosta Tuf-Flex 880. Amer. Hoechst Corp./Plastics Div.
Chem. Descrip.: Polystyrene
Category/Applications: Extra high impact polystyrene for sheet extrusion and critical thermoforming applications, incl. internal packaging articles and cottage cheese tubs
Form: Solid
Gen'l. Prop.: Sp. Gr.: 1.04 (D792); Soften. Pt. (Vicat): 97 C (D1525)
Mech. Prop.: Tens. Str.: 21 MPa (D-638); Tens. Elong.: 50% (D638); Tens. Mod.: 1800 MPa (D638); Impact Str.: 161 J/m (injection molded, D256); Hardness (Rockwell): L 50 (D785)
Therm. Prop.: Distort. Temp.: 82 C (@ 264 psi, D648)
Ref.: Bulletin No. I-23-78

Fosta Tuf-Flex 929. Amer. Hoechst Corp./Plastics Div.
Chem. Descrip.: Polystyrene
Category/Applications: High flow, fast set, med. impact material for use in housewares and packaging applications; esp. suitable for thin wall sections; also used for structural foam molding
Form: Solid
Gen'l. Prop.: Sp. Gr.: 1.04 (D792); Soften. Pt. (Vicat): 92 C (D1525)
Mech. Prop.: Tens. Str.: 28 MPa (D-638); Tens. Elong.: 25% (D638); Tens. Mod.: 2346 MPa (D638); Impact Str.: 51 J/m notch (injection molded, D-256); Hardness (Rockwell): M 40 (D-785)
Therm. Prop.: Distort. Temp.: 74 C (@ 264 psi, D648)
Ref.: Bulletin No. I-26-78

Fostafoam Type 3775. Amer. Hoechst Corp./Plastics Div.
Chem. Descrip.: Expandable polystyrene beads
Category/Applications: May be pre-expanded to ≈ 30 times their original size, are easily molded with fast cycles into blocks or billets, and exhibit excellent dimensional stability; molded blocks may be saw-cut or hot-wire cut for use as flotation billets or in other specialized applications, also for use as doll or beanbag filler material
Form: Beads
All properties as molded, at ≈ 1.0 lb/ft^3 density:
Mech. Prop.: Tens. Str.: 25–35 psi (D-1623); Flex. Str.: 25–35 psi (C203); Compr. Str.: 8–13 psi (@ 5% deformation, D1621)
Therm. Prop.: Conduct.: 0.26 Btu/h/ft^2/F/in. (@ 75 F, C177); Coeff. of Linear Exp.: 3.5×10^{-5} in./in./F (D-696)
Ref.: Bulletin No. F-2-78

Fostafoam Type 4775. Amer. Hoechst Corp./Plastics Div.
Chem. Descrip.: Expandable polystyrene beads
Category/Applications: Med.-size beads which can be preexpanded to ≈ 30 times their original size, are easily molded with fast cycles and superior gloss into a variety of shapes, and exhibit excellent dimensional stability; used for custom packaging for such items as instruments and glassware, and for molding toys, ice buckets, etc.
Form: Beads
All properties as molded, @ ≈ 1.5 lb/ft^3 density:

Mech. Prop.: Tens. Str.: 40–60 psi (D-1623); Flex. Str.: 40–60 psi (C203); Compr. Str.: 18–22 psi (@ 5% deformation, D1621)
Therm. Prop.: Conduct.: 0.25 Btu/h/ft^2/F/in. (@ 75 F, C177); Coeff. of Linear Exp.: 3.5 X 10^{-5} in./in./F (D-696)
Ref.: Bulletin No. F-1-78

Fostafoam Type 5775. Amer. Hoechst Corp./Plastics Div.
Chem. Descrip.: Expandable polystyrene beads
Category/Applications: See Fostafoam Type 4775; Type 5775 is the smallest bead size, and is designed to alleviate fill problems in molds with thin sections, sharp corners, or intricate designs; exhibits outstanding cycle advantages in thick sections
Form: Beads
All properties as molded, @ ≈ 1.5 lb/ft^3 density:
Mech. Prop.: Tens. Str.: 40–60 psi (D-1623); Flex. Str.: 40–60 psi (C203); Compr. Str.: 18–22 psi (@ 5% deformation, D1621)
Therm. Prop.: Conduct.: 0.25 Btu/h/ft^2/F/in. (@ 75 F, C177); Coeff. of Linear Exp.: 3.5 X 10^{-5} in./in./F (D-696)
Ref.: Bulletin F-3-78

Fostalite. Amer. Hoechst Corp./Plastics Div.
Chem. Descrip.: Light stable polystyrene
Category/Applications: Light stable polystyrene specially formulated for fluorescent lighting louvers and panels, with good resistance to yellowing, dimensional stability, light weight, and good processability
Form: Pellets
Gen'l. Prop.: Sp. Gr.: 1.05 (D792); Soften. Pt. (Vicat): 106 C (D1525)
Mech. Prop.: Tens. Str.: 52 MPa (@ break, D638); Tens. Elong.: 2.5% (@ break, D638); Tens. Mod.: 3100 MPa (D-638); Flex. Str.: 86 MPa (D790); Impact Str. (Izod): 21 J/m (D256); Hardness (Rockwell): M 75 (D785)
Therm. Prop.: Distort. Temp.: 91 C (@ 264 psi, D648)
Elec. Prop.: Dissip. Factor: 0.0003 ($60 \cdot 10^6$ Hz, D150); Dielec. Str.: 22.6 KV/mm (S/T, D149); Dielec. Const.: 2.6 ($60 \cdot 10^6$ Hz, D150); Vol. Resist.: > 10^{14} ohm-cm (D257)
Toxicity/Handling: Product does not absorb moisture; however, store in a cool, dry place; should moisture be found it may be removed by drying in an air-circulating oven for 1–2 h @ 160 F
Ref.: Bulletin SLS-4-79

Fostarene 50. Amer. Hoechst Corp./Plastics Div.
Chem. Descrip.: High heat polystyrene
Category/Applications: Molding material with high heat resistance, high strength properties, and good clarity; used for film, monofilaments, battery cases, battery caps, radio cabinets, cut glass housewares, trays, various medical disposable items
Form: Solid; Color: Clear and colors
Gen'l. Prop.: Sp. Gr.: 1.05 g/cc (D792); Soften. Pt. (Vicat): 107 C (D1525)
Mech. Prop.: Tens. Str.: 52 MPa (D638); Tens. Elong.: 2.5% (@ break, D638); Tens. Mod.: 3100 MPa (D638); Impact Str.: 21 J/m notch (injection molded, D256); Hardness (Rockwell): M 75 (D-785)
Therm. Prop.: Distort. Temp.: 93 C (@ 264 psi, D648)
Ref.: Bulletin No. S-11-79

Fostarene 57. Amer. Hoechst Corp./Plastics Div.
Chem. Descrip.: HIgh heat polystyrene
Category/Applications: High m.w. crystal polystyrene with flow and strength properties designed to optimize its use in extruded foam sheet; used in meat and poultry trays, egg cartons, and fast food packaging items
Form: Crystalline
Gen'l. Prop.: Sp. Gr.: 1.05 g/cc (D792); Soften. Pt. (Vicat): 107 C (D1525)
Mech. Prop.: Tens. Str.: 53 MPa (D638); Tens. Elong.: 2.5% (@ break, D638); Tens. Mod.: 3100 MPa (D638); Impact Str.: 21 J/m (injection molded, D256); Hardness (Rockwell): M 75 (D785)
Therm. Prop.: Distort. Temp.: 93 C (@ 264 psi, D648)
Ref.: Bulletin No. S-10-79

Fostarene 58. Amer. Hoechst Corp./Plastics Div.
Chem. Descrip.: High heat polystyrene
Category/Applications: High m.w. polystyrene designed for extrusion applications where extra strength is required; used in oriented film extrusion; foam extrusion for use in meat tray, egg carton, and other packaging requiring toughness; injection molding for disposable cutlery and other various sectioned parts; good clarity
Form: Solid
Gen'l. Prop.: Sp. Gr.: 1.05 g/cc (D792); Soften. Pt. (Vicat): 107 C (D1525)
Mech. Prop.: Tens. Str.: 53 MPa (D638); Tens. Elong.: 2.5% (@ break, D638); Tens. Mod.: 3100 MPa (D638); Impact Str.: 21 J/m (injection molded, D256); Hardness (Rockwell): M 75 (D785)
Therm. Prop.: Distort. Temp.: 93 C (@ 264 psi, D648)
Ref.: Bulletin No. S-9-78

Fostarene 817. Amer. Hoechst Corp./Plastics Div.
Chem. Descrip.: High flow polystyrene
Category/Applications: Molding material with easy flow and fast set properties, esp. suitable for thin section applications and difficult mold design; used in containers, wall tile, thin section housewares, thin section packaging, novelties, vials, disposable tumblers, and various disposable medical components
Form: Solid
Gen'l. Prop.: Sp. Gr.: 1.05 (D792); Soften. Pt. (Vicat): 92 C (D1525)
Mech. Prop.: Tens. Str.: 39 MPa (D638); Tens. Elong.: 1.5% (@ break, D638); Tens. Mod.: 2760 MPa (D638); Impact Str.: 19 J/m notch (injection molded, D256); Hardness (Rockwell): M 75 (D-785)
Therm. Prop.: Distort. Temp.: 79 C (@ 264 psi, D648)
Ref.: Bulletin No. S-12-79

FPC 450. Firestone Plastics Co.
Chem. Descrip.: Vinyl chloride copolymer
Category/Applications: Intermediate m.w. resin for use in solution resin applications; compatible with vinyl plasticizers, pigments, and stabilizers; used in strippable coatings, printing inks, and protective coatings
Form: Powder
Gen'l. Prop.: Sol.: Sol. in ketones; will tolerate aromatic hydrocarbons; Sp. Gr.: 1.36; Visc. (Relative): 1.85 (1% in cyclohexanone); Stability: Film is resistant to moisture, grease, chemicals, etc.
Ref.: Data sheet

FPC 454. Firestone Plastics Co.
Chem. Descrip.: Vinyl chloride copolymer resin
Category/Applications: Very low m.w.

resin for solution application; adheres to FPC 470 primed surfaces and to porous substrates, e.g., paper and cloth; used for chemical resistant, flexible coatings, etc.; film is nonflammable, transparent, odorless, tasteless, and resistant to water, moisture vapor, chemicals, oils, greases, etc.

Form: Powder

Gen'l. Prop.: Sol.: Excellent sol. with resultant high solids in vinyl type solvents (ketones, esters, chlorinated hydrocarbons); tolerates a substantial quantity of toluene and/or xylene, or small quantity of aliphatic diluents; Sp. Gr.: 1.36; Visc. (Relative): 1.45 (1% in cyclohexanone)

Ref.: Data sheet

FPC 461. Firestone Plastics Co.

Chem. Descrip.: Fluorine-containing vinyl chloride copolymer

Category/Applications: Vinyl solution resin with excellent heat and light stability; can also be used as a dry processing material (e.g., molding, extrusion, and calendering); not compatible with other solution vinyl copolymer resins

Gen'l. Prop.: Sol.: Sol. in toluene, xylene, chlorinated hydrocarbons, ketones, esters; Sp. Gr.: 1.70; Visc. (Relative): 1.50

Ref.: Data sheet

FPC 470. Firestone Plastics Co.

Chem. Descrip.: Vinyl chloride terpolymer

Category/Applications: Solution resin with alkyd and extender resin compatibility and excellent adhesion to metal, vinyl, and alkyd-type surfaces; used in coatings, e.g., as top coat over old paint surfaces, vinyl floor coverings, wash primers, and conventional alkyd-type primers

Form: Powder

Gen'l. Prop.: Sol.: Sol. in ketones, esters, chlorinated hydrocarbons; tolerates toluene and xylene; Sp. Gr.: 1.31; Visc. (Relative): 1.35 (1% in cyclohexanone)

Ref.: Data sheet

FPC 471. Firestone Plastics Co.

Chem. Descrip.: Vinyl chloride copolymer

Category/Applications: Solution resin for coating applications requiring high vinyl resin solids; forms rigid or flexible films with good resistance to water, moisture vapor, chemicals, and weather; good adhesion to porous surfaces, e.g., paper, cloth, and adhesion to FPC 470-primed metal surfaces

Form: Powder

Gen'l. Prop.: Sol.: Sol. in ketones, esters; sol. in toluene or xylene with heat and high shear mixing; limited tolerance of aliphatic hydrocarbons and alcohols; Sp. Gr.: 1.31; Visc. (Relative): 1.45 (1% in cyclohexanone); Stability: Film has excellent light stability

Ref.: Data sheet

FPC 481. Firestone Plastics Co.

Chem. Descrip.: Vinyl chloride copolymer

Category/Applications: High m.w. solution resin for printing inks, strippable coatings, protective coatings, and decorative coatings; compatible with plasticizers, stabilizers, and all pigment types; films have good adhesion to vinyl surfaces and porous substrates, and high tensile strength and elongation

Form: Powder

Gen'l. Prop.: Sol.: Sol. in ketones; tolerates moderate proportions of aromatic solvents when ketones present; Sp. Gr.: 1.35; Visc. (Relative): 1.92 (1% cyclohexanone); Stability: Films are resistant to water, moisture, grease, chemicals

Ref.: Data sheet

FPC 497. Firestone Plastics Co.
Chem. Descrip.: Vinyl chloride copolymer
Category/Applications: Med. m.w. solution resin used in vinyl paints; adheres to porous surfaces and to primed metal surfaces
Form: Powder
Gen'l. Prop.: Sol.: Sol. in ketones; tolerates xylene and toluene; Sp. Gr.: 1.36; Visc. (Relative): 1.55 (1% in cyclohexanone); Stability: Film has resistance to water, moisture vapor, chemicals, grease
Ref.: Data sheet

FPC 605. Firestone Plastics Co.
Chem. Descrip.: PVC resin
Category/Applications: Low m.w. general purpose dispersion resin used in roll coating, spraying, strand coating, dip coating, slush molding, rotation casting (molding), cellular vinyls, sealants, and organosols; also used in cast cellular vinyls using azo type chemical blowing agents
Gen'l. Prop.: Sp. Gr.: 1.4; Visc. (Brookfield): 30–60
Mech. Prop.: Tens. Str.: 350 psi (ultimate, when fused at 120 C, D412)
Elec. Prop.: Vol. Resist.: 6.6×10^9 ohm-cm (@ R.T., D257-66)
Ref.: Data sheet

FPC 654. Firestone Plastics Co.
Chem. Descrip.: High m.w. PVC resin
Category/Applications: General purpose dispersion resin used in strand coating, dip coating, slush molding, rotational casting (molding), sealants, and organosols
Gen'l. Prop.: Sp. Gr.: 1.4; Visc. (Brookfield): 30–60
Mech. Prop.: Tens. Str.: 450 psi (ultimate, when fused at 120 C, D412)
Elec. Prop.: Vol. Resist.: 7.6×10^9 ohm-cm (@ R.T., D257-66)
Ref.: Data sheet

FPC 654H. Firestone Plastics Co.
Chem. Descrip.: High m.w. PVC resin
Category/Applications: General purpose dispersion resin used in roll coating, spraying, strand coating, dip coating, slush molding, rotational casting (molding), cellular vinyls, sealants, and organosols; fast gelling time
Gen'l. Prop.: Sp. Gr.: 1.4; Visc. (Brookfield): 30–60
Mech. Prop.: Tens. Str.: 450 psi (ultimate, when fused at 120 C, D412)
Elec. Prop.: Vol. Resist.: 7.6×10^9 *ohm-cm (@*
Ref.: Data sheet

FPC 965. Firestone Plastics Co.
Chem. Descrip.: Extremely low m.w. PVC homopolymer resin
Category/Applications: Used as a modifying resin with homopolymer and copolymer resins; suitable for operations having a minimum of energy available to generate fusion
Form: Particles
Gen'l. Prop.: Sp. Gr.: 1.40 (D792); Density (Bulk): 0.60 g/cc (D1895/B); Visc. (Inherent): 0.52 (D1243)
Ref.: Data sheet

FPC 1238. Firestone Plastics Co.
Chem. Descrip.: PVC comp.
Category/Applications: General purpose extrusion comp. meeting requirements for 70 durometer weatherstrip; used for weatherstripping and profile extrusion
Form: Opaque, filled solid; Color: Natural, black, white, gray
Gen'l. Prop.: Sp. Gr.: 1.37 (D792-66)
Mech. Prop.: Tens. Str.: 1600 psi (D-412-68); Tens. Elong.: 360% (ultimate,

D412-68); Tens. Mod.: 730 psi (@ 100% elongation, D412-68); Hardness (Shore Durometer): 71A (instantaneous, D-2240-68)
Ref.: Data sheet

FPC 1239. Firestone Plastics Co.
Chem. Descrip.: PVC comp.
Category/Applications: General purpose extrusion comp. meeting requirements for 80 durometer weatherstrip; used for weatherstripping and profile extrusion
Form: Opaque, filled solid; Color: Natural, black, white, gray
Gen'l. Prop.: Sp. Gr.: 1.39 (D792-66)
Mech. Prop.: Tens. Str.: 1870 psi (D-412-68); Tens. Elong.: 310% (ultimate, D412-68); Tens. Mod.: 840 psi (@ 100% elong., D412-68); Hardness (Shore Durometer): 77A (instantaneous, D2240-68)
Ref.: Data sheet

FPC 1240. Firestone Plastics Co.
Chem. Descrip.: PVC comp.
Category/Applications: Used for weatherstripping and general purpose profile extrusion
Gen'l. Prop.: Sp. Gr.: 1.40 (D792)
Mech. Prop.: Tens. Prop. (75 mil plaque, Die "C"): Tens. Str.: 2400 psi (D412); Tens. Elong.: 280% (ultimate, D412); Tens. Mod.: 1370 psi (@ 100% elong., D412); Hardness (Shore Durometer): 89A (instantaneous, D2240)
Ref.: Data sheet

FPC 1265C. Firestone Plastics Co.
Chem. Descrip.: PVC comp.
Category/Applications: Extrusion comp. for exterior fence coating
Form: Filled, opaque solid; Color: Natural
All properties—75 mil plaque:
Gen'l. Prop.: Sp. Gr.: 1.31 (D792-66)
Mech. Prop.: Tens. Str.: 3140 psi (D-412-68); Tens. Elong.: 240% (D412-68); Tens. Mod.: 2420 psi (@ 100% elong., D412-68); Hardness (Shore Durometer): 94A (instantaneous, D2240-68)
Ref.: Data sheet

FPC 1359. Firestone Plastics Co.
Chem. Descrip.: PVC comp.
Category/Applications: Profile extrusion comp.
Form: Filled, opaque solid; Color: Natural, black, gray
All properties—75 mil plaque:
Gen'l. Prop.: Sp. Gr.: 1.35 (D792-66)
Mech. Prop.: Tens. Str.: 1070 psi (D-412-68); Tens. Elong.: 250% (ultimate, D412-68); Tens. Mod.: 520 psi (@ 100% elong., D412-68); Hardness (Shore Durometer): 65A (instantaneous, D2240-75)
Ref.: Data sheet

FPC 1363A. Firestone Plastics Co.
Chem. Descrip.: PVC comp.
Category/Applications: Injection molding comp. recommended for shoe soles
Form: Filled comp.
Gen'l. Prop.: Sp. Gr.: 1.24 (D792-66)
Mech. Prop.: Tens. Str.: 1900 psi (D-412-68); Tens. Elong.: 330% (ultimate, D412-68); Tens. Mod.: 880 psi (@ 100% elong., D412-68); Hardness (Shore Durometer): 75 (instantaneous, D2240-75)
Ref.: Data sheet

FPC 1379. Firestone Plastics Co.
Chem. Descrip.: PVC comp.
Category/Applications: Profile extrusion comp.
Form: Filled, opaque solid; Color: Natural, black, gray
All properties—75 mil plaque:
Gen'l. Prop.: Sp. Gr.: 1.36 (D792-66)
Mech. Prop.: Tens. Str.: 1490 psi (D-412-68); Tens. Elong.: 290% (ultimate, D412-68); Tens. Mod.: 670 psi (@ 100%

elong., D412-68); Hardness (Shore Durometer): 70A (instantaneous, D2240-75)
Ref.: Data sheet

FPC 1380. Firestone Plastics Co.
Chem. Descrip.: PVC comp.
Category/Applications: Profile extrusion comp.
Form: Filled, opaque solid; Color: Natural, black, gray
All properties–75 mil plaque:
Gen'l. Prop.: Sp. Gr.: 1.37 (D792-66)
Mech. Prop.: Tens. Str.: 1760 psi (D-412-68); Tens. Elong.: 250% (ultimate, D412-68); Tens. Mod.: 960 psi (@ 100% elong., D412-68); Hardness (Shore Durometer): 75A (instantaneous, D2240-75)
Ref.: Data sheet

FPC 1380A. Firestone Plastics Co.
Chem. Descrip.: PVC comp.
Category/Applications: Profile extrusion comp.
Form: Filled, opaque solid
All properties–75 mil plaque:
Gen'l. Prop.: Sp. Gr.: 1.37 (D792)
Mech. Prop.: Tens. Str.: 1270 psi (D-412); Tens. Elong.: 240% (ultimate, D-412); Tens. Mod.: 680 psi (@ 100% elong., D412); Hardness (Shore Durometer): 75A (instantaneous, D2240)
Ref.: Data sheet

FPC 1381A. Firestone Plastics Co.
Chem. Descrip.: PVC comp.
Category/Applications: Profile extrusion comp.
Form: Filled, opaque solid; Color: Natural, black, gray
All properties–75 mil plaque:
Gen'l. Prop.: Sp. Gr.: 1.39 (D792-66)
Mech. Prop.: Tens. Str.: 2890 psi (D-412-68); Tens. Elong.: 300% (ultimate, D412-68); Tens. Mod.: 1770 psi (@ 100% elong., D412-68); Hardness (Shore Durometer): 93A (instantaneous, D-2240-75)
Ref.: Data sheet

FPC 1415. Firestone Plastics Co.
Chem. Descrip.: PVC comp.
Category/Applications: Injection molding comp. used for shoe soles
Form: Filled solid
Gen'l. Prop.: Sp. Gr.: 1.21 (D792-66)
Mech. Prop.: Tens. Str.: 1400 psi (D-412-68); Tens. Elong.: 320% (ultimate, D412-68); Tens. Mod.: 600 psi (@ 100% elong., D412-68); Hardness (Shore Durometer): 65 (instantaneous, D2240-75)
Ref.: Data sheet

FPC 1418. Firestone Plastics Co.
Chem. Descrip.: PVC comp.
Category/Applications: Extrusion comp. used for automotive battery jumper cables
Form: Filled, opaque solid
All properties–75 mil plaque:
Gen'l. Prop.: Sp. Gr.: 1.25 (D792-66)
Mech. Prop.: Tens. Str.: 2000 psi (D-412-68); Tens. Elong.: 300% (ultimate, D412-68); Tens. Mod.: 960 psi (@ 100% elong., D412-68); Hardness (Shore Durometer): 84A (instantaneous, D2240-75)
Ref.: Data sheet

FPC 1419. Firestone Plastics Co.
Chem. Descrip.: PVC comp.
Category/Applications: Injection molding comp. for shoe soling, etc.
Form: Filled opaque solid
Gen'l. Prop.: Sp. Gr.: 1.24 (D792)
Mech. Prop.: Tens. Prop. (75 mil plaque): Tens. Str.: 2100 psi (D412); Tens. Elong.: 370% (ultimate, D412); Tens. Mod.: 790 psi (@ 100% elong., D412); Hardness (Shore Durometer): 73A (instantaneous, D2240)
Ref.: Data sheet

FPC 1430. Firestone Plastics Co.
Chem. Descrip.: PVC comp.
Category/Applications: Injection molding comp. for shoe soling, ice skate soling, athletic shoe soling, etc.
Form: Filled, opaque solid
Gen'l. Prop.: Sp. Gr.: 1.24 (D792)
Mech. Prop.: Tens. Prop. (75 mil plaque): Tens. Str.: 1800 psi (D412); Tens. Elong.: 300% (ultimate, D412); Tens. Mod.: 950 psi (@ 100% elong., D412); Hardness (Shore Durometer): 80A (instantaneous, D2240)
Ref.: Data sheet

FPC 1433. Firestone Plastics Co.
Chem. Descrip.: PVC comp.
Category/Applications: Rubber-like injection molding comp. with good traction, excellent low temp. flexibility; used for shoe soles
All properties–75 mil plaque:
Gen'l. Prop.: Sp. Gr.: 1.18 (D792-66)
Mech. Prop.: Tens. Str.: 1520 psi (D-412-68); Tens. Elong.: 370% (ultimate, D412-68); Tens. Mod.: 510 psi (@ 100% elong., D412-68); Hardness (Shore Durometer): 62A (instantaneous, D2240-75)
Ref.: Data sheet

FPC 4296. Firestone Plastics Co.
Chem. Descrip.: Vinyl copolymer resin
Category/Applications: High m.w. solution resin used in resilient floor coverings, rigid sheet applications, permanent and strippable coatings, and in free films
Gen'l. Prop.: Sol.: Sol. in ketones, ketone-aromatic blends (e.g., ketone/toluene 7/3); Sp. Gr.: 1.36 (D792-66); Density (Bulk): 0.76 g/cc(D1895-69/B); Visc. (Relative): 1.88 (1% in cyclohexanone)
Ref.: Data sheet

FPC 4301. Firestone Plastics Co.
Chem. Descrip.: Med. m.w. vinyl chloride/vinyl acetate copolymer resin
Category/Applications: Resin compatible with plasticizers, stabilizers, etc. for the production of vinyl asbestos floor tiles exhibiting excellent heat and light stability, and high resistance to ammonia, soaps, detergents, and mineral and weak organic acids
Form: Particles (80–200 mesh)
Gen'l. Prop.: Sp. Gr.: 1.36 (D792-66); Density (Bulk): 0.77 g/cc (D1895-69/B); Visc. (Inherent): 0.47 (D1243-66)
Ref.: Data sheet

FPC 4414. Firestone Plastics Co.
Chem. Descrip.: Low m.w. vinyl chloride/vinyl acetate copolymer resin
Category/Applications: See FPC 4301
Form: Particles (80–200 mesh)
Gen'l. Prop.: Sp. Gr.: 1.36 (D792-66); Density (Bulk): 0.73 g/cc (D1895-69/B); Visc. (Inherent): 0.42 (D1243-66)
Ref.: Data sheet

FPC 5000B. Firestone Plastics Co.
Chem. Descrip.: PVC comp.
Category/Applications: Flame retardant extrusion comp. meeting requirements for naval shipboard special purpose jacket
Gen'l. Prop.: Sp. Gr.: 1.29 (D792); Stability: 100% retention of tensile and 90% retention of elong. after heat aging 7 days @ 136 C
Mech. Prop.: Tens. Prop. (50 mil plaque–Die C): Tens. Str.: 2800 psi (D-412); Tens. Elong.: 320% (ultimate, D-412); Tens. Mod.: 1600 psi (@ 100% elong., D412); Hardness (Shore Durometer): 95A (instantaneous, D2240)
Ref.: Data sheet

FPC 5002. Firestone Plastics Co.
Chem. Descrip.: PVC comp.
Category/Applications: Flame retardant extrusion comp. for 105 C jacketing, power cable jacket
Gen'l. Prop.: Sp. Gr.: 1.42 (D792); Stability: 80% retention of elong. after heat aging 7 days @ 136 C
Mech. Prop.: Tens. Prop. (75 mil plaque–Die C): Tens. Str.: 3200 psi (D-412); Tens. Elong.: 270% (ultimate, D-412); Tens. Mod.: 2600 psi (@ 100% elong., D412); Hardness (Shore Durometer): 90C (instantaneous, D2240)
Ref.: Data sheet

FPC 5004. Firestone Plastics Co.
Chem. Descrip.: PVC comp.
Category/Applications: General purpose jacketing comp. with good low temp. and physical properties for outdoor use, e.g., outdoor protective jackets for wire and cable
Gen'l. Prop.: Sp. Gr.: 1.37 (D792); Stability: 89% retention of elong. on heat aging 7 days @ 100 C (75 mil plaque, Die C, D573)
Mech. Prop.: Tens. Prop. (75 mil plaque–Die C): Tens. Str.: 2200 psi (D-412); Tens. Elong.: 275% (ultimate, D-412); Tens. Mod.: 1550 psi (@ 100% elong., D412); Hardness (Shore Durometer): 82A (instantaneous, D2240)
Elec. Prop. (75 mil plaque–Die C): Dielec. Const.: 5.75 (D150)
Ref.: Data sheet

FPC 5005. Firestone Plastics Co.
Chem. Descrip.: PVC comp.
Category/Applications: Extrusion comp. for thin wall (8 mil) 105 C appliance wire insulation
Gen'l. Prop.: Sp. Gr.: 1.34 (D792); Stability: 80% retention of elong. on heat aging 60 days @ 113 C (20 AWG, 8 mil wall)
Mech. Prop.: Tens. Prop. (20 AWG, 8 mil wall): Tens. Str.: 2565 psi (D412); Tens. Elong.: 180% (ultimate, D412); Tens. Mod.: 2100 psi (@ 100% elong., D412); Hardness (Shore Durometer): 94A (instantaneous, D2240)
Elec. Prop.: Vol. Resist.: 1.5 X 10^{13} ohm-cm (60 C, D257)
Ref.: Data sheet

FPC 5006A. Firestone Plastics Co.
Chem. Descrip.: PVC comp.
Category/Applications: Flame retardant, low temp. outdoor jacketing comp. for REA PE 7 parallel conductor drop wire
Gen'l. Prop.: Sp. Gr.: 1.39 (D792); Stability: 88% retention of elong. on heat aging 7 days @ 100 C (75 mil plaque–Die C, D573)
Mech. Prop.: Tens. Prop. (75 mil plaque–Die C): Tens. Str.: 2100 psi (D-412); Tens. Elong.: 320% (ultimate, D-412); Tens. Mod.: 1325 psi (@ 100% elong., D412); Hardness (Shore Durometer): 83A (instantaneous, D2240)
Elec. Prop.: Dielec. Const.: 5.75 (D150)
Ref.: Data sheet

FPC 5008. Firestone Plastics Co.
Chem. Descrip.: PVC comp.
Category/Applications: Extrusion comp. for low temp. protective jackets for wire and cable
Gen'l. Prop.: Sp. Gr.: 1.35 (D792); Stability: 80% retention of elong. on heat aging 7 days @ 113 C (75 mil plaque, UL 758)
Mech. Prop.: Tens. Prop. (75 mil plaque): Tens. Str.: 1850 psi (D412); Tens. Elong.: 290% (ultimate, D412); Tens. Mod.: 900 psi (@ 100% elong., D-412); Hardness (Shore Durometer): 80A (instantaneous, D2240)
Ref.: Data sheet

FPC 5009. Firestone Plastics Co.
Chem. Descrip.: PVC comp.
Category/Applications: Flame retardant extrusion comp. for 75 C and 90 C tray cable jacketing; suitable for an "all color" sunlight resistant rating
Gen'l. Prop.: Sp. Gr.: 1.39 (D792); Stability (50 mil plaque–Die B): 95% retention of tensile and 90% retention of elong. after sunlight (UL 62, 720 h, Weatherometer); 95% retention of elong. on heat aging 10 days @ 100 C (D573)
Mech. Prop.: Tens. Prop. (50 mil plaque–Die B): Tens. Str.: 2750 psi (D-412); Tens. Elong.: 310% (ultimate, D-412); Tens. Mod.: 1650 psi (@ 100% elong., D412); Hardness (Shore Durometer): 93A (instantaneous, D2240)
Ref.: Data sheet

FPC 5060. Firestone Plastics Co.
Chem. Descrip.: PVC comp.
Category/Applications: Low-cost comp. meeting requirements for 60 C TW insulation
Gen'l. Prop.: Sp. Gr.: 1.38 (D792-66); Stability: 84% retention of elong. on heat aging 7 days @ 100 C (14 AWG, 1/32 in. wall)
Mech. Prop.: Tens. Str.: 2350 psi (14 AWG, 1/32 in. wall, D412-68); Tens. Elong.: 250% (ultimate, D412-68); Tens. Mod.: 1350 psi (@ 100% elong., D412-68); Hardness (Shore Durometer): 88A (instantaneous, D2240-68)
Elec. Prop.: Vol. Resist.: 1.5×10^{13} ohm-cm (60 C, D257-66)
Ref.: Data sheet

FPC 5105. Firestone Plastics Co.
Chem. Descrip.: PVC comp.
Category/Applications: Extrusion comp. for use as 105 C appliance wire insulation at wall thickness ⩾ 15 mil
Gen'l. Prop.: Sp. Gr.: 1.34 (D792); Stability: 75% retention of elong. on heat aging 7 days @ 136 C (20 AWG, 15 mil)
Mech. Prop.: Tens. Str.: 3125 psi (20 AWG, 15 mil, D412); Tens. Elong.: 290% (ultimate, D412); Tens. Mod.: 1750 psi (@ 100% elong., D412); Hardness (Shore Durometer): 93A (instantaneous, D-2240)
Elec. Prop.: Vol. Resist.: 10^{13} ohm-cm (50 mil plaque, 60 C, D257)
Ref.: Data sheet

FPC 5125A. Firestone Plastics Co.
Chem. Descrip.: PVC comp.
Category/Applications: Extrusion comp. for under-hood automotive wiring at 30 mil wall or greater
Gen'l. Prop.: Sp. Gr.: 1.63 (D792); Stability: 66% retention of elong. on heat aging 7 days @ 158 C (20 AWG, 0.03 in. wall)
Mech. Prop.: Tens. Prop. (20 AWG, 0.03 in. wall): Tens. Str.: 3200 psi (D-412); Tens. Elong.: 240% (ultimate, D-412); Tens. Mod.: 2200 psi (@ 100% elong., D412); Hardness (Shore Durometer): 62D (instantaneous, D2240)
Elec. Prop.: Vol. Resist.: 2.0×10^{11} ohm-cm (50 mil plaque, 60 C, D257)
Ref.: Data sheet

FPC 5160. Firestone Plastics Co.
Chem. Descrip.: PVC comp.
Category/Applications: Semirigid 60 C comp. with high mechanical strength, wide extrusion latitude, gloss, toughness, and flexibility; used in TW applications
Gen'l. Prop.: Sp. Gr.: 1.38 (D792-66); Stability: 90% retention of elong. on heat aging 7 days @ 100 C (14 AWG, 1/32 in. wall)
Mech. Prop.: Tens. Str.: 3000 psi (14 AWG, 1/32 in. wall, D412-68); Tens. Elong.: 275% (ultimate, D412-68); Tens. Mod.: 2100 psi (@ 100% elong., D412-68); Hardness (Shore Durometer): 96A

(instantaneous, D2240-68)
Elec. Prop.: Vol. Resist.: 5.0 X 10^{13} ohm-cm (60 C, D257-66)
Ref.: Data sheet

FPC 5175. Firestone Plastics Co.
Chem. Descrip.: PVC comp.
Category/Applications: Extrusion comp. for THW-MTW, 90 C AWM-600V, 75 C oil resistant; 90 C-AWM 1000V applications
Gen'l. Prop.: Sp. Gr.: 1.37 (D792-66); Stability: 84% retention of elong. on heat aging 7 days @ 121 C (14 AWG, 3/64 in. wall)
Mech. Prop.: Tens. Str.: 2600 psi (14 AWG, 3/64 in. wall, D412-68); Tens. Elong.: 260% (ultimate, D412-68); Tens. Mod.: 1750 psi (@ 100% elong., D412-68); Hardness (Shore Durometer): 96A (instantaneous, D2240-68)
Elec. Prop.: Vol. Resist.: 1.5 X 10^{13} ohm-cm (60 C, D257-66)
Ref.: Data sheet

FPC 5264. Firestone Plastics Co.
Chem. Descrip.: PVC comp.
Category/Applications: Protective jacket comp. for the manufacture of nonmetallic cable having superior abrasion resistance and low temp. properties
Gen'l. Prop.: Sp. Gr.: 1.44 (D792-66); Stability: 80% retention of elong. on heat aging 7 days @ 100 C (20 mil plaque)
Mech. Prop.: Tens. Str.: 2140 psi (20 mil plaque, D412-68); Tens. Elong.: 220% (ultimate, D412-68); Tens. Mod.: 1330 psi (@ 100% elong., D412-68); Hardness (Shore Durometer): 86A (instantaneous, D2240-68)
Elec. Prop.: Vol. Resist.: 10^{12} ohm-cm (60 C, D257-66)
Ref.: Data sheet

FPC 5290. Firestone Plastics Co.
Chem. Descrip.: PVC comp.
Category/Applications: Extrusion comp. for 90 C appliance wire, radiant heating cable
Gen'l. Prop.: Sp. Gr.: 1.36 (D792-66); Stability: 80% retention of elong. on heat aging 7 days @ 121 C (14 AWG, 1/32 in. wall, UL 758)
Mech. Prop.: Tens. Str.: 2400 psi (14 AWG, 1/32 in. wall, D412-68); Tens. Elong.: 220% (ultimate, D412-68); Tens. Mod.: 1800 psi (@ 100% elong., D412-68); Hardness (Shore Durometer): 92A (instantaneous, D2240-68)
Elec. Prop.: Vol. Resist.: 1.5 X 10^{13} ohm-cm (60 C, D257-66)
Ref.: Data sheet

FPC 5305. Firestone Plastics Co.
Chem. Descrip.: PVC comp.
Category/Applications: Extrusion comp. for use as 30 mil wall and heavier appliance wire insulation where low taste and odor are required
Gen'l. Prop.: Sp. Gr.: 1.37 (D792); Stability: 85% retention of elong. on heat aging 7 days @ 136 C (14 AWG, 30 mil wall)
Mech. Prop.: Tens. Prop. (14 AWG, 30 mil wall): Tens. Str.: 2900 psi (D412); Tens. Elong.: 275% (ultimate, D412); Tens. Mod.: 2000 psi (@ 100% elong., D412); Hardness (Shore Durometer): 91A (instantaneous, D2240)
Elec. Prop.: Vol. Resist.: 5 X 10^{12} ohm-cm (50 mil plaque, 60 C, D257)
Ref.: Data sheet

FPC 5360. Firestone Plastics Co.
Chem. Descrip.: PVC comp.
Category/Applications: Extrusion comp. for TW, T, 60 C in oil, and 80 C appliance wire use at a 30 mil wall and heavier

Gen'l. Prop.: Sp. Gr.: 1.37 (D792); Stability: 90% retention of elong. on heat aging 7 days @ 100 C (14 AWG, 30 mil wall, UL 83)
Mech. Prop.: Tens. Str.: 2600 psi (14 AWG, 30 mil wall, D412); Tens. Elong.: 280% (ultimate, D412); Tens. Mod.: 1800 psi (@ 100% elong., D412); Hardness (Shore Durometer): 92A (instantaneous, D2240)
Elec. Prop.: Vol. Resist.: 1.5×10^{13} ohm-cm (50 mil plaque, 60 C, D257)
Ref.: Data sheet

FPC 5390. Firestone Plastics Co.
Chem. Descrip.: PVC comp.
Category/Applications: Extrusion comp. for 90 C SPT compound insulation where polystyrene migration resistance is required
Gen'l. Prop.: Sp. Gr.: 1.43 (D792); Stability: 85% retention of elong. on heat aging 7 days @ 121 C (14 AWG, 30 mil wall)
Mech. Prop.: Tens. Prop. (14 AWG, 30 mil wall): Tens. Str.: 2300 psi (D412); Tens. Elong.: 300% (ultimate, D412); Tens. Mod.: 1300 psi (@ 100% elong., D412); Hardness (Shore Durometer): 87A (instantaneous, D2240)
Elec. Prop.: Vol. Resist.: 7×10^{12} ohm-cm (50 mil plaque, 60 C, D257)
Ref.: Data sheet

FPC 5439A. Firestone Plastics Co.
Chem. Descrip.: PVC comp.
Category/Applications: Extrusion comp. for automotive jumper cable jacketing
Form: Filled opaque solid; Color: Natural, black
Gen'l. Prop.: Sp. Gr.: 1.24 (D792)
Mech. Prop.: Tens. Prop. (75 mil plaque): Tens. Str.: 2150 psi (D412); Tens. Elong.: 315% (ultimate, D412); Tens. Mod.: 1020 psi (@ 100% elong., D412); Hardness (Shore Durometer): 75 (instantaneous, D2240)
Ref.: Data sheet

FPC 5440. Firestone Plastics Co.
Chem. Descrip.: PVC comp.
Category/Applications: Extrusion comp. for REA PE 20 station wire jacketing
Gen'l. Prop.: Sp. Gr.: 1.36 (D792); Stability: 84% retention of elong. on heat aging 7 days @ 100 C (75 mil plaque–Die C)
Mech. Prop.: Tens. Prop. (75 mil plaque–Die C): Tens. Str.: 2200 psi (D-412); Tens. Elong.: 320% (ultimate, D-412); Tens. Mod.: 1075 psi (@ 100% elong., D412); Hardness (Shore Durometer): 81A (instantaneous, D2240)
Elec. Prop.: Dielec. Const.: 5.24 (D150)
Ref.: Data sheet

FPC 5460. Firestone Plastics Co.
Chem. Descrip.: PVC comp.
Category/Applications: Extrusion comp. used for flexible cord and appliance wire insulation at 0.03 in. wall thickness; max. temp. rating is 80 C
Gen'l. Prop.: Sp. Gr.: 1.36 (D792); Stability: 90% retention of elong. on heat aging 7 days @ 100 C (14 AWG, 0.03 in. wall)
Mech. Prop.: Tens. Prop. (14 AWG, 0.03 in. wall): Tens. Str.: 2100 psi (D-412); Tens. Elong.: 280% (ultimate, D-412); Tens. Mod.: 1250 psi (@ 100% elong., D412); Hardness (Shore Durometer): 85A (instantaneous, D2240)
Elec. Prop.: Vol. Resist.: 5.0×10^{12} ohm-cm (50 mil plaque, 60 C, D257)
Ref.: Data sheet

FPC 5605. Firestone Plastics Co.
Chem. Descrip.: PVC comp.
Category/Applications: Extrusion comp. used as 105 C SPT insulation where

polystyrene migration resistance is required
Gen'l. Prop.: Sp. Gr.: 1.46 (D792); Stability: 85% retention of elong. on heat aging 7 days @ 136 C (14 AWG, 30 mil wall)
Mech. Prop.: Tens. Prop. (14 AWG, 30 mil wall): Tens. Str.: 2500 psi (D412); Tens. Elong.: 280% (ultimate, D412); Tens. Mod.: 1500 psi (@ 100% elong., D412); Hardness (Shore Durometer): 90A (instantaneous, D2240)
Elec. Prop.: Vol. Resist.: 10^{12} ohm-cm (50 mil plaque, 60 C, D257)
Ref.: Data sheet

FPC 5760. Firestone Plastics Co.
Chem. Descrip.: PVC comp.
Category/Applications: Extrusion comp. for multiple line wire insulation and jacket
Gen'l. Prop.: Sp. Gr.: 1.29 (D792); Stability: 80% retention of elong. on heat aging 7 days @ 100 C (75 mil plaque)
Mech. Prop.: Tens. Prop. (75 mil plaque): Tens. Str.: 2800 psi (D412); Tens. Elong.: 290% (ultimate, D412); Tens. Mod.: 1400 psi (@ 100% elong., D412); Hardness (Shore Durometer): 88A (instantaneous, D2240)
Elec. Prop.: Vol. Resist.: 3.0×10^{12} ohm-cm (50 mil plaque, 60 C, D257)
Ref.: Data sheet

FPC 5760B. Firestone Plastics Co.
Chem. Descrip.: PVC comp.
Category/Applications: Extrusion comp. for multiple line wire insulation and direct burial jacket
Gen'l. Prop.: Sp. Gr.: 1.32 (D792); Stability: 97% retention of elong. on heat aging 7 days @ 100 C (75 mil plaque)
Mech. Prop.: Tens. Prop. (75 mil plaque): Tens. Str.: 2700 psi (D412); Tens. Elong.: 310% (ultimate, D412); Tens. Mod.: 1610 psi (@ 100% elong., D412); Hardness (Shore Durometer): 92A (instantaneous, D2240)
Elec. Prop.: Vol. Resist.: 2.0×10^{11} ohm-cm (50 mil plaque, 60 C, D257)
Ref.: Data sheet

FPC 5790. Firestone Plastics Co.
Chem. Descrip.: PVC comp.
Category/Applications: Extrusion comp. for THHN/THWN applications
Gen'l. Prop.: Sp. Gr.: 1.38 (D792); Stability: 77% retention of elong. on heat aging 7 days @ 121 C (15 mil wall)
Mech. Prop.: Tens. Prop. (15 mil wall): Tens. Str.: 2920 psi (D412); Tens. Elong.: 190% (ultimate, D412); Tens. Mod.: 2370 psi (@ 100% elong., D412); Hardness (Shore Durometer): 61D (instantaneous, D2240)
Elec. Prop.: Vol. Resist.: 10^{13} ohm-cm (50 mil plaque, 60 C, D257)
Ref.: Data sheet

FPC 5805. Firestone Plastics Co.
Chem. Descrip.: PVC comp.
Category/Applications: Extrusion comp. for 30 mil and heavier UL appliance wire insulation, 90 C CSA TR 64 radio device insulation, and 30 mil and heavier CSA equipment wiring
Gen'l. Prop.: Sp. Gr.: 1.39 (D792); Stability: 74% retention of elong. on heat aging 7 days @ 136 C (20 AWG, 30 mil wall)
Mech. Prop.: Tens. Str.: 3000 psi (20 AWG, 30 mil wall, D412); Tens. Elong.: 290% (ultimate, D412); Tens. Mod.: 1800 psi (@ 100% elong., D412); Hardness (Shore Durometer): 93A (instantaneous, D2240)
Elec. Prop.: Vol. Resist.: 3.5×10^{13} ohm-cm (50 mil plaque, 60 C, D257)
Ref.: Data sheet

FPC 6337. Firestone Plastics Co.
Chem. Descrip.: High m.w. PVC resin
Category/Applications: General purpose dispersion resin for roll coating, spraying, strand coating, dip coating, slush molding, rotational casting (molding), cellular vinyls, sealants, and organosols
Gen'l. Prop.: Sp. Gr.: 1.4; Visc. (Brookfield): 30–50; Stability: Excellent visc. stability
Mech. Prop.: Tens. Str.: 470 psi (ultimate, when fused at 120 C, D412)
Elec. Prop.: Vol. Resist.: 1.57×10^{11} ohm-cm (@ R.T., D257-66)
Ref.: Data sheet

FPC 6338. Firestone Plastics Co.
Chem. Descrip.: Med. m.w. vinyl chloride/vinyl acetate copolymer resin
Category/Applications: Dispersion resin for processing at low fusion temps.; used in coatings for polyolefins and other heat-sensitive fabrics, vinyl foams, dip coatings, spray coatings, rotational castings, slush molding, and specialty applications (e.g., vinyl stencil inks)
Gen'l. Prop.: Sp. Gr.: 1.38; Visc. (Brookfield): 30–50
Mech. Prop.: Tens. Str.: 600 psi (ultimate, when fused at 120 C, D412)
Elec. Prop.: Vol. Resist.: 2.19×10^{11} ohm-cm (@ R.T., D257-66)
Ref.: Data sheet

FPC 6366. Firestone Plastics Co.
Chem. Descrip.: Med. m.w. PVC resin
Category/Applications: See FPC 6337; also used in cast cellular vinyls using azo type chemical blowing agents
Gen'l. Prop.: Sp. Gr.: 1.4; Visc. (Brookfield): 30–60
Mech. Prop.: Tens. Str.: 450 psi (ultimate, when fused at 120 C, D412)
Elec. Prop.: Vol. Resist.: 8.1×10^{9} ohm-cm (@ R.T., D257-66)
Ref.: Data sheet

FPC 6411. Firestone Plastics Co.
Chem. Descrip.: High m.w. PVC resin
Category/Applications: Dispersion resin with exceptional high gloss and clarity for use in clear wear layers in resilient flooring products, coated fabrics, and coil coatings
Gen'l. Prop.: Sp. Gr.: 1.4; Visc. (Brookfield): 40–60
Mech. Prop.: Tens. Str.: 450 psi (ultimate, when fused at 120 C, D412)
Elec. Prop.: Vol. Resist.: 6.83×10^{10} ohm-cm (@ R.T., D257-66)
Ref.: Data sheet

FPC 6422. Firestone Plastics Co.
Chem. Descrip.: High m.w. PVC resin
Category/Applications: See FPC 6411
Gen'l. Prop.: Sp. Gr.: 1.4; Visc. (Brookfield): 30–50
Mech. Prop.: Tens. Str.: 400 psi (ultimate, when fused at 120 C, D412)
Ref.: Data sheet

FPC 6854. Firestone Plastics Co.
Chem. Descrip.: High m.w. PVC resin
Category/Applications: See FPC 6337; 6854 has exceptional air release characteristics and fast gelling time; esp. suitable in foaming and other operations requiring fast physical property development
Gen'l. Prop.: Sp. Gr.: 1.4; Visc. (Brookfield): 50–80 poise (20 rpm)
Mech. Prop.: Tens. Str.: 500 psi (ultimate, when fused at 120 C, D412)
Elec. Prop.: Vol. Resist.: 7.6×10^{9} ohm-cm (@ R.T., D257-66)
Ref.: Data sheet

FPC 6866. Firestone Plastics Co.
Chem. Descrip.: Med. m.w. PVC resin
Category/Applications: See FPC 6366
Gen'l. Prop.: Sp. Gr.: 1.4; Visc. (Brookfield): 35–60 poise (20 rpm)
Mech. Prop.: Tens. Str.: 400 psi (ulti-

mate, when fused at 120 C, D412)
Elec. Prop.: Vol. Resist.: 8.1 $\times$ 10^9 ohm-cm (@ R.T., D257-66)
Ref.: Data sheet

FPC 9269A. Firestone Plastics Co.
Chem. Descrip.: High m.w. PVC homopolymer resin
Category/Applications: Resin with excellent electrical properties and plasticizer sorption characteristics with monomeric and/or combined polymeric systems; used in dry processing and calendering operations, e.g., wire and cable; compounds based on this resin have excellent physical properties, good heat and light stability, and ease of processability
Form: Particles (80–200 mesh)
Gen'l. Prop.: Sp. Gr.: 1.4 (D792-66); Density (Bulk): 0.48 g/cc (D1895-69/B); Visc. (Inherent): 1.07 (D1243-66)
Ref.: Data sheet

FPC 9275. Firestone Plastics Co.
Chem. Descrip.: PVC resin
Category/Applications: Resin used in cold blending operations and products requiring high plasticizer levels; not recommended for applications requiring extruder densification because of its fine particle size
Form: Particles (140–200 mesh)
Gen'l. Prop.: Sp. Gr.: 1.4 (D792-66); Density (Bulk): 0.38 g/cc (D1895-69/B); Visc. (Inherent): 1.24 (D1243-66)
Ref.: Data sheet

FPC 9282. Firestone Plastics Co.
Chem. Descrip.: Low m.w. PVC homopolymer resin
Category/Applications: Lowest m.w. PVC resin normally recommended in dry processing and calendering operations; used where a minimum of energy is available for processing due to equipment or product limitations; has excellent clarity and fuses rapidly; also as a modifier with other PVC resins, and for extrusion of rigid PVC foam profiles
Form: Particles (100–200 mesh)
Gen'l. Prop.: Sp. Gr.: 1.4 (D792); Density (Bulk): 0.60 g/cc (D1895/B); Visc. (Inherent): 0.65 (D1243)
Ref.: Data sheet

FPC 9290. Firestone Plastics Co.
Chem. Descrip.: Low m.w. PVC homopolymer resin
Category/Applications: Resin used in dry processing and calendering operations, and for injection and blow molding compounds; suitable for operations where a minimum of heat is available for processing or where other ingred. are heat sensitive
Form: Particles (100–200 mesh)
Gen'l. Prop.: Sp. Gr.: 1.4 (D792); Density (Bulk): 0.50 g/cc (D1895/B); Visc. (Inherent): 0.80 (D1243)
Ref.: Data sheet

FPC 9300. Firestone Plastics Co.
Chem. Descrip.: Med. m.w. PVC homopolymer resin
Category/Applications: Resin used in dry processing and calendering operations; controlled particle size facilitates handling in bulk systems; plasticizer sorption properties permit fast blend cycles and high outputs from mills, banburys, and extruders; products based on this resin have good color, clarity, surface luster, and low gel count
Form: Particles (100–200 mesh)
Gen'l. Prop.: Sp. Gr.: 1.4 (D792-66); Density (Bulk): 0.5 g/cc (D1895-69/B); Visc. (Inherent): 0.96 (D1243-66); Stability: Good heat stability during processing and aging
Ref.: Data sheet

FPC 9326. Firestone Plastics Co.
Chem. Descrip.: Med. m.w. PVC homopolymer resin
Category/Applications: Resin for rigid PVC applications yielding products with good color and high surface luster; esp. suitable for dry blend extrusion due to narrow particle size distribution, low fines, and high bulk density; limited plasticizer sorption properties
Form: Particles (100–200 mesh)
Gen'l. Prop.: Sp. Gr.: 1.4 (D792); Density (Bulk): 0.56 g/cc (D1895/B); Visc. (Inherent): 0.97 (D1243); Stability: Good fusion and heat stability
Ref.: Data sheet

FPC 9339. Firestone Plastics Co.
Chem. Descrip.: Ultra high m.w. PVC homopolymer resin
Category/Applications: Resin for use in calendering and dry process extrusion equipment yielding products with superior physical properties, color and clarity, and surface luster
Form: Particles (80–200 mesh)
Gen'l. Prop.: Sp. Gr.: 1.4 (D792); Density (Bulk): 0.48 g/cc (D1895/B); Visc. (Inherent): 1.33 (D1243)
Ref.: Data sheet

FPC 9418. Firestone Plastics Co.
Chem. Descrip.: Med. m.w. PVC homopolymer resin
Category/Applications: Resin for rigid PVC applications; suitable for dry blend extrusion due to narrow particle size distribution, low fines, and high bulk density; finished products have good color and high surface luster; used in single and multi-screw extrusion of PVC siding, pipe and conduit, profiles, etc.
Form: Particles (100–200 mesh)
Gen'l. Prop.: Sp. Gr.: 1.4 (D792); Density (Bulk): 0.56 g/cc (D1895/B); Visc. (Inherent): 0.91 (D1243)
Ref.: Data sheet

Fulton 404. Liquid Nitrogen Processing Corp.
Chem. Descrip.: Modified TFE and acetal resin comp.
Category/Applications: Thermoplastic comp. suited for moving parts that require excellent frictional properties, resistance to wear, and ease of fabrication; used in bearings and sliding-contact parts, e.g., gears, cams, valve seats, stem packing, thrust washers, seals, etc. in the textile, printing, automotive, appliance, and furniture industries
Form: Pellets
Gen'l. Prop.: Sp. Gr.: 1.52 (D792); Sp. Vol.: 17.8 in.^{3}/lb; Density: 0.0559 lb/in.^{3}; Stability: Finished products are resistant to corrosive and excessively moist atmospheres
Mech. Prop.: Tens. Str.: 6500 psi (D-638); Tens. Elong.: 7% (D638); Tens. Mod.: 390,000 psi (D638); Flex. Str.: 10,000 psi (D790); Flex. Mod.: 350,000 psi (D790); Compr. Str.: 1800 psi (1% deformation, D695); Shear Str.: 6500 psi (D732); Impact Str. (Izod): 0.6 ft lb/in. notch (D256); Hardness (Rockwell): M 60 (D785)
Therm. Prop.: Conduct.: 1.4 Btu/h/ft^2/F/in. (Cenco-Fitch); Distort. Temp.: 210 F (@ 264 psi, D648); Coeff. of Linear Exp.: 5.4 × 10^{-5} in./in./F (D-696)
Ref.: Product data 201-1-277

Fulton 441. Liquid Nitrogen Processing Corp.
Chem. Descrip.: Silicone-lubricated acetal
Category/Applications: Used for med. duty service in cost-sensitive bearing applications
Gen'l. Prop.: Sp. Gr.: 1.40 (D792); Sp. Vol.: 19.6 in.^{3}/lb (D792)
Mech. Prop.: Tens. Str.: 8000 psi (D-

638); Tens. Elong.: 35% (D638); Flex. Str.: 10,000 psi (D790); Flex. Mod.: 300,000 psi (D790); Impact Str. (Izod): 1.6 ft lb/in. notch (D256)
Therm. Prop.: Conduct.: 1.6 Btu/h/ft^2/F/in. (C177); Distort. Temp.: 230 F (@ 264 psi, D648); Coeff. of Linear Exp.: 5.0 $\times$ 10^{-5} in./in./F (D696); Flamm.: 94HB
Ref.: Product data 201-4-1076

Fulton 441D. Liquid Nitrogen Processing Corp.
Chem. Descrip.: Silicone-lubricated acetal
Category/Applications: See Fulton 441
Gen'l. Prop.: Sp. Gr.: 1.42 (D792); Sp. Vol.: 19.6 in.^{3}/lb
Mech. Prop.: Tens. Str.: 9500 psi (D-638); Tens. Elong.: 40% (D638); Flex. Str.: 14,000 psi (D790); Flex. Mod.: 410,000 psi (D790); Impact Str. (Izod): 1.2 ft lb/in. notch (D256)
Therm. Prop.: Conduct.: 1.7 Btu/h/ft^2/F/in. (C177); Distort. Temp.: 240 F (@ 264 psi, D648); Coeff. of Linear Exp.: 5.4 $\times$ 10^{-5} in./in./F (D696); Flamm.: 94HB
Ref.: Product data 201-4-1076

G

Gafite 1432F. GAF Corp.

Chem. Descrip.: Polyester resin (polybutylene terephthalate base)–15% glass reinforced, flame retarded

Category/Applications: Thermoplastic molding comp. having high strength and rigidity, low moisture absorption, rapid molding cycles, smooth, glossy surface, excellent elec. properties; used in automotive (under-the-hood and exterior body) parts, electrical/electronic components, appliance parts and housings, hardware, business machines, hydraulic controls

Form: Pellets; Color: Natural and colors

All properties @ 23 C:

Gen'l. Prop.: Sp. Gr.: 1.58 (D792); Stability: Excellent resistance to organic solvents, incl. aliphatic hydrocarbons, gasoline, carbon tetrachloride, alcohols, ketones, ethylene glycol, fluorinated hydrocarbons, oils, and fats; resistant to salt sol'ns., soap sol'ns., and dilute acids and bases; attacked by strong acids and bases; excellent dimensional stability under load

Mech. Prop.: Tens. Str.: 940 kg/cm^2 (@ yield, D638); Tens. Elong.: 6% (@ break, D638); Flex. Str.: 1600 kg/cm^2 (D790); Flex. Mod.: 56,000 kg/cm^2 (D790); Impact Str. (Izod): 5.4 cm-kg/cm notch (D256); Hardness (Rockwell): M 90 (D-785)

Therm. Prop.: Distort. Temp.: 212 C (@ 4.6 kg/cm^2, D648); Flamm.: V-O (1/32 in., UL94)

Elec. Prop.: Dissip. Factor: 0.002 (1 KHz, D150); Dielec. Str.: 16.2 KV/mm (S/T, 2.4 mm, D149); Dielec. Const.: 3.48 (1 KHz, D150); Vol. Resist.: > 10^{16} ohm-cm (D257); Arc Resist.: 85 s (D495)

Std. Pkgs.: 50-lb to truck-load quantities

Ref.: Bulletin 2308-001 R3 3M-88

Gafite 1432Z. GAF Corp.

Chem. Descrip.: Polyester resin (polybutylene terephthalate base)–15% glass-reinforced

Category/Applications & Form: See Gafite 1432F

All properties @ 23 C:

Gen'l. Prop.: Sp. Gr.: 1.42 (D792); Stability: See Gafite 1432F

Mech. Prop.: Tens. Str.: 880 kg/cm^2 (@ yield, D638); Tens. Elong.: 6% (@ break, D638); Flex. Str.: 1600 kg/cm^2 (D790); Flex. Mod.: 47,000 kg/cm^2 (D790); Impact Str. (Izod): 6.0 cm-kg/cm notch (D256); Hardness (Rockwell): M 90 (D785)

Therm. Prop.: Distort. Temp.: 212 C (@ 4.6 kg/cm^2, D648); Flamm.: HB (UL94)

Elec. Prop.: Dissip. Factor: 0.002 (1

KHz, D150); Dielec. Str.: 17.7 KV/mm (S/T, 2.5 mm, D149); Dielec. Const.: 3.38 (1 KHz, D150); Vol. Resist.: > 10^{16} ohm-cm (D257); Arc Resist.: 125 s (D495)
Std. Pkgs.: 50-lb to truck-load quantities
Ref.: Bulletin 2308-001 R3 3M-88

Gafite 1452F. GAF Corp.
Chem. Descrip.: Polyester resin (polybutylene terephthalate base)–26% glass-reinforced, flame-retarded
Category/Applications & Form: See Gafite 1432F
All properties @ 23 C:
Gen'l. Prop.: Sp. Gr.: 1.68 (D792); Stability: See Gafite 1432F
Mech. Prop.: Tens. Str.: 1200 kg/cm^2 (@ yield, D638); Tens. Elong.: 2% (@ break, D638); Flex. Str.: 1810 kg/cm^2 (D790); Flex. Mod.: 85,000 kg/cm^2 (D-790); Impact Str. (Izod): 7.0 cm-kg/cm notch (D256); Hardness (Rockwell): M 90 (D785)
Therm. Prop.: Distort. Temp.: 217 C (@ 4.6 kg/cm^2, D648); Flamm.: V-O (1/32 in., UL94)
Elec. Prop.: Dissip. Factor: 0.002 (1 KHz, D150); Dielec. Str.: 15.8 KV/mm (S/T, 2.5 mm, D149); Dielec. Const.: 3.78 (1 KHz, D150); Vol. Resist.: > 10^{16} ohm-cm (D257); Arc Resist.: 125 s (D495)
Std. Pkgs.: 50-lb to truck-load quantities
Ref.: Bulletin 2308-001·R3 3M-88

Gafite 1462Z. GAF Corp.
Chem. Descrip.: Polyester resin (polybutylene terephthalate base)–30% glass-reinforced
Category/Applications & Form: See Gafite 1432F
All properties @ 23 C:
Gen'l. Prop.: Sp. Gr.: 1.52 (D792); Stability: See Gafite 1432F
Mech. Prop.: Tens. Str.: 1300 kg/cm^2 (@ yield, D638); Tens. Elong.: 4% (@ break, D638); Flex. Str.: 2020 kg/cm^2 (D790); Flex. Mod.: 85,000 kg/cm^2 (D-790); Impact Str. (Izod): 9.0 cm-kg/cm notch (D256); Hardness (Rockwell): M 90 (D785)
Therm. Prop.: Distort. Temp.: 217 C (@ 4.6 kg/cm^2, D648); Flamm.: HB (UL94)
Elec. Prop.: Dissip. Factor: 0.002 (1 KHz, D150); Dielec. Str.: 20.5 KV/mm (S/T, 2.5 mm, D149); Dielec. Const.: 3.66 (1 KHz, D150); Vol. Resist.: > 10^{16} ohm-cm (D257); Arc Resist.: 130 s (D495)
Std. Pkgs.: 50-lb to truck-load quantities
Ref.: Bulletin 2308-001 R3 3M-88

Gafite 1600A, 1602Z. GAF Corp.
Chem. Descrip.: Polyester resin (polybutylene terephthalate base)–unreinforced
Category/Applications & Form: See Gafite 1432F
All properties @ 23 C:
Gen'l. Prop.: Sp. Gr.: 1.31 (D792); Stability: See Gafite 1432F
Mech. Prop.: Tens. Str.: 560 kg/cm^2 (@ yield, D638); Tens. Elong.: > 200/ (@ break, D638); Flex. Str.: 850 kg/cm^2 (D790); Flex. Mod.: 23,000 kg/cm^2 (D-790); Impact Str. (Izod): 5.4 cm-kg/cm notch (D256); Hardness (Rockwell): M 72 (D785)
Therm. Prop.: Distort. Temp.: 154 C (@ 4.6 kg/cm^2, D648); Flamm.: HB (UL94)
Elec. Prop.: Dissip. Factor: 0.001 (1 KHz, D150); Dielec. Str.: 15.8 KV/mm (S/T, 2.5 mm, D149); Dielec. Const.: 3.24 (1 KHz, D150); Vol. Resist.: > 10^{16} ohm-cm (D257); Arc Resist.: 190 s (D495)
Std. Pkgs.: 50-lb to truck-load quantities
Ref.: Bulletin 2308-001 R3 3M-88

Gafite 1602F. GAF Corp.
Chem. Descrip.: Polyester resin (polybutylene terephthalate base)–unreinforced, flame retarded
Category/Applications & Form: See Gafite 1432F
All properties @ 23 C:
Gen'l. Prop.: Sp. Gr.: 1.44 (D792); Stability: See Gafite 1432F
Mech. Prop.: Tens. Str.: 560 kg/cm^2 (@ yield, D638); Tens. Elong.: 25% (@ break, D638); Flex. Str.: 910 kg/cm^2 (D790); Flex. Mod.: 26,000 kg/cm^2 (D790); Impact Str. (Izod): 4.5 cm-kg/cm notch (D256); Hardness (Rockwell): M 81 (D785)
Therm. Prop.: Distort. Temp.: 166 C (@ 4.6 kg/cm^2, D648); Flamm.: V-O (1/32 in., UL94)
Elec. Prop.: Dissip. Factor: 0.002 (1 KHz, D150); Dielec. Str.: 13.8 KV/mm (S/T, 2.5 mm, D149); Dielec. Const.: 3.24 (1 KHz, D150); Vol. Resist.: > 10^{16} ohm-cm (D257); Arc Resist.: 63 s (D495)
Std. Pkgs.: 50-lb to truck-load quantities
Ref.: Bulletin 2308-001 R3 3M-88

Gafite 1642Z. GAF Corp.
Chem. Descrip.: Polyester resin (polybutylene terephthalate base)–20% glass-reinforced, flexible)
Category/Applications & Form: See Gafite 1432F
All properties @ 23 C:
Gen'l. Prop.: Sp. Gr.: 1.45 (D792); Stability: See Gafite 1432F
Mech. Prop.: Tens. Str.: 1100 kg/cm^2 (@ yield, D638); Tens. Elong.: 5% (@ break, D638); Flex. Str.: 1770 kg/cm^2 (D790); Flex. Mod.: 64,000 kg/cm^2 (D-790); Impact Str. (Izod): 9.2 cm-kg/cm notch (D256); Hardness (Rockwell): M 89 (D785)
Therm. Prop.: Distort. Temp.: 213 C (@ 4.6 kg/cm^2, D648); Flamm.: HB (UL94)
Elec. Prop.: Dissip. Factor: 0.002 (1 KHz, D150); Dielec. Str.: 19 KV/mm (S/T, 2.5 mm, D149); Dielec. Const.: 3.45 (1 KHz, D150); Vol. Resist.: > 10^{16} ohm-cm (D257); Arc Resist.: 130 s (D495)
Std. Pkgs.: 50-lb to truck-load quantities
Ref.: Bulletin 2308-001 R3 3M-88

Gafite 1662Z. GAF Corp.
Chem. Descrip.: Polyester resin (polybutylene terephthalate base)–30% glass-reinforced, flexible
Category/Applications & Form: See Gafite 1432F
All properties @ 23 C:
Gen'l. Prop.: Sp. Gr.: 1.52 (D792); Stability: See Gafite 1432F
Mech. Prop.: Tens. Str.: 1300 kg/cm^2 (@ yield, D638); Tens. Elong.: 4% (@ break, D638); Flex. Str.: 2020 kg/cm^2 (D790); Flex. Mod.: 85,000 kg/cm^2 (D790); Impact Str. (Izod): 10.9 cm-kg/cm notch (D256); Hardness (Rockwell): M 90 (D785)
Therm. Prop.: Distort. Temp.: 217 C (@ 4.6 kg/cm^2, D648); Flamm.: HB (UL94)
Elec. Prop.: Dissip. Factor: 0.002 (1 KHz, D150); Dielec. Str.: 20.9 KV/mm (S/T, 2.5 mm, D149); Dielec. Const.: 3.69 (1 KHz, D150); Vol. Resist.: > 10^{16} ohm-cm (D257); Arc Resist.: 130 s (D495)
Std. Pkgs.: 50-lb to truck-load quantities
Ref.: Bulletin 2308-001 R3 3M-88

Gafite LW X-4424R. GAF Corp.
Chem. Descrip.: Low-warp polyester (polybutylene terephthalate) comp.–impact-modified, 40% glass spheres/fiberglass, flexible
Category/Applications: Thermoplastic molding comp. with fast molding cycles, and good electrical insu-

lation used for large automotive exterior parts, e.g., side and rear panels, louvres, deck lids, fender extensions, camper housings, in business machines, and in other low warpage applications
All properties @ 23 C:
Gen'l. Prop.: Sp. Gr.: 1.55 (D792); Stability: Excellent chemical- and wear-resistance; good dimensional stability under load
Mech. Prop.: Tens. Str.: 560 kg/cm^2 (@ yield, D638); Tens. Elong.: 3% (@ break, D638); Flex. Str.: 920 kg/cm^2 (D790); Flex. Mod.: 42,500 kg/cm^2 (D790); Impact Str. (Izod): 7.6 cm-kg/cm notch (D256); Hardness (Rockwell): M 33 (D785)
Therm. Prop.: Distort. Temp.: 165 C (@ 184 kg/cm^2, D648)
Elec. Prop.: Dielec. Str.: 490 V/mil (S/T, 1/8 in., D149); Dielec. Const.: 4.26 (1000 Hz, D150); Vol. Resist.: > 10^{15} ohm-cm (D257); Arc Resist.: 115 s (D-495)
Ref.: Bulletin 2308-018/R

Gafite LW X-4424S. GAF Corp.
Chem. Descrip.: Low-warp polyester (polybutylene terephthalate) comp.—impact-modified, 26% glass spheres/fiberglass, flexible
Category/Applications: See Gafite LW X-4424R
All properties @ 23 C:
Gen'l. Prop.: Sp. Gr.: 1.45 (D792); Stability: See Gafite LW X-4424R
Mech. Prop.: Tens. Str.: 560 kg/cm^2 (@ yield, D638); Tens. Elong.: 3% (@ break, D638); Flex. Str.: 920 kg/cm^2 (D790); Flex. Mod.: 42,500 kg/cm^2 (D790); Impact Str.: 7.6 cm-kg/cm notch (D-256); Hardness (Rockwell): M 33 (D785)
Therm. Prop.: Distort. Temp.: 168 C (@ 184 kg/cm^2, D648)
Elec. Prop.: Dielec. Str.: 460 V/mil (S/T, 1/8 in., D149); Dielec. Const.: 3.86 (1000 Hz, D150); Vol. Resist.: > 10^{15} ohm-cm (D257); Arc Resist.: 125 s (D495)
Ref.: Bulletin 2308-018/R

Gafite LW X-4612R. GAF Corp.
Chem. Descrip.: Low-warp polyester (polybutylene terephthalate) comp.—impact-modified, very high impact strength grade
Category/Applications: See Gafite LW X-4424R
All properties @ 23 C:
Gen'l. Prop.: Sp. Gr.: 1.30 (D792); Stability: See Gafite LW X-4424R
Mech. Prop.: Tens. Str.: 520 kg/cm^2 (@ yield, D638); Tens. Elong.: 6% (@ break, D638); Flex. Str.: 870 kg/cm^2 (D790); Flex. Mod.: 28,000 kg/cm^2 (D790); Impact Str. (Izod): 18.4 cm-kg/cm notch (D256); Hardness (Rockwell): M 41 (D-785)
Therm. Prop.: Distort. Temp.: 77 C (@ 184 kg/cm^2, D648)
Elec. Prop.: Dielec. Str.: 400 V/mil (S/T, 1/8 in., D149); Dielec. Const.: 3.61 (1000 Hz, D150); Vol. Resist.: > 10^{15} ohm-cm (D257); Arc Resist.: 130 s (D495)
Ref.: Bulletin 2308-018/R

Gafite LW X-5632Z. GAF Corp.
Chem. Descrip.: Low-warp polyester (polybutylene terephthalate) comp.—fiberglass/mica-reinforced, stiff grade
Category/Applications: See Gafite LW X-4424R
All properties @ 23 C:
Gen'l. Prop.: Sp. Gr.: 1.56 (D792); Stability: See Gafite LW X-4424R
Mech. Prop.: Tens. Str.: 860 kg/cm^2 (@ yield, D638); Tens. Elong.: 3% (@ break, D638); Flex. Str.: 1420 kg/cm^2 (D790); Flex. Mod.: 85,000 kg/cm^2 (D790); Im-

pact Str. (Izod): 4.9 cm-kg/cm notch (D256); Hardness (Rockwell): M 75 (D-785)
Therm. Prop.: Distort. Temp.: 181 C (@ 184 kg/cm^2, D648)
Elec. Prop.: Dielec. Str.: 600 V/mil (S/T, 1/8 in., D149); Dielec. Const.: 3.81 (1000 Hz, D150); Vol. Resist.: > 10^{15} ohm-cm (D257); Arc Resist.: 110 s (D-495)
Ref.: Bulletin 2308-018/R

Gafite LW X-6443R. GAF Corp.
Chem. Descrip.: Low-warp polyester (polybutylene terephthalate) comp.—impact-modified, mica-reinforced
Category/Applications: See Gafite LW X-4424R
All properties @ 23 C:
Gen'l. Prop.: Sp. Gr.: 1.57 (D792); Stability: See Gafite LW X-4424R
Mech. Prop.: Tens. Str.: 825 kg/cm^2 (@ yield, D638); Tens. Elong.: 3% (@ break, D638); Flex. Str.: 1200 kg/cm^2 (D790); Flex. Mod.: 83,000 kg/cm^2 (D790); Impact Str. (Izod): 10.3 cm-kg/cm notch (D256); Hardness (Rockwell): M 57 (D-785)
Therm. Prop.: Distort. Temp.: 195 C (@ 184 kg/cm^2, D648)
Elec. Prop.: Dielec. Str.: 600 V/mil (S/T, 1/8 in., D149); Dielec. Const.: 4.18 (1000 Hz, D150); Vol. Resist.: > 10^{15} ohm-cm (D257); Arc Resist.: 130 s (D-495)
Ref.: Bulletin 2308-018/R

Gafite LW X-7443R. GAF Corp.
Chem. Descrip.: Low-warp polyester (polybutylene terephthalate) comp.—impact-modified, mica-reinforced, improved stiffness grade
Category/Applications: See Gafite LW X-4424R
All properties @ 23 C:
Gen'l. Prop.: Sp. Gr.: 1.59 (D792); Stability: See Gafite LW X-4424R
Mech. Prop.: Tens. Str.: 800 kg/cm^2 (@ yield, D638); Tens. Elong.: 3% (@ break, D638); Flex. Str.: 1300 kg/cm^2 (D790); Flex. Mod.: 86,000 kg/cm^2 (D790); Impact Str. (Izod): 9.7 cm-kg/cm notch (D256); Hardness (Rockwell): M 50 (D-785)
Therm. Prop.: Distort. Temp.: 165 C (@ 184 kg/cm^2, D648)
Elec. Prop.: Dielec. Str.: 600 V/mil (S/T, 1/8 in., D149); Dielec. Const.: 4.15 (1000 Hz, D150); Vol. Resist.: > 10^{15} ohm-cm (D257); Arc Resist.: 125 s (D-495)
Ref.: Bulletin 2308-018/R

Gafquat 734 Polymer. GAF Corp.
Chem. Descrip.: Vinylpyrrolidone/dimethylaminoethylmethacrylate copolymer; med. m.w.
Category/Applications: Film-forming polymer for superior hair and skin care products; adsorbed on hair giving improved strength, thickness, manageability, high luster, and exceptional smoothness; for formulation of hair conditioners, rinses, sprays, shampoos, dyes, and semipermanents; also used in deodorants, antiperspirants, shaving preparations, antiseptics, toilet soaps, skin creams, sunburn remedies
Form: Liquid
Composition: 50% alcoholic sol'n.
Ref.: Catalogue 2303-036

Gafquat 755N Polymer. GAF Corp.
Chem. Descrip.: Vinylpyrrolidone/dimethylaminoethylmethacrylate copolymer; high m.w.
Category/Applications: See Gafquat 734 Polymer
Form: Liquid
Composition: 20% aq. sol'n.
Ref.: Catalogue 2303-036

Gaftex CK-160 Conc. GAF Corp.
Chem. Descrip.: Polyether (nonionic)
Category/Applications: Multicolor differential dyeing assistant for continuous dyeing of carpets on Kuster machines; used as antiprecipitant and leveling agent; yields excellent contrast; applicable in beck and hosiery dyeing
Form: Liquid
Ref.: Catalogue 2303-036

Gaftex COM-154, COM-154 Conc. GAF Corp.
Chem. Descrip.: Polyether (nonionic)
Category/Applications: Wetter and dispersant for cationic and disperse dyes on acrylics; compatibilizes cationic and disperse dyes; provides clean, bright shades in cross dyeings; emulsifies residual oil and lubricant in the dyebath
Form: Liquid
Ref.: Catalogue 2303-036

Gaftex DN-159 Conc. GAF Corp.
Chem. Descrip.: Polyether (nonionic)
Category/Applications: Leveling agent for disperse dyes on nylon; highly effective in preventing shaded dyeings due to temp. variations in dyeing equipment; promotes uniform dye transfer and allows full color development; low foaming biodegradable lubricant
Form: Liquid
Ref.: Catalogue 2303-036

Gaftex LN-110. GAF Corp.
Chem. Descrip.: Polyamide polyether (modified nonionic)
Category/Applications: Leveling-migrating agent in atmospheric and pressure dyeing of nylon yarn, piece goods, and carpeting; compatibilizes anionic/cationic dyes in a single bath; highly effective in repair dyeing
Form: Liquid
Ref.: Catalogue 2303-036

Gaftex PT. GAF Corp.
Chem. Descrip.: Modified copolymer of methyl vinyl ether and maleic anhydride
Category/Applications: Thickener in aq. media; esp. effective in textile print paste systems
Form: Powder
Ref.: Catalogue 2303-036

Ganex P and V Polymers. GAF Corp.
Chem. Descrip.: Alkylated vinylpyrrolidone polymers
Category/Applications: Polymers of varying hydrophobic/hydrophilic characteristics, exhibiting a high degree of surface activity; Ganex P-904 is a water-soluble dispersant in aq. ag chemical formulations or pigment suspensions; Ganex V polymers are sol. in mineral oil, organic solvents, and other polymers; used in cosmetics and toiletries as moisture barriers, adhesives, protective colloids, and microencapsulating resins; function as dispersants for pigments and solubilizers for dyes; employed as petroleum additives, as sludge and detergent dispersants, and pour point depressants; protective colloids in coatings and suspending aids in polymerization; synthetic sizing additives, dye receptors for synthetic fibers, and dyeing assistants for anionic/cationic systems; anti-redeposition agents in drycleaning
Form: Liquids to waxy solids
Ref.: Catalogue 2303-036

Gantrez AN-119, AN-139, AN-149, AN-169, AN-179. GAF Corp.
Chem. Descrip.: Poly(methyl vinyl ether/maleic anhydride)
Category/Applications: Water-sol. copolymer with thickening, dispersing, and stabilizing properties; produces clear films of high tensile and cohesive strength; compatible with plasticizers;

protective colloids, particle size regulators, and mechanical stabilizers in emulsion polymerization of vinyl acetate and in suspension polymerization of vinyl chloride; gelling agents, thickeners, emulsion stabilizers, explosive stabilizers, anticorrosion coatings, and suspending aids; used in adhesives, detergents, textiles
Form: Powder; AN-119 is a low-visc. type; AN-139 and AN-149 are med.-visc.; AN-169 and AN-179 are high-visc.
Gen'l. Prop.: Sol.: Water-sol.; sol. in acid, caustic, and certain organic solvents; can be insolubilized by crosslinking with polyfunctional compounds such as polyols and polyamines
Ref.: Catalogue 2303-036

Gantrez AN-8194. GAF Corp.
Chem. Descrip.: Poly(octadecyl vinyl ether/maleic anhydride)
Category/Applications: Hydrocarbon-sol. copolymer that forms waxy films with good water resistance and thermal properties; compatible with silicone release agents and modifies release level for pressure-sensitive labels and tapes; used in adhesives, coatings, and water-repellent coatings for porous and nonporous substrates
Form: Liquid
Composition: 40% toluene sol'n.
Ref.: Catalogue 2303-036

Gantrez B-773. GAF Corp.
Chem. Descrip.: Poly(vinyl isobutyl ether)
Category/Applications: Tacky polymer with excellent adhesion to plastic, metal, and coated surfaces; used in a variety of adhesives systems; also as plasticizer and leveling agent for surface coatings
Form: Liquid
Composition: 70% in hexane sol'n.
Gen'l. Prop.: Sol.: Sol. in aromatic, aliphatic, and chlorinated hydrocarbons
Ref.: Catalogue 2303-036

Gantrez ES-225. GAF Corp.
Chem. Descrip.: Monoethyl ester of poly(methyl vinyl ether/maleic acid)
Category/Applications: Polymer forming tough, clear, glossy films which dry tack-free and exhibit good adhesion and moisture resistance; films have enteric and controlled-release properties; compatible with various plasticizers and modifiers; used in cosmetics (film-formers in hair sprays and grooming aids), coatings (film-formers in shoe polishes, decorative and protective sprays, paper and leather coatings; temporary protective metal and plastic coatings; enteric and sustained-release coating agents for tablets; film-formers in aerosol and spray bandages)
Form: Clear, viscous liquid
Composition: 50% ethanol sol'n.
Gen'l. Prop.: Sol. [1 g resin (100% solids) in 9 g solvent]: Sol. in ethanol, isopropanol, diethylene glycol, tetrahydrofuran, ethylene glycol monomethyl ether, butyl Carbitol, acetone, cyclohexanone, dioxane; water-sol.; Sp. Gr.: 0.983; Density: 8.18 lb/gal; Acid No.: 275–300 (100% solids)
Toxicity/Handling: Gantrez ES monoesters have the same flammability characteristics as the alcohol vehicles in which the polymers are supplied; relatively nontoxic by the single dose oral route and are neither ocular nor primary skin irritants
Ref.: Bulletin 9642-052, catalogue 2303-036

Gantrez ES-335. GAF Corp.
Chem. Descrip.: Monoisopropyl ester of poly(methyl vinyl ether/maleic acid)
Category/Applications & Form: See

Gantrez ES-225
Composition: 50% isopropanol sol'n.
Gen'l. Prop.: Sol.: Sol. in esters, ketones, glycol ethers, alcohol; insol. in acidic water; sol. in alkali; Sp. Gr.: 0.957; Density: 7.98 lb/gal; Acid No.: 250–280 (100% solids)
Toxicity/Handling: See Gantrez ES-225
Ref.: Bulletin 9642-052, catalogue 2303-036

Gantrez ES-425. GAF Corp.
Chem. Descrip.: Monobutyl ester of poly(methyl vinyl ether/maleic acid)
Category/Applications & Form: See Gantrez ES-225
Composition: 50% ethanol sol'n.
Gen'l. Prop.: Sol. [1 g resin (100% solids) in 9 g solvent]: Sol. in ethanol, isopropanol, diethylene glycol, tetrahydrofuran, ethylene glycol monomethyl ether, butyl Carbitol, ethyl acetate, acetone, cyclohexanone, dioxane; water-sol.; Sp. Gr.: 0.977; Density: 8.13 lb/gal; Acid No.: 235–265 (100% solids)
Toxicity/Handling: See Gantrez ES-225
Ref.: Bulletin 9642-052, catalogue 2303-036

Gantrez ES-435. GAF Corp.
Chem. Descrip.: Monobutyl ester of poly(methyl vinyl ether/maleic acid)
Category/Applications & Form: See Gantrez ES-225
Composition: 50% isopropanol sol'n.
Gen'l. Prop.: Sol.: See Gantrez ES-335; Sp. Gr.: 0.962; Density: 8.02 lb/gal; Acid No.: 245–275 (100% solids)
Toxicity/Handling: See Gantrez ES-225
Ref.: Bulletin 9642-052, catalogue 2303-036

Gantrez M-154, M-555, M-556, M-574. GAF Corp.
Chem. Descrip.: Poly(methyl vinyl ether)
Category/Applications: Polymer which functions as tackifier, binder, and plasticizer; used in printing inks, textile sizes and finishes, latex modification
Form: Liquids
Composition: 50% aq. sol'n. (M-154); 50% toluene sol'n.(M-555, M-556); 70% toluene sol'n. (M-574)
Gen'l. Prop.: Sol.: Sol. in water and diverse organic solvents; thermally reversible solubility in aq. systems
Ref.: Catalogue 2303-036

Gantrez S-95. GAF Corp.
Chem. Descrip.: Poly(methyl vinyl ether/maleic acid)
Category/Applications: Hydrolyzed low m.w. polymer; water-sol. polyelectrolyte similar to Gantrez AN series; rapid cold-water solubility over entire pH range
Form: Powder
Gen'l. Prop.: pH: ≈ 2 (5% aq. sol'n.)
Ref.: Catalogue 2303-036

Gantrez S-97. GAF Corp.
Chem. Descrip.: Poly(methyl vinyl ether/maleic acid)
Category/Applications: Hydrolyzed high m.w. polymer; see Gantrez S-95
Form: Powder
Gen'l. Prop.: pH: ≈ 2 (5% aq. sol'n.)
Ref.: Catalogue 2303-036

Gel 151. Isochem Resins Co.
Category/Applications: Two-part thixotropic dips; service temps. –45 to 165 C; mix ratio: 14 phr of Hardener 16
Form: Thixotropic gel; Color: Black
Gen'l. Prop.: Sp. Gr.: 1.24
Mech. Prop.: Tens. Str.: 7600 psi; Compr. Str.: 8700 psi; Hardness (Shore D): 90
Therm. Prop.: Conduct.: 8.2×10^{-4} Cal/s cm^2 C/cm; Distort. Temp.: 138 C; Coeff. of Linear Exp.: 3.8×10^{-5}/C

Elec. Prop.: Dissip. Factor: 0.001 (100 KC @ 27 C); Dielec. Str.: 488 V/mil; Dielec. Const.: 4.65 (100 KC @ 27 C); Vol. Resist.: 10^{16} ohm-cm (27 C)
Curing Charact.: Pot Life: 4 h (1 lb mass @ 27 C); Suggested Cure: 90 min @ 100 C
Ref.: Bulletin

Gel Trans Flex. Isochem Resins Co.
Category/Applications: Flexible dielectric gel and dipping comp.; coating printed circuit boards; cuttable and repairable; top thermal shock properties; mix ratio: 50 phr of Hardener TF
Form: Color: Clear
Gen'l. Prop.: Sp. Gr.: 1.01; Visc.: 6000 cps (mixed, 27 C)
Mech. Prop.: Tens. Str.: 2200 psi; Compr. Str.: 3300 psi
Therm. Prop.: Conduct.: 4.5×10^{-4} Cal/s cm^2 C/cm; Coeff. of Linear Exp.: 6.2×10^{-5}/C
Elec. Prop.: Dissip. Factor: 0.035 (100 KC, 27 C); Dielec. Str.: 310 V/mil; Dielec. Const.: 3.1 (100 KC, 27 C); Vol. Resist.: 10^{13} ohm-cm (27 C)
Curing Charact.: Pot Life: 2 h (1 lb mass @ 27 C); Suggested Cure: R.T. or mild heat
Ref.: Bulletin

Genal GF-7000 Series. General Electric Co.
Chem. Descrip.: Asbestos-free, glass-filled phenolic comp.
Category/Applications: Molding comp. offering excellent dimensional stability, chemical resistance, and heat resistance up to 500 F, and lower creep under load than glass-filled thermoplastics; used in automotive parts (disc brake components, solenoid covers, accessory housings, engine components, transmission components, commutators, etc.), electrical components (rotors, coils, high-energy ignition system components), electrical applications (circuit breakers, thermostat componentry, connectors, relays, light diffusers or baffles, printed circuit boards, sockets, switches), portable and major appliances, motor support housings, brush holders, etc.
Gen'l. Prop.: Stability: Excellent dimensional stability, heat resistance, and resistance to gasoline, ethylene glycol, engine and brake fluids, oils, lubricants, and engine degreasers
Ref.: Pamphlet GF-5

Genal GF-7011. General Electric Co.
Chem. Descrip.: Asbestos-free, glass-filled phenolic comp.
Category/Applications: Low glass content material developed esp. for metal replacement; its coeff. of linear thermal expansion approximates steel and nonferrous metals to meet the stringent application requirements where a GF-7011 part is assembled and mated with precision machined metal parts; see also Genal GF-7000 Series
Gen'l. Prop.: Sp. Gr.: 2.0
Mech. Prop.: Tens. Str.: 6500 psi; Flex. Str.: 11,000 psi; Flex. Mod.: 3.1×10^6 psi; Compr. Str.: 28,000 psi; Impact Str. (Izod): 0.35 ft lb/in. notch; Hardness (Rockwell): M 105
Therm. Prop.: Distort. Temp.: 380 F (@ 264 psi); Coeff. of Linear Exp.: 0.83×10^{-5} in./in./F; Flamm.: V-O (1/16 in., UL94)
Elec. Prop.: Dissip. Factor: 0.03 (dry, 60 cycle); Dielec. Str.: 350 V/mil (60 cycle, 25 C, S/T); Dielec. Const.: 6.5 (Dry, 60 cycle); Vol. Resist.: 10^{11} ohm-cm; Arc Resist.: 180 s
Ref.: Pamphlet GF-5

Genal GF-7021, GF-7021P. General Electric Co.

Chem. Descrip.: Asbestos-free, glass-filled phenolic comp.

Category/Applications: Med. glass-content material designed to replace die-cast metals and long-fiber asbestos compounds; esp. suitable for use in commutators, flatiron skirts, motor housings, and transmission components; see also Genal GF-7000 Series

All properties–compression molded (GF-7021) and injection molded (GF-7021P):

Gen'l. Prop.: Sp. Gr.: 1.74 (both)

Mech. Prop. (both): Tens. Str.: 9000 psi; Flex. Str.: 12,500 psi; Flex. Mod.: 2 X 10^6 psi; Compr. Str.: 26,000 psi; Impact Str. (Izod): 0.45 ft lb/in. notch; Hardness (Rockwell): M 115

Therm. Prop.: Distort. Temp.: 390 F (both, @ 264 psi); Coeff. of Linear Exp.: 1.39 X 10^{-5} in./in./F (GF-7021P); Flamm.: V-O (both, 1/16 in., UL94)

Elec. Prop. (both): Dissip. Factor: 0.03 (dry, 60 cycle); Dielec. Str.: 360 V/mil (60 cycle, 25 C, S/T); Dielec. Const.: 6.5 (dry, 60 cycle); Vol. Resist.: 6 X 10^{11} ohm-cm; Arc Resist.: 182 s

Ref.: Pamphlet GF-5

Genal GF-7031P. General Electric Co.

Chem. Descrip.: Asbestos-free, glass-filled phenolic comp.

Category/Applications: High glass-content material developed specifically for connectors, structural motor supports, and automotive components; offers improved impact and flex. strength with minimal change in sp. gr.; see also Genal GF-7000 Series

Gen'l. Prop.: Sp. Gr.: 1.70

Mech. Prop.: Tens. Str.: 9000 psi; Flex. Str.: 15,000 psi; Flex. Mod.: 1.8 X 10^6 psi; Compr. Str.: 25,000 psi; Impact Str. (Izod): 0.45 ft lb/in. notch; Hardness (Rockwell): M 108

Therm. Prop.: Distort. Temp.: 420 F (@ 264 psi); Coeff. of Linear Exp.: 1.1 X 10^{-5} in./in./F; Flamm.: V-O (1/16 in., UL94)

Elec. Prop.: Dissip. Factor: 0.03 (dry, 60 cycle); Dielec. Str.: 350 V/mil (60 cycle, 25 C, S/T); Dielec. Const.: 6.0 (dry, 60 cycle); Vol. Resist.: 2 X 10^{12} ohm-cm; Arc Resist.: 182 s

Ref.: Pamphlet GF-5

Genal MX582. General Electric Co.

Chem. Descrip.: Asbestos-free phenolic

Category/Applications: High heat, improved impact molding comp. offering a UL continuous use rating of 180 C, fast-cycling moldability with excellent mold release characteristics, and excellent surface finish; used as a mechanical interface with metal (e.g., bearing retainer), in consumer applications (e.g., automotive ashtray), hand-held electrical appliances (iron plate), fog lamps, diesel engine gears, appliance bases, etc.

Gen'l. Prop.: Sp. Gr.: 1.56 (D792A); Density: 0.65 g/cc (apparent, D1895); Stability: Excellent dimensional stability, high-heat endurance, and chemical resistance

Mech. Prop.: Tens. Str.: 7500 psi (D-651); Flex. Str.: 12,300 psi (D790); Flex. Mod.: 1.2 X 10^6 psi (D790); Compr. Str.: 16,000 psi (D695); Impact Str. (Izod): 0.6 ft lb/in. notch (D256A); Hardness (Rockwell): M 85 (D785A)

Therm. Prop.: Distort. Temp.: 500 F + (@ 264 psi, D648); Flamm.: V-O (UL-94)

Elec. Prop.: Dielec. Str.: 395 V/mil (dry, 60 Hz, S/T, 40 h, 23 C, 50% r.h., D149);

Vol. Resist.: 10^{11} ohm-cm (D257); Arc Resist.: 132 s (D495)
Ref.: Pamphlet GF-20-7

Genal P4300. General Electric Co.
Chem. Descrip.: Asbestos-free E grade of phenolic
Category/Applications: See Genal 4300-E, P4300E
Form: Pellets
Gen'l. Prop.: Sp. Gr.: 1.57 (D792A)
Mech. Prop.: Tens. Str.: 6000 psi min. (D651); Flex. Str.: 10,000 psi min. (D-790); Flex. Mod.: 10^6 psi (D790); Compr. Str.: 30,000 psi min. (D695); Impact Str. (Izod): 0.36 ft lb/in. notch (D256A); Hardness (Rockwell): 105 (D-785A)
Therm. Prop.: Distort. Temp.: 360 F (@ 264 psi, D648); Flamm.: NB (D635)
Elec. Prop.: Dissip. Factor: 0.22 (dry, 60 cycles, D150); Dielec. Str.: 400 (25 C, S/T, D149); Dielec. Const.: 10.6 (dry, 60 cycles, D150); Arc Resist.: 0184 s (D495); Insul. Resist.: 800,000 megohms (D257)
Ref.: Pamphlet CDC-491

Genal P4301. General Electric Co.
Chem. Descrip.: Asbestos-free E grade of phenolic
Category/Applications: See Genal 4300-E, P4300E
Form: Pellets
Gen'l. Prop.: Sp. Gr.: 1.50 (D792A)
Mech. Prop.: Tens. Str.: 5500 psi min. (D651); Flex. Str.: 12,000 psi min. (D-790); Flex. Mod.: 10^6 psi (D790); Compr. Str.: 29,000 psi min. (D695); Impact Str. (Izod): 0.32 ft lb/in. notch (D256A); Hardness (Rockwell): 106 (D-785A)
Therm. Prop.: Distort. Temp.: 370 F (@ 264 psi, D648); Flamm.: NB (D635)
Elec. Prop.: Dissip. Factor: 0.25 (dry, 60 cycles, D150); Dielec. Str.: 210 (25 C, S/T, D149); Dielec. Const.: 22.5 (dry, 60 cycles, D150); Arc Resist.: 183 s (D-495); Insul. Resist.: 800,000 megohms (D257)
Ref.: Pamphlet CDC-491

Genal P4050. General Electric Co.
Chem. Descrip.: Asbestos-free E grade of phenolic
Category/Applications: See Genal 4300-E, P4300E
Form: Pellets
Gen'l. Prop.: Sp. Gr.: 1.39 (D792A)
Mech. Prop.: Tens. Str.: 6500 psi min. (D651); Flex. Str.: 9000 psi min. (D-790); Flex. Mod.: 10^6 psi (D790); Compr. Str.: 27,000 psi min. (D695); Impact Str. (Izod): 0.30 ft lb/in. notch (D256A); Hardness (Rockwell): 110 (D-785A)
Therm. Prop.: Distort. Temp.: 335 F (@ 264 psi, D648); Flamm.: NB (D635)
Elec. Prop.: Dissip. Factor: 0.20 (dry, 60 cycles, D150); Dielec. Str.: 340 (25 C, S/T, D149); Dielec. Const.: 7.8 (dry, 60 cycles, D150); Insul. Resist.: 330,000 megohms (D257)
Ref.: Pamphlet CDC-491

Genal 4300E, P4300E. General Electric Co.
Chem. Descrip.: Asbestos-free E grade of phenolic
Category/Applications: Thermosetting, screw-injection molding material offering good mechanical properties, UL acceptance, heat resistance, dimensional stability, and chemical resistance at low cost; used in appliances (handles, knobs, timing motor housings, blower wheels, can opener fronts, heater bases,

lamp housings), automotive (transmission components, power brake components, ignition system parts, wiring systems, commutators, fan spacers, rotors, fuse blocks, coil towers, solenoid covers, ash trays, various light and instrument housings, etc.), and electrical-electronic applications (circuit breakers, switches, thermostat components, relays, receptacles, terminal blocks, stiffener boards, meter bases, coil bobbins, IC carriers, etc.)

Gen'l. Prop.: Sp. Gr.: 1.55 (D792A)

Mech. Prop.: Tens. Str.: 8000 psi (D-651); Flex. Str.: 11,000 psi (D790); Flex. Mod.: 10^6 psi (D790); Compr. Str.: 30,000 psi (D695); Impact Str. (Izod): 0.30 ft lb/in. notch (D256A)

Therm. Prop.: Distort. Temp.: 380 F (@ 264 psi, D648); Flamm.: SE-O (UL94)

Elec. Prop.: Dielec. Str.: 350 V/mil (60 cps, 25 C, S/T, D149); Arc Resist.: 180 s (D495)

Ref.: Pamphlet CDC-491

Genal 4301E, P4301E. General Electric Co.

Chem. Descrip.: Asbestos-free E grade of phenolic

Category/Applications: See Genal 4300-E, P4300E

Gen'l. Prop.: Sp. Gr.: 1.50 (D792A)

Mech. Prop.: Tens. Str.: 8500 psi (D-651); Flex. Str.: 11,000 psi (D790); Flex. Mod.: 10^6 psi (D790); Compr. Str.: 30,000 psi (D695); Impact Str. (Izod): 0.32 ft lb/in. notch (D256A)

Therm. Prop.: Distort. Temp.: 370 F (@ 264 psi, D648); Flamm.: SE-O (UL94)

Elec. Prop.: Dielec. Str.: 250 V/mil (60 cps, 25 C, S/T, D149); Arc Resist.: 150 s (D495)

Ref.: Pamphlet CDC-491

Genal 12553E, 12983E. General Electric Co.

Chem. Descrip.: Asbestos-free E grade of phenolic

Category/Applications: See Genal 4300-E, P4300E

Form: Granular

Gen'l. Prop.: Sp. Gr.: 1.52, 1.55 respectively (D792A)

Mech. Prop.: Tens. Str.: 6000 psi (both, D651); Flex. Str.: 10,000 and 9000 psi respectively (D790); Flex. Mod.: 10^6 psi (both, D790); Compr. Str.: 30,000 psi (both, D695); Impact Str. (Izod): 0.32 and 0.34 ft lb/in. notch respectively (D-256A)

Therm. Prop.: Distort. Temp.: 365 and 380 F (@ 264 psi, D648); Flamm.: SE-O (both, UL94)

Elec. Prop.: Dielec. Str.: 250 and 350 V/mil respectively (60 cps, 25 C, S/T, D-149); Arc Resist.: 150 and 180 s respectively (D495)

Ref.: Pamphlet CDC-491

Genu Carrageenan. Hercules Inc.

Chem. Descrip.: Refined carrageenan

Category/Applications: Emulsifier, stabilizer, thickener, or gelling agent in foods, pharmaceuticals, and cosmetics

Form: Powder

Ref.: Products catalogue

Genu Pectins. Hercules Inc.

Chem. Descrip.: High-methoxy and low-methoxy purified natural hydrocolloid derived from citrus peels; consists chiefly of partially methoxylated polygalacturonic acid

Category/Applications: Used primarily as a gelling agent for jams, jellies, and other foods in a controlled form and as a viscosity builder, protective colloid,

and stabilizer for the food systems, pharmaceutical, and cosmetic industries
Form: Powder; Color: Light cream to grayish; Odor and Flavor: None
Ref.: Products catalogue

Geon 92, 99. BF Goodrich Co.
Chem. Descrip.: Vinyl resin—super EP type
Form: Particles (100% through 30 mesh)
Gen'l. Prop.: Visc. (Inherent): 1.02, 1.37 respectively (D1243-66)
Ref.: Bulletin

Geon 100X122. BF Goodrich Co.
Chem. Descrip.: Vinyl blending resin
Gen'l. Prop.: Sp. Gr.: ≈ 1.40; Visc. (Inherent): 0.85 (D1243-66)
Mech. Prop. (20-mil film fused 5 min @ 149 C, Die C, 500 mm/min): Tens. Str.: 8.4 MPa; Tens. Elong.: 183%
Ref.: Bulletin

Geon 101F-4. BF Goodrich Co.
Chem. Descrip.: Vinyl blending resin
Gen'l. Prop.: Sp. Gr.: ≈ 1.40; Visc. (Inherent): 1.10 (D1243-66)
Mech. Prop. (20-mil film fused 5 min @ 149 C, Die C, 500 mm/min): Tens. Str.: 8.3 MPa; Tens. Elong.: 137%
Ref.: Bulletin

Geon 102EP F-5, 103EP, 103EP F-76, 110X334. BF Goodrich Co.
Chem. Descrip.: Vinyl resin
Category/Applications: General purpose applications
Form: Particles (100% through 40 mesh)
Gen'l. Prop.: Visc. (Inherent): 1.13, 1.02, 0.92, 0.68 respectively (D1243-66)
Ref.: Bulletin

Geon 120X241. BF Goodrich Co.
Chem. Descrip.: Vinyl dispersion resin
Category/Applications: Foam applications
Gen'l. Prop.: Sp. Gr.: ≈ 1.40; Visc. (Inherent): 0.73 (D1243-66)
Mech. Prop. (0.51-mm film fused 5 min @ 135 C, Die C, 500 mm/min): Tens. Str.: 5.2 MPa; Tens. Elong.: 80%
Ref.: Bulletin

Geon 120X251. BF Goodrich Co.
Chem. Descrip.: Vinyl dispersion resin
Category/Applications: Foam applications
Gen'l. Prop.: Sp. Gr.: ≈ 1.40; Visc. (Inherent): 1.00 (D1243-66)
Mech. Prop. (0.51-mm film fused 5 min @ 135 C, Die C, 500 mm/min): Tens. Str.: 6.9 MPa; Tens. Elong.: 100%
Ref.: Bulletin

Geon 120X271. BF Goodrich Co.
Chem. Descrip.: Vinyl dispersion resin
Category/Applications: Specialty resin
Gen'l. Prop.: Sp. Gr.: ≈ 1.40; Visc. (Inherent): 1.40 (D1243-66)
Mech. Prop. (0.51-mm film fused 5 min @ 135 C, Die C, 500 mm/min): Tens. Str.: 5.2 MPa; Tens. Elong.: 80%
Ref.: Bulletin

Geon 121. BF Goodrich Co.
Chem. Descrip.: Vinyl dispersion resin
Category/Applications: General purpose applications
Gen'l. Prop.: Sp. Gr.: ≈ 1.40; Visc. (Inherent): 1.20 (D1243-66)
Mech. Prop. (0.51-mm film fused 5 min @ 135 C, Die C, 500 mm/min): Tens. Str.: 6.9 MPa; Tens. Elong.: 100%
Ref.: Bulletin

Geon 124. BF Goodrich Co.
Chem. Descrip.: Vinyl dispersion resin
Category/Applications: Foam applications
Gen'l. Prop.: Sp. Gr.: ≈ 1.40; Visc. (Inherent): 0.89 (D1243-66)
Mech. Prop. (0.51-mm film fused 5 min @ 135 C, Die C, 500 mm/min): Tens. Str.: 6.5 MPa; Tens. Elong.: 100%
Ref.: Bulletin

Geon 124F-1. BF Goodrich Co.
Chem. Descrip.: Vinyl dispersion resin
Category/Applications: Foam applications
Gen'l. Prop.: Sp. Gr.: ≈ 1.40; Visc. (Inherent): 0.92 (D1243-66)
Mech. Prop. (0.51-mm film fused 5 min @ 135 C, Die C, 500 mm/min): Tens. Str.: 5.4 MPa; Tens. Elong.: 90%
Ref.: Bulletin

Geon 128. BF Goodrich Co.
Chem. Descrip.: Vinyl dispersion resin
Category/Applications: General purpose applications
Gen'l. Prop.: Sp. Gr.: ≈ 1.40; Visc. (Inherent): 1.20 (D1243-66)
Mech. Prop. (0.51-mm film fused 5 min @ 135 C, Die C, 500 mm/min): Tens. Str.: 6.9 MPa; Tens. Elong.: 80%
Ref.: Bulletin

Geon 130X17. BF Goodrich Co.
Chem. Descrip.: Vinyl dispersion resin
Category/Applications: Specialty resin
Gen'l. Prop.: Sp. Gr.: ≈ 1.39; Visc. (Inherent): 1.40 (D1243-66)
Mech. Prop. (0.51-mm film fused 5 min @ 149 C, Die C, 500 mm/min): Tens. Str.: 8.0 MPa; Tens. Elong.: 190%
Ref.:Bu lletin

Geon 135. BF Goodrich Co.
Chem. Descrip.: Vinyl dispersion resin
Category/Applicaitons: Low fusion resin
Gen'l. Prop.: Sp. Gr.: ≈ 1.39; Visc. (Inherent): 1.12 (D1243-66)
Mech. Prop. (0.51-mm film fused 5 min @ 135 C, Die C, 500 mm/min): Tens. Str.: 11.0 MPa; Tens. Elong.: 200%
Ref.: Bulletin

Geon 140X30. BF Goodrich Co.
Chem. Descrip.: Vinyl blending resin
Gen'l. Prop.: Sp. Gr.: ≈ 1.40; Visc. (Inherent): 0.74 (D1243-66)
Mech. Prop. (20-mil film fused 5 min @ 149 C, Die C, 500 mm/min): Tens. Str.: 10.0 MPa; Tens. Elong.: 220%
Ref.: Bulletin

Geon 140X31. BF Goodrich Co.
Chem. Descrip.: Vinyl blending resin
Category/Applications: Powder coating resin which dry blends well in high-intensity mixers; used in the electrostatic spray and fluidized bed coating market; its vinyl powder compounds readily melt and flow into durable, continuous coatings when heated to 204–232 C
Form: Fine particles with narrow size distribution
Gen'l. Prop.: Sp. Gr.: ≈ 1.40; Visc. (Inherent): 0.84 (D1243-66)
Mech. Prop. (20-mil film fused 5 min @ 149 C, Die C, 500 mm/min): Tens. Str.: 11.7 MPa; Tens. Elong.: 223%
Ref.: Bulletin

Geon 140X32. BF Goodrich Co.
Chem. Descrip.: Vinyl blending resin
Category/Applications & Form: See Geon 140X31

Gen'l. Prop.: Sp. Gr.: ≈ 1.40; Visc. (Inherent): 0.76 (D1243-66)
Mech. Prop. (20-mil film fused 5 min @ 149 C, Die C, 500 mm/min): Tens. Str.: 11.4 MPa; Tens. Elong.: 215%
Ref.: Bulletin

Geon 2042. BF Goodrich Co.
Chem. Descrip.: Vinyl
Category/Applications: Wire and cable insulation; suggested for T, TW, TW 60 C oil, type UA flexible conduit
Form: Color: Ivory, B. gray
Gen'l. Prop.: Sp. Gr.: 1.31 ± 0.02 (40-mil molded sheet, D792)
Mech. Prop. (40-mil molded sheet): Tens. Str.: 17.23 MPa (D412); Tens. Elong.: 300% (D412); Tens. Mod.: 10.34 MPa (@ 100% elong., D412); Hardness: 87A ± 3 (instantaneous, D-2240)
Elec. Prop.: Dielec. Str.: 27,580 V/mm (14 AWG 2/64 in. wall, S/T, D149); Dielec. Const.: 6.5 (14 AWG 2/64 in. wall, 1000 Hz, 1 day @ 30 C, UL83); Vol. Resist.: 2.5 × 10^{12} ohm-cm (75-mil slab, 95 C, D257)
Ref.: Bulletin

Geon 3007. BF Goodrich Co.
Chem. Descrip.: CPVC comp.
Category/Applications: Extrusion pipe applications
Form: Cube
All properties as compression molded:
Gen'l. Prop.: Sp. Gr.: 1.55 ± 0.02 (D-792)
Mech. Prop.: Tens. Str.: 58 MPa (23 C, D638); Tens. Mod.: 2919 MPa (23 C, D-638); Flex. Str.: 108 MPa (23 C, D790); Flex. Mod.: 2939 MPa (23 C, D790); Impact Str. (Izod): 1.6 J/cm notch (23 C, D256); Hardness (Rockwell): R 120 (A, D785)
Therm. Prop.: Conduct.: 0.137 W/m•K (C177); Distort. Temp.: 105 C (annealed, @ 1.8 MPa, D648); Coeff. of Linear Exp.: 6.8 × 10^5 cm/cm/C (D696)
Elec. Prop.: Dissip. Factor: 0.019 (60 Hz, D150); Dielec. Str.: 51,220 V/mm (D149); Dielec. Const.: 3.4 (10^3 Hz, D-150)
Ref.: Bulletin

Geon 3010. BF Goodrich Co.
Chem. Descrip.: CPVC comp.
Category/Applications: Molding fittings
Form: Cube
All properties as compression molded:
Gen'l. Prop.: Sp. Gr.: 1.52 ± 0.02 (D-792)
Mech. Prop.: Tens. Str.: 57 MPa (23 C, D638); Tens. Mod.: 3312 MPa (23 C, D638); Flex. Str.: 110 MPa (23 C, D-790); Flex. Mod.: 2760 MPa (23 C, D-790); Impact Str. (Izod): 1.1 J/cm notch (23 C, D256); Hardness (Rockwell): R 118 (A, D785)
Therm. Prop.: Distort. Temp.: 100 C (annealed, @ 1.8 MPa, D648); Coeff. of Linear Exp.: 6.8 × 10^5 cm/cm/C (D696)
Elec. Prop.: Dissip. Factor: 0.019 (60 Hz, D150); Dielec. Str.: 51,220 V/mm (D149); Dielec. Const.: 3.4 (10^3 Hz, D150)
Ref.: Bulletin

Geon 7082. BF Goodrich Co.
Chem. Descrip.: Rigid vinyl comp.
Category/Applications: Normal-impact extrusion comp. for exterior building products; profile applications
Form: Cube
Gen'l. Prop.: Sp. Gr.: 1.46 ± 0.02 (D-792)
Mech. Prop.: Tens. Str.: 49 MPa (26 C, D638); Tens. Mod.: 2829 MPa (D638); Flex. Str.: 112 MPa (D790); Compr.

Str.: 70 MPa (D695); Impact Str. (Izod): 0.95 J/cm notch (23 C, D256); Hardness (Durometer D): 78 ± 3 (D785)
Therm. Prop.: Conduct.: 0.19 W/m•K, C (C177); Distort. Temp.: 72 C (@ 1.82 MPa, D648); Coeff. of Linear Exp.: 6.3 × 10^{-5} cm/cm/C (D696)
Elec. Prop.: Dielec. Str.: 28,920 V/mm (normal, D149); Vol. Resist.: 7.55 × 10^{8} megohm-cm (normal, D257); Arc Resist.: 6 s (to track, D495)
Ref.: Bulletin

Geon 7084. BF Goodrich Co.
Chem. Descrip.: Rigid vinyl comp.
Category/Applications: See Geon 7082
Form: Cube
Gen'l. Prop.: Sp. Gr.: 1.46 ± 0.02 (D-792)
Mech. Prop.: Tens. Str.: 50 MPa (26 C, D638); Tens. Mod.: 2898 MPa (D638); Flex. Str.: 100 MPa (D790); Compr. Str.: 66 MPa (D695); Impact Str. (Izod): 0.74 J/cm notch (23 C, D256); Hardness (Durometer D): 80 ± 3 (D785)
Therm. Prop.: Conduct.: 0.14 W/m•K, C (C177); Distort. Temp.: 71 C (@ 1.82 MPa, D648); Coeff. of Linear Exp.: 5.2 × 10^{-5} cm/cm/C (D696)
Ref.: Bulletin

Geon 8610. BF Goodrich Co.
Chem. Descrip.: Vinyl
Category/Applications: Wire and cable insulation; suggested for T, TW, TW 60 C oil
Form: Color: Natural
Gen'l. Prop.: Sp. Gr.: 1.39 ± 0.02 (40-mil molded sheet, D792)
Mech. Prop. (40-mil molded sheet): Tens. Str.: 13.78 MPa (D412); Tens. Elong.: 270% (D412); Tens. Mod.: 8.96 MPa (@ 100% elong., D412); Hardness: 86A ± 3 (instantaneous, D2240)
Elec. Prop.: Dielec. Str.: 29,550 V/mm (14 AWG 2/64 in. wall, S/T, D149); Dielec. Const.: 6.8 (14 AWG 2/64 in. wall, 1000 Hz, 1 day @ 30 C, UL83); Vol. Resist.: 2.2 × 10^{12} ohm-cm (75-mil slab, 95 C, D257)
Ref.: Bulletin

Geon 8700A. BF Goodrich Co.
Chem. Descrip.: Rigid vinyl comp.
Category/Applications: Extrusion comp. for interior building products; profile applications
Form: Cube; Color: Natural
Gen'l. Prop.: Sp. Gr.: 1.39 ± 0.02 (D-792)
Mech. Prop.: Tens. Str.: 43 MPa (D638); Tens. Mod.: 2691 MPa (D638); Flex. Str.: 76 MPa (D790); Flex. Mod.: 2415 MPa (D790); Compr. Str.: 56 MPa (D-695); Impact Str. (Izod): 9.0 J/cm notch (23 C, D256); Hardness (Rockwell): R 111 (D785)
Therm. Prop.: Conduct.: 0.16 W/m C (D177); Distort. Temp.: 73 C (@ 1.82 MPa, D648); Coeff. of Linear Exp.: 6.6 × 10^{5} cm/cm/C (D696); Flamm.: V-O (UL94)
Elec. Prop.: Dissip. Factor: 2.9 (60 Hz, 30 C, D150); Dielec. Str.: 27,186 V/mm (normal, D149); Dielec. Const.: 3.9 (60 Hz, 30 C, D150); Vol. Resist.: 5.1 × 10^{8} megohm-cm (normal, D257); Arc Resist.: 6 s (to track, D495)
Ref.: Bulletin

Geon 8702. BF Goodrich Co.
Chem. Descrip.: Rigid vinyl comp.
Category/Applications: See Geon 8700A
Form: Cube; Color: Natural
Gen'l. Prop.: Sp. Gr.: 1.39 ± 0.02 (D-792)
Mech. Prop.: Tens. Str.: 37 MPa (D638); Tens. Mod.: 2412 MPa (D638); Flex. Str.: 96 MPa (D790); Flex. Mod.: 3101 MPa (D790); Compr. Str.: 50 MPa (D-695); Impact Str. (Izod): 11 J/cm notch (23 C, D256); Hardness (Rockwell): R

107 (D785)
Therm. Prop.: Conduct.: 0.16 W/m C (D177); Distort. Temp.: 73.5 C (@ 1.82 MPa, D648); Coeff. of Linear Exp.: 5.8 $\times 10^5$ cm/cm/C (D696)
Elec. Prop.: Dissip. Factor: 2.9 (60 Hz, 30 C, D150); Dielec. Str.: 27,186 V/mm (normal, D149); Dielec. Const.: 3.9 (60 Hz, 30 C, D150)
Ref.: Bulletin

Geon 8714. BF Goodrich Co.
Chem. Descrip.: Vinyl comp.
Category/Applications: Pipe extrusion comp.; NSF-approved 2110 for potable water
Form: Cube
Gen'l. Prop.: Sp. Gr.: 1.35 ± 0.02 (D-792)
Mech. Prop.: Tens. Str.: 43 MPa (D638); Tens. Mod.: 2373 MPa (D638); Flex. Str.: 80 MPa (D790); Flex. Mod.: 2653 MPa (D790); Impact Str. (Izod): 7.95 J/cm (23 C, D256); Hardness (Rockwell): R 107 (D785)
Therm. Prop.: Distort. Temp.: 72.6 C (@ 1.8 MPa, D648)
Ref.: Bulletin

Geon 8720. BF Goodrich Co.
Chem. Descrip.: Vinyl
Category/Applications: Wire and cable sheath; coaxial cable, noncontaminating jacket
Form: Color: Natural, gray, black
Gen'l. Prop.: Sp. Gr.: 1.20 ± 0.02 (40-mil molded sheet, D792)
Mech. Prop. (40-mil molded sheet): Tens. Str.: 11.71 MPa (D412); Tens. Elong.: 380% (D412); Tens. Mod.: 6.23 MPa (@ 100% elong., D412); Hardness: 84A ± 3 (instantaneous, D2240)
Elec. Prop.: Dielec. Str.: 23,640 V/mm (D149); Dielec. Const.: 11.2 (1000 Hz, 1 day @ 30 C, UL83); Vol. Resist.: 0.004 $\times 10^{11}$ ohm-cm (D257)
Ref.: Bulletin

Geon 8730. BF Goodrich Co.
Chem. Descrip.: Vinyl
Category/Applications: Wire and cable insulation; suggested for T, TW, TW 60 C oil, THW 60 C oil, 90 C MTW with 3/64 in. wall
Form: Color: Natural, black
Gen'l. Prop.: Sp. Gr.: 1.37 ± 0.02 (40-mil molded sheet, D792)
Mech. Prop. (40-mil molded sheet): Tens. Str.: 16.54 MPa (D412); Tens. Elong.: 275% (D412); Tens. Mod.: 10.68 MPa (@ 100% elong., D412); Hardness: 74C ± 3 (instantaneous, D-2240)
Elec. Prop.: Dielec. Str.: 29,550 V/mm (14 AWG 2/64 in. wall, S/T, D149); Dielec. Const.: 6.2 (14 AWG 2/64 in. wall, 1000 Hz, 1 day @ 30 C, UL83); Vol. Resist.: 1.4 $\times 10^{12}$ ohm-cm (75-mil slab, 95 C, D257)
Ref.: Bulletin

Geon 8737. BF Goodrich Co.
Chem. Descrip.: Vinyl
Category/Applications: Wire and cable sheath; REA PE 7, PE 73, MIL-C-915E (black) 80 C jacket
Form: Color: Natural, flame retarded black
Gen'l. Prop.: Sp. Gr.: 1.40 ± 0.02 (40-mil molded sheet, D792)
Mech. Prop. (40-mil molded sheet): Tens. Str.: 13.78 MPa (D412); Tens. Elong.: 300% (D412); Tens. Mod.: 7.58 MPa (@ 100% elong., D412); Hardness: 82A ± 3 (instantaneous, D2240)
Elec. Prop.: Dielec. Str.: 23,640 V/mm (D149); Dielec. Const.: 7.1 (1000 Hz,

1 day @ 30 C, UL83); Vol. Resist.: 5.2 $\times 10^{11}$ ohm-cm (D257)
Ref.: Bulletin

Geon 8745. BF Goodrich Co.
Chem. Descrip.: Vinyl
Category/Applications: Wire and cable sheath; 105 C jacket SPT 1, 2, 3, class 12, oil resistant and oil proof cord
Form: Color: Natural
Gen'l. Prop.: Sp. Gr.: 1.37 ± 0.02 (40-mil molded sheet, D792)
Mech. Prop. (40-mil molded sheet): Tens. Str.: 13.78 MPa (D412); Tens. Elong.: 340% (D412), Tens. Mod.: 8.96 MPa (@ 100% elong., D412); Hardness: 85A ± 3 (instantaneous, D2240)
Elec. Prop.: Dielec. Str.: 31,520 V/mm (D149); Dielec. Const.: 7.2 (1000 Hz, 1 day @ 30 C, UL83); Vol. Resist.: 0.136 $\times 10^{11}$ ohm-cm (D257)
Ref.: Bulletin

Geon 8750. BF Goodrich Co.
Chem. Descrip.: Rigid vinyl comp.
Category/Applications: Extrusion comp. for interior building products; sheet and profile applications
Form: Cube; Color: Natural
Gen'l. Prop.: Sp. Gr.: 1.42 ± 0.02 (D-792)
Mech. Prop.: Tens. Str.: 52 MPa (D638); Tens. Mod.: 3714 MPa (D638); Flex. Str.: 90 MPa (D790); Flex. Mod.: 3243 MPa (D790); Compr. Str.: 66 MPa (D-695); Impact Str. (Izod): 0.5 J/cm notch (23 C, D256); Hardness (Rockwell): R 118 (D785)
Therm. Prop.: Conduct.: 0.16 W/m C (D177); Distort. Temp.: 74 C (@ 1.82 MPa, D648); Coeff. of Linear Exp.: 5.7 $\times 10^{5}$ cm/cm/C (D696)
Elec. Prop.: Dissip. Factor: 1.2 (60 Hz, 30 C, D150); Dielec. Str.: 55,160 V/mm (normal, D149); Dielec. Const.: 3.7 (60 Hz, 30 C, D150)
Ref.: Bulletin

Geon 8761. BF Goodrich Co.
Chem. Descrip.: Vinyl comp.
Category/Applications: Pipe extrusion comp.; NSF-approved 1120 for potable water, TI-G1 for DWV
Form: Cube
Gen'l. Prop.: Sp. Gr.: 1.38 ± 0.02 (D-792)
Mech. Prop.: Tens. Str.: 51 MPa (D638); Tens. Mod.: 2862 MPa (D638); Flex. Str.: 105 MPa (D790); Flex. Mod.: 2860 MPa (D790); Impact Str.: 0.42 J/cm (23 C, D256); Hardness (Rockwell): R 113 (D785)
Therm. Prop.: Distort. Temp.: 70 C (@ 1.8 MPa, D648)
Ref.: Bulletin

Geon 8800. BF Goodrich Co.
Chem. Descrip.: Vinyl
Category/Applications: Wire and cable insulation; suggested for THHN-THWN, 105 C AWM, MIL-W-15878, MIL-W-5086, MIL-W-76 applications
Form: Color: Transparent, natural
Gen'l. Prop.: Sp. Gr.: 1.35 ± 0.02 (40-mil molded sheet, D792)
Mech. Prop. (40-mil molded sheet): Tens. Str.: 20.67 MPa (D412); Tens. Elong.: 250% (D412); Tens. Mod.: 20.40 MPa (@ 100% elong., D412); Hardness: 79C ± 3 (instantaneous, D-2240)
Elec. Prop.: Dielec. Str.: 27,580 V/mm (14 AWG 2/64 in. wall, S/T, D149); Dielec. Const.: 4.8 (14 AWG 2/64 in. wall, 1000 Hz, 1 day @ 30 C, UL83); Vol. Resist.: 1.8 $\times 10^{12}$ ohm-cm (75-mil slab, 95 C, D257)
Ref.: Bulletin

Geon 8801-021. BF Goodrich Co.
Chem. Descrip.: Vinyl
Category/Applications: Wire and cable insulation; suggested for T, TW, TW 60 C oil, NM jacket, 80 C AWM with 2/64 in. wall applications
Form: Color: Natural; also avail. in black and beige
Gen'l. Prop.: Sp. Gr.: 1.37 ± 0.02 (40-mil molded sheet, D792)
Mech. Prop. (40-mil molded sheet): Tens. Str.: 17.23 MPa (D412); Tens. Elong.: 300% (D412); Tens. Mod.: 11.71 MPa (@ 100% elong., D412); Hardness: 90A ± 3 (instantaneous, D-2240)
Elec. Prop.: Dielec. Str.: 33,490 V/mm (14 AWG 2/64 in. wall, S/T, D149); Dielec. Const.: 4.8 (14 AWG 2/64 in. wall, 1000 Hz, 1 day @ 30 C, UL83); Vol. Resist.: 2.1 × 10^{12} ohm-cm (75-mil slab, 95 C, D257)
Ref.: Bulletin

Geon 8801-025. BF Goodrich Co.
Chem. Descrip.: Vinyl
Category/Applications: Wire and cable insulation; suggested for T, TW, TW 60 C oil, NM jacket 80 C AWM, 75 C THW applications
Form: Color: Natural, flame-retarded natural, black
Gen'l. Prop.: Sp. Gr.: 1.37 ± 0.02 (40-mil molded sheet, D792)
Mech. Prop. (40-mil molded sheet): Tens. Str.: 17.23 MPa (D412); Tens. Elong.: 300% (D412); Tens. Mod.: 11.71 MPa (@ 100% elong., D412); Hardness: 90A ± 3 (instantaneous, D-2240)
Elec. Prop.: Dielec. Const.: 5.0 (14 AWG 2/64 in. wall, 1000 Hz, 1 day @ 75 C, UL83)
Ref.: Bulletin

Geon 8803. BF Goodrich Co.
Chem. Descrip.: Vinyl
Category/Applications: Telephone cable insulation and jacket
Form: Color: Natural, ivory, white, black, beige
Gen'l. Prop.: Sp. Gr.: 1.32 ± 0.02 (40-mil molded sheet, D792)
Mech. Prop. (40-mil molded sheet): Tens. Str.: 19.29 MPa (D412); Tens. Elong.: 300% (D412); Tens. Mod.: 10.34 MPa (@ 100% elong., D412); Hardness: 86A ± 3 (instantaneous, D-2240)
Elec. Prop.: Dielec. Str.: 19,700 V/mm (14 AWG 2/64 in. wall, S/T, D149); Dielec. Const.: 6.0 (14 AWG 2/64 in. wall, 1000 Hz, 1 day @ 30 C, UL83); Vol. Resist.: 0.2 × 10^{12} ohm-cm (75-mil slab, 95 C, D257)
Ref.: Bulletin

Geon 8804. BF Goodrich Co.
Chem. Descrip.: Vinyl
Category/Applications: Wire and cable sheath; general purpose jacket, CB, CATV coaxial cable
Form: Color: Natural, gray, black, red, and flame-retarded natural and black
Gen'l. Prop.: Sp. Gr.: 1.35 ± 0.02 (40-mil molded sheet, D792)
Mech. Prop. (40-mil molded sheet): Tens. Str.: 13.78 MPa (D412); Tens. Elong.: 300% (D412); Tens. Mod.: 11.71 MPa (@ 100% elong., D412); Hardness: 83A ± 3 (instantaneous, D-2240)
Elec. Prop.: Dielec. Str.: 23,640 V/mm (D149); Dielec. Const.: 8.0 (1000 Hz, 1 day @ 30 C, UL83); Vol. Resist.: 3.7 × 10^{11} ohm-cm (D257)
Ref.: Bulletin

Geon 8806. BF Goodrich Co.
Chem. Descrip.: Vinyl
Category/Applications: Wire and cable

sheath; coil cords, telephone handset cords
Form: Color: Natural, black, beige
All properties–20 AWG 1/64 in. wall:
Gen'l. Prop.: Sp. Gr.: 1.25 ± 0.02 (D-792)
Mech. Prop.: Tens. Str.: 15.16 MPa (D-412); Tens. Elong.: 375% (D412); Tens. Mod.: 6.58 MPa (@ 100% elong., D412); Hardness: 78A ± 3 (instantaneous, D-2240)
Elec. Prop.: Dielec. Str.: 19,700 V/mm (D149); Dielec. Const.: 9.5 (1000 Hz, 1 day @ 30 C, UL83); Vol. Resist.: 8.0 X 10^8 ohm-cm (D257)
Ref.: Bulletin

Geon 8812. BF Goodrich Co.
Chem. Descrip.: Vinyl comp.
Category/Applications: General purpose flexible molding and extrusion comp.
Gen'l. Prop.: Sp. Gr.: 1.24 ± 0.02 (D-792)
Mech. Prop.: Tens. Str.: 11.3 MPa (D-412); Tens. Elong.: 400% (D412); Tens. Mod.: 3.4 MPa (@ 100% elong., D412); Hardness (Durometer A): 65A ± 3 (D-2240)
Elec. Prop.: Dielec. Str.: 27,600 V/mm (D149)
Ref.: Bulletin

Geon 8813. BF Goodrich Co.
Chem. Descrip.: Vinyl comp.
Category/Applications: General purpose flexible molding and extrusion comp.
Gen'l. Prop.: Sp. Gr.: 1.27 ± 0.02 (D-792)
Mech. Prop.: Tens. Str.: 13.1 MPa (D-412); Tens. Elong.: 380% (D412); Tens. Mod.: 5.5 MPa (@ 100% elong., D412); Hardness (Durometer A): 75A ± 3 (D-2240)
Elec. Prop.: Dielec. Str.: 30,000 V/mm (D149)
Ref.: Bulletin

Geon 8814. BF Goodrich Co.
Chem. Descrip.: Vinyl comp.
Category/Applications: General purpose flexible molding and extrusion comp.
Gen'l. Prop.: Sp. Gr.: 1.32 ± 0.02 (D-792)
Mech. Prop.: Tens. Str.: 16.5 MPa (D-412); Tens. Elong.: 310% (D412); Tens. Mod.: 9.0 MPa (@ 100% elong., D412); Hardness (Durometer A): 85A ± 3 (D-2240)
Elec. Prop.: Dielec. Str.: 34,000 V/mm (D149)
Ref.: Bulletin

Geon 8815. BF Goodrich Co.
Chem. Descrip.: Vinyl comp.
Category/Applications: General purpose flexible molding and extrusion comp.
Gen'l. Prop.: Sp. Gr.: 1.34 ± 0.02 (D-792)
Mech. Prop.: Tens. Str.: 19.3 MPa (D-412); Tens. Elong.: 290% (D412); Tens. Mod.: 12.4 MPa (@ 100% elong., D412); Hardness (Durometer A): 95A ± 3 (D-2240)
Elec. Prop.: Dielec. Str.: 35,600 V/mm (D149)
Ref.: Bulletin

Geon 8825. BF Goodrich Co.
Chem. Descrip.: Vinyl
Category/Applications: Wire and cable insulation; 80 C AWM
Form: Color: Natural, white, black
Gen'l. Prop.: Sp. Gr.: 1.37 ± 0.02 (40-mil molded sheet, D792)
Mech. Prop. (40-mil molded sheet): Tens. Str.: 15.85 MPa (D412); Tens. Elong.: 275% (D412); Tens. Mod.: 11.02 MPa (@ 100% elong., D412); Hardness: 86A ± 3 (instantaneous, D-2240)
Elec. Prop.: Dielec. Str.: 19,700 V/mm (14 AWG 2/64 in. wall, S/T, D149); Dielec. Const.: 6.6 (14 AWG 2/64 in.

wall, 1000 Hz, 1 day @ 75 C, UL83); Vol. Resist.: 1.5 X 10^{12} ohm-cm (75-mil slab, 95 C, D257)
Ref.: Bulletin

Geon 8830. BF Goodrich Co.
Chem. Descrip.: Vinyl
Category/Applications: Wire and cable insulation; suggested for 80 C AWM
Form: Color: Natural, black
All properties—20 AWG 1/64 in. wall:
Gen'l. Prop.: Sp. Gr.: 1.34 ± 0.02 (D-792)
Mech. Prop.: Tens. Str.: 23.08 MPa (D-412); Tens. Elong.: 250% (D412); Tens. Mod.: 20.67 MPa (@ 100% elong., D-412); Hardness: 68D ± 3 (instantaneous, D2240)
Elec. Prop.: Dielec. Str.: 25,610 V/mm (S/T, D149); Dielec. Const.: 4.7 (1000 Hz, 1 day @ 75 C, UL83); Vol. Resist.: 0.08 X 10^{12} ohm-cm (95 C, D257)
Ref.: Bulletin

Geon 8841. BF Goodrich Co.
Chem. Descrip.: Vinyl
Category/Applications: Wire and cable insulation for automotive applications
Form: Color: Natural
All properties—20 AWG 1/64 in. wall:
Gen'l. Prop.: Sp. Gr.: 1.36 ± 0.02 (D-792)
Mech. Prop.: Tens. Str.: 20.67 MPa (D-412); Tens. Elong.: 280% (D412); Tens. Mod.: 15.16 MPa (@ 100% elong., D-412); Hardness: 80C ± 3 (D2240)
Elec. Prop.: Dielec. Str.: 25,610 V/mm (S/T, D149); Vol. Resist.: 0.02 X 10^{12} ohm-cm (95 C, D257)
Ref.: Bulletin

Geon 8843. BF Goodrich Co.
Chem. Descrip.: Vinyl
Category/Applications: Wire and cable insulation; low odor 105 C AWM applications
Form: Color: Transparent, natural, flame-retarded natural, white, black
Gen'l. Prop.: Sp. Gr.: 1.35 ± 0.02 (40-mil molded sheet, D792)
Mech. Prop. (40-mil molded sheet): Tens. Str.: 18.95 MPa (D412); Tens. Elong.: 275% (D412); Tens. Mod.: 13.78 MPa (@ 100% elong., D412); Hardness: 92A ± 3 (instantaneous, D2240)
Elec. Prop.: Dielec. Str.: 23,640 V/mm (14 AWG 2/64 in. wall, S/T, D149); Dielec. Const.: 5.9 (14 AWG 2/64 in. wall, 1000 Hz, 1 day @ 75 C, UL83); Vol. Resist.: 0.14 X 10^{12} ohm-cm (75-mil slab, 95 C, D257)
Ref.: Bulletin

Geon 8852. BF Goodrich Co.
Chem. Descrip.: Vinyl
Category/Applications: Wire and cable insulation; NM and UF applications
Form: Color: Natural, black
Gen'l. Prop.: Sp. Gr.: 1.37 ± 0.02 (40-mil molded sheet, D792)
Mech. Prop. (40-mil molded sheet): Tens. Str.: 22.05 MPa (D412); Tens. Elong.: 275% (D412); Tens. Mod.: 13.78 MPa (@ 100% elong., D412); Hardness: 83C ± 3 (instantaneous, D-2240)
Elec. Prop.: Dielec. Str.: 27,580 V/mm (14 AWG 2/64 in. wall, S/T, D149); Dielec. Const.: 4.9 (14 AWG 2/64 in. wall, 1000 Hz, 1 day @ 30 C, UL83); Vol. Resist.: 0.9 X 10^{12} ohm-cm (75-mil slab, 95 C, D257)
Ref.: Bulletin

Geon 8857. BF Goodrich Co.
Chem. Descrip.: Vinyl comp.
Category/Applications: Nonmigrating, flexible molding and extruding comp.
Gen'l. Prop.: Sp. Gr.: 1.42 ± 0.02 (D-792)
Mech. Prop.: Tens. Str.: 10.4 MPa (D-

412); Tens. Elong.: 270% (D412); Tens. Mod.: 4.5 MPa (@ 100% elong., D412); Hardness (Durometer A): 70A ± 3 (D-2240)
Ref.: Bulletin

Geon 8863. BF Goodrich Co.
Chem. Descrip.: Vinyl
Category/Applications: Wire and cable sheath; low-temp., oil-resistant jacket, missile cable
Form: Color: Black and flame-retarded gray and black
Gen'l. Prop.: Sp. Gr.: 1.22 ± 0.02 (40-mil molded sheet, D792)
Mech. Prop. (40-mil molded sheet): Tens. Str.: 14.47 MPa (D412); Tens. Elong.: 325% (D412); Tens. Mod.: 8.96 MPa (@ 100% elong., D412); Hardness: 82A ± 3 (instantaneous, D2240)
Elec. Prop.: Dielec. Str.: 23,640 V/mm (D149); Dielec. Const.: 10.5 (1000 Hz, 1 day @ 30 C, UL83); Vol. Resist.: 2.0 $\times 10^8$ ohm-cm (D257)
Ref.: Bulletin

Geon 8870. BF Goodrich Co.
Chem. Descrip.: Vinyl
Category/Applications: Wire and cable sheath; –40 to 105 C jacket
Form: Color: Natural, gray, black
Gen'l. Prop.: Sp. Gr.: 1.26 ± 0.02 (40-mil molded sheet, D792)
Mech. Prop. (40-mil molded sheet): Tens. Str.: 13.78 MPa (D412); Tens. Elong.: 300% (D412); Tens. Mod.: 6.89 MPa (@ 100% elong., D412), Hardness: 78A ± 3 (instantaneous, D2240)
Elec. Prop.: Dielec. Str.: 27,580 V/mm (D149); Dielec. Const.: 7.0 (1000 Hz, 1 day @ 30 C, UL83); Vol. Resist.: 1.5 $\times 10^9$ ohm-cm (D257)
Ref.: Bulletin

Geon 8880. BF Goodrich Co.
Chem. Descrip.: Vinyl
Category/Applications: Wire and cable sheath, noncontaminating to polyethylene power cable jacket, MIL-C-915E special purpose jacket
Form: Color: Flame-retarded black
Gen'l. Prop.: Sp. Gr.: 1.32 ± 0.02 (40-mil molded sheet, D792)
Mech. Prop. (40-mil molded sheet): Tens. Str.: 15.16 MPa (D412); Tens. Elong.: 300% (D412); Tens. Mod.: 9.65 MPa (@ 100% elong., D412); Hardness: 68C ± 3 (instantaneous, D2240)
Elec. Prop.: Dielec. Str.: 31,520 V/mm (D149); Dielec. Const.: 7.1 (1000 Hz, 1 day @ 30 C, UL83); Vol. Resist.: 5.0 $\times 10^9$ ohm-cm (D257)
Ref.: Bulletin

Geon 8883. BF Goodrich Co.
Chem. Descrip.: Vinyl comp.
Category/Applications: High-gloss, flexible extrusion, molding comp.
Gen'l. Prop.: Sp. Gr.: 1.26 ± 0.02 (D-792)
Mech. Prop.: Tens. Str.: 19.3 MPa (D-412); Tens. Elong.: 300% (D412); Tens. Mod.: 14.8 MPa (@ 100% elong., D412); Hardness (Durometer A): 90A ± 3 (D-2240)
Ref.: Bulletin

Geon 8884. BF Goodrich Co.
Chem. Descrip.: Vinyl
Category/Applications: Wire and cable insulation; 75 C THW, 90 C AWM
Form: Color: Natural, flame-retarded natural, black
Gen'l. Prop.: Sp. Gr.: 1.30 ± 0.02 (40-mil molded sheet, D412)
Mech. Prop. (40-mil molded sheet): Tens. Str.: 17.91 MPa (D412); Tens. Elong.: 300% (D412); Tens. Mod.:

13.09 MPa (@ 100% elong., D412); Hardness: 80C ± 3 (instantaneous, D-2240)
Elec. Prop.: Dielec. Str.: 31,520 V/mm (14 AWG 2/64 in. wall, S/T, D149); Dielec. Const.: 6.5 (14 AWG 2/64 in. wall, 1000 Hz, 1 day @ 75 C, UL83); Vol. Resist.: 1.5 × 10^{12} ohm-cm (75-mil slab, 95 C, D257)
Ref.: Bulletin

Geon 8891. BF Goodrich Co.
Chem. Descrip.: Vinyl
Category/Applications: Wire and cable insulation; 105 C AWM; MIL-W-16878
Form: Color: Natural, flame-retarded natural and black, red
Gen'l. Prop.: Sp. Gr.: 1.32 ± 0.02 (40-mil molded sheet, D792)
Mech. Prop. (40-mil molded sheet): Tens. Str.: 17.23 MPa (D412); Tens. Elong.: 250% (D412); Tens. Mod.: 13.78 MPa (@ 100% elong., D412); Hardness: 81C ± 3 (instantaneous, D2240)
Elec. Prop.: Vol. Resist.: 1.5 × 10^{12} ohm-cm (75-mil slab, 95 C, D257)
Ref.: Bulletin

Geon 8896. BF Goodrich Co.
Chem. Descrip.: Vinyl
Category/Applications: Wire and cable insulation; SPT cord
Form: Color: Natural, white, gray
Gen'l. Prop.: Sp. Gr.: 1.37 ± 0.02 (40-mil molded sheet, D792)
Mech. Prop. (40-mil molded sheet): Tens. Str.: 14.47 MPa (D412); Tens. Elong.: 275% (D412); Tens. Mod.: 6.89 MPa (@ 100% elong., D412); Hardness: 82A ± 3 (instantaneous, D2240)
Elec. Prop.: Dielec. Str.: 31,520 V/mm (14 AWG 2/64 in. wall, S/T, D149); Dielec. Const.: 7.1 (14 AWG 2/64 in. wall, 1000 Hz, 1 day @ 30 C, UL83); Vol. Resist.: 0.5 × 10^{12} ohm-cm (75-mil slab, 95 C, D257)
Ref.: Bulletin

Geon 8898. BF Goodrich Co.
Chem. Descrip.: Vinyl
Category/Applications: Wire and cable insulation; TW, THW, THHN, THWN, 105 C AWM
Form: Color: Flame-retarded natural and black
All proeprties—14 AWG 15-mil wall:
Gen'l. Prop.: Sp. Gr.: 1.35 ± 0.02 (D-792)
Mech. Prop.: Tens. Str.: 19.29 MPa (D-412); Tens. Elong.: 275% (D412); Tens. Mod.: 15.85 MPa (@ 100% elong., D-412); Hardness: 81C ± 3 (instantaneous, D2240)
Elec. Prop.: Dielec. Str.: 17,730 V/mm (S/T, D149); Dielec. Const.: 5.02 (60 Hz, 1 day @ 30 C, UL83); Vol. Resist.: 1.5 × 10^{12} ohm-cm (95 C, D257)
Ref.: Bulletin

Geon 8918. BF Goodrich Co.
Chem. Descrip.: Vinyl
Category/Applications: Wire and cable insulation; retractile cords
Form:Color: Natural, white, black
Gen'l. Prop.: Sp. Gr.: 1.26 ± 0.02 (40-mil molded sheet, D792)
Mech. Prop. (40-mil molded sheet): Tens. Str.: 10.34 MPa (D412); Tens. Elong.: 400% (D412); Tens. Mod.: 4.13 MPa (@ 100% elong., D412); Hardness: 66A ± 3 (instantaneous, D2240)
Elec. Prop.: Dielec. Str.: 17,730 V/mm (14 AWG 2/64 in. wall, S/T, D149); Dielec. Const.: 8.5 (14 AWG 2/64 in. wall, 1000 Hz, 1 day @ 30 C, UL83); Vol. Resist.: 1.8 × 10^{12} ohm-cm (75-mil slab, 95 C, D257)
Ref.: Bulletin

Geon 82461. BF Goodrich Co.
Chem. Descrip.: Vinyl
Category/Applications: Wire and cable insulation; 105 C plug molding
Form: Color: Black
Gen'l. Prop.: Sp. Gr.: 1.31 ± 0.02 (40-mil molded sheet, D792)
Mech. Prop. (40-mil molded sheet): Tens. Str.: 11.37 MPa (D412); Tens. Elong.: 375% (D412); Tens. Mod.: 4.96 MPa (@ 100% elong., D412); Hardness: 72A ± 3 (instantaneous, D2240)
Ref.: Bulletin

Geon 82662. BF Goodrich Co.
Chem. Descrip.: Rigid vinyl comp.
Category/Applications: See Geon 8700A
Form: Cube; Color: Natural
Gen'l. Prop.: Sp. Gr.: 1.37 ± 0.02 (D-792)
Mech. Prop.: Tens. Str.: 47 MPa (D638); Tens. Mod.: 3514 MPa (D638); Flex. Str.: 90 MPa (D790); Flex. Mod.: 3307 MPa (D790); Compr. Str.: 66 MPa (D-695); Impact Str. (Izod): 1.0 J/cm notch (23 C, D256); Hardness (Rockwell): R 117 (D785)
Therm. Prop.: Conduct.: 0.16 W/m C (D177); Distort. Temp.: 73.5 C (@ 1.82 MPa, D648); Coeff. of Linear Exp.: 5.7 $\times 10^5$ cm/cm/C (D696)
Elec. Prop.: Dissip. Factor: 1.2 (60 Hz, 30 C, D150); Dielec. Str.: 55,160 V/mm (normal, D149); Dielec. Const.: 3.7 (60 Hz, 30 C, D150)
Ref.: Bulletin

Geon 82718. BF Goodrich Co.
Chem. Descrip.: Vinyl
Category/Applications: Wire and cable insulation; 90 C insulation, low odor 90 C plug molding
Form: Color: Black
Gen'l. Prop.: Sp. Gr.: 1.29 ± 0.02 (40-mil molded sheet, D792)
Mech. Prop. (40-mil molded sheet): Tens. Str.: 12.40 MPa (D412); Tens. Elong.: 350% (D412); Hardness: 80A ± 3 (instantaneous, D2240)
Elec. Prop.: Vol. Resist.: 0.1 $\times 10^{12}$ ohm-cm (75-mil slab, 95 C, D257)
Ref.: Bulletin

Geon 82882. BF Goodrich Co.
Chem. Descrip.: Vinyl
Category/Applications: Wire and cable insulation; 60 C insulation, REA PE 220 switchboard insulation
Form: Color: Natural
Gen'l. Prop.: Sp. Gr.: 1.31 ± 0.02 (40-mil molded sheet, D792)
Mech. Prop. (40-mil molded sheet): Tens. Str.: 21.70 MPa (D412); Tens. Elong.: 300% (D412); Tens. Mod.: 15.85 MPa (@ 100% elong., D412); Hardness: 83C ± 3 (instantaneous, D-2240)
Elec. Prop.: Dielec. Str.: 35,460 V/mm (14 AWG 2/64 in. wall, S/T, D149); Dielec. Const.: 4.3 (14 AWG 2/64 in. wall, 1000 Hz, 1 day @ 30 C, UL83); Vol. Resist.: 0.04 $\times 10^{12}$ ohm-cm (75-mil slab, 95 C, D257)
Ref.: Bulletin

Geon 83403. BF Goodrich Co.
Chem. Descrip.: Vinyl comp.
Category/Applications: FDA-acceptable, general purpose, flexible molding and extrusion comp.
Form: Color: Transparent
Gen'l. Prop.: Sp. Gr.: 1.19 ± 0.02 (D-792)
Mech. Prop.: Tens. Str.: 11.0 MPa (D-412); Tens. Elong.: 425% (D412); Tens. Mod.: 4.5 MPa (@ 100% elong., D412); Hardness (Durometer A): 65A ± 3 (D-2240)
Ref.: Bulletin

Geon 83457. BF Goodrich Co.
Chem. Descrip.: Vinyl comp.
Category/Applications: General purpose flexible molding and extrusion comp.
Form: Color: Transparent
Gen'l. Prop.: Sp. Gr.: 1.22 ± 0.02 (D-792)
Mech. Prop.: Tens. Str.: 17.3 MPa (D-412); Tens. Elong.: 300% (D412); Tens. Mod.: 8.2 MPa (@ 100% elong., D412); Hardness (Durometer A): 78A ± 3 (D-2240)
Ref.: Bulletin

Geon 83718. BF Goodrich Co.
Chem. Descrip.: Vinyl comp.
Category/Applications: Flexible molding and extruding comp.; weatherstrip
Gen'l. Prop.: Sp. Gr.: 1.38 ± 0.02 (D-792)
Mech. Prop.: Tens. Str.: 12.1 MPa (D-412); Tens. Elong.: 275% (D412); Tens. Mod.: 6.2 MPa (@ 100% elong., D412); Hardness (Durometer A): 80A ± 3 (D-2240)
Ref.: Bulletin

Geon 83741. BF Goodrich Co.
Chem. Descrip.: Vinyl comp.
Category/Applications: Flexible molding and extruding comp.; weatherstrip
Gen'l. Prop.: Sp. Gr.: 1.37 ± 0.02 (D-792)
Mech. Prop.: Tens. Str.: 10.7 MPa (D-412); Tens. Elong.: 310% (D412); Tens. Mod.: 4.7 MPa (@ 100% elong., D412); Hardness (Durometer A): 72A ± 3 (D-2240)
Ref.: Bulletin

Geon 83794. BF Goodrich Co.
Chem. Descrip.: Vinyl comp.
Category/Applications: Flexible molding and extruding comp.; weatherstrip
Gen'l. Prop.: Sp. Gr.: 1.42 ± 0.02 (D-792)
Mech. Prop.: Tens. Str.: 15.2 MPa (D-412); Tens. Elong.: 280% (D412); Tens. Mod.: 8.3 MPa (@ 100% elong., D412); Hardness (Durometer A): 90A ± 3 (D-2240)
Ref.: Bulletin

Geon 84059. BF Goodrich Co.
Chem. Descrip.: Vinyl
Category/Applications: Wire and cable sheath; switchboard cable jacket, REA PE 220 for indoor use
Form: Color: Natural, gray
Gen'l. Prop.: Sp. Gr.: 1.37 ± 0.02 (40-mil molded sheet, D792)
Mech. Prop. (40-mil molded sheet): Tens. Str.: 15.16 MPa (D412); Tens. Elong.: 300%; Hardness: 85A ± 3 (instantaneous, D2240)
Ref.: Bulletin

Geon 84203. BF Goodrich Co.
Chem. Descrip.: Vinyl
Category/Applications: Wire and cable insulation; UL 224 90 C nonheat shrinkable tubing
Form: Color: Transparent
All properties–20 AWG 1/64 in. wall:
Gen'l. Prop.: Sp. Gr.: 1.21 ± 0.02 (D-792)
Mech. Prop.: Tens. Str.: 18.60 MPa (D-412); Tens. Elong.: 325% (D412); Tens. Mod.: 11.02 MPa (@ 100% elong., D-412); Hardness: 91A ± 3 (instantaneous, D2240)
Ref.: Bulletin

Geon 84295. BF Goodrich Co.
Chem. Descrip.: Vinyl
Category/Applications: Wire and cable sheath; for applications requiring low

chlorine content jacket
Form: Color: Flame-retarded black
Gen'l. Prop.: Sp. Gr.: 1.55 ± 0.02 (40-mil molded sheet, D792)
Mech. Prop. (40-mil molded sheet): Tens. Str.: 11.02 MPa (D412); Tens. Elong.: 275% (D412); Tens. Mod.: 7.23 MPa (@ 100% elong., D412); Hardness: 88A ± 3 (instantaneous, D2240)
Elec. Prop.: Dielec. Str.: 27,580 V/mm (D149); Dielec. Const.: 6.8 (1000 Hz, 1 day @ 30 C, UL83); Vol. Resist.: 2.6×10^{11} ohm-cm (D257)
Ref.: Bulletin

Geon 84319. BF Goodrich Co.
Chem. Descrip.: Vinyl
Category/Applications: Wire and cable sheath; REA PE 20 station wire jacket, REA PE 220 for outdoor jacket
Form: Color: Flame-retarded natural, gray, black, and beige
Gen'l. Prop.: Sp. Gr.: 1.29 ± 0.02 (40-mil molded sheet, D792)
Mech. Prop. (40-mil molded sheet): Tens. Str.: 17.23 MPa (D412); Tens. Elong.: 350% (D412); Tens. Mod.: 8.96 MPa (@ 100% elong., D412); Hardness: 83A ± 3 (instantaneous, D2240)
Ref.: Bulletin

Geon 84343. BF Goodrich Co.
Chem. Descrip.: Vinyl
Category/Applications: Wire and cable sheath; MIL-C-915 repeated flex jacket, low-temp. jacket
Form: Color: Flame-retarded natural, gray, black
Gen'l. Prop.: Sp. Gr.: 1.25 ± 0.02 (40-mil molded sheet, D792)
Mech. Prop. (40-mil molded sheet): Tens. Str.: 13.78 MPa (D412); Tens. Elong.: 350% (D412); Tens. Mod.: 5.86 MPa (@ 100% elong., D412); Hardness: 76A ± 3 (instantaneous, D2240)
Ref.: Bulletin

Geon 84350. BF Goodrich Co.
Chem. Descrip.: Vinyl
Category/Applications: Wire and cable sheath; telephone modular cord jacket
Form: Color: Flame-retarded transparent
Gen'l. Prop.: Sp. Gr.: 1.23 ± 0.02 (40-mil molded sheet, D792)
Mech. Prop. (40-mil molded sheet): Tens. Str.: 13.78 MPa (D412); Tens. Elong.: 325% (D412); Tens. Mod.: 6.89 MPa (@ 100% elong., D412); Hardness: 80A ± 3 (instantaneous, D2240)
Ref.: Bulletin

Geon 84360. BF Goodrich Co.
Chem. Descrip.: Vinyl
Category/Applications: Wire and cable sheath; low cost coaxial cable, CATV, and general purpose jacket
Form: Color: Black
Gen'l. Prop.: Sp. Gr.: 1.41 ± 0.02 (40-mil molded sheet, D792)
Mech. Prop. (40-mil molded sheet): Tens. Str.: 12.92 MPa (D412); Tens. Elong.: 290% (D412); Tens. Mod.: 7.23 MPa (@ 100% elong., D412); Hardness: 83A ± 3 (instantaneous, D2240)
Ref.: Bulletin

Geon 84862. BF Goodrich Co.
Chem. Descrip.: Rigid vinyl comp.
Category/Applications: See Geon 84857
Form: Powder
Gen'l. Prop.: Sp. Gr.: 1.46 ± 0.02 (D-792)
Mech. Prop.: Tens. Str.: 44 MPa (26 C, D638); Tens. Mod.: 3105 MPa (D638); Flex. Str.: 81 MPa (D790); Flex. Mod.: 3284 MPa (D790); Impact Str. (Izod): 4.24 J/cm notch (23 C, D256); Hardness (Durometer D): 78 ± 3 (D785)
Therm. Prop.: Distort. Temp.: 73 C (@ 1.82 MPa, D648); Coeff. of Linear Exp.: 6.3×10^{-5} cm/cm/C (D696)
Ref.: Bulletin

Geon 85483. BF Goodrich Co.
Chem. Descrip.: Vinyl comp.
Category/Applications: Pipe extrusion comp.; NSF-approved 1120 for potable water, T I-G1 for DWV
Form: Powder
Gen'l. Prop.: Sp. Gr.: 1.38 ± 0.02 (D-792)
Mech. Prop.: Tens. Str.: 50 MPa (D638); Tens. Mod.: 2792 MPa (D638); Flex. Str.: 93 MPa (D790); Flex. Mod.: 2932 MPa (D790); Impact Str. (Izod): 1.16 J/cm (23 C, D256); Hardness (Rockwell): R 113 (D785)
Therm. Prop.: Distort. Temp.: 72 C (@ 1.8 MPa, D648)
Ref.: Bulletin

Geon 85555. BF Goodrich Co.
Chem. Descrip.: Vinyl comp.
Category/Applications: Pipe extrusion comp.; NSF-approved 1120 for potable water, TI-G1 for DWV
Form: Cube
Gen'l. Prop.: Sp. Gr.: 1.38 ± 0.02 (D-792)
Mech. Prop.: Tens. Str.: 50 MPa (D638); Tens. Mod.: 2792 MPa (D638); Flex. Str.: 93 MPa (D790); Flex. Mod.: 2932 MPa (D790); Impact Str. (Izod): 1.16 J/cm (23 C, D256); Hardness (Rockwell): R 113 (D785)
Therm. Prop.: Distort. Temp.: 72 C (@ 1.8 MPa, D648)
Ref.: Bulletin

Geon 85672. BF Goodrich Co.
Chem. Descrip.: Vinyl comp.
Category/Applications: Pipe extrusion comp.; NSF-approved 2116 for potable water
Form: Cube
Gen'l. Prop.: Sp. Gr.: 1.35 ± 0.02 (D-792)
Mech. Prop.: Tens. Str.: 48 MPa (D638); Tens. Mod.: 2827 MPa (D638); Flex. Str.: 85 MPa (D790); Flex. Mod.: 2771 MPa (D790); Impact Str. (Izod): 9.01 J/cm (23 C, D256); Hardness (Rockwell): R 113 (D785)
Therm. Prop.: Distort. Temp.: 72.7 C (@ 1.8 MPa, D648)
Ref.: Bulletin

Geon 85690. BF Goodrich Co.
Chem. Descrip.: Vinyl comp.
Category/Applications: Pipe fitting comp.; NSF-approved T II-G1 for potable water
Form: Cube
Gen'l. Prop.: Sp. Gr.: 1.35 ± 0.02 (D-792)
Mech. Prop.: Tens. Str.: 47 MPa (D638); Tens. Mod.: 2736 MPa (D638); Flex. Str.: 94 MPa (D790); Impact Str. (Izod): 7.36 J/cm (23 C, D256); Hardness (Rockwell): R 111 (D785)
Therm. Prop.: Distort. Temp.: 75 C (@ 1.8 MPa, D648)
Ref.: Bulletin

Geon 85707. BF Goodrich Co.
Chem. Descrip.: Vinyl comp.
Category/Applications: See Geon 85483
Form: Powder
Gen'l. Prop.: Sp. Gr.: 1.40 ± 0.02 (D-792)
Mech. Prop.: Tens. Str.: 52 MPa (D638); Tens. Mod.: 3211 MPa (D638); Flex. Str.: 101 MPa (D790); Flex. Mod.: 3204 MPa (D790); Impact Str. (Izod): 0.39 J/cm (23 C, D256); Hardness (Rockwell): R 117 (D785)
Therm. Prop.: Distort. Temp.: 16.6 C (@ 1.8 MPa, D648)
Ref.: Bulletin

Geon 85721. BF Goodrich Co.
Chem. Descrip.: Cellular rigid PVC comp.
Category/Applications: Extrusion comp.

for interior applications
Gen'l. Prop.: 0.50–0.60 g/cc (expanded, D792)
Mech. Prop.: Tens. Str.: 10–12 MPa (D-638); Tens. Elong.: 38–40% (D638); Flex. Str.: 23–32 MPa (D790); Flex. Mod.: 1035–1380 MPa (D790); Impact Str. (Izod): 0.6 J/cm notch (D256); Hardness (Durometer D): 40–50 (D-2240)
Therm. Prop.: Distort. Temp.: 57–58 C (@ 1.8 MPa, D648); Coeff. of Linear Exp.: 6.3 × 10^5 cm/cm/C (D696)
Ref.: Bulletin

Geon 85781. BF Goodrich Co.
Chem. Descrip.: Vinyl comp.
Category/Applications: Pipe extrusion comp.; NSF-approved 1120 for potable water, TI-G1 for DWV
Form: Cube
Gen'l. Prop.: Sp. Gr.: 1.40 ± 0.02 (D-792)
Mech. Prop.: Tens. Str.: 51 MPa (D638); Tens. Mod.: 2932 MPa (D638); Flex. Str.: 91 MPa (D790); Flex. Mod.: 3135 MPa (D790); Impact Str. (Izod): 0.46 J/cm (23 C, D256); Hardness (Rockwell): R 115 (D785)
Therm. Prop.: Distort. Temp.: 74.9 C (@ 1.8 MPa, D648)
Ref.: Bulletin

Geon 85796. BF Goodrich Co.
Chem. Descrip.: Vinyl
Category/Applications: Pipe fitting comp.; NSF-approved for T1-G1 for potable water, DWV
Form: Cube
Gen'l. Prop.: Sp. Gr.: 1.41 ± 0.02 (D-792)
Mech. Prop.: Tens. Str.: 49 MPa (D638); Tens. Mod.: 3180 MPa (D638); Flex. Str.: 90 MPa (D790); Impact Str. (Izod): 0.71 J/cm (23 C, D256); Hardness (Rockwell): R 112 (D785) .
Therm. Prop.: Distort. Temp.: 73 C (@ 1.8 MPa, D648)
Ref.: Bulletin

Geon 85856. BF Goodrich Co.
Chem. Descrip.: Rigid vinyl comp.
Category/Applications: Injection molding comp. for exterior applications
Gen'l. Prop.: Sp. Gr.:1.45 ± 0.02 (D-792)
Mech. Prop.: Tens. Str.: 44.4 MPa (D-638); Tens. Elong.: 125% (D638); Tens. Mod.: 3169 MPa (D638); Flex. Str.: 69.0 MPa (D790); Flex. Mod.: 2997 MPa (D790); Impact Str. (Izod): 6.9 J/cm notch (23 C, D256); Hardness (Durometer D): 80 (D2240)
Therm. Prop.: Distort. Temp.: 71 C (@ 1.8 MPa, D648); Coeff. of Linear Exp.: 11.7 × 10^5 cm/cm/C (D696)
Ref.: Bulletin

Geon 85857. BF Goodrich Co.
Chem. Descrip.: Rigid vinyl comp.
Category/Applications: Impact-modified extrusion comp. for exterior building products; sheet and profile applications
Form: Cube
Gen'l. Prop.: Sp. Gr.: 1.46 ± 0.02 (D-792)
Mech. Prop.: Tens. Str.: 43 MPa (26 C, D638); Tens. Mod.: 2446 MPa (D638); Flex. Str.: 78 MPa (D790); Flex. Mod.: 2825 MPa (D790); Impact Str. (Izod): 2.01 J/cm notch (23 C, D256); Hardness (Durometer D): 82 ± 3 (D785)
Therm. Prop.: Distort. Temp.: 73 C (@ 1.82 MPa, D648); Coeff. of Linear Exp.: 5.9 × 10^{-5} cm/cm/C (D696)
Ref.: Bulletin

Geon 85858. BF Goodrich Co.
Chem. Descrip.: Cellular rigid PVC comp.
Category/Applications: Extrusion comp.

for exterior applications
Gen'l. Prop.: Sp. Gr.: 0.70–0.80 g/cc (expanded, D792)
Mech. Prop.: Tens. Str.: 17–21 MPa (D-638); Tens. Elong.: 60–65% (D638); Flex. Str.: 36–39 MPa (D790); Flex. Mod.: 1378–1723 MPa (D790); Impact Str. (Izod): 0.7 J/cm notch (D256); Hardness (Durometer D): 55 (D2240)
Therm. Prop.: Distort. Temp.: 66 C (@ 1.8 MPa, D648)
Ref.: Bulletin

Geon 86100, 86101, and 86103. BF Goodrich Co.
Chem. Descrip.: Vinyl comp.
Category/Applications: Low-gloss grade for coextrusion capstock
Form: Cube
All properties–0.08 mm film
Gen'l. Prop.: Sp. Gr.: 1.43 ± 0.02 (D-792-60T)
Mech. Prop.: Tens. Str.: 20.7 MPa (with grain, yield, D882-75); Tens. Elong.: 220% (with grain, rupture, D882-61T)
Elec. Prop.: Dielec. Str.: 99,446 V/mm (D149-64T); Dielec. Const.: 3.4 (100 cps, D150-65T); Vol. Resist.: 2.91 ohm-$cm^3 \times 10^5$ (D257-61)
Ref.: Bulletin

Geon 87001. BF Goodrich Co.
Chem. Descrip.: Cellular rigid PVC comp.
Category/Applications: See Geon 85721
Gen'l. Prop.: Sp. Gr.: 0.70–0.80 g/cc (expanded, D792)
Mech. Prop.: Tens. Str.: 14–17 MPa (D-638); Tens. Elong.: 40–45 MPa (D638); Flex. Str.: 31–35 MPa (D790); Flex. Mod.: 1380–1725 MPa (D790); Impact Str. (Izod): 0.6–0.7 J/cm notch (D256); Hardness (Durometer D): 55 (D2240)
Therm. Prop.: Distort. Temp.: 65 C (@ 1.8 MPa, D648)
Ref.: Bulletin

Geon 87237. BF Goodrich Co.
Chem. Descrip.: Rigid vinyl comp., lead-stabilized
Category/Applications: Injection molding comp.
Gen'l. Prop.: Sp. Gr.: 1.40 ± 0.02 (D-792)
Mech. Prop.: Tens. Str.: 41.4 MPa (D-638); Tens. Elong.: 125% (D638); Tens. Mod.: 2530 MPa (D638); Flex. Str.: 79.3 MPa (D790); Flex. Mod.: 2792 MPa (D790); Impact Str. (Izod): 0.58 J/cm notch (23 C, D256); Hardness (Durometer D): 78 (D2240)
Therm. Prop.: Distort. Temp.: 68 C (@ 1.8 MPa, D648)
Ref.: Bulletin

Geon 87238. BF Goodrich Co.
Chem. Descrip.: Rigid vinyl comp., lead-stabilized
Category/Applications: Injection molding comp.
Gen'l. Prop.: Sp. Gr.: 1.35 ± 0.02 (D-792)
Mech. Prop.: Tens. Str.: 40.7 MPa (D-638); Tens. Elong.: 125% (D638); Tens. Mod.: 2241 MPa (D638); Flex. Str.: 75.8 MPa (D790); Flex. Mod.: 2565 MPa (D790); Impact Str. (Izod): 8.0 J/cm notch (23 C, D256); Hardness (Durometer D): 77 (D2240)
Therm. Prop.: Distort. Temp.: 67 C (@ 1.8 MPa, D648)
Ref.: Bulletin

Geon 87239. BF Goodrich Co.
Chem. Descrip.: Rigid vinyl comp., tin-stabilized
Category/Applications: Injection molding comp.
Gen'l. Prop.: Sp. Gr.: 1.38 ± 0.02 (D-792)
Mech. Prop.: Tens. Str.: 46.2 MPa (D-638); Tens. Elong.: 125% (D638); Tens.

Mod.: 2827 MPa (D638); Flex. Str.: 89.0 MPa (D790); Flex. Mod.: 3227 MPa (D790); Impact Str. (Izod): 0.4 J/cm notch (23 C, D256); Hardness (Durometer D): 78 (D2240)
Therm. Prop.: Distort. Temp.: 68 C (@ 1.8 MPa, D648)
Ref.: Bulletin

Geon 87241. BF Goodrich Co.
Chem. Descrip.: Rigid vinyl comp., tin-stabilized
Category/Applications: Injection molding comp.
Gen'l. Prop.: Sp. Gr.: 1.35 ± 0.02 (D-792)
Mech. Prop.: Tens. Str.: 42.7 MPa (D-638); Tens. Elong.: 100% (D638); Tens. Mod.: 2517 MPa (D638); Flex. Str.: 79.3 MPa (D790); Flex. Mod.: 2737 MPa (D790); Impact Str. (Izod): 6.4 J/cm notch (23 C, D256); Hardness (Durometer D): 80 (D2240)
Therm. Prop.: Distort. Temp.: 68 C (@ 1.8 MPa, D648)
Ref.: Bulletin

Geon 87242. BF Goodrich Co.
Chem. Descrip.: Rigid vinyl comp.
Category/Applications: Injection molding comp.
Form: Color: Transparent
Gen'l. Prop.: Sp. Gr.: 1.35 ± 0.02 (D-792)
Mech. Prop.: Tens. Str.: 52.7 MPa (D-638); Tens. Elong.: 145% (D638); Tens. Mod.: 2859 MPa (D638); Flex. Str.: 92.5 MPa (D790); Flex. Mod.: 3204 MPa (D790); Impact Str. (Izod): 1.2 J/cm notch (23 C, D256); Hardness (Durometer D): 81 (D2240)
Therm. Prop.: Distort. Temp.: 67 C (@ 1.8 MPa, D648)
Ref.: Bulletin

Geon 87255. BF Goodrich Co.
Chem. Descrip.: Rigid vinyl comp.
Category/Applications: See Geon 8700A
Form: Cube; Color: Natural
Gen'l. Prop.: Sp. Gr.: 1.34 ± 0.02 (D-792)
Mech. Prop.: Tens. Str.: 50 MPa (D638); Tens. Mod.: 2898 MPa (D638); Flex. Str.: 90 MPa (D790); Flex. Mod.: 2963 MPa (D790); Impact Str. (Izod): 0.7 J/cm notch (23 C, D256); Hardness (Rockwell): R 112 (D785)
Therm. Prop.: Distort. Temp.: 66 C (@ 1.82 MPa, D648); Coeff. of Linear Exp.: 6.8×10^5 cm/cm/C (D696)
Ref.: Bulletin

Geon 87256. BF Goodrich Co.
Chem. Descrip.: Rigid vinyl comp.
Category/Applications: See Geon 8700A
Form: Cube; Color: Natural
Gen'l. Prop.: Sp. Gr.: 1.41 ± 0.02 (D-792)
Mech. Prop.: Tens. Str.: 45 MPa (D638); Tens. Mod.: 2756 MPa (D638); Flex. Str.: 88 MPa (D790); Flex. Mod.: 3231 MPa (D790); Impact Str. (Izod): 5.3 J/cm notch (23 C, D256); Hardness (Rockwell): R 113 (D785)
Therm. Prop.: Distort. Temp.: 73.5 C (@ 1.82 MPa, D648); Coeff. of Linear Exp.: 6.0×10^5 cm/cm/C (D696)
Ref.: Bulletin

Geon 87262. BF Goodrich Co.
Chem. Descrip.: Rigid vinyl comp.
Category/Applications: See Geon 8700A
Form: Cube; Color: Natural
Gen'l. Prop.: Sp. Gr.: 1.39 ± 0.02 (D-792)
Mech. Prop.: Tens. Str.: 42 MPa (D638); Tens. Mod.: 2308 MPa (D638); Flex. Str.: 78 MPa (D790); Flex. Mod.: 2598 MPa (D790); Compr. Str.: 56 MPa (D-695); Impact Str. (Izod): 7.3 J/cm notch

(23 C, D256); Hardness (Rockwell): R 111 (D785)
Therm. Prop.: Conduct.: 0.16 W/m C (D177); Distort. Temp.: 77 C (@ 1.82 MPa, D648); Coeff. of Linear Exp.: 6.6 × 10^5 cm/cm/C (D696)
Ref.: Bulletin

Geon 88933. BF Goodrich Co.
Chem. Descrip.: CPVC comp.
Category/Applications: Molding and custom extrusion
Form: Cube
All properties as compression molded:
Gen'l. Prop.: Sp. Gr.: 1.52 ± 0.02 (D-792)
Mech. Prop.: Tens. Str.: 8200 psi (60 C, D638); Tens. Mod.: 405,000 psi (23 C, D638); Flex. Str.: 14,500 psi (23 C, D-790); Flex. Mod.: 387,000 psi (23 C, D-790); Impact Str. (Izod): 2.3 ft lb/in. notch (23 C, D256); Hardness (Rockwell): R 119 (A, D785)
Therm. Prop.: Distort. Temp.: 212 F (annealed, @ 1.8 MPa, D648); Coeff. of Linear Exp.: 3.9 × 10^5 in./in./F (D696)
Ref.: Bulletin

Geon 88934. BF Goodrich Co.
Chem. Descrip.: CPVC comp.
Category/Applications: Extrusion pipe and profile applications
Form: Cube
All properties as compression molded:
Gen'l. Prop.: Sp. Gr.: 1.57 ± 0.02 (D-792)
Mech. Prop.: Tens. Str.: 8400 psi (23 C, D638); Tens. Mod.: 405,000 psi (23 C, D638); Flex. Str.: 15,000 psi (23 C, D-790); Flex. Mod.: 395,000 psi (23 C, D-790); Impact Str. (Izod): 2.0 ft lb/in. notch (23 C, D256); Hardness (Rockwell): R 119 (A, D785)
Therm. Prop.: Distort. Temp.: 216 F (annealed, @ 1.8 MPa, D648); Coeff. of Linear Exp.: 3.6 × 10^5 in./in./F (D696)
Ref.: Bulletin

Geon CPVC. BF Goodrich Co.
Chem. Descrip.: Chlorinated polyvinyl chloride
Category/Applications: Lightweight vinyl material with good heat resistance (withstands temps. 60 F higher than ordinary vinyls), flame retardant characteristic, and electrical resistance; used in process piping, valves, tanks, ductwork, protective covers for electrical conductors, tubing for residential hot and cold water supply piping
Gen'l. Prop.: Stability: Good chemical resistance
Mech. Prop.: Remarkably strong at high temps., good impact strength and toughness
Ref.: Product information

Geon Latexes. BF Goodrich Co.
Chem. Descrip.: Vinyl chloride latices
Category/Applications: Latices for use in paper and textile coatings, fiber and fabric sizes, nonwoven fabric binding, release coatings, leather finishing, and adhesives; coatings are wear and water-resistant, with good gloss and toughness; heat sealable
Form: Latices range from hard, unplasticized resin systems to relatively soft, vinyl chloride/acrylic copolymer systems
Gen'l. Prop.: Stability: Good stability
Therm. Prop.: Flamm.: Flame retardant
Ref.: Product information

Geon Vinyls. BF Goodrich Co.
Chem. Descrip.: Polyvinyl chloride vinyls available as resins, compounds, and latices

Category/Applications: Thermoplastic, strong vinyls with high tensile and impact strengths, and broad range of impact, flexural strengths, and hardnesses; resistant to corrosion, weather, and to harsh chemicals; good electrical insulator; processed by extruding, molding, dipping, fluid-bed coating, calendering; used in rigid compounds (rigid vinyl siding, windows, floor coverings, piping, automotive, appliance parts, toys, etc.), flexible compounds (vinyl weather-proof clothing, wire and cable insulation, upholstery, sheeting and molded parts, gasketing, mechanical parts, etc.), dispersion resins (textile coatings, flooring, carpeting, paints, paper, packaging, adhesives, industrial coatings, molded automobile parts, tool handles, etc.)
Form: Various; Color: Broad range of colors available
Gen'l. Prop.: Stability: Chemically inert; resistant to chemicals, water, oil, and most solvents
Therm. Prop.: Flamm.: Can be flame-retardant
Ref.: Product information

Gillab 990C. M.C. Gill Corp.
Chem. Descrip.: Overlay laminate of special chemically-resistant polyester resin and fiberglass
Category/Applications: Strong, flat laminate with outstanding resistance to chemicals, organic acids, strong bases, solvents, water, and heat; radioactive decontaminable; can be laminated on various core materials (e.g., plywood, particle board, honeycomb and asbestos sheeting); no surface porosity (not stained by ink, dyes, blood, or rust); highly resistant to abrasion; used for plastic counter and laboratory tops, etc.
Form: Standard 1/8-in. sheet in various sizes (other thicknesses avail.); Color: Gray (other colors avail.)
Composition: 70 ± 5% resin content
Gen'l. Prop.: Density: 0.50 lb/ft^2 (0.06 laminate)
Mech. Prop.: Tens. Str.: 16,000 psi; Flex. Str.: 18,500 psi; Flex. Mod.: 10^6 psi; Compr. Str.: 24,000 psi; Shear Str.: 9400 psi; Hardness (Barcol): 45
Therm. Prop.: Coeff. of Linear Exp.: 1.5–3.0 × 10^{-5} C; Flamm.: Self-extinguishing in 2 s; will not support combustion
Elec. Prop.: Dielec. Str.: 350–500 V/mil (S/T); Arc Resist.: 60–120 s
Ref.: Bulletin

Gillfab 1000. M.C. Gill Corp.
Chem. Descrip.: Acrylic resin, fiberglass cloth, flat sheet laminate
Category/Applications: Used for acrylic canopy edge attachments; bonded to the acrylic canopy to provide reinforcement strength for bolting the canopy to the airframe
Form: Any thickness sheet by standard cloth widths with lengths up to 168 in.; Color: Natural, clear light green
Composition: 47 ± 5% resin content
All properties—1/8-in. laminate:
Gen'l. Prop.: Sp. Gr.: 1.8 ± 0.1 (Fed Std 406-5011)
Mech. Prop.: Tens. Str.: 243 MPa (Fed Std 406-1011); Flex. Str.: 207 MPa (Fed Std 406-1031); Hardness (Barcol): 35 min.
Therm. Prop.: Flamm. (Burn Rate): 2.79 cm/min (Fed Std 406-2021)
Ref.: Bulletin section B, pg. 1

Gillfab 1007. M.C. Gill Corp.
Chem. Descrip.: Flat sheet laminate, phenolic resin and Refrasil glass cloth
Category/Applications: Laminate with continuous high-temp. resistance for use

in firewalls on aircraft, heat shields on missiles, and similar high-temp. applications
Form: Sheet; Color: Natural
Composition: 30 ± 3% resin content
Gen'l. Prop.: Sp. Gr.: 1.68 min. (Fed Std 406-5011); Density: 1682 kg/m^3
Mech. Prop.: Tens. Str.: 103 MPa (Fed Std 406-1011); Flex. Str.: 152 MPa (Fed Std 406-1031); Flex. Mod.: 21 MPa (Fed Std 406-1031); Compr. Str.: 172 MPa (Fed Std 406-1021); Shear Str.: 34.5 MPa (double shear, Fed Std 406-1041); Hardness (Barcol): 65 min. (Model 934-1)
Ref.: Bulletin section L, pg. 15

Gillfab 1009. M.C. Gill Corp.
Chem. Descrip.: Hot pressed, acrylic resin, nylon fabric flat sheet laminate
Category/Applications: Designed as a reinforcement to be bonded to acrylic enclosures and canopies for bolting to the airframe
Form: Sheet; Color: Natural (other colors avail.)
Composition: 55 ± 5% resin content
Gen'l. Prop.: Sp. Gr.: 1.2 (Fed Std 406-5011); Density: 3.9 kg/m^2 (0.32-cm laminate)
Mech. Prop.: Tens. Str.: 131 MPa (Fed Std 406-1011); Tens. Mod.: 3500 MPa (Fed Std 406-1011); Flex. Str.: 104 MPa (Fed Std 406-1031); Flex. Mod.: 2800 MPa (Fed Std 406-1031); Shear Str.: 104 MPa (Fed Std 406-1041), Hardness (Barcol): 30 min. (Model 934-1)
Ref.: Bulletin section B, pg. 5

Gillfab 1017. M.C. Gill Corp.
Chem. Descrip.: Silicone, fiberglass flat sheet laminate
Category/Applications: Laminate with good electrical and mechanical properties @ 500 F service; used for structural aircraft parts
Form: Sheet; Color: Natural
Composition: 32 ± 3% resin content
Gen'l. Prop.: Sp. Gr.: 1.61–1.8 (Fed Std 406-5011); Density: 2.78 kg/m^2 (0.06-in. laminate)
Mech. Prop.: Tens. Str.: 207 MPa (Fed Std 406-1011); Flex. Str.: 242 MPa (Fed Std 406-1031); Flex. Mod.: 17,300 MPa (Fed Std 406-1031); Compr. Str.: 124 MPa (Fed Std 406-1021); Hardness (Barcol): 55 min. (Model 934-1)
Therm. Prop.: Coeff. of Linear Exp.: 0.8 10^{-5}/C (D696); Flamm.: 1 s to extinguish (Fed Std 406-2021)
Elec. Prop.: Dielec. Str.: 100 V/mil (S/T, D149); Dielec. Const.: 4.03 (Type II, 1 MC, D150); Arc Resist.: 250 s (D495)
Ref.: Bulletin section L, pg. 3

Gillfab 1045. M.C. Gill Corp.
Chem. Descrip.: Epoxy resin, fiberglass cloth, flat sheet laminate
Category/Applications: Self-extinguishing laminate which retains considerable strength at elevated temps.; suitable for continuous service @ 300 F; used in structural aircraft parts, for NEMA electrical insulation laminates
Form: Sheet; Color: Natural, light green
Composition: 34 ± 3% resin content (wt.)
Gen'l. Prop.: Sp. Gr.: 1.85 (@ 3% resin content, Fed Std 406-5011); Density: 2.93 kg/m^2 (0.06-in. laminate, Fed Std 406-5011)
Mech. Prop.: Tens. Str.: 324 MPa (Fed Std 406-1011); Flex. Str.: 552 MPa (warp, Fed Std 406-1031); Flex. Mod.: 23,000 MPa (Fed Std 406-1031); Compr. Str.: 311 MPa (Fed Std 406-1021); Impact Str. (Izod): 15 ft lb/in. notch (lengthwise, 48 h @ 50 C); Hardness (Barcol): 60 min. (Model 934-1)

Therm. Prop.: Coeff. of Linear Exp.: 1.1–3.5 10^{-5} in./in./C (D696); Flamm.: self-extinguishing in 2 s (Fed Std 406-2021)
Elec. Prop.: Dissip. Factor: 0.012 (1 MC, 23 C, 50% r.h.); Dielec. Const.: 5.1 (1 MC, 23 C, 50% r.h.); Vol. Resist.: 10^8 megohm-cm (96 h, 35 C, 90% r.h.)
Ref.: Bulletin section L, pg. 6

Gillfab 1053. M.C. Gill Corp.
Chem. Descrip.: Epoxy resin, fiberglass, flat sheet laminate
Category/Applications: Laminate for general purpose electrical insulation; used in aircraft structures and ducts for electrical insulation
Form: Sheet; Color: Natural
Composition: 35 ± 5% resin content
All properties—1/8-in. laminate:
Gen'l. Prop.: Sp. Gr.: 1.85 ± 0.1 (Fed Std 406-5011)
Mech. Prop.: Flex. Str.: 483 MPa (dry); Flex. Mod.: 15,900 MPa (dry); Impact Str. (Izod): 1.4 kgm/2.54 cm notch (Fed Std 406-1071); Hardness (Barcol): 60 min. (Model 934-1)
Elec. Prop.: Dissip. Factor: 0.018 (1 MC, 23 C, 50% r.h., D150); Dielec. Str.: 56 KV (parallel, S/S, D149); Dielec. Const.: 5.1 (1 MC, 23 C, 50% r.h., D150); Vol. Resist.: 10^8 megohm-cm (23 C, after 24 h in water, D257)
Ref.: Bulletin section F, pg. 5

Gillfab 1060. M.C. Gill Corp.
Chem. Descrip.: Kevlar-reinforced polyester laminate
Category/Applications: Laminate with high bullet and shrapnel stopping capability, used for personnel shielding in military and civilian applications; has max. bullet stopping to weight ratio
Form: Sheet
Composition: 25–27% resin content
Gen'l. Prop.: Sp. Gr.: 1.26 ± 0.08
Mech. Prop.: Hardness (Barcol): 45 min.
Ref.: Bulletin section N, pg. 21

Gillfab 1094. M.C. Gill Corp.
Chem. Descrip.: Rigid polyester-glass flat sheet laminate
Category/Applications: Used as a bullet-stopping material where ricochets must be avoided and/or weight is critical, e.g., cabin doors in aircraft, judges benches, bank teller booths, and inside lining of armored cars; when surfaced with a thick section of porcelain, used as personnel protecting armor on helicopters and other military aircraft
Form: Sheet
Composition: 25 ± 5% resin content
All properties—1/8-in. laminate:
Gen'l. Prop.: Sp. Gr.: 1.95; Density: 1.3 lb/ft^2
Mech. Prop.: Tens. Str.: 152 MPa (Fed Std 406-1011); Flex. Str.: 207 MPa (warp, Fed Std 406-1031); Hardness (Barcol): 45 min. (Model 934-1)
Ref.: Bulletin section N, pg. 14

Gillfab 1109. M.C. Gill Corp.
Chem. Descrip.: Epoxy, nylon-reinforced flat sheet laminate
Category/Applications: Laminate with good decorative and interlaminar peel characteristics, for use where the flexibility and strength of a nylon/epoxy laminate are desired
Form: Sheet; Color: Royal blue
All properties—0.76-mm laminate:
Gen'l. Prop.: Density: 0.97 kg/m^2 ± 10%
Mech. Prop.: Tens. Str.: 186 MPa (Fed Std 406-1011); Tens. Mod.: 3200 MPa (Fed Std 406-1011); Flex. Str.: 138 MPa (Fed Std 406-1031); Flex. Mod.: 2500 MPa (Fed Std 406-1031); Hardness (Barcol): 25 min.
Ref.: Bulletin section V, pg. 9

Gillfab 1338. M.C. Gill Corp.
Chem. Descrip.: Synthetic fiber (Nomex) reinforced, flame-retardant polyester, flat sheet laminate
Category/Applications: Laminate meeting FAA flammability requirements, with high impact resistance and 1/3 lower weight than glass-reinforced laminates; used as equipment liners in Lockheed L-1011 cargo compartments
Form: Sheet; Color: Natural cream
Composition: 40 ± 5% resin content
Gen'l. Prop.: Sp. Gr.: 1.25 ± 0.08 (Fed Std 406-5011); Density: 0.63 kg/m^2 (0.51 mm)
Mech. Prop.: Tens. Str.: 110 MPa (Fed Std 406-1011); Tens. Mod.: 4000 MPa (Fed Std 406-1011); Flex. Str.: 193 MPa (Fed Std 406-1031); Flex. Mod.: 6900 MPa (Fed Std 406-1031); Hardness (Barcol): 40 min. (Model 934-1)
Ref.: Bulletin section D, pg. 6

Gillfab 1366. M.C. Gill Corp.
Chem. Descrip.: Fiberglass-reinforced, flame-ratardant polyester, flat sheet laminate
Category/Applications: Laminate with highest impact resistance to weight ratio of any fiberglass cargo liner; original equipment liner in Lockheed L-1011 cargo compartments
Form: Sheet; Color: White
Gen'l. Prop.: Sp. Gr.: 1.9 ± 0.1 (Fed Std 406-5011); Density: 2.92 kg/m^2 ± 10% (0.06-in. laminate)
Mech. Prop.: Tens. Str.: 311 MPa (Fed Std 406-1011); Tens. Mod.: 17,000 MPa (Fed Std 406-1051); Flex. Str.: 242 MPa (Fed Std 406-1031); Flex. Mod.: 14,000 MPa (Fed STd 406-1031)
Therm. Prop.: Flamm.: 1 s to extinguish (FAR 25.853)
Ref.: Bulletin section D, pg. 7

Gillfab 4003. M.C. Gill Corp.
Chem. Descrip.: Flat, honeycomb sandwich panel with phenolic fiberglass skins
Category/Applications: Laminate with good dimensional stability and light weight; can be varied to exhibit high mechanical strength or high temp. resistance depending on the type of skins used and the core material
Form: Sheet
Gen'l. Prop.: Density: 5.27 kg/m^2
Mech. Prop.: Tens. Str.: 2.76 MPa (flatwise, Mil-Std-401B)
Therm. Prop.: Flamm.: Meets self-extinguishing requirements of FAR25.853, Appendix F
Ref.: Bulletin section R, pg. 4

Gillfab 4020. M.C. Gill Corp.
Chem. Descrip.: Phenolic-reinforced plastic honeycomb sandwich panel
Category/Applications: High-strength, good-rigidity, and short duration heat-resistant (up to 500 F) panel for portable landing fields
Form: Panel; Color: Natural
Ref.: Bulletin section R, pg. 7

Gillfab 4037. M.C. Gill Corp.
Chem. Descrip.: Phenolic-fiberglass skin and fiberglass honeycomb core sandwich panel
Category/Applications: Panel with high bonding strength adhesive, heat resistance to 350 F for short periods, and to 180 F for extended periods, and good moisture resistance; used in aircraft bulkheads, ceiling panels, etc.
Form: Panel
Mech. Prop. (0.015 in. Gillfab 1042 skins with 0.25 in. NP ¼ in. 4.0 lb/ft^3 core): Tens. Str.: 3.105 MPa (flatwise, Mil-Std-401B); Compr. Str.: 2.76 MPa (Mil-Std-401B)

Therm. Prop.: Flamm.: Meets requirements for FAR25.853
Ref.: Bulletin section R, pg. 14

Gillfab 5019. M.C. Gill Corp.
Chem. Descrip.: Polyurethane foam core, fiberglass phenolic skin sandwich panel with epoxy adhesive
Category/Applications: Used as abrasion-resistant sidewalls in liquor carts by some airlines
Mech. Prop.: Tens. Str.: 34.5 kPa (flatwise, Mil-Std-401B)
Therm. Prop.: Conduct.: 0.3 Btu/h/ft^2/F/in.; Flamm.: 15 s extinguishing time (FAR25.853)
Ref.: Bulletin section R, pg. 41

Gillfab 5045. M.C. Gill Corp.
Chem. Descrip.: Rigid PVC closed cell foam core, Gillfab 1088, epoxy fiberglass sheet skin sandwich panel
Category/Applications: Light weight, corrosion- and puncture-resistant panel for partitions, refrigerated storage boxes, and other interior requirements of aircraft
Form: Panel; Color: Natural (greenish brown)
Gen'l. Prop.: Density: 4.1 kg/m^2
Mech. Prop.: Compr. Str.: 2.584 MPa (flatwise)
Therm. Prop.: Flamm.: 5 s extingusising time (FAR25.853)
Ref.: Bulletin section R, pg. 39

Gilliner 1066. M.C. Gill Corp.
Chem. Descrip.: Polyester resin, fiberglass flat sheet laminate
Category/Applications: Self-extinguishing, high puncture, wear, shear, and scuff resistant laminate used as a liner for commercial aircraft baggage or cargo compartments
Form: Sheet; Color: White
Composition: 37 ± 5% resin content
Gen'l. Prop.: Density: 2.98 kg/m^2 (0.06-in. laminate)
Mech. Prop.: Tens. Str.: 262 MPa (Fed Std 406-1011); Tens. Mod.: 6200 MPa (Fed Std 406-1041); Flex. Str.: 124 MP MPa (Fed Std 406-1031); Flex. Mod.: 8300 MPa (Fed Std 406-1031); Shear Str.: 124 MPa (Fed Std 406-1041); Hardness (Barcol): 30 min. (Model 934-1)
Therm. Prop.: Coeff. of Linear Exp.: 1.5–3.0 × 10^{-5} /C (D696); Flamm.: 2 s to self-ext. (FAR25.853)
Elec. Prop.: Dielec. Str.: 350–500 V/mil (S/T, D149); Arc Resist.: 60–120 s (D-495)
Ref.: Bulletin section D, pg. 8

Glitter Adhesive #5048. Polymer Research Corp. of America
Chem. Descrip.: Aqueous nonionic emulsion
Category/Applications: Thickened emulsion suitable for adhering glitter to natural or synthetic blended fabrics (100% cotton and cotton/polyester blends)
Form: Thickened pasty emulsion; Color: White
Composition: 43.2% solids
Gen'l. Prop.: 9.0 lb/gal
Curing Charact.: Recommended curing cycle–4 min @ 325 F
Toxicity/Handling: Store @ R.T.
Std. Pkgs.: 30- and 55-gal fiber drums
Ref.: Tech. bulletin

Good-rite Antioxidants. BF Goodrich Co.
Chem. Descrip.: Polymer chemical
Category/Applications: Stabilizer for polypropylene and polyethylene, EPDM, nitrile, epichlorohydrin, and polyurethanes; ultraviolet protection and light stability, high-temp. protection; resists

water extraction; prolonged color retention; thermal stability in processing; FDA-acceptable
Toxicity/Handling: Relatively nontoxic, nonirritating
Ref.: Product information

Good-rite Latexes. BF Goodrich Co.
Chem. Descrip.: Copolymer and vinyl pyridine latexes
Category/Applications: Latex used for adhesives, coatings, saturants, tire cord adhesives, and printing inks for fabrics
Ref.: Product information

Good-rite Polyacrylates. BF Goodrich Co.
Chem. Descrip.: Polyacrylic resin
Category/Applications: Resin for use as dispersant for pigments and fillers; prevents scale and nonbiological fouling in cooling water; sludge conditioner in boiler water treatment; latex thickener; used in drilling muds, pigment dispersions, and detergents
Gen'l. Prop.: Sol.: Infinitely dilutable in water; Stability: Resists hydrolysis and bacterial attack; compatible in variety of aqueous dispersions
Ref.: Product information

Guar Gum. Hercules Inc.
Chem. Descrip.: Refined guar gum
Category/Applications: Thickener and stabilizer for food applications; used to improve mouthfeel, retain moisture, retard crystal growth, or as a binder or lubricant; used in frozen dairy desserts, pet foods, baked goods, salad dressings, sauces, breakfast cereals, soups, and gravies; special grades avail. excelling in acid stability and whiteness and having minimal odor and taste
Form: Powder (avail. in coarse to fine grinds)
Gen'l. Prop.: Visc.: 1000–> 6000 cps
Ref.: Products catalogue